Surveying Solved Problems

Fifth Edition

Jan Van Sickle, PhD, PLS

Report Errors for This Book

PPI is grateful to every reader who notifies us of a possible error. Your feedback allows us to improve the quality and accuracy of our products. Report errata at **ppi2pass.com**.

SURVEYING SOLVED PROBLEMS
Fifth Edition

Current release of this edition: 3

Release History

date	edition number	revision number	update
May 2019	5	1	New edition. Code updates.
May 2020	5	2	Minor corrections.
Sep 2020	5	3	Minor corrections.

Printed in the United States of America.

PPI
ppi2pass.com

ISBN: 978-1-59126-655-6

Table of Contents

Preface

Over the past 40 years, I have seen the everyday work of surveyors change from transit and tape to the global positioning system (GPS). Calculation tools have evolved from logarithmic tables to personal computers. In surveying today, you have to run as fast as you can to keep up with new methods and technologies. As technology grows more complex, the requirements for licensure become more challenging. As a result, it is imperative that you spend the time to adequately prepare for your exam.

The problems in *Surveying Solved Problems* cover a broad range of the surveying methods and knowledge areas essential for modern practice. They are from the "real world" of surveying. From total stations to GPS, and taping corrections to hydrography, the topics included in this book are broad.

For this fifth edition, I have written new problems covering the following topics: land development; BLM *Manual of Surveying Instructions*; FEMA requirements; risk management procedures; ALTA/NSPS Land Title Surveys; route surveys for alignments and utilities; surveys to establish new parcels, lots, or units; as-built/record drawing surveys; consultation services.

It is my sincere hope that you will find this book helpful, whether you are a student or a seasoned surveyor.

Jan Van Sickle, PhD, PLS

Acknowledgments

For this fifth edition and the previous edition of this book, I would like to thank George M. Cole, PhD, PE, PLS, and Peter Boniface, PhD, PLS, for ensuring the accuracy of the content. For this book's third edition, I would like to thank George M. Cole, PhD, PE, PLS, for technically reviewing the new material. For the first and second editions of this book, I would like to express my appreciation to J. Hardwick Butler, PE, of Middle Georgia College for his expert review of the material in this book. Such a review is particularly difficult when the subjects involved are so wide-ranging.

The help of Daniel Hoekstra, PLS, was also invaluable, not only because he allowed me to be involved in the first GPS survey of the Grand Canyon, but also for his review of the GPS data in the first two editions of this book.

My sincere appreciation goes to PPI's product development and implementation team, including Megan Synnestvedt, senior product manager; Meghan Finley, content manager; Tyler Hayes, senior copy editor; Bradley Burch, production editor; Beth Christmas, editorial project manager; Tom Bergstrom, technical illustrator and cover designer; Richard Iriye, typesetter; Cathy Schrott, production services manager; and Grace Wong, director of editorial operations. These individuals showed genuine concern about the details of this book. Real commitment to accuracy is hard to find.

Finally, I want to thank my father, Norman Van Sickle, PE, PLS, who was an inspiration to me. He was an example of what a land surveyor should know and, more importantly, an example of what a land surveyor should be. It is to him that this book is dedicated.

The problems and solutions in this book have all been carefully prepared and reviewed to ensure that they are clear and free of mistakes. However, if you suspect you've found an error, please let me know by submitting your comment through PPI's website, **ppi2pass.com**. I'll review your comment and make any necessary changes to the next printing of this book.

Jan Van Sickle, PhD, PLS

Introduction

ABOUT THIS BOOK

Surveying Solved Problems is designed to prepare you for the National Council of Examiners for Engineering and Surveying (NCEES) Fundamentals of Surveying (FS) exam and the Principles and Practice of Surveying (PS) exam. The included problems cover a broad range of subject areas relevant to surveying. These problems are both conceptual and practical, and they are representative of the type and difficulty of the problems you will encounter on the FS and PS exams.

Problems are grouped by topic into 17 chapters. Each chapter is further organized by subtopics covering many of the FS and PS exam specifications. Within each subtopic, problems are ordered from easiest to most challenging. Like the actual exam, each of the problems has four answer options, labeled (A), (B), (C), and (D). If the options are numerical, they will be displayed in increasing value. One of the answer options is correct (or, will be "most nearly" correct). The remaining answer options are incorrect and may consist of one or more "logical distractors," the term used by NCEES to designate incorrect options that result from simple mathematical errors, such as failing to square a term in an equation, or more serious errors, such as using the wrong equation.

The quantitative solutions in this book are presented step-by-step to help you follow the logical development of the solving approach and to provide examples of how you may want to solve similar problems on the exam. The qualitative solutions explain the correct answer and present related supportive information, such as surveying laws, regulations, and history.

HOW TO USE THIS BOOK

The chapters may be worked in any order, but it is recommended that you work only one chapter at a time. Problems within a chapter are categorized by subtopic, and they should be worked in order as they proceed from the simplest to the most difficult. Before attempting to solve problems, review the topics covered and identify those topics with which you're less familiar. Then, work a few of these problems to assess your general understanding, and identify your strengths and weaknesses. Try to solve problems without looking at the solutions in this book. Then, you can use the solutions to check your work or for guidance in solving problems you found difficult. Every step needed to arrive at the final answer, whether it be a function of logic or of calculation (and sometimes both), is provided so that you can determine which topics need further study.

Once you've identified which topics will require more study, locate relevant resource materials (such as those provided in the previous section), and work the problems in one subject area at a time. You should also work problems using the appropriate NCEES *Handbook*, as it will be the only resource you'll be able to use during your exam.

After you have worked all the problems with which you're least familiar, move on to the problems whose subject areas you are more familiar with. Follow the same procedure as previously described until you have worked all problems.

Remember that the value of a collection of problems such as this does not lie in its ability to guide your preparation. Rather, the value is in giving you an opportunity to bring together all your knowledge and to practice your problem-solving skills.

THE FS AND PS EXAMS

The FS and PS exams are standardized tests prepared by the NCEES to ensure that only qualified people are legally allowed to practice as surveyors. To ensure the reliability and validity of the tests, the FS and PS exams are based on input from committees of professional surveyors and educators throughout the United States.

The FS exam tests general entry-level surveying principles you are expected to have gained through academic study. In most states, passing the FS exam is a requirement for registration as a surveyor-in-training or surveyor intern. The PS exam tests your ability to apply those principles to the kinds of problems typically encountered in professional practice. Passing the PS exam is required in most states for full licensure as a professional surveyor. This may be necessary if your company requires licensure for employment or advancement, if your state requires registration before you may use the title "Surveyor," if you wish to be an independent consultant, or for other reasons. The PS exam is usually administered in conjunction with an exam on surveying practices and regulations specific to the state administering the exam.

Although both the FS and PS exams are prepared by NCEES, they are administered under the direction of the licensing boards of the various states, which usually also require you to have attained a certain level of education and (for the PS exam) experience before taking the exams. For information regarding these requirements in the state in which you plan to become licensed, and to apply for licensure, contact that state's board. Current addresses and phone numbers for each state board may be obtained at PPI's website, **ppi2pass.com**.

THE FS EXAM

Structure

The FS exam is a computer-based test that concentrates on the fundamentals and basics of surveying. You may take it at any Pearson VUE test center.

The exam contains 110 problems. Only one problem is given onscreen at a time. The exam is not adaptive (i.e., your response to one problem has no bearing on the next problem you are given); even if you answer the first five mathematics problems correctly, you'll still have to answer the sixth problem.

Your exam will include a limited (and unknown) number of problems that will not be scored and will not have an impact on your results (known as "pretest items"). NCEES does this to evaluate potential problems for future exams. You won't know which problems are pretest items. They are not identifiable and are randomly distributed throughout the exam.

The FS exam is 6 hours long and includes an 8-minute tutorial, a 25-minute break, and 2 minutes to sign a nondisclosure agreement. The total time you'll have to answer the exam problems is 5 hours and 20 minutes, which works out to slightly less than 3 minutes per problem. However, the exam does not pace you. You may spend as much time as you like on each question. You may can take less than a 25-minute break, if you like, but you cannot work through the break, and the break time cannot be added to the time you have to complete your exam. You may also leave your seat for personal reasons, but the "clock" does not stop for your absence.

You may work through the problems in any sequence. If you want to go back and check your answers before you submit them for grading, you may. However, once you submit your answers, you are not able to go back and review them. Unanswered problems are scored the same as problems answered incorrectly, so you should use the last few minutes of your exam time to guess at all unanswered problems.

The NCEES Nondisclosure Agreement

At the beginning of the FS exam, a nondisclosure agreement will appear on the screen. To begin the exam, you must accept the agreement within 2 minutes. If you do not accept within 2 minutes, your exam appointment will end, and you will forfeit your appointment and exam fees. The nondisclosure agreement is discussed in the section titled "Subversion After the Exam." The nondisclosure agreement, as stated in the *NCEES Examinee Guide*, is as follows.

> This exam is confidential and secure, owned and copyrighted by NCEES, and protected by the laws of the United States and elsewhere. It is made available to you, the examinee, solely for valid assessment and licensing purposes. To take this exam, you must agree not to disclose, publish, reproduce, or transmit this exam, in whole or in part, in any form or by any means, oral or written, electronic or mechanical, for any purpose, without the prior express written permission of NCEES. This includes agreeing not to post or disclose any test questions or answers from this exam, in whole or in part, on any websites, online forums, or chat rooms, or in any other electronic transmissions, at any time.

Your Exam Is Unique

The exam that you take will not be exactly the same exam taken by the person sitting next to you. NCEES says that, for each examinee, its computer-based testing (CBT) system randomly selects different but equivalent problems from its database using a linear, on-the-fly (LOFT) algorithm. Each examinee will have a unique exam, and all exams will be of equivalent difficulty.

The Exam Interface

The onscreen exam interface contains only minimal navigational tools. Onscreen navigation is limited to selecting an answer, advancing to the next problem, going back to the previous problem, and flagging the current problem for later review. The interface also includes a timer, the current problem number (e.g., 45 of 110), a pop-up scientific calculator, and access to an onscreen version of the *FS Reference Handbook*.

During the exam, you can advance through the problems in sequence, but you cannot jump to any specific problem, whether or not it has been flagged. After you have completed the last problem in a session, however, the navigation capabilities change, and you are permitted to review problems in any sequence and navigate to flagged problems.

Knowledge Areas and Problem Distribution

The FS exam includes problems in 13 knowledge areas. Each area is listed as follows, with the number of problems relating to this area that you can expect to see on your exam and a list of topics included in this area.

1. **Mathematics (13–20 problems):** algebra, trigonometry, and basic geometry; spherical trigonometry; linear algebra and matrix theory; analytic geometry and calculus

2. **Basic sciences (5–8 problems):** geology, dendrology, cartography, and environmental sciences
3. **Spatial data acquisition and reduction (6–9 problems):** vertical measurement, distance measurement, angle measurement, unit conversions, redundancy, knowledge and utilization of instruments and methods, and understanding of historical methods and instruments
4. **Survey computations and computer applications (19–29 problems):** coordinate geometry, traverse closure and adjustment, area, volume, horizontal and vertical curves, spirals, and spreadsheets
5. **Statistics and adjustments (6–9 problems):** mean, median, and mode; variance and standard deviation; error analysis; least squares adjustment; measurement and positional tolerance; and relative, network, and positional accuracy
6. **Geodesy (5–8 problems):** basic theory, satellite positioning, gravity, coordinate systems, datums, and map projections
7. **Boundary and cadastral survey law (13–20 problems):** controlling elements, gathering and identifying evidence, records research, legal descriptions, case law, riparian rights, public land survey system, metes and bounds, simultaneously created parcels, easements and encumbrances
8. **Photogrammetry and remote sensing (4–6 problems):** interpretation and analysis, project and flight planning, quality control, ground control, and LiDAR
9. **Survey processes and methods (11–17 problems):** land development (principles, standards, and regulations); boundary location; mapping, cartography, and topography; construction; riparian surveys; route surveying; and control surveys
10. **Geographic information systems (GIS) (5–8 problems):** feature collection and integration, database concepts and design, accuracy and use, and metadata
11. **Graphical communication and mapping (6–9 problems):** plans and specifications; contours and slopes; scales; planimetric features and symbols; land forms; digital terrain modeling and digital elevation modeling; and survey maps, plats, drawings, and reports
12. **Professional communication (4–6 problems):** oral and written communication; alternative forms of communication; documentation and recordkeeping
13. **Business concepts (3–5 problems):** contracts, liability and risk management, financial practices, leadership and management principles, personnel management principles, project planning and design, ethics, and safety

Exam content is subject to change. Consult PPI's website (**ppi2pass.com**) for current specifications.

THE PS EXAM

Structure

The PS exam is also a computer-based, closed-book exam. There are 100 problems on the exam. A PS testing session is 7 hours long, which includes 2 minutes for completion of a nondisclosure agreement, 8 minutes for a tutorial, and 6 hours for actually taking the exam. The exam is divided into two sections, each with approximately 50 problems, and an optional 50-minute scheduled break between the sections.

Knowledge Areas and Problem Distribution

The PS exam includes problems in five knowledge areas. Those areas are listed here with their subtopics and the number of problems to expect in each knowledge area.

1. Legal principles (18–27 problems)

Principles of Evidence: how to search for data and for physical evidence to evaluate data; how to evaluate data; parol evidence; prescriptive rights; adverse possession; acquiescence; controlling elements; easement rights

Common Law Boundary Principles: historical and current common law principles; riparian and littoral rights; sovereign rights, including both navigable waters and eminent domain; sovereign land grants

Sequential and Simultaneous Conveyance Concepts: types of conveyances; junior/senior rights; record and physical evidence

Legal Descriptions for Real Property Transactions: preparation and interpretation of legal descriptions; controlling elements and how they impact the description; unwritten rights and how they impact the description; encumbrances and how they impact the description; easements and how they impact the description

Evidence for the Perpetuation of the U.S. PLSS

2. Professional survey practices (22–33 problems)

Public/Private Record Sources: resources for private and public records; local public records indexing and filing system; local survey office records

Documentation, Supervision, and Clear Communication of Field Procedures: field surveying techniques; field surveying practices; data collection protocols

GPS/GNSS Including Satellite Constellations, Static GPS, RTK, PPP, and Virtual Networks

Surveying Principles and Computations: technical computations; applicable software

Monumentation Standards: applicable monumentation criteria; monument types

Land Development Solutions: regulatory land development criteria; construction criteria; land development implementation procedures

Survey Maps/Plats/Reports: technical communications by schematic, platting, and mapping processes and procedures; communication options

GIS: GIS spatial databases and metadata; datums and projections related to GIS

3. Standards and specifications (8–12 problems)

BLM Manual of Surveying Instructions

ALTA/NSPS Land Title Survey Standards: current ALTA/NSPS Land Title Survey Standards; state statutes regarding boundary surveys in conjunction with ALTA/NSPS Land Title Surveys

FEMA Requirements: FEMA specifications and instructions; horizontal and vertical datums related to FEMA flood zones; current FEMA elevation certificate; FEMA Flood Insurance Study

4. Business practices (13–19 problems)

General Business Practices and Procedures: project planning and project management; deliverables; costs, budgets, and contracts; types of surveys; site features and conditions; scope of services; appropriate equipment and instruments

Risk Management Procedures: safety procedures; QA/QC methods; risk management in contracts; insurance needs and requirements; potential liabilities

Professional Conduct

Communication with Clients, Staff, Related Professions, and the Public: different forms of communications; appropriate type of communication to convey concepts; related professions and their impact on client needs and deliverables

5. Areas of practice (24–36 problems)

ALTA/NSPS Land Title Surveys: legal documents, such as deeds, easements, and agreements; zoning information as applied to ALTA/NSPS Land Title Surveys; title insurance commitment letters and policies; underground features as applied to ALTA/NSPS Land Title Surveys

Control Networks and Geodetic Network Surveys: datums and reference frames relative to control networks; differences between local datums and geodetic datums; equipment appropriate for control surveys; the Federal Geographic Data Committee Geospatial Positioning Accuracy Standards; the National Geospatial Programs (NGP) Standards and Specifications—Digital Data Standards

Construction Surveys: construction plan reading; construction calculations including slopes, grades, and plan details; construction techniques and activities; horizontal and vertical positioning relative to a plan or datum

Boundary Surveys: physical boundary evidence; boundary reconciliations; historical measurement accuracy, equipment, and techniques; legal principles related to boundary surveys

Route Surveys for Alignments and Utilities: route alignment stationing practices; reading and interpreting roadway and utility plans

Topographic: topographic/planimetric mapping and control standards; interpretation, reconciliation, and adjustment of topographic survey data; QA/QC procedures as applied to topographic surveys; ground, hydrographic, and remote sensing equipment; the U.S. National Map Accuracy Standards as applied to topographic surveys; tools and techniques required to perform hydrographic, bathymetric, and remote sensing surveys; nomenclature related to utilities

Surveys to Establish New Parcels, Lots, or Units: types of subdivisions; platting; condominiums and associations; deed restrictions and restrictive covenants; zoning and subdivision ordinances

As-Built/Record Drawing Surveys: as-built/record drawing calculations including slopes, grades, and plan details; as-built/record drawing techniques and activities; horizontal and vertical as-built/record drawing positions relative to a plan or datum

Consultation Services: site topography and slope for development purposes; site access for development purposes; zoning standards related to new projects; floodplains as related to land development

Note: Exam content is subject to change. Consult PPI's website (**ppi2pass.com**) for current specifications.

TYPICAL PROBLEM FORMAT FOR THE FS AND PS EXAMS

The multiple-choice problems on the FS exam are typically short, straightforward, and designed to test your knowledge of the fundamentals of surveying and mapping.

The PS exam is designed to test your ability to apply surveying fundamentals to typical problems encountered in surveying practice. For example, a series of land descriptions from deeds might be provided, followed by problems requiring you to analyze the descriptions, establish certain boundaries or corner positions, and treat encroachments.

Most of the problems in both exams will be in the traditional multiple-choice exam, with a problem statement followed by four answer options. But, with the new CBT format, the exams may include alternative item type (AIT) problems, such as multiple correct problems, fill-in-the-blank problems, point-and-click (on points on a graphic) problems, and drag-and-drop problems for ranking or labeling items.

Both the FS and PS exams test in customary U.S. units. Therefore, the majority of this book also utilizes U.S. units.

EXAM SCORING

Neither the FS exam nor the PS exam is graded on a curve, since a certain minimum competency must be demonstrated to safeguard the public welfare. Nevertheless, the tests may vary slightly in difficulty, depending upon the problems selected for a particular exam. Therefore, problems are reviewed by committees of practicing surveyors before the exams. These committees evaluate the difficulty of each problem in order to develop a "standard of minimum competency," or recommended passing score for each exam. However, the individual state boards have the authority to determine the passing score in their respective states. Credit is given for each correct answer, and no points are deducted for incorrect answers.

USE OF CALCULATORS AND COMPUTERS IN THE EXAMS

The exams require use of a scientific calculator. However, it may not be obvious that you should also bring a spare calculator with you. It would be unfortunate not to be able to finish because your calculator was dropped or stolen or stopped working for some unknown reason.

NCEES has banned communicating and text-editing calculators from the exam site. Only select types of calculators are permitted. Check the current list of permissible devices at PPI's website (**ppi2pass.com**). All the listed calculators have enough functionality for the exam.

The exams have not been optimized for any particular brand or type of calculator. In fact, for most calculations, a $15 scientific calculator will produce results as satisfactory as those from a $200 calculator. There are definite benefits to having built-in statistical functions, graphing, unit-conversion, and equation-solving capabilities. However, these benefits are not so great as to give anyone an unfair advantage.

You may not share calculators with other examinees. Be sure to take your calculator with you whenever you leave the exam room for any length of time.

Laptop computers are not permitted in the exam. You may not use a walkie-talkie, cell phone, or other communications device during the exam.

THE NCEES REFERENCE HANDBOOKS

Both the FS and PS exam are "closed book." No references may be used except for either the *FS Reference Handbook* or *PS Reference Handbook*, as appropriate. Your *Handbook* will be made available in computer format on a split-screen, to allow viewing of the exam problems together with a searchable version of the *Handbook*. The search function can find only precise terms (e.g., searching for "non-annual compounding" will not locate "nonannual compounding").

Whichever exam you are preparing for, you should download a copy of the appropriate reference handbook from the NCEES website and use it in your studies. Become familiar with its contents and organization so that during the exam you can find equations and tables quickly when you need them. You may download and print out either handbook for your personal use, but you may not take your personal copy to the exam.

The *PS Reference Handbook* contains the following statement.

> The *Handbook* does not contain all the information required to answer every question on the exam. Some of the basic theories, conversions, formulas, and definitions examinees are expected to know have not been included in the supplied references. When appropriate, NCEES will provide information in the question statement itself to assist you in solving the problem.

Some basic formulas and conversion factors not in the handbooks may be needed to finish the exam. As a result, to be well prepared, you should know many of the basic formulas for both exams.

CHEATING AND EXAM SUBVERSION

The proctors are well trained to ensure that cheating does not occur. Obviously, you should not talk to other examinees during the exam, nor should you pass notes back and forth. To prevent discussion, the number of people permitted to use the restrooms at the same time will typically be limited.

The NCEES regularly reuses good problems from previous exams. Therefore, exam security is a serious issue with NCEES, which goes to great lengths to prevent copying of problems. You may not copy the problems in any manner.

The proctors are especially concerned about exam subversion, which generally means any activity that might invalidate the exam or the exam process. The most common form of exam subversion involves trying to copy exam problems for future use.

PREPARING FOR YOUR EXAM

Plan Your Approach

You should consider preparation for the FS or PS exam to be a long-term project and plan carefully. The exams are both comprehensive and fast paced; rapid recall, discipline, stamina, and mastery of the subject areas covered are all essential to success. Development of these qualities may require months of preparation in addition to the years of academic study and work practice you needed to qualify. Therefore, it is important to plan your preparation for the exam as you would plan for a large surveying and mapping project.

These steps can help prepare you.

1. Review the list of subject areas earlier in this Introduction to gain insight into the nature and content of the exams.
2. For future reference, prepare a concise outline as you work through each area.
3. Your review should be on a rigorous schedule to help you develop the discipline and stamina necessary to do well on the exams.
4. Take a practice exam, such as the *Fundamentals of Surveying Practice Exam* or *Principles and Practice of Surveying Practice Exam* (both available from PPI), to evaluate your readiness for the exams.
5. Work on any weak areas revealed by the practice exam.
6. Conduct a final review of your notes.

Learning to use your time wisely is one of the most important things that you can do during your review. You will undoubtedly encounter review problems that take much longer than you expect. You may cause some delays yourself by spending too much time looking through the *Handbook* for the information you need. Other problems will just entail too much work. Learning to recognize such situations more quickly will help you make intelligent decisions during the exams.

Additional Reference Material

You will find that this book is an excellent starting point for preparing for your exam. However, additional references may be helpful, especially in areas in which you are uncomfortable. There are countless available texts that cover the various topics in depth. Listed here are several personal favorites that offer coverage of the areas to be tested on the exams. Edition numbers have been omitted since new editions are often issued. Use the most recent edition available.

Legal Principles

Brown, Curtis A., Walter G. Robillard and Donald A. Wilson. *Evidence and Procedures for Boundary Location.* New York: John Wiley & Sons.

Cole, George M. and Donald Wilson. *Land Tenure, Boundary Surveying and Cadastral Systems.* Boca Raton, FL: CRC Press.

Cole, George M. *Water Boundaries.* New York: John Wiley & Sons.

Robillard, Walter G. and Donald A. Wilson. *Brown's Boundary Control and Legal Principles.* New York: John Wiley & Sons.

Measurement and Computation Theory and Practice

Bureau of Land Management and United States Department of the Interior. *Manual of Surveying Instructions – For the Survey of the Public Lands of the United States.* Washington D.C.: Government Printing Office.

Ghilani, Charles D. *Adjustment Computations: Spatial Data Analysis.* New York, NY: John Wiley & Sons.

Smith, James R. *Introduction to Geodesy.* New York, NY: John Wiley & Sons.

Stem, James E. *State Plane Coordinate System of 1983.* Silver Springs, MD: NOAA Manual NOS NGS.

Van Sickle, Jan. *GPS for Land Surveyors.* Boca Raton, FL: CRC Press.

Geographic Information Systems and Photogrammetry

Clarke, Keith C. *Getting Started with Geographic Information Systems.* Upper Saddle River, NJ: Prentice Hall.

Land Development

Colley, Barbara C. *Practical Manual of Land Development.* New York, NY: McGraw-Hill.

Dewberry. *Land Development Handbook.* New York, NY. McGraw-Hill.

Practice Problems

Cole, George M. *Fundamentals of Surveying Practice Exam* Professional Publications, Inc.

Cole, George M. *Principles and Practice of Surveying Practice Exam,* Professional Publications, Inc.

Last-Minute Preparation

A week or so before your exam, conduct an intensive review of the outlines you prepared during your study. Arrange for child care and transportation. Since the exam does not always start or end at the designated time, make sure such arrangements are flexible. If convenient, visit the exam site ahead of time to locate the building, parking areas, exam rooms, and restrooms.

Take a backup calculator to your exam. If your spare calculator is not the same type as your primary one, spend some time familiarizing yourself with it. Make sure that you have correct replacement batteries for both calculators. In addition, you should prepare a kit of items to take to the exam.

Take the day before the exam off from work to relax. If you live far from the exam site, consider getting a hotel room in which to spend the night. Do not attempt to cram the night before the exam. Calculate your wake-up time, and set two alarms. Select and lay out your clothing and breakfast items, and make sure that you have gas in your car and money in your wallet.

TAKING YOUR EXAM

What to Take to Your Exam

Your exam kit should contain items you need, such as your photo ID and your calculator, as well as items pertaining to your personal comfort. However, you may bring only certain items into the testing room, and the list of permitted items is different for the FS and PS exams.

You will need your photo ID in order to be admitted to the test site for either exam. Your ID must be government issued and must include

- your name
- your date of birth
- a recognizable photo of you
- your signature (except for U.S. military IDs)
- an expiration date (not past)

Your exam kit might contain the following items. Some of these, however, you may not bring into the testing room and must leave in a small locker that will be provided to you. You may access the locker during the break between sessions, but not during either session.

- an acceptable form of photo ID (essential)
- a printed copy of your appointment confirmation letter (strongly recommended)
- your primary calculator, with fresh batteries installed
- your backup calculator (left in locker)
- spare batteries for both calculators (left in locker)
- eyeglasses (case left in locker)
- eyeglasses repair kit, including a small screwdriver for fixing glasses or removing batteries from your calculator (left in locker)
- contact lens wetting solution (left in locker)
- a light sweater or jacket (pockets empty)
- cough drops (unwrapped and not in a bottle or container; a clear plastic bag is acceptable)
- aspirin or other pills (unwrapped and not in a bottle or container; a clear plastic bag is acceptable)
- eyedrops
- a pillow or cushion (if you need one in order to sit comfortably through the exam)
- several dollars in loose change (left in locker)
- an extra set of car keys (left in locker)
- something to eat during the break (left in locker)

You will be provided with earplugs, noise-cancelling headphones, tissues, and a reusable booklet and marker for scratch work, so you don't have to bring your own; if you do, you will not be permitted to bring them into the testing room. You must leave your wallet, purse, wristwatch, car keys, cell phone, and other personal items in your locker.

You may also bring essential medicines, medical devices, and mobility devices. These items will be inspected visually before you may bring them into the testing room. These include

- bandages
- braces (neck, back, wrist, leg, or ankle)
- canes
- crutches
- casts, slings, and other injury-related items that cannot be removed
- eye patches
- handheld magnifying glasses (not electronic; case left in locker)
- hearing aids or cochlear implants
- inhalers
- insulin pumps (or other medical devices attached to your body)
- medical alert bracelets
- medical/surgical face masks
- motorized scooters or chairs
- oxygen tanks
- walkers
- wheelchairs

For medical-related items not on this list, you must get approval in advance of your exam day.

What to Do at the Exam

Arrive at least 30 minutes before your exam is scheduled to begin. This will allow you to find a convenient parking place, get to the exam room, and calm down. Be prepared, though, to find that the exam room is not open or ready at the designated time.

On both the FS and PS exams, every problem is worth the same number of points, so it is a good idea to answer all of the problems that you can within a reasonable amount of time before attempting to solve problems that will take a disproportionate amount of time. If time allows, you can go back to those difficult problems.

Many points are lost due to carelessness. Therefore, it is a good idea to read each problem twice before solving. Check to make sure that you used all of the given data and done the appropriate conversion of units. While the exam problems are not tricky, you may find the results of commonly made mistakes are represented among the available answer choices. Thus, just because there is an answer matching your results does not mean that you have obtained the correct results.

Credit is given for correct answers, but no credit is deducted for wrong answers. Therefore, it is in your best interest to answer every question. It is a good idea to use the last ten minutes of the exam to guess at any remaining unsolved multiple-choice problems. You will be successful with about 25% of your guesses, and those points will more than make up for the few points you might earn by working during the last ten minutes.

After Your Exam

People react quite differently to the exam experience. Some people are energized and need to unwind by talking with other examinees, describing every detail of their experience and dissecting every exam problem. However, most people are completely exhausted and need a lot of quiet space and a hot tub in which to soak and sulk. Since everyone who took the exam has seen it, you will not be violating your "oath of silence" if you talk about the details with other examinees. It is difficult not to ask how someone else approached a problem that had you completely stumped. However, it is also very disquieting to think you did well on a problem, only to have someone else tell you where you went wrong.

Waiting for your exam results is its own form of mental torture. There is no predictable pattern to the release of the results. Exam results are not released by NCEES to all states simultaneously. They are not released alphabetically by state or examinee name. The people who failed are not notified first or last. Your co-worker might receive their notification today, and you might have to wait another three weeks. It all depends on when the entire process is complete. Some states have to have the results approved at a board meeting. Some prepare certificates before sending out notifications. Some states are more highly automated than others. The number of examinees also varies by state, as do numerous other factors. Therefore, you just have to wait patiently.

You will typically receive your results within 7 to 10 days. Your licensing board will contact you with your results. If you passed the exam, you will receive a letter that states you passed. If you failed, you will receive notice of this and get a diagnostic report that shows your strengths and weaknesses.

Now that you know all there is to know about the exams and about how to prepare for them, the rest is up to you. Plan your approach, and get to work. The very best of luck to you!

About the Author

Jan Van Sickle, PhD, PLS, has been a licensed professional land surveyor for over 30 years and is currently licensed in seven states. Dr. Van Sickle's diverse experience includes a variety of surveying, geomatics, and GIS projects. He has published three GIS and GPS textbooks and has advised many companies, including Microsoft, Jeppesen, and Hess, in geospatial matters. Dr. Van Sickle served as the GIS director at Quest Communications, where he led the effort to build the GIS spatial database for the company's worldwide network. He is a senior lecturer at The Pennsylvania State University and has been a featured speaker at many conferences.

Surveying Mathematics

UNIT DEFINITIONS AND CONVERSIONS

PROBLEM 1

A surveyor receives a request from a client on a control project to report the results in state plane coordinates, expressed in meters. The project is in a state that has adopted the U.S. survey foot as its standard. In such a state, which relationship is correct?

(A) 1 ft = 0.3048000 m

(B) 1 ft = 0.3048006 m

(C) 1 m = 3.2808000 ft

(D) 1 m = 3.2808399 ft

PROBLEM 2

A surveyor making measurements to guide a welder in the installation of a sill on a building wall must provide a 1.40 ft minimum clearance along the property line. However, the welder requires the measurement in feet and inches. What is 1.40 ft expressed to the nearest eighth of an inch?

(A) 1 ft, $3\frac{1}{8}$ in

(B) 1 ft, $3\frac{5}{8}$ in

(C) 1 ft, $4\frac{3}{8}$ in

(D) 1 ft, $4\frac{3}{4}$ in

PROBLEM 3

A surveyor staking a circular arc 100 ft long finds that the plans show the curve has a degree of curvature of one. The surveyor is about to calculate the length of the radius when the party chief says, "It is simply 100 times 1 radian." Which of the following relationships correctly describes 1 radian?

(A) $1 \text{ radian} = \dfrac{\pi}{360^\circ}$

(B) $1 \text{ radian} = \dfrac{180^\circ}{\pi}$

(C) $1 \text{ radian} = 57^\circ 29' 58''$

(D) $1 \text{ radian} = \dfrac{\pi}{3200 \text{ mils}}$

PROBLEM 4

A property owner from overseas writes to a surveyor with instructions concerning a survey. One of the instructions says, "Please be certain that the property has at least 4 hectares." What is a hectare?

(A) a large shade tree with gray bark

(B) a type of concrete monument

(C) a category of easements

(D) a measure of area

PROBLEM 5

Which numerical value is most nearly equal to $(81)^{-1/4}$?

(A) -3

(B) $\dfrac{1}{3}$

(C) 3

(D) 6

PROBLEM 6

While shopping for an inexpensive theodolite, a surveyor finds that the instrument with the lowest price has a circle that is divided into more graduations than usual. The merchant explains that it is a European instrument and the circle is divided into grads. Which relationship correctly describes a grad?

(A) 0.9 grads = 01°

(B) 1 grad = 00°54′

(C) 1.5 grads = 01°

(D) 100 grads = 10°

PROBLEM 7

An old deed describes two corner monuments as being 18 rods apart. The monuments are recovered and found to be 300 ft apart. What is the difference between the two measurements?

(A) 0.25 rod

(B) 1 yd

(C) 1 m

(D) 2 ft

PROBLEM 8

While preparing to set the parts per million on an EDM, the surveyor discovers that the only chart available is calibrated in degrees Celsius. The thermometer indicates 68° Fahrenheit. What is the corresponding temperature in degrees Celsius?

(A) 6°C

(B) 20°C

(C) 60°C

(D) 90°C

PROBLEM 9

While working down a section line from a witness corner monument toward a standard corner in the Public Land Surveying System, a surveyor finds that the record indicates a measurement of 15 chains, 64 links. Which value most nearly corresponds to that length?

(A) 65 rods

(B) 310 m

(C) 340 yd

(D) 510 ft, 3 in

PROBLEM 10

What is the sum of the following fractions of an inch?

$$\frac{1}{4}\text{ in},\ \frac{1}{16}\text{ in},\ \frac{17}{32}\text{ in},\ \frac{5}{8}\text{ in},\ \frac{1}{32}\text{ in}$$

(A) $\frac{6}{16}$ in

(B) 1 in

(C) $1\frac{1}{2}$ in

(D) $1\frac{3}{4}$ in

TRIGONOMETRY

PROBLEM 11

In the Cartesian coordinate system, the cosine and secant functions are positive in two of the four quadrants. In which two quadrants does this occur?

(A) the first and second quadrants

(B) the third and fourth quadrants

(C) the first and fourth quadrants

(D) the second and third quadrants

PROBLEM 12

In an equilateral triangle, the sine of half of any one of the interior angles is most nearly

(A) 0.500

(B) 0.700

(C) 0.900

(D) 1.000

PROBLEM 13

The sine of an angle A is equal to which of the following expressions?

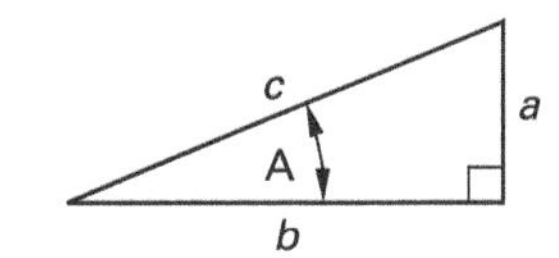

(A) $1 - \csc A$

(B) $\dfrac{\tan A}{\csc A}$

(C) $1 - \cos^2 A$

(D) $\pm\sqrt{1 - \cos^2 A}$

PROBLEM 14

An instrument is 377.14 ft, measured horizontally from the center of the base of a high-power transmission line tower. The reading at the center, on a vertical rod, taken with the telescope of the instrument level is 6.32 ft. The vertical angle measured to the top of the tower is 11°58′35″. What is the height of the tower?

(A) 74 ft

(B) 80 ft

(C) 84 ft

(D) 86 ft

PROBLEM 15

The chord of the arc subtended by 5°03′25″ will vary in length with the radius. Which of the following chord lengths is most correct for the indicated radius?

(A) 22 ft at 500 ft

(B) 44 ft at 1000 ft

(C) 180 ft at 2000 ft

(D) 350 ft at 4000 ft

PROBLEM 16

If the cosine of an angle can be expressed by the fraction $^{12}/_{13}$, which fraction would correctly represent the sine of the same angle?

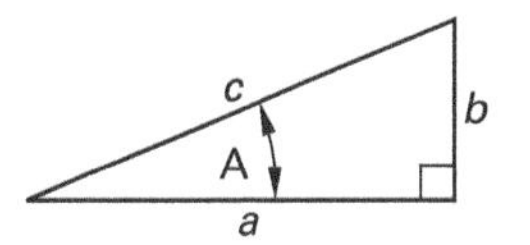

(A) $\dfrac{1}{12}$

(B) $\dfrac{5}{13}$

(C) $\dfrac{18}{12}$

(D) $\dfrac{12}{5}$

PROBLEM 17

In the triangle shown, which expression is equivalent to $h \csc C$?

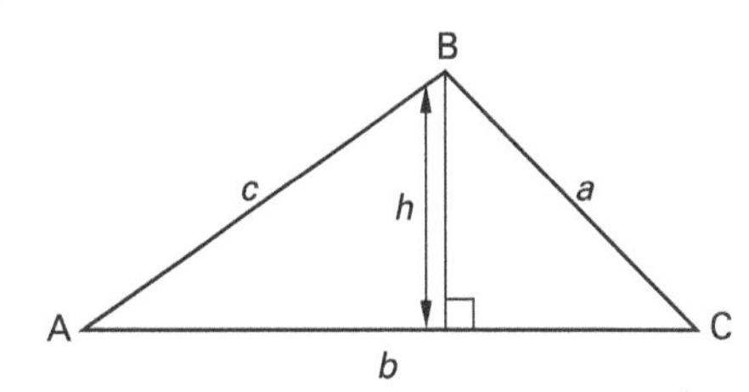

(A) a

(B) b

(C) a^2

(D) h

PROBLEM 18

The beginning of a circular curve (PC) occurs at station 12+37.48. The instrument, occupying station 11+00, backsights station 10+00 and turns an angle to the right of 242°15′18″ to the center of the curve.

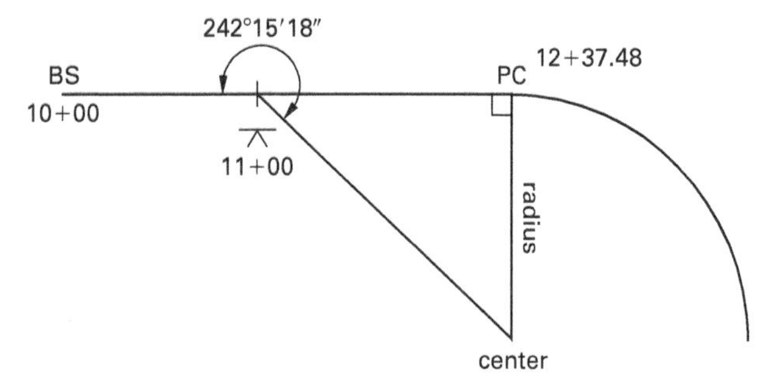

The radius of the circular curve is most nearly

(A) 64 ft

(B) 120 ft

(C) 260 ft

(D) 450 ft

PROBLEM 19

The circle shown has a radius of one unit and a delta angle, Δ, of 90°. Which expression correctly describes the length of line AB?

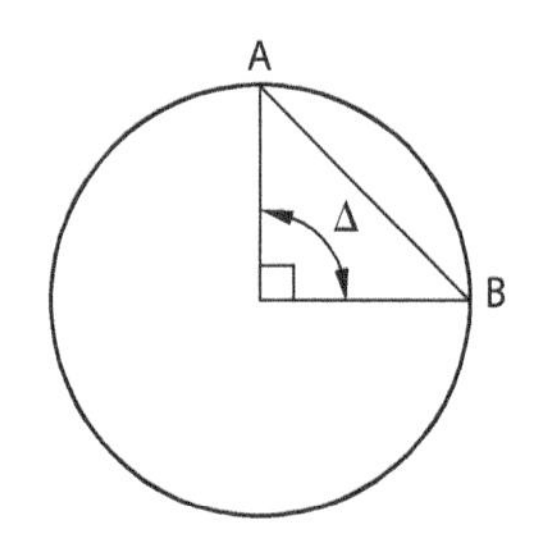

(A) $\dfrac{\Delta}{2}$

(B) $\cos\dfrac{\Delta}{2}$

(C) $\tan\Delta$

(D) $\csc\dfrac{\Delta}{2}$

PROBLEM 20

If the sine of A in a right triangle is x, what is the cotangent of A in terms of x?

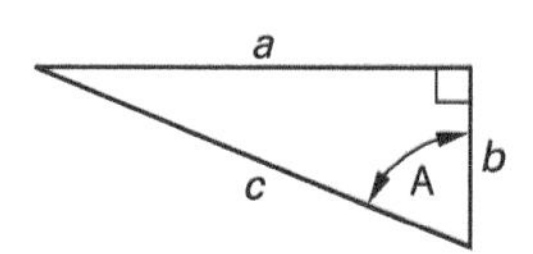

(A) $(1-\cos \mathrm{A})x$

(B) $\dfrac{b}{cx}$

(C) $\left(\dfrac{c}{a}\right)x$

(D) $(\tan \mathrm{A})x$

PROBLEM 21

One of the fundamental tools used to solve oblique triangles is the law of cosines. Which of the following expressions is a correct form of the law of cosines?

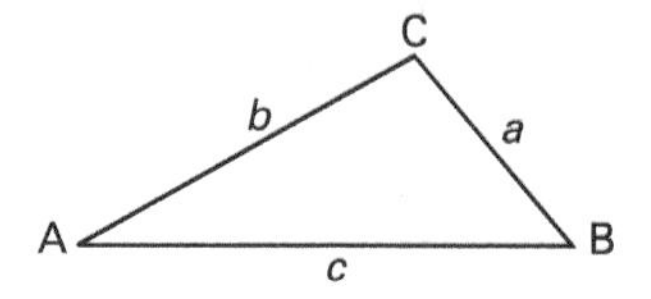

(A) $a^2 = b^2 + c^2 - 2ac\cos \mathrm{A}$

(B) $\cos b = \cos a \cos c + \sin b \sin c$

(C) $\cos \mathrm{A} = \dfrac{b^2 + c^2 - a^2}{2bc}$

(D) $\cos \mathrm{A} = \dfrac{\cot \mathrm{A}}{\csc \mathrm{A}}$

PROBLEM 22

A part of an old deed reads, "to an iron buggy axle; thence N85°15′E, 18 chains and 86 links to a wooden stake; thence S02°30′E, 16 chains and 98 links to a 3 in diameter iron pipe." The wooden stake has decayed. Current measurements yield the following coordinates.

iron buggy axle	7631.413N
	5723.610E
3 in iron pipe	6615.016N
	7012.813E

The decision was made to define the position of the decayed wooden stake by relying on the distance alone. Which coordinate would correspond to the position of the stake under these conditions?

(A) 6401.105N
5912.738E

(B) 6512.043N
5774.562E

(C) 7734.634N
6964.083E

(D) 8223.869N
7509.105E

PROBLEM 23

No point along a stationed line may be closer than 15 ft to an existing building. With the instrument occupying station 0+00 and backsighting station 0+40, an angle of 339°07′44″ is turned to the building corner. Then, with the instrument occupying station 0+40 and backsighting station 0+00, an angle of 104°25′09″ is turned to the same building corner. What is the minimum distance between the line and the building corner?

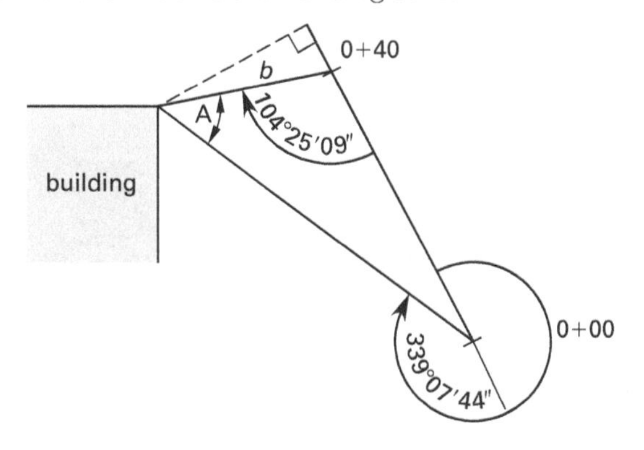

(A) 16.91 ft

(B) 17.46 ft

(C) 17.99 ft

(D) 20.00 ft

PROBLEM 24

A three-sided gore is found between adjoining tracts. One side of the gore is 471.67 ft, and the second side is 450.83 ft. The angle between these two sides is 05°12′40″. What is the length of the third side?

(A) 20.84 ft

(B) 40.95 ft

(C) 42.84 ft

(D) 46.82 ft

PROBLEM 25

After all measurements for a survey have been completed, it is important to determine the height of the top of a tower with respect to the instrument used to observe it. The only available information is the actual height of the tower, 157 ft, and the two vertical angles observed. The vertical angle to the base of the tower is 01°46′52″. The vertical angle to the top of the tower is 09°50′33″. What is the vertical distance from the instrument to the top of the tower? Ignore the effects of the curvature and refraction in the solution.

(A) 188.08 ft

(B) 191.29 ft

(C) 191.38 ft

(D) 205.34 ft

PROBLEM 26

Two sides of a triangle are 83.40 ft and 95.20 ft, respectively. The angle between the two sides is 51°15′00″.

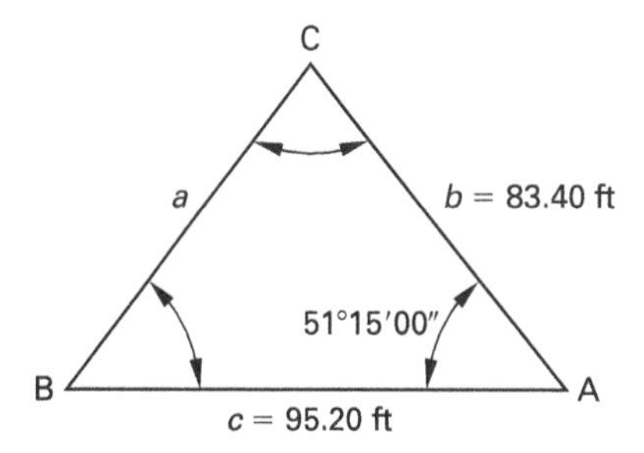

Which values correctly represent the remaining three elements of the triangle?

(A) 77.97 ft, 56°31′56″, 72°13′04″

(B) 77.97 ft, 56°31′56″, 72°15′07″

(C) 78.23 ft, 56°31′56″, 72°13′04″

(D) 85.46 ft, 49°35′15″, 79°09′45″

PROBLEM 27

x is the length of one side of triangle ABC. The altitude of the triangle, CD, is $x - 4$ ft. It bisects angle ACB and the base AB as well. AB is equal to x.

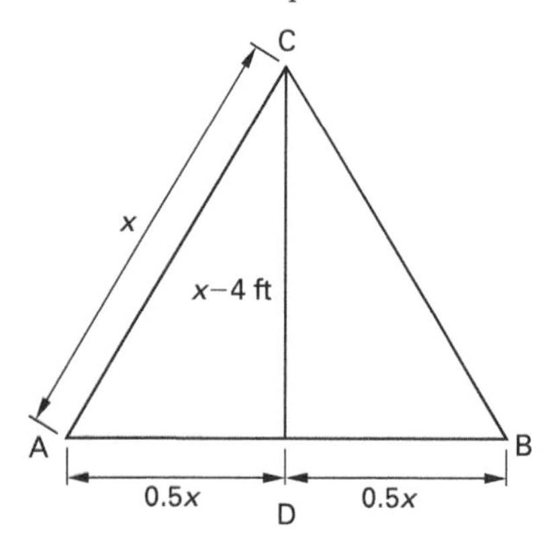

What is the length of x in ft?

(A) 12.42 ft

(B) 18.76 ft

(C) 29.85 ft

(D) 35.33 ft

PROBLEM 28

The triangle shown contains 1.5 acres. It is an oblique triangle in which angle B is 102°03′50″, angle A is 47°56′10″, and angle C is 30°00′00″.

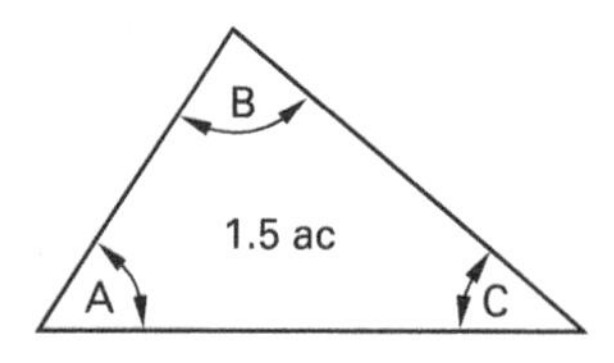

What is the length of the side opposite angle C?

(A) 300.00 ft

(B) 375.99 ft

(C) 460.43 ft

(D) 586.75 ft

PROBLEM 29

Two angles of a triangle are 79°59′ and 44°41′, respectively. The side opposite the angle 44°41′ is 568.00 ft.

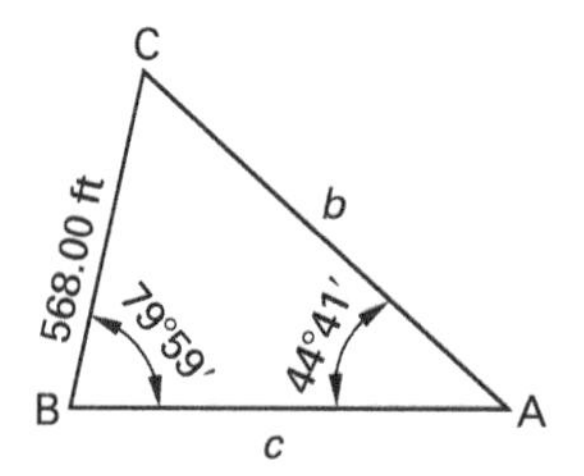

What are the lengths of the other two sides?

(A) 493.35 ft and 584.52 ft

(B) 537.09 ft and 543.77 ft

(C) 795.44 ft and 664.35 ft

(D) 823.33 ft and 774.52 ft

PROBLEM 30

Two sides of a triangle are 82.00 and 73.00, respectively, while the angle between them is 71°38′. What are the other two angles of the triangle?

(A) 42°51′ and 65°31′

(B) 48°12′ and 60°10′

(C) 49°35′ and 58°47′

(D) 51°32′ and 56°50′

HORIZONTAL CURVES

PROBLEM 31

The following six problems refer to the illustration shown.

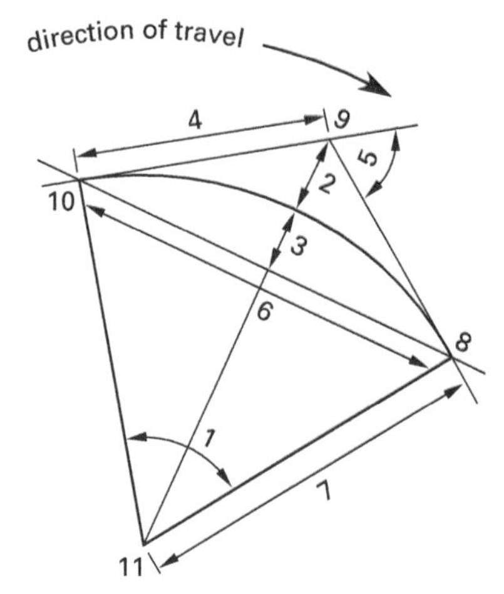

31.1. The distance labeled 2 is known by which name?

(A) middle ordinate

(B) long chord

(C) external distance

(D) semitangent

31.2. The angle labeled 1 is equal to which of the following?

(A) 360° minus the deflection per foot of the length

(B) 2π radians

(C) the angle labeled 5 in the illustration

(D) one-half of the angle labeled 5 in the illustration

31.3. The distance labeled 3 is known by which name?

(A) middle ordinate

(B) point of curvature

(C) long chord

(D) tangent

31.4. The distance labeled 4 is known by which name?

(A) semitangent

(B) back tangent

(C) radius

(D) both A and B

31.5. The length of the curve from the point labeled 10 to the point labeled 8 is correctly indicated by which mathematical expression?

(A) $\dfrac{RL}{2}$

(B) $2R\sin\dfrac{I}{2}$

(C) $R\left(1-\cos\dfrac{I}{2}\right)$

(D) $\dfrac{2\pi RI}{360°}$

31.6. Which of the following abbreviations might properly be used for the points labeled 10, 11, and 8, respectively?

(A) PC, CEN, and PT

(B) PI, RP, and PT

(C) PC, RP, and PI

(D) PT, CEN, and PC

PROBLEM 32

The long chord of a horizontal circular curve is mathematically represented by which expression?

(A) $\text{LC} = R\tan\dfrac{I}{2}$

(B) $\text{LC} = R\cos\dfrac{I}{2}$

(C) $\text{LC} = \dfrac{\pi RI}{180°}$

(D) $\text{LC} = 2R\sin\dfrac{I}{2}$

PROBLEM 33

The middle ordinate of a horizontal circular curve is mathematically represented by which expression?

(A) $\text{middle ordinate} = R\left(\sec\dfrac{I}{2}-1\right)$

(B) $\text{middle ordinate} = R\left(1-\cos\dfrac{I}{2}\right)$

(C) $\text{middle ordinate} = R\sec\dfrac{I}{2}$

(D) $\text{middle ordinate} = R\tan\dfrac{I}{2}$

PROBLEM 34

By the highway definition of the term, what would be the degree of curve, to the nearest minute, of a circular curve with a radius of 450.00 ft?

(A) 12°44′

(B) 12°45′

(C) 13°00′

(D) 15°53′

PROBLEM 35

The radius of a circular curve with a chord defined degree of curvature of 4° and a central angle, I, of 20° would most nearly be

(A) 720 ft

(B) 870 ft

(C) 1100 ft

(D) 1400 ft

PROBLEM 36

A circular curve has a radius of 100.00 ft and a central angle of 20°. Beginning at the PC, the tangent distance, TD, and the tangent offset, TO, to the point of tangency would most nearly be

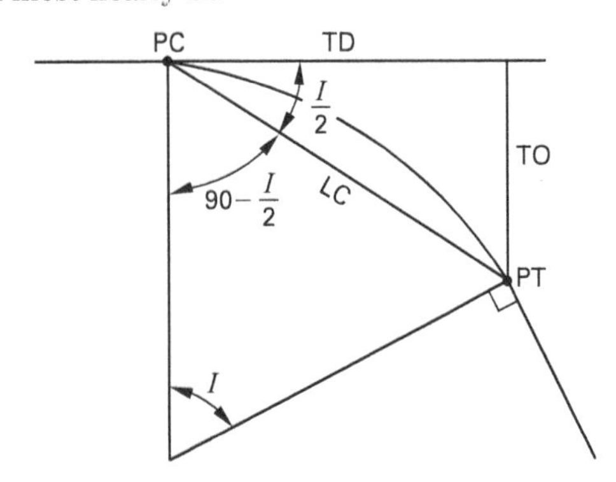

(A) TD = 34.2 ft, TO = 6.03 ft

(B) TD = 35.4 ft, TO = 5.79 ft

(C) TD = 44.1 ft, TO = 15.7 ft

(D) TD = 50.0 ft, TO = 25.1 ft

PROBLEM 37

A circular curve with a radius of 750.00 ft, a central angle of 70°, and its PC at station 13+30 must be staked by deflection angles and chords. What is the deflection angle and chord length to station 14+00?

(A) 01°23′32″ and 69.95 ft

(B) 02°40′26″ and 69.97 ft

(C) 03°14′42″ and 67.54 ft

(D) 03°15′12″ and 68.32 ft

PROBLEM 38

The PC of a circular curve is at station 214+23.61, the central angle of the curve is 42°15′, and the radius is 1438.92 ft. What is the station of the PT of the curve?

(A) 224+60.79

(B) 224+84.67

(C) 225+08.55

(D) 228+62.53

PROBLEM 39

The following two problems refer to the illustration shown.

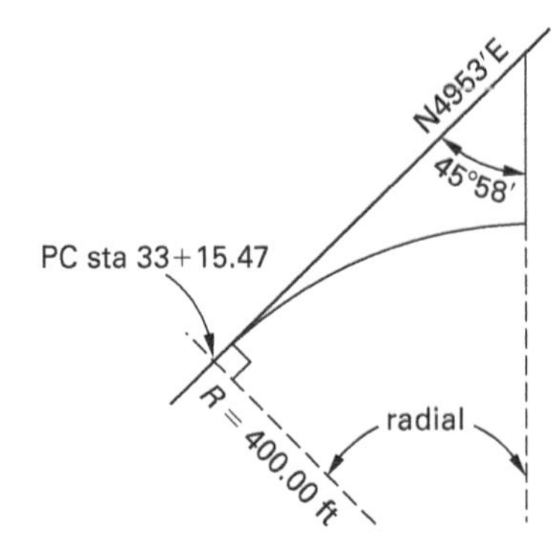

39.1. What is the station of the intersection of the curve and the line on the right side of the illustration?

(A) 35+45.91

(B) 35+92.07

(C) 36+02.38

(D) 36+22.88

39.2. What is the external distance from the midpoint of the curve to the tangent line?

(A) 30.98 ft

(B) 31.46 ft

(C) 32.21 ft

(D) 33.45 ft

PROBLEM 40

Which of the descriptions below correctly defines a reverse curve?

(A) two circular curves lying on opposite sides of a common tangent

(B) a variable radius curve

(C) two curves with radii of different lengths

(D) two curves with equal central angles

PROBLEM 41

Which of the descriptions below correctly defines a compound curve?

(A) two circular curves with equal radii

(B) two circular curves with different radii lying on opposite sides of a common tangent

(C) two circular curves with different radii lying on the same side of a common tangent

(D) two variable radius curves

PROBLEM 42

The abbreviations PRC and PCC are correctly defined by which of the following descriptions?

(A) PRC = point of reverse curvature
PCC = point of constant curvature

(B) PRC = point to resume curvature
PCC = point of compound curvature

(C) PRC = point of reverse curvature
PCC = point of compound curvature

(D) PRC = point of a reverse compound
PCC = point of compound curvature

PROBLEM 43

A line connecting the PC with the PT of a particular reverse curve passes through the PRC. Which of the following statements describes the characteristics of this reverse curve?

(A) The radii are equal.

(B) The central angles are equal.

(C) The long chords are equal.

(D) The line described from the PC to the PT passes through the PRC in all reverse curves.

PROBLEM 44

The following five problems refer to the illustration shown and the accompanying statement.

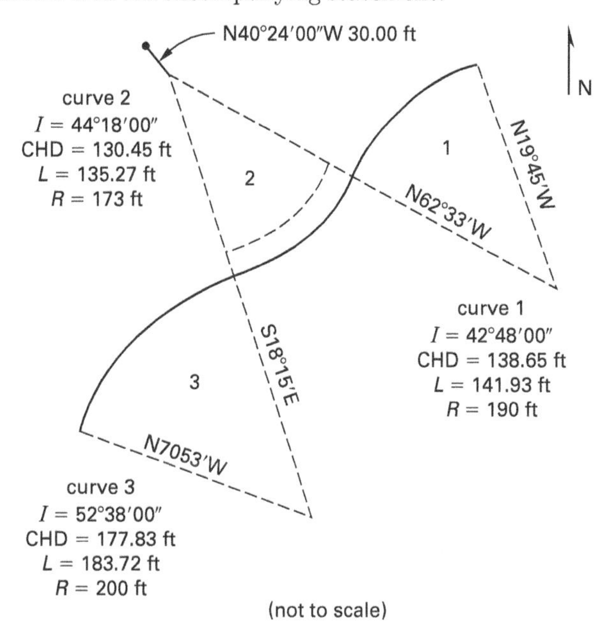

The curving centerline of a street must be changed. The center of curve 2 will be moved N40°24′00″W by 30.00 ft, but its radius will not change. The centers of the other two curves will remain unchanged as well.

44.1. Following the alteration, which list represents the central angles of the curves?

(A) 1 = 41°08′30″
2 = 41°01′24″
3 = 51°00′54″

(B) 1 = 41°48′08″
2 = 42°00′15″
3 = 53°38′23″

(C) 1 = 42°35′39″
2 = 44°18′25″
3 = 51°52′02″

(D) 1 = 42°40′15″
2 = 40°31′00″
3 = 52°35′50″

44.2. Following the alteration, which list correctly represents the revised radius of each curve?

(A) 1 = 190.00 ft
2 = 173.00 ft
3 = 200.00 ft

(B) 1 = 190.00 ft
2 = 200.93 ft
3 = 200.00 ft

(C) 1 = 204.62 ft
2 = 186.31 ft
3 = 214.98 ft

(D) 1 = 217.95 ft
2 = 173.00 ft
3 = 227.96 ft

44.3. Following the alteration, which list correctly represents the long chord of each curve?

(A) 1 = 153.16 ft
2 = 121.24 ft
3 = 196.33 ft

(B) 1 = 154.66 ft
2 = 123.88 ft
3 = 202.96 ft

(C) 1 = 155.32 ft
2 = 115.69 ft
3 = 202.96 ft

(D) 1 = 156.48 ft
2 = 123.88 ft
3 = 135.27 ft

44.4. Following the alteration, which list correctly represents the arc length of each curve?

(A) 1 = 143.67 ft
2 = 127.37 ft
3 = 231.07 ft

(B) 1 = 153.14 ft
2 = 121.25 ft
3 = 196.33 ft

(C) 1 = 155.03 ft
2 = 128.93 ft
3 = 200.86 ft

(D) 1 = 156.50 ft
2 = 123.87 ft
3 = 202.97 ft

44.5. Following the alteration, which list correctly represents the bearing of the long chord of each curve?

(A) 1 = S48°51′00″W
2 = S49°36′00″W
3 = S45°26′00″W

(B) 1 = S49°15′54″W
2 = S49°36′51″W
3 = S45°01′48″W

(C) 1 = S49°40′45″W
2 = S49°37′18″W
3 = S44°37′44″W

(D) 1 = S49°51′25″W
2 = S49°26′09″W
3 = S44°59′43″W

PROBLEM 45

The following three problems refer to the illustration shown.

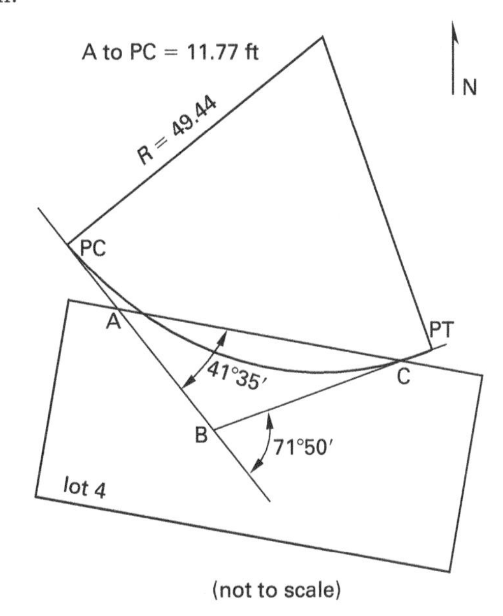

(not to scale)

45.1. The circular curve shown is the southerly edge of a proposed right-of-way. The encroachment of the curve onto lot 4 is not acceptable, so the curve must be realigned. A compound curve will be constructed; both curves will be tangent to the northern boundary of lot 4. The first curve of the compound curve will begin at the PC, and point A will be its PI. What will be the radius of the first curve of the compound curve?

(A) 31.00 ft

(B) 47.35 ft

(C) 49.44 ft

(D) 52.33 ft

45.2. Point C is to be the PI of the second curve. The second curve in the compound curve begins at the PCC at the end of the first curve. Using the tangent distance from the PCC to point C and the prolongation of the original tangent as shown in the following illustration, what is the radius of the second curve?

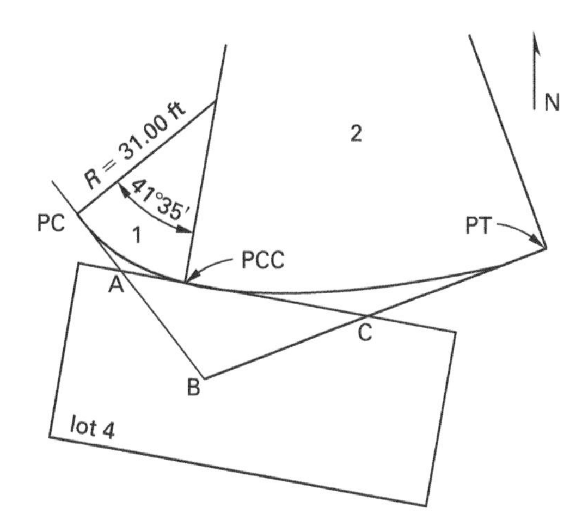

(A) 121.35 ft

(B) 122.30 ft

(C) 123.25 ft

(D) 124.20 ft

45.3. Which list correctly represents the lengths of the long chords of the first and the second curves, respectively?

(A) 1 = 20.18 ft
2 = 50.43 ft

(B) 1 = 20.18 ft
2 = 64.21 ft

(C) 1 = 22.01 ft
2 = 64.81 ft

(D) 1 = 23.89 ft
2 = 64.81 ft

PROBLEM 46

Which of the following statements concerning plane horizontal circular curves is true?

(A) Any three points not in a straight line define a unique circular curve.

(B) Any two points not in a straight line define a unique circular curve.

(C) Any four points not in a straight line define a unique circular curve.

(D) All of the above are true.

PROBLEM 47

Which of the following statements regarding the perpendicular bisectors of all chords of a circle is correct?

(A) They will divide the circle into sectors of equal area.

(B) They will be perpendicular to each other.

(C) They will be equal to twice the length of the chord they bisect.

(D) They will intersect at the center of the circle.

PROBLEM 48

The following three problems refer to the illustration shown.

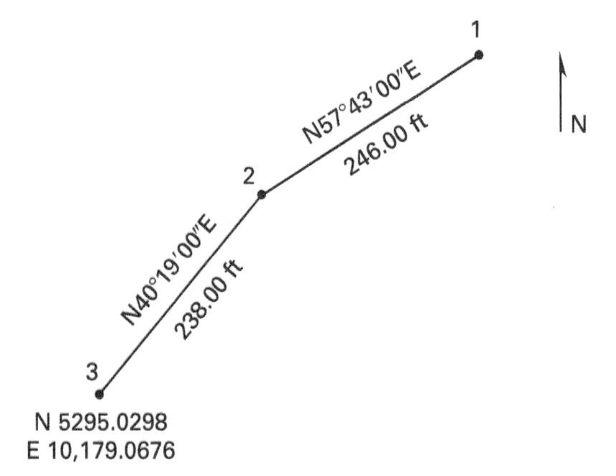

48.1. What is the radius of the curve that passes through points 1, 2, and 3?

(A) 790.43 ft

(B) 791.04 ft

(C) 799.94 ft

(D) 800.00 ft

48.2. What is the length of the arc of the curve in the illustration?

(A) 239.21 ft

(B) 358.82 ft

(C) 478.43 ft

(D) 485.86 ft

48.3. What is the bearing of the long chord and its length for the curve in the illustration?

(A) N40°19′00″E and 478.43 ft

(B) N49°09′42″E and 478.43 ft

(C) N49°09′42″E and 485.86 ft

(D) N57°43′00″E and 485.86 ft

AREAS BOUNDED BY CURVES

PROBLEM 49

A portion of the area within a circle is called a *sector.* Which of the following statements best defines the meaning of the word?

(A) the area bounded by the long chord and the arc of a curve

(B) the area bounded by the semitangents and the arc of a curve

(C) the area bounded by two radii and the arc of a curve

(D) the area bounded by two radii and the semitangents of a curve

PROBLEM 50

Which of the following mathematical expressions correctly describes the area within a circle?

(A) $2\pi R$

(B) πD

(C) πR^2

(D) $\pi^2 R$

PROBLEM 51

What is the area, in acres, of a sector with a central angle of 33° within a circle that has a radius of 1400.00 ft?

(A) 10.75 ac

(B) 12.96 ac

(C) 13.00 ac

(D) 14.35 ac

PROBLEM 52

A portion of the area within a circle is known as the *segment.* Which of the following statements best defines the term?

(A) the area bounded by two radii and the included long chord

(B) the area bounded by two radii and the attendant semitangents

(C) the area bounded by the arc of a curve and its two semitangents

(D) the area bounded by the arc of a curve and its long chord

PROBLEM 53

What is the area, in square feet, of the segment included in a curve with a radius of 52.46 ft and a central angle of 23°42′?

(A) 16 ft^2

(B) 21 ft^2

(C) 270 ft^2

(D) 280 ft^2

PROBLEM 54

The area between a circular curve and its tangents is sometimes called the *fillet.*

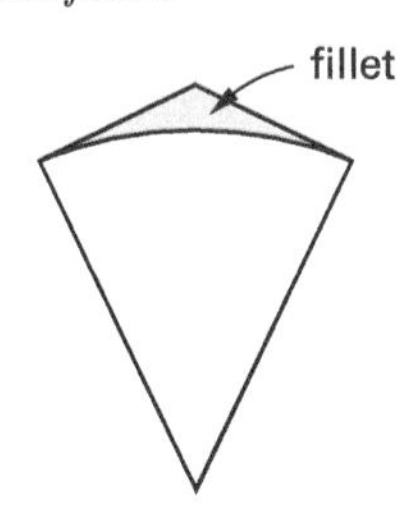

Which of the following mathematical expressions correctly defines the area of a fillet?

(A) RT

(B) $\frac{LR}{2}$

(C) $R\left(T-\frac{L}{2}\right)$

(D) $\frac{I\pi R^2}{360°}$

PROBLEM 55

A horizontal circular curve with a radius of 270.00 ft and a central angle of 55° will have a fillet of which of the following areas?

(A) 2506.88 ft^2

(B) 2643.91 ft^2

(C) 2921.75 ft^2

(D) 2959.20 ft^2

PROBLEM 56

The following illustration represents two concentric circular curves. The outermost curve has a radius of 39.73 ft, and both curves have a central angle of 53°29′. What is the area between the two arcs?

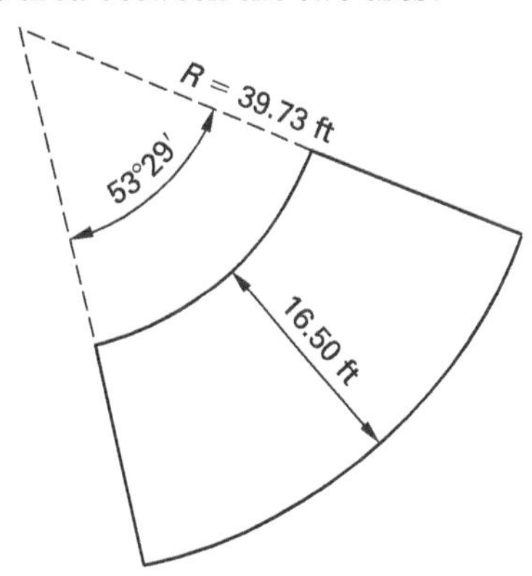

(A) 484.85 ft^2

(B) 487.31 ft^2

(C) 499.95 ft^2

(D) 500.00 ft^2

VERTICAL CURVES

PROBLEM 57

What is the curve that will result from the cutting of a right circular cone by a plane parallel with the side of the cone?

(A) a loxodrome

(B) an ellipse

(C) a catenary

(D) a parabola

PROBLEM 58

Which of the following statements is true of a vertical curve?

(A) The center of the curve is always either the highest or lowest point of the curve.

(B) The tangents are always of equal length.

(C) The tangent offsets always vary as the square of their distances from the point of tangency.

(D) The central angle is always 90°.

PROBLEM 59

A particular vertical curve passes over the summit of a hill. The grade of its first tangent is positive and the grade of its second tangent is negative. The algebraic difference between the two grades divided by the total length of the curve will provide which of the following values?

(A) the average slope of the tangents

(B) the total change in grade from the beginning of the vertical curve to its ending

(C) the rate at which the grade changes along the vertical curve

(D) the square of the tangent offset at the PVI

PROBLEM 60

If r is used to stand for the rate of change for a vertical curve and x for the distance from the point of tangency, which of the following expressions would yield the tangent offset for any station along the curve?

(A) $\left(\frac{r}{2}\right)x^2$

(B) rx

(C) $r^2\left(\frac{x}{2}\right)$

(D) $\frac{2r}{x}$

PROBLEM 61

Besides its elevation, what is unique about the highest point on a summit type of vertical curve?

(A) The slope of a tangent to the highest point on a summit vertical curve is zero.

(B) The tangent offset is the largest to the highest point on a summit vertical curve.

(C) The highest point on a summit vertical curve always occurs precisely beneath the PVI.

(D) Both A and B are true.

PROBLEM 62

A symmetrical vertical curve may be said to have a long chord—a straight line connecting the PVC with the PVT. The PVI occurs at the intersection of the lines of constant grade that are the tangents of the vertical curve. Which statement applies to these elements?

(A) The vertical distance from the midpoint of the long chord to the midpoint of the curve is equal to the vertical distance from the midpoint of the curve to the PVI.

(B) The highest point of a vertical curve occurs at the midpoint.

(C) The grade on the curve at the PVI is always zero.

(D) The station of the PVI is dependent on the grade of the long chord.

PROBLEM 63

If x is the distance along the vertical curve from the PVC to the high or low point, g_1 is the initial grade, and r is the rate of change, which mathematical expression is true?

(A) $x = \frac{-g_1}{r}$

(B) $-g_1x = r$

(C) $\frac{r}{x} = g_1$

(D) $-g_1r = x$

PROBLEM 64

The following five problems refer to the illustration shown.

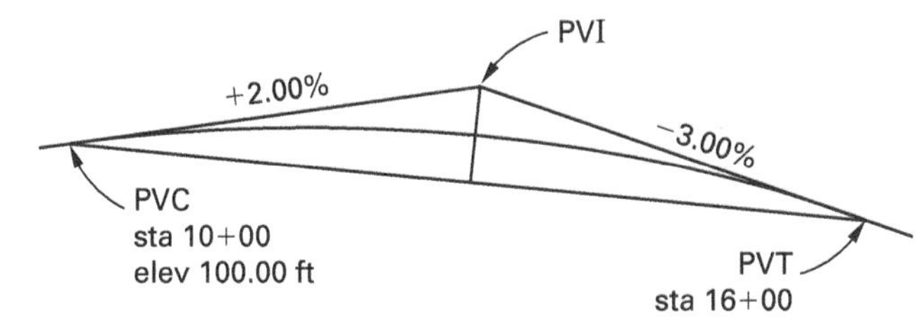

64.1. Referring to the symmetrical vertical curve, what is the elevation of the PVI, PVT, and the midpoint of the long chord?

(A) PVI = 100.60 ft
PVT = 99.70 ft
chord midpoint = 99.85 ft

(B) PVI = 106.00 ft
PVT = 97.00 ft
chord midpoint = 98.50 ft

(C) PVI = 122.20 ft
PVT = 88.90 ft
chord midpoint = 90.80 ft

(D) PVI = 160.00 ft
PVT = 70.00 ft
chord midpoint = 85.00 ft

64.2. Referring to the same symmetrical vertical curve, what is the largest tangent offset distance from the tangent to the curve anywhere along its length?

(A) 1.67 ft
(B) 2.60 ft
(C) 3.75 ft
(D) 4.90 ft

64.3. Referring to the same symmetrical vertical curve, what is the tangent offset at station 12+50?

(A) 0.10 ft
(B) 0.42 ft
(C) 0.94 ft
(D) 2.60 ft

64.4. What is the station of the highest point of the vertical curve?

(A) 10+03
(B) 10+50
(C) 12+40
(D) 13+12

64.5. What is the elevation of the highest point on the vertical curve?

(A) 100.90 ft
(B) 101.90 ft
(C) 102.25 ft
(D) 102.40 ft

PROBLEM 65

The following three problems refer to the vertical curve described.

A surveyor must stake a symmetrical vertical curve where an entering grade of +0.80% meets an exiting grade of −0.40% at station 90+00, which has an elevation of 100.00 ft. The maximum allowable change in grade per 100.00 ft station is −0.2%.

65.1. What is the length, in stations, of the vertical curve described?

(A) 5.5 sta
(B) 6.0 sta
(C) 8.2 sta
(D) 9.1 sta

65.2. What is the elevation of 92+00 on the vertical curve described?

(A) 98.30 ft
(B) 98.80 ft
(C) 99.10 ft
(D) 99.20 ft

65.3. What is the highest point on the curve previously described?

(A) 99.10 ft
(B) 99.20 ft
(C) 99.43 ft
(D) 100.00 ft

PROBLEM 66

The following two problems refer to the illustration shown.

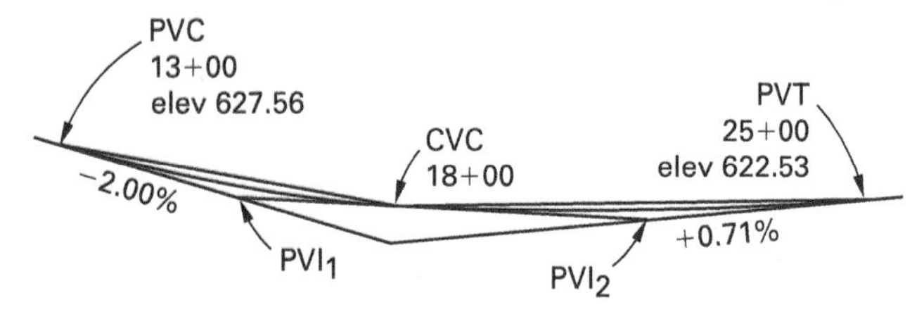

66.1. What is the elevation of the point of compound vertical curvature, CVC?

(A) 620.96 ft

(B) 621.51 ft

(C) 622.98 ft

(D) 625.72 ft

66.2. Referring to the same asymmetrical vertical curve, what is the elevation of sta 20+60 on the curve?

(A) 620.96 ft

(B) 621.51 ft

(C) 622.53 ft

(D) 627.56 ft

AREAS

PROBLEM 67

Which of the four-sided figures described contains an area of 1 ac?

(A) a square that is 209.10 ft on a side

(B) a rectangle that is 2 chains by 5 chains

(C) a square that is 10 chains on a side

(D) both B and C

PROBLEM 68

A property owner wishes to have a lot created from his property that is exactly square and contains 3 ac. What should be the dimension of one side of the square, in feet?

(A) 208.71 ft

(B) 290.81 ft

(C) 295.16 ft

(D) 361.50 ft

PROBLEM 69

What is the area, in acres, of the trapezoid shown?

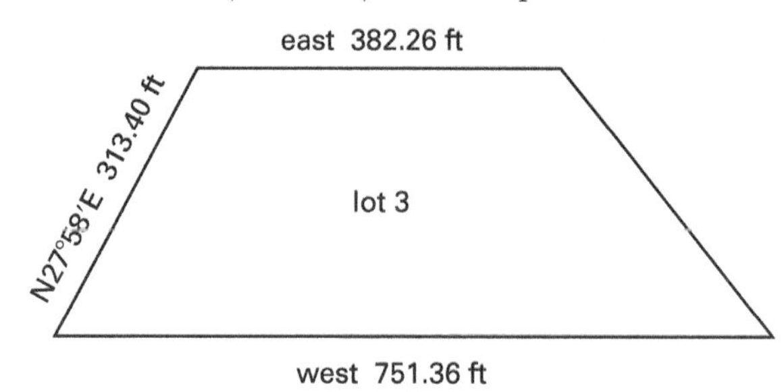

(A) 1.7 ac

(B) 3.0 ac

(C) 3.6 ac

(D) 4.2 ac

PROBLEM 70

A large lot is to be subdivided as indicated. The owner has stipulated that the parcel labeled 1 must contain 6 ac.

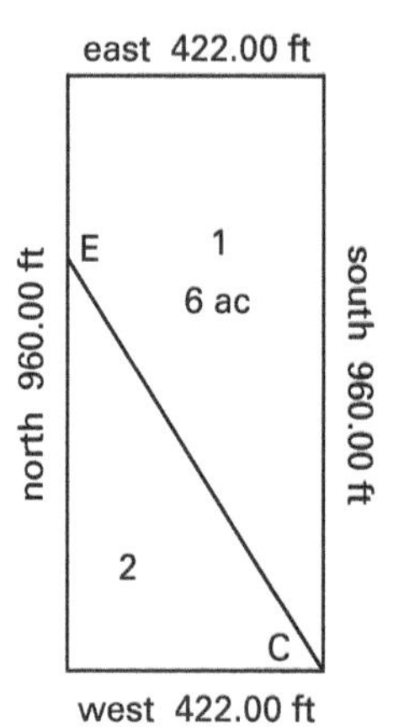

What is the length of line EC?

(A) 502.12 ft

(B) 681.33 ft

(C) 763.98 ft

(D) 801.38 ft

PROBLEM 71

The area of triangle XYZ contains an eighth of the area within triangle XAB.

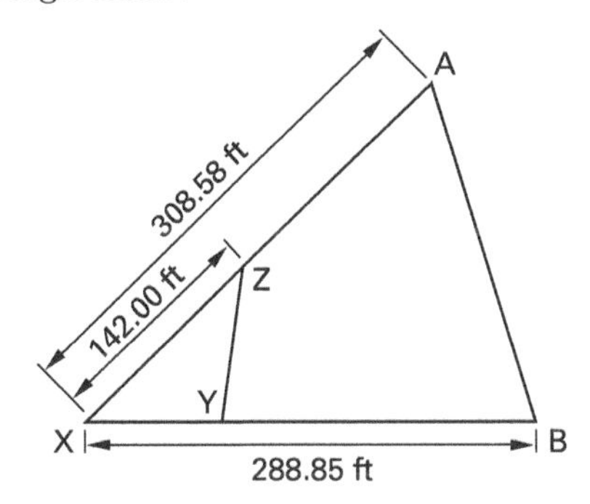

What is the length of line XY?

(A) 53.47 ft

(B) 78.46 ft

(C) 95.21 ft

(D) 103.45 ft

PROBLEM 72

A right-of-way, 100.00 ft wide, crosses a lot as shown in the following illustration.

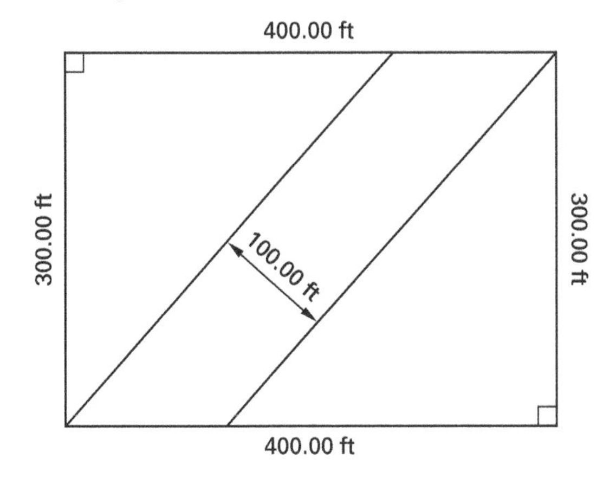

What is the area, in acres, within the right-of-way?

(A) 0.73 ac

(B) 0.88 ac

(C) 0.92 ac

(D) 1.01 ac

PROBLEM 73

In order to determine the area of a parcel that adjoins a small lake, a line was established nearly parallel to the lake's shore and stationed from 0+00 to 1+75, at intervals of 25.00 ft. Measurements perpendicular to the line were taken from each station to the lake's shore.

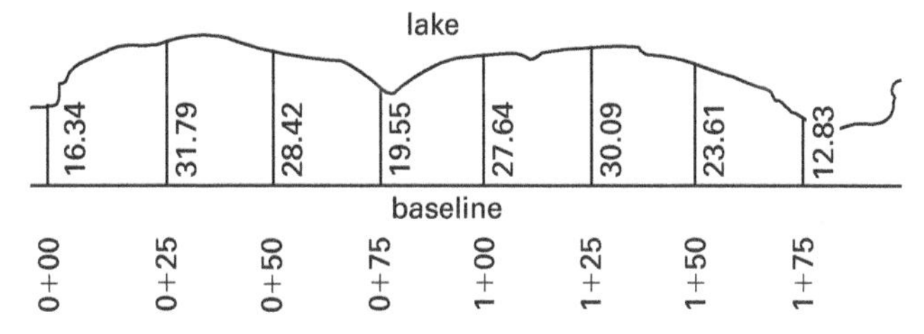

From the measurements given, what is the area, in acres, between the line and the lake?

(A) 0.08 ac

(B) 0.10 ac

(C) 0.22 ac

(D) 0.35 ac

PROBLEM 74

To find the area between a straight fence and the bottom of a dry gulch, stationing was established along the fence and perpendicular offsets measured from them to the bottom of the ravine. All of the given distances were measured to the nearest foot.

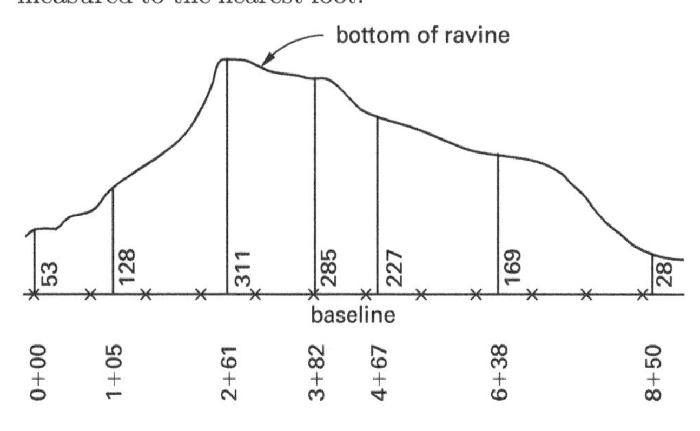

What is the area, in acres, between the fence and the bottom of the ravine?

(A) 1.9 ac

(B) 2.7 ac

(C) 3.2 ac

(D) 3.6 ac

VOLUMES

PROBLEM 75

What quantity, in cubic feet, is contained within 1 yd^3?

(A) 6 ft^3

(B) 9 ft^3

(C) 14 ft^3

(D) 27 ft^3

PROBLEM 76

Which of the following methods will consistently produce a more accurate volume computation than the others?

(A) DMD method

(B) coordinate method

(C) average end area method

(D) prismoidal method

PROBLEM 77

Which of the following characteristics of a route survey would be likely to increase the difference between the volumes calculated using the prismoidal method and the same volumes calculated using the average end area method?

(A) Consecutive cross sections, or end areas, change shape and area very little along the route.

(B) Consecutive cross sections change shape and area dramatically along the route.

(C) The ground over which the route passes is quite irregular and the cross sections are taken at shorter than usual intervals.

(D) Both B and C are true.

PROBLEM 78

Which of the following statements characterizes a fairly consistent difference between volumes calculated by the average end area method and those calculated by the prismoidal method?

(A) Volumes calculated by the prismoidal method are consistently larger than those calculated by the average end area method.

(B) Volumes calculated by the prismoidal method are consistently smaller than those calculated by the average end area method.

(C) Volumes calculated by the prismoidal method are less reliable than those calculated by the average end area method, but are more frequently used.

(D) Volumes calculated by the average end area method are only an approximation of the actual quantity of earth, but those calculated by the prismoidal method are not.

PROBLEM 79

A storage pile of crushed rock is 100.00 ft square at its base and 50.00 ft square on the top. The pile is 25.00 ft high. What is the volume of rock in the pile?

(A) 5208 yd^3

(B) 5401 yd^3

(C) 6944 yd^3

(D) 7593 yd^3

PROBLEM 80

Two level cross sections a full station (100 ft) apart have center heights of 3.60 ft and 4.80 ft, respectively. Both cross sections are in cut and the base width of the designed roadway is 44.00 ft. The side slopes are $1\frac{1}{2}$:1.

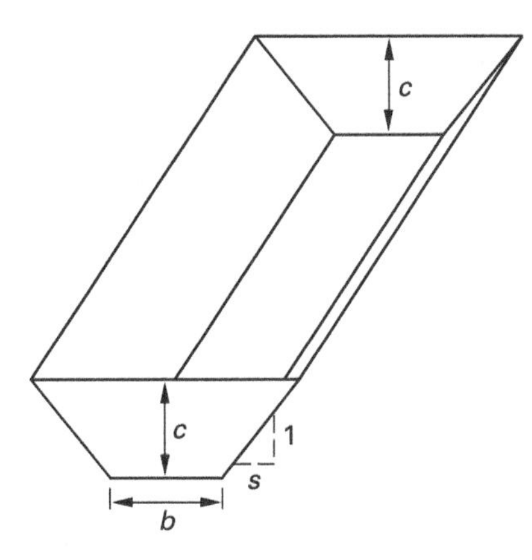

What is most nearly the volume between these cross sections calculated by the average end area method?

(A) 780 yd^3

(B) 900 yd^3

(C) 950 yd^3

(D) 1000 yd^3

PROBLEM 81

The following four problems refer to the following illustration and the data returned from the setting of slope stakes along a proposed road. All dimensions are in feet.

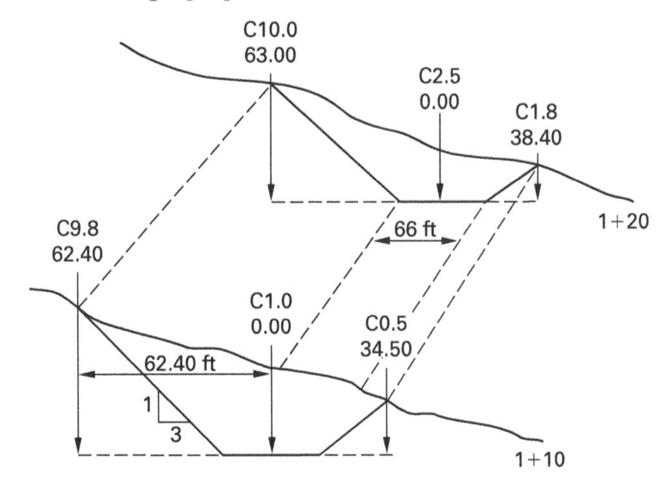

width of the roadbed = 66 ft
side slope ratio = 3:1

1+10	C9.8	C1.0	C0.5
	62.40	0.00	34.50
1+20	C10.0	C2.5	C1.8
	63.00	0.00	38.40

81.1. Referring to the data, what is the cross-sectional area indicated at stations 1+10 and 1+20, respectively?

(A) 193.56 ft^2 and 245.13 ft^2

(B) 218.40 ft^2 and 321.45 ft^2

(C) 270.99 ft^2 and 363.04 ft^2

(D) 285.93 ft^2 and 308.71 ft^2

81.2. Using the average end area method, what is most nearly the volume between the cross sections at 1+10 and 1+20?

(A) 97 yd^3

(B) 100 yd^3

(C) 110 yd^3

(D) 120 yd^3

81.3. Using the prismoidal method, what is the volume between the cross sections at 1+10 and 1+20?

(A) 98.9 yd^3

(B) 99.8 yd^3

(C) 100.2 yd^3

(D) 101.1 yd^3

81.4. What is the prismoidal correction that would be applied to the volume found by the average end area to approximate the volume found by the prismoidal method?

(A) 0.0 yd^3

(B) 0.1 yd^3

(C) 0.2 yd^3

(D) 0.4 yd^3

ADJUSTMENTS

PROBLEM 82

The calculation of a closed traverse reveals that the closing coordinates of the station from which the traverse began do not match the coordinates assigned to it at the beginning of the calculation. The coordinates are in feet.

N 5000.000
E 10,000.000 (coordinates assigned at the beginning)
N 5000.324
E 9999.871 (closing coordinates at the same station)

Which of the following answers correctly represents the linear error of closure of the traverse and its direction?

(A) 0.349 ft, N21°42′36″W

(B) 0.453 ft, N31°29′12″E

(C) 0.698 ft, N68°17′24″E

(D) 0.929 ft, N20°31′46″W

PROBLEM 83

When a traverse is adjusted by the compass rule, which of the following statements correctly defines the relationship between the correction applied to the departure of one course and the total correction in departure?

(A) It is in the same proportion as that of the length of that course to the entire length of the traverse.

(B) It is in the same proportion as that of the azimuth of that course to the sum of the interior angles of the entire traverse.

(C) It is inversely proportional to the sum of the squares of the lengths of the entire traverse.

(D) It is in the same proportion as that of the length of that course to the arithmetic sum of all the departures.

PROBLEM 84

The sum of the interior angles of a closed plane traverse of 12 sides has a theoretically correct value. What is that value?

(A) 1620°

(B) 1800°

(C) 1980°

(D) 2160°

PROBLEM 85

The sum of the exterior angles of a closed plane traverse with eight sides has a theoretically correct value. What is that value?

(A) 1080°

(B) 1440°

(C) 1620°

(D) 1800°

PROBLEM 86

Which of the following statements is an assumption appropriate to the compass rule, but not to the transit rule?

(A) It is impossible to exactly determine the magnitude of the random errors in the angles and of a traverse.

(B) The angular measurements are more precise than the distance measurements.

(C) The systematic errors in a traverse have been, as much as is possible, eliminated from the measurements before any adjustment is undertaken.

(D) The effect of the errors in angular measurements is equal to the effect of the errors in the distance measurements.

PROBLEM 87

While turning six complete sets of angles, direct and reverse, from a triangulation station with a one-second optical theodolite, a surveyor incremented the plate at the start of each new set. The first set backsight was near 0° on the plate, the next set began near 30°, the next near 60°, and so on in increments of 30° until all six sets were complete. Which statement best describes the primary reason for the incrementation?

(A) The incrementation was a correction for refraction and parallax.

(B) The incrementation was an effort to ensure honest note keeping.

(C) The incrementation was an effort to minimize the effect of pointing errors at the backsight station.

(D) The incrementation was an effort to minimize the effect of the systematic error in the plate of the instrument.

PROBLEM 88

Which of the following statements is NOT a basis for the adjustment of surveys?

(A) Measurements of the same quantity may agree within very close limits and still be quite inaccurate because of the presence of uncompensated systematic errors.

(B) The effect of random errors can be eliminated from a survey by a properly applied adjustment.

(C) Errors and mistakes are distinctly different in surveying. Mistakes may be entirely avoided, but errors may not.

(D) In some cases, a distribution of the errors in a survey based on the knowledge of the field conditions at the time the measurements were made is superior to a mathematically structured adjustment.

PROBLEM 89

The interior angles of a closed traverse of six sides are given.

118°15′58″
93°59′01″
23°44′10″
269°21′00″
98°00′39″
116°38′12″

Each angle has been measured with equal precision. Which of the following angles would be the correct adjusted value for the first angle?

(A) 118°15′48″

(B) 118°15′58″

(C) 118°16′08″

(D) 118°16′15″

PROBLEM 90

What relative error is the closest to that which would be the result of an angular error of 20″?

(A) 1 part in 5000

(B) 1 part in 10,000

(C) 1 part in 20,000

(D) 1 part in 100,000

PROBLEM 91

In surveying, the words *precision* and *accuracy* are not synonymous. Which of the following statements correctly defines the differences between the two terms?

(A) Precision is the degree of agreement between a measurement and the true value; accuracy is the degree of agreement between repetitive measurements of the same value.

(B) Accuracy refers to the final result of a measurement; precision refers to the procedure by which the result was obtained.

(C) Higher precision in an operation tends to indicate reduced random errors; higher accuracy tends to indicate reduced systematic errors.

(D) Both B and C are true.

PROBLEM 92

The formula, $\sqrt{\sum v^2}/\sqrt{n-1} = 1\sigma$, where $\sum v^2$ is the sum of the squares of the residuals and n is the number of observations, yields the standard deviation of a series of measurements of a single quantity. What is most nearly the percent of probability associated with the standard deviation?

(A) 68%

(B) 90%

(C) 95%

(D) 100%

PROBLEM 93

If a series of measurements is made of the same quantity, which of the following values is most likely to be closest to the true value?

(A) the median

(B) the mode

(C) the mean

(D) the residual

PROBLEM 94

Point Allen and point 100 are two adjacent points on a seven-sided unadjusted closed traverse. The state plane coordinates of the points, in feet, are shown.

Allen
N 2,125,651.942
E 738,869.199

100
N 2,132,478.378
E 745,786.469

The sum of the lengths of the traverse, which begins at point Allen, is 63,954.250 ft. The first of the points along the traverse is point 100 and the closing coordinate at point Allen is

N 2,125,652.070
E 738,869.068

The error of closure at point Allen is +0.128 ft in the northing and −0.131 ft in the easting. The unadjusted distance between point Allen and point 100 is 9718.480 ft, and the angles of the traverse are balanced.

Using the compass rule, what are the adjusted coordinates of point 100?

(A) N 2,132,478.319
E 745,786.402

(B) N 2,132,478.359
E 745,786.489

(C) N 2,132,478.398
E 745,786.449

(D) N 2,132,478.412
E 745,786.489

PROBLEM 95

Three NGS stations, Jackson, Nill, and Red Peak, form a triangle that encloses an area of 131.24 mi². The interior angles of this triangle are 52°28′15″, 50°15′10″, and 77°16′32″. Presuming 1 second of spherical excess is accumulated when a figure encloses 75.60 mi², which angles represent the correctly adjusted values?

(A) 52°28′15.42″, 50°15′10.42″, and 77°16′32.42″

(B) 52°28′16″, 50°15′11″, and 77°16′33″

(C) 52°28′16.33″, 50°15′11.33″, and 77°16′33.33″

(D) 52°28′16.58″, 50°15′11.58″, and 77°16′33.58″

DIRECTIONS

PROBLEM 96

What is the south azimuth of a line with a bearing of S10°59′01″E?

(A) 10°59′01″

(B) 79°00′59″

(C) 169°00′59″

(D) 349°00′59″

PROBLEM 97

Occupying point B at the easterly end of a line AB that bears S89°56′15″E, a deflection angle of 00°04′15″ is turned to the left to point C. Occupying point C, a deflection angle of 00°00′35″ is turned to the right to point D. What is the azimuth from north of line CD?

(A) 89°56′15″

(B) 89°59′55″

(C) 90°00′00″

(D) 90°00′05″

PROBLEM 98

Line EF bears N65°14′02″E and line FG bears S 75°23′10″E. What is the deflection angle to the right at point F if the direction of the survey is from west to east?

(A) 10°09′08″

(B) 39°22′48″

(C) 50°37′12″

(D) 140°37′12″

PROBLEM 99

A surveyor retracing a boundary that was originally established in 1886 using magnetic bearings finds that a particular line has a magnetic bearing of record of N01°20′W, when the magnetic declination was 0°31′E. Today the magnetic declination is 0°12′W, and the true bearing of the same line is N00°09′W. Which pair of angles correctly reflects the difference between the two magnetic bearings of the line and the two true bearings of the line, respectively?

(A) 00°52′ (difference between the magnetic bearings)
00°40′ (difference between the true-bearings)

(B) 01°17′ (difference between the magnetic bearings)
00°59′ (difference between the true bearings)

(C) 01°23′ (difference between the magnetic bearings)
00°40′ (difference between the true bearings)

(D) 01°23′ (difference between the magnetic bearings)
00°59′ (difference between the true bearings)

PROBLEM 100

The grid azimuth, from north, from station Morrison to station Boulder is 348°32′24.0″. The convergence angle (i.e., gamma angle), γ, in the Lambert projection at Morrison is +00°10′53.2″. What is the geodetic azimuth, from north, from Morrison to Boulder?

(A) 348°21′30.8″

(B) 348°32′24.0″

(C) 348°40′21.8″

(D) 348°43′17.2″

PROBLEM 101

The geodetic azimuth, from south, from station Adams to station Douglas is 25°17′49.11″. If the convergence angle, γ, at station Adams is +00°34′37.61″, what is the grid azimuth, from north, from Douglas to Adams?

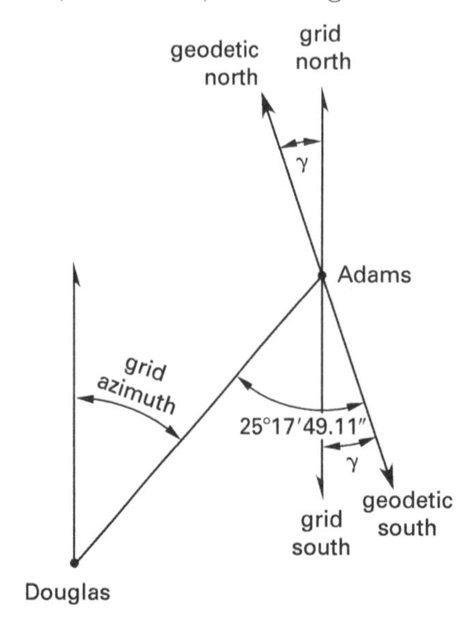

(A) 24°43′11.50″

(B) 25°52′26.72″

(C) 204°43′11.50″

(D) 205°52′26.72″

PROBLEM 102

The deviation of the vertical is an effect that influences the difference between an astronomic azimuth and the geodetic azimuth at a point. Which of the following statements describes an aspect of the deviation of the vertical?

(A) It is the second term correction applied to lines over 5 mi long.

(B) It is an orthometric height.

(C) It is the small angle between a line that is normal to the spheroid through a point and a line through the same point that is normal to the geoid.

(D) It is the angle between an azimuth to Polaris at elongation and an azimuth to Polaris at culmination.

SLOPE INTERCEPTS

PROBLEM 103

Which statement describes the typical position of a slope stake?

(A) Slope stakes are set at the intersection of the profile grade and the natural ground.

(B) Slope stakes are set at the intersection of the natural ground and the side slopes.

(C) Slope stakes are set at each PVC on the centerline of a roadway.

(D) Slope stakes are set at the beginning of a superelevated curve.

PROBLEM 104

A levee built on level ground is parallel to a river channel. The width of the levee on the top is 20 ft. The side slope on the river side is 3:1, and the side slope on the land side is 2:1. The levee is 15 ft high at its center. What is the width of the base of the levee?

(A) 50 ft

(B) 75 ft

(C) 95 ft

(D) 100 ft

PROBLEM 105

The following field data have been used to set a slope stake. The width of the roadway is 60 ft and the side slopes are $1\frac{1}{2}$:1. The elevation of the subgrade is 609.45 ft.

$$\text{HI} = 618.23 \text{ ft}$$
$$\text{ground rod} = 7.3 \text{ ft}$$

What is most nearly the distance from the center of the roadway to the slope stake?

(A) 30 ft

(B) 32 ft

(C) 34 ft

(D) 36 ft

PROBLEM 106

Which of the following best defines the term *angle of repose*?

(A) the angle between the normal of one centerline and the normal of an intersecting centerline

(B) an angle of elevation or depression along a vertical curve

(C) the vertical angle between the true horizon and the perspective center

(D) the angle between the horizontal plane and the slope of the material in a loosely compacted embankment

SOLUTION 1

The key to this question is the adoption of the survey foot rather than the international foot. Some states have adopted the international foot, which is defined by the wavelength of light emitted when krypton gas is electrically excited. Choices (A) and (D) are correct relationships for describing the international foot.

$$\begin{aligned} 1\ \text{ft} &= 12\ \text{in} \\ 1\ \text{m} &= 39.37\ \text{U.S. in} \\ 1\ \text{U.S. survey ft} &= \frac{12\ \text{in}}{39.37\ \frac{\text{in}}{\text{m}}} \\ &= 0.3048006\ \text{m} \end{aligned}$$

The answer is (B).

SOLUTION 2

The number of whole feet remains the same (1 ft). To convert 0.40 ft into inches, use a proportion.

$$\begin{aligned} \frac{40}{100} &= \frac{x}{12\ \text{in}} \\ x &= (12\ \text{in})\left(\frac{40}{100}\right) \\ &= 4.8\ \text{in} \end{aligned}$$

Since the answer must be in eighths of an inch, the next step is to convert 8/10 of an inch into eighths of an inch.

$$\begin{aligned} \frac{8}{10} &= \frac{x}{8} \\ x &= (8)\left(\frac{8}{10}\right) \\ &= 6.4 \\ \frac{6.4}{8}\ \text{in} &= \frac{3.2}{4}\ \text{in} \quad \left[\text{round to } \tfrac{3}{4}\ \text{in}\right] \end{aligned}$$

The final result, to the nearest eighth of an inch, is 1 ft, $4\frac{3}{4}$ in.

The answer is (D).

SOLUTION 3

A radian is the angle at the center of a circle subtended by an arc with a length equal to the circle's radius. The radian is a unit convenient in calculating circular curves that are expressed in degrees of curve. One radian is equal to $180^\circ/\pi$, or $57^\circ 17' 44.8''$.

The answer is (B).

SOLUTION 4

A hectare is a measure of land area in the metric system. The term is built on the word *are*, which is 100 m^2, with the prefix *hect*, which indicates multiplication by 100. Therefore, a hectare is equal to 10,000 m^2, or 2.471 ac.

The answer is (D).

SOLUTION 5

The expression $(81)^{-1/4}$ may be expressed as the inverse of the fourth root of 81.

$$\begin{aligned} (81)^{-1/4} &= \frac{1}{(81)^{1/4}} = \frac{1}{\sqrt[4]{81}} \\ &= 1/3 \end{aligned}$$

The answer is (B).

SOLUTION 6

A grad, also known as *grade*, is one hundredth of a right angle. There are 400 grads in a circle, so the grad is a smaller increment than a degree. The degree is a unit of the sexagesimal system and the grad is a unit of the centesimal system. One grad is 0.9 of a degree, or $00^\circ 54'$.

The answer is (B).

SOLUTION 7

Early surveyors used a *rod*, otherwise known as a *pole* or a *perch*, whose length was approximately $16\frac{1}{2}$ ft. While these units are now considered archaic, they are still found in the public record. Eighteen rods is equivalent to 297 ft. If the current measurement is 300 ft, the difference is 3 ft, or 1 yd.

The answer is (B).

SOLUTION 8

According to the Celsius (or Centigrade) temperature scale, water freezes at 0° and boils at 100°. The corresponding freezing and boiling points on the Fahrenheit scale are 32° and 212°, respectively. The conversion from one to the other can be expressed as

$$\begin{aligned} {}^\circ\text{F} &= \frac{9}{5}{}^\circ\text{C} + 32^\circ \\ {}^\circ\text{C} &= \left(\frac{5}{9}\right)({}^\circ\text{F} - 32^\circ) \end{aligned}$$

To find 68°F in degrees Celsius,

$$
\begin{aligned}
{}^\circ\mathrm{C} &= \left(\frac{5}{9}\right)(68^\circ - 32^\circ) \\
&= \left(\frac{5}{9}\right)(36^\circ) \\
&= 20^\circ\mathrm{C}
\end{aligned}
$$

The answer is (B).

SOLUTION 9

In the public land survey system, the unit defined by Gunter's chain remains the standard, even though the instrument is long out of use. The chain is 66 ft long and is divided into 100 links. Therefore, the measurement can be expressed as

$$
\begin{aligned}
15 \text{ chains, } 64 \text{ links} &= 15.64 \text{ chains} \\
(15.64 \text{ chains})\left(66\ \frac{\text{ft}}{\text{chain}}\right) &= 1032.24 \text{ ft} \\
(1032.24 \text{ ft})\left(\frac{1 \text{ yd}}{3 \text{ ft}}\right) &= 344.08 \text{ yd} \quad (340 \text{ yd})
\end{aligned}
$$

The answer is (C).

SOLUTION 10

Express each fraction in 32nds, then add them together.

$$
\frac{8}{32} + \frac{2}{32} + \frac{17}{32} + \frac{20}{32} + \frac{1}{32} = \frac{48}{32} = 1\tfrac{16}{32} = 1\tfrac{1}{2} \text{ in}
$$

The answer is (C).

SOLUTION 11

All trigonometric functions are positive in the first quadrant. In the second quadrant, all are negative except two: the sine and the cosecant functions. Only two functions are positive in the third quadrant: the tangent and cotangent. In the fourth quadrant, the cosine and secant functions are positive; all others are negative.

The answer is (C).

SOLUTION 12

In an equilateral triangle, all sides and all angles are equal. Since the sum of the angles must be 180°, each angle of an equilateral triangle must be one-third of 180°, or 60°. The sine of 30° is 0.500.

The answer is (A).

SOLUTION 13

In a right triangle with legs represented by a and b, and the hypotenuse by c, the Pythagorean theorem indicates

$$a^2 + b^2 = c^2$$

Also,

$$
\begin{aligned}
\sin \mathrm{A} &= \frac{a}{c} \\
\cos \mathrm{A} &= \frac{b}{c}
\end{aligned}
$$

Using the formula $a^2 + b^2 = c^2$, reduce the right side of the equation to 1 by dividing every term by c^2.

$$
\begin{aligned}
\frac{a^2}{c^2} + \frac{b^2}{c^2} &= 1 \\
\left(\frac{a}{c}\right)^2 + \left(\frac{b}{c}\right)^2 &= 1 \\
\sin^2 \mathrm{A} + \cos^2 \mathrm{A} &= 1
\end{aligned}
$$

Finally, isolate sin A on the left side of the equation.

$$
\begin{aligned}
\sin^2 \mathrm{A} &= 1 - \cos^2 \mathrm{A} \\
\sin \mathrm{A} &= \pm\sqrt{1 - \cos^2 \mathrm{A}}
\end{aligned}
$$

The answer is (D).

SOLUTION 14

To find the height of the tower, it is helpful to sketch the problem parameters.

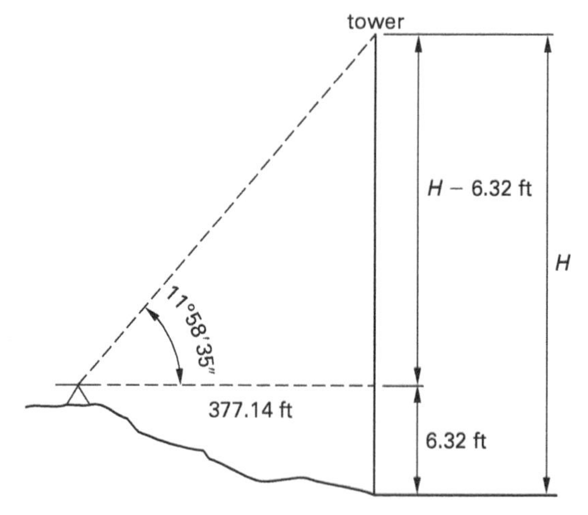

$$
\begin{aligned}
\tan 11^\circ 58' 35'' &= \frac{H - 6.32 \text{ ft}}{377.14 \text{ ft}} \\
H - 6.32 \text{ ft} &= (\tan 11^\circ 58' 35'')(377.14 \text{ ft}) \\
H &= 80.00 \text{ ft} + 6.32 \text{ ft} \\
&= 86.32 \text{ ft} \quad (86 \text{ ft})
\end{aligned}
$$

The answer is (D).

SOLUTION 15

First, sketch the problem parameters.

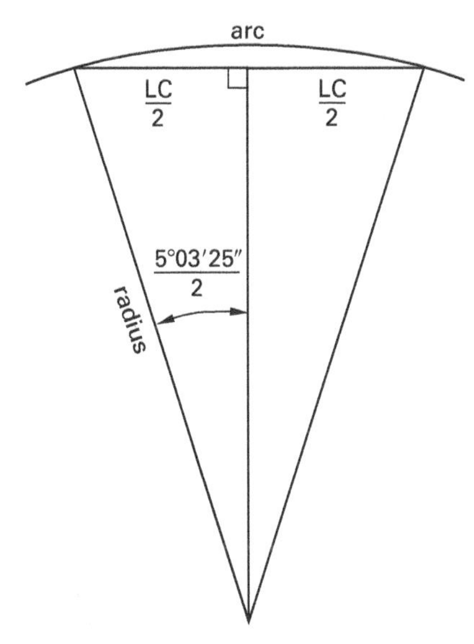

Dividing the angle 5°03′25″ in half allows the problem to be handled with two identical right triangles. The hypotenuse of each of the triangles is the radius. The distance to the center of the long chord is one leg of the triangle. One-half of the length of the long chord is the other leg.

$$\begin{aligned}\sin\frac{5^\circ03'25''}{2} &= \frac{\frac{\text{LC}}{2}}{\text{radius}}\\ \frac{\text{LC}}{2} &= (\sin 2^\circ31'42.5'')(4000\text{ ft})\\ &= 176.46\text{ ft}\\ \text{LC} &= 352.93\text{ ft}\quad(350\text{ ft at }4000\text{ ft})\end{aligned}$$

The answer is (D).

SOLUTION 16

Since 12/13 is the cosine, let $a = 12$ and $c = 13$. According to the Pythagorean theorem,

$$\begin{aligned}a^2 + b^2 &= c^2\\ a^2 - c^2 &= b^2\\ (12)^2 - (13)^2 &= b^2\\ b^2 &= 144 - 169 = 25\\ b &= \sqrt{25} = 5\end{aligned}$$

Therefore, the sine of the angle is $\frac{5}{13}$.

The answer is (B).

SOLUTION 17

The cosecant of angle C is indicated by a/h. When the fraction is multiplied by h, the result is a.

$$\left(\frac{a}{h}\right)h = a$$

The answer is (A).

SOLUTION 18

The distance from the station occupied by the instrument to the beginning of the circular curve, PC, is 137.48 ft, found by subtraction of the stations.

$$\begin{array}{r}12{+}37.48\\ -11{+}00.00\\ \hline 1{+}37.48\end{array} = 137.48\text{ ft}$$

The radius forms a right angle to the tangent at the PC. Since the backsight was taken down station, 180° is subtracted from the turned angle, leaving 62°15′18″ as the interior angle of a right triangle. The radius is found by solving the right triangle.

$$\begin{aligned}\tan 62^\circ15'18'' &= \frac{\text{radius}}{137.48\text{ ft}}\\ \text{radius} &= (137.48\text{ ft})(\tan 62^\circ15'18'')\\ &= 261.36\text{ ft}\quad(260\text{ ft})\end{aligned}$$

The answer is (C).

SOLUTION 19

Since the angles at A and B must each be 45° and the radius is 1, the length of AB is the cosecant of 45°, or $\csc\Delta/2$.

The answer is (D).

SOLUTION 20

The ratio that expresses the cotangent of angle A is b/a. Since the sine (a/c) is called x, its reciprocal, the cosecant (c/a), can be written as $1/x$. Multiply the cosine by the cosecant to find the cotangent.

$$\left(\frac{b}{c}\right)\left(\frac{c}{a}\right) = \frac{b}{a}$$

Substituting $(1/x)$ for (c/a),

$$\left(\frac{b}{c}\right)\left(\frac{1}{x}\right) = \frac{b}{cx}$$

$$\cot \mathrm{A} = \frac{b}{cx}$$

Expressed another way,

$$(\cos \mathrm{A})(\cos \mathrm{A}) = \cot \mathrm{A}$$

The answer is (B).

SOLUTION 21

The law of cosines can be expressed in several different forms. The correct form for this problem is one that is convenient in solving triangles where all of the sides and none of the angles are known, which is

$$\cos \mathrm{A} = \frac{b^2 + c^2 - a^2}{2bc}$$

The answer is (C).

SOLUTION 22

The triangle described by the three points is devoid of angles, but all three distances are available. The distance and bearing from the pipe to the axle are found by inverse, and the two other distances are taken from the deed. These descriptions can be multiplied by 66 to convert them to feet. A form of the law of cosines can be used to discover the missing angles from the available distances.

$$\cos \mathrm{A} = \frac{b^2 + c^2 - a^2}{2bc}$$

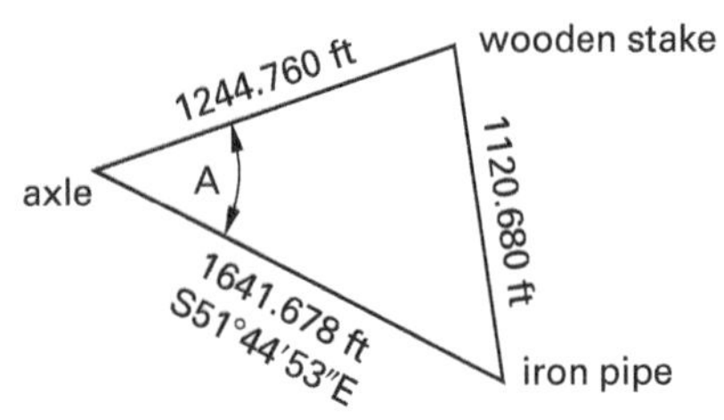

$$\begin{aligned}\cos \mathrm{A} &= \frac{(1244.760)^2 + (1641.678)^2 - (1120.680)^2}{(2)(1244.760)(1641.678)} \\ &= 0.73125 \\ \mathrm{A} &= 43°00'31.4''\end{aligned}$$

The azimuth from the axle to the pipe is

$$180° - 51°44'53'' = 128°15'07''$$

Therefore, the azimuth from the axle to the stake is

$$128°15'07'' - 43°00'31'' = 85°14'36''$$

The component distances from the axle to the stake are

$$(1244.760 \text{ ft})\sin 85°14'36'' = 1240.473 \text{ ft}$$

$$(1244.760 \text{ ft})\cos 85°14'36'' = 103.221 \text{ ft}$$

To find the coordinates of the stake, add the component distances from the axle to the stake to the coordinates of the axle.

$$7631.413 \text{ N} + 103.221 \text{ ft} = 7734.634\text{N}$$

$$5723.610 \text{ E} + 1240.473 \text{ ft} = 6964.083\text{E}$$

The answer is (C).

SOLUTION 23

In the oblique triangle shown in the problem, the distance along the line between the two stations is 40 ft. The two interior angles are 20°52′16″ and 104°25′09″. First, find the third angle, A.

$$\begin{aligned}\mathrm{A} &= 20°52'16'' + 104°25'09'' - 180°00'00'' \\ &= 125°17'25'' - 180°00'00'' \\ &= 54°42'35''\end{aligned}$$

Then use the law of sines to find b, the shortest side of the triangle.

$$\frac{a}{\sin \mathrm{A}} = \frac{b}{\sin \mathrm{B}}$$

$$\frac{40.00 \text{ ft}}{\sin 54°42'35''} = \frac{b}{\sin 20°52'16''}$$

$$b = (\sin 20°52'16'')\left(\frac{40.00 \text{ ft}}{\sin 54°42'35''}\right)$$

$$\frac{14.25 \text{ ft}}{\sin 54°42'35''} = 17.46 \text{ ft}$$

However, this distance is not the minimum distance between the line and the building corner. The minimum distance must be along a line perpendicular to the stationed line. One side of the right triangle including that perpendicular is the distance found, 17.46 ft. One angle of the same right triangle must be the supplement of the second turned angle.

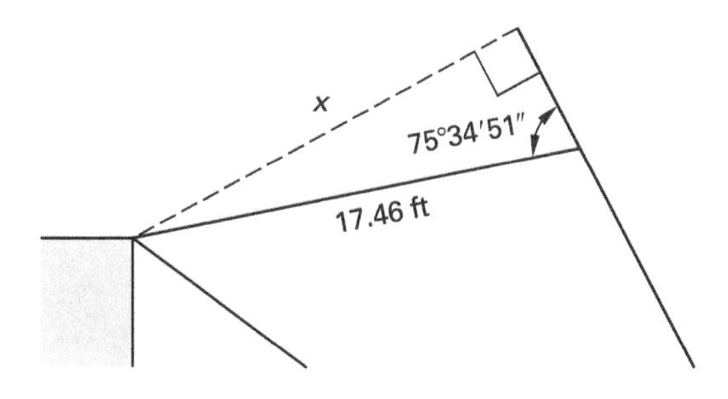

$$180°00'00'' - 104°25'09'' = 75°34'51''$$
$$\sin 75°34'51'' = \frac{x}{17.46 \text{ ft}}$$
$$x = (\sin 75°34'51'')(17.46 \text{ ft}) = 16.91 \text{ ft}$$

The answer is (A).

SOLUTION 24

This problem can be solved using the law of cosines in its most familiar form.

$$\begin{aligned} a^2 &= b^2 + c^2 - 2bc\cos \text{A} \\ &= (450.83 \text{ ft})^2 + (471.67 \text{ ft})^2 \\ &\quad -(2)(450.83 \text{ ft})(471.67 \text{ ft})(\cos 05°12'40'') \\ &= (425{,}720.28 \text{ ft}) - (425{,}285.97 \text{ ft})(\cos 05°12'40'') \\ &= 2192.10 \text{ ft} \\ a &= \sqrt{2192.10 \text{ ft}} \\ &= 46.82 \text{ ft} \end{aligned}$$

The answer is (D).

SOLUTION 25

The law of sines is useful in solving this problem. Two triangles are created by the problem's situation.

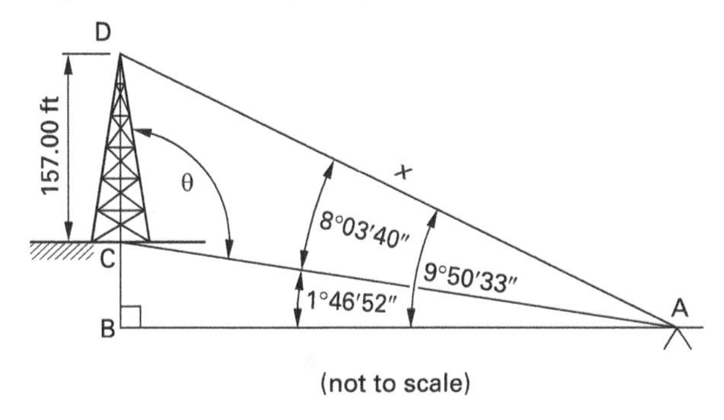

(not to scale)

Two of the angles of triangle ABC are known: 90°00′00″ and 01°46′52″.

Find angle ACB.

$$90°00'00'' + 01°46'52'' = 91°46'52''$$
$$\text{ACB} = 180°00'00'' - 91°46'52'' = 88°13'08''$$

Find angle ACD.

$$\theta = \text{ACD} = 180°00'00'' - 88°13'08'' = 91°46'52''$$

Find the length of x.

$$\frac{a}{\sin \text{A}} = \frac{b}{\sin \text{B}}$$
$$\frac{\text{DC}}{\sin \text{DAC}} = \frac{x}{\sin \theta}$$
$$\frac{157.00 \text{ ft}}{\sin 08°03'40''} = \frac{x}{\sin 91°46'52''}$$
$$x = (\sin 91°46'52'')\left(\frac{157.00 \text{ ft}}{\sin 08°03'40''}\right) = 1119.05 \text{ ft}$$

Finally, find the length of side DB.

$$\text{DB} = (\sin 09°50'32'')(1119.05 \text{ ft}) = 191.29 \text{ ft}$$

The answer is (B).

SOLUTION 26

The side opposite the given angle can be found by the law of cosines, and the remaining sides and angles by the law of sines.

Using the law of cosines,

$$\begin{aligned} a^2 &= b^2 + c^2 - 2bc\cos \text{A} \\ &= (83.40 \text{ ft})^2 + (95.20 \text{ ft})^2 \\ &\quad -(2)(83.40 \text{ ft})(95.20 \text{ ft})(\cos 51°15'00'') \\ &= 16{,}018.60 \text{ ft}^2 - (15{,}879.36 \text{ ft}^2)(\cos 51°15'00'') \\ &= 6079.34 \text{ ft}^2 \\ a &= \sqrt{6079.34 \text{ ft}^2} \\ &= 77.97 \text{ ft} \end{aligned}$$

Using the law of sines,

$$\frac{\sin A}{a} = \frac{\sin B}{b}$$

$$\frac{\sin 51°15'00''}{77.97 \text{ ft}} = \frac{\sin B}{83.40 \text{ ft}}$$

$$\sin B = (83.40 \text{ ft})\left(\frac{\sin 51°15'00''}{77.97 \text{ ft}}\right) = 0.8342$$

$$B = 56°31'56''$$

$$\frac{\sin A}{a} = \frac{\sin C}{c}$$

$$\frac{\sin 51°15'00''}{77.97 \text{ ft}} = \frac{\sin C}{95.20 \text{ ft}}$$

$$\sin C = (95.20 \text{ ft})\left(\frac{\sin 51°15'00''}{77.97 \text{ ft}}\right) = 0.9522$$

$$C = 72°13'04''$$

The answer is (A).

SOLUTION 27

First, find the angle at A.

$$\cos A = \frac{0.5x}{x} = 0.5$$

$$A = 60°$$

Then use the sine of 60° to find the solution.

$$\sin 60° = \frac{x - 4 \text{ ft}}{x}$$

$$4 \text{ ft} = x - x\sin 60° = 1x - 0.866x = 0.1340x$$

$$x = \frac{4 \text{ ft}}{0.1340} = 29.85 \text{ ft}$$

The answer is (C).

SOLUTION 28

The area, A, of the triangle can be found from the formula

$$A = \frac{c^2 \sin A \sin B}{2 \sin C}$$

$$(2\sin C)A = (c^2)(\sin A)(\sin B)$$

$$c^2 = \frac{(2\sin C)A}{(\sin A)(\sin B)} = \frac{(2\sin 30°00'00'')(65{,}340 \text{ ft}^2)}{(\sin 47°56'10'')(\sin 102°03'50'')} = \frac{65{,}340 \text{ ft}^2}{0.7260} = 90{,}000.00 \text{ ft}^2$$

$$c = \sqrt{90{,}000.00 \text{ ft}^2} = 300.00 \text{ ft}$$

The answer is (A).

SOLUTION 29

Use the law of sines to find the remaining sides.

$$\frac{a}{\sin A} = \frac{b}{\sin B}$$

$$\frac{568.00 \text{ ft}}{\sin 44°41'} = \frac{b}{\sin 79°59'}$$

$$b = (\sin 79°59')\left(\frac{568.00 \text{ ft}}{\sin 44°41'}\right) = \frac{559.342 \text{ ft}}{\sin 44°41'} = 795.44 \text{ ft}$$

$$79°59' + 44°41' = 124°40'$$

$$180°00' - 124°40' = 55°20'$$

$$\frac{568.00 \text{ ft}}{\sin 44°41'} = \frac{c}{\sin 55°20'}$$

$$c = (\sin 55°20')\left(\frac{568.00 \text{ ft}}{\sin 44°41'}\right) = \frac{467.166 \text{ ft}}{\sin 44°41'} = 664.35 \text{ ft}$$

The answer is (C).

SOLUTION 30

Use the law of cosines to find the missing side of the triangle, and the law of sines to find the remaining angles.

$$\begin{aligned} a^2 &= b^2 + c^2 - 2bc\cos A \\ &= (82.00\text{ ft})^2 + (73.00\text{ ft})^2 \\ &\quad -(2)(82.00\text{ ft})(73.00\text{ ft})(\cos 71°38') \\ &= 12{,}053\text{ ft}^2 - 3772.34\text{ ft}^2 \\ &= 8280.66\text{ ft}^2 \\ a &= 91.00\text{ ft} \end{aligned}$$

$$\frac{\sin A}{a} = \frac{\sin B}{b}$$

$$\frac{\sin 71°38'}{91.00\text{ ft}} = \frac{\sin B}{82.00\text{ ft}}$$

$$\sin B = (82.00\text{ ft})\left(\frac{\sin 71°38'}{91.00\text{ ft}}\right)$$

$$\begin{aligned} B &= 58°47' \\ C &= 180°00'00'' - (58°47' + 71°38') \\ &= 49°35' \end{aligned}$$

The answer is (C).

SOLUTION 31

31.1. The distance from the PI to the midpoint of the curve is known as the *external distance.*

The answer is (C).

31.2. The angle labeled 5 is usually known as the *angle of intersection*; it is equal to the angle labeled 1, which is usually known as the *delta angle.*

The answer is (C).

31.3. The distance from the midpoint of the long chord to the middle of the curve is known as the *middle ordinate.*

The answer is (A).

31.4. The terms *semitangent* and *back tangent* both can be properly applied to the line in question. The latter term is dependent on the direction of travel of the work.

The answer is (D).

31.5. The length of the curve, L, is a portion of the circumference, which can be expressed as $2\pi R$. The central angle, I, is the same portion of the whole circle. Therefore,

$$\frac{L}{2\pi R} = \frac{I}{360°}$$

$$L = \frac{2\pi RI}{360°}$$

The answer is (D).

31.6. The abbreviations stand for the point of curvature (PC), center of the curve (CEN), and point of tangency (PT).

The answer is (A).

SOLUTION 32

The right side of each of the answer choice's expressions describes one element of a circular curve, but only choice (D) correctly represents the long chord.

The answer is (D).

SOLUTION 33

The middle ordinate can also be expressed by R versin $I/2$, which is equivalent to

$$R\left(1 - \cos\frac{I}{2}\right)$$

The answer is (B).

SOLUTION 34

The highway definition of the degree of curve is that the central angle, I, subtends an arc of 100 ft. This is in contrast to the railroad definition of the term, which states that the central angle subtends a chord of 100 ft. The known values in the problem are the radius, 450.00 ft, and the arc length, 100.00 ft.

$$\text{arc length} = L = 100.00\text{ ft} = \frac{\pi RI}{180°}$$

$$\begin{aligned} 100.00\text{ ft} &= \frac{\pi(450.00\text{ ft})I}{180°} \\ (180°)(100.00) &= 1413.72I \\ 18{,}000.00 &= 1413.72I \\ I &= \frac{18{,}000.00}{1413.72} \\ &= 12.732 \\ &= 12°44' \end{aligned}$$

Another method of performing this calculation is to recognize that the degree of curve, by arc definition, is always equivalent to 5729.58 divided by the radius.

$$\begin{aligned} I &= \frac{5729.58}{R} \\ &= \frac{5729.58}{450.00} \\ &= 12.732 \\ &= 12°44' \end{aligned}$$

The answer is (A).

SOLUTION 35

An angle of 4° subtends a chord of 100.00 ft. To find the radius of the curve, use the formula for the long chord.

$$\begin{aligned} \text{LC} &= 2R\sin\frac{I}{2} \\ 100.00\text{ ft} &= 2R\sin\frac{4°}{2} \\ &= 2R\sin 2° \end{aligned}$$

$$\begin{aligned} 2R &= \frac{100.00\text{ ft}}{\sin 2°} \\ &= 2865.37 \\ R &= \frac{2865.37}{2} \\ &= 1432.69\text{ ft} \end{aligned}$$

Another method of performing this calculation is to use the following formula.

$$R = \frac{50\text{ ft}}{\sin\frac{D}{2}}$$

R is the radius in feet, and D is the degree of curve.

$$\begin{aligned} R &= \frac{50\text{ ft}}{\sin\frac{4°}{2}} \\ &= \frac{50\text{ ft}}{0.034899} \\ &= 1432.69\text{ ft} \quad (1400\text{ ft}) \end{aligned}$$

The answer is (D).

SOLUTION 36

The best way to begin this solution is to find the length of the chord.

$$\begin{aligned} \text{LC} &= 2R\sin\frac{I}{2} \\ &= (2)(100.00\text{ ft})\sin\frac{20°}{2} \\ &= (200.00\text{ ft})(\sin 10°) \\ &= 34.73\text{ ft} \end{aligned}$$

As shown in the problem's illustration, the deflection angle is one half of the central angle.

$$\begin{aligned} \text{TD} &= (\text{cosine of the deflection angle})(\text{chord}) \\ &= (\cos 10°)(34.73\text{ ft}) \\ &= 34.2\text{ ft} \end{aligned}$$

The tangent offset, TO, is also found using the deflection angle.

$$\begin{aligned} \text{TO} &= (\text{sine of the deflection angle})(\text{chord}) \\ &= (\sin 10°)(34.73\text{ ft}) \\ &= 6.03\text{ ft} \end{aligned}$$

The answer is (A).

SOLUTION 37

The arc length desired is 70.00 ft.

$$\begin{aligned} \text{deflection angle} &= \frac{180°L}{2\pi R} = \frac{(180°)(70.00\text{ ft})}{2\pi(750.00\text{ ft})} \\ &= \frac{12{,}600.00}{4712.39} \\ &= 02.6738° \\ &= 02°40'26'' \end{aligned}$$

The deflection angle can be used to find the chord length.

$$\begin{aligned} \text{LC} &= 2R\sin\frac{I}{2} \\ &= 2R\sin(\text{deflection angle}) \\ &= (2)(750.00\text{ ft})(\sin 02°40'26'') \\ &= (1500\text{ ft})(\sin 02°40'26'') \\ &= 69.97\text{ ft} \end{aligned}$$

The answer is (B).

SOLUTION 38

The stationing of a route generally follows the arc of any included curves. To solve the problem, determine the length of the arc of the curve.

$$\begin{aligned}\text{arc length} = L &= \frac{\pi RI}{180°} \\ &= \frac{\pi(1438.92 \text{ ft})(42°15')}{180°} \\ &= \frac{60{,}794.37\pi}{180°} \\ &= \frac{190{,}991.15}{180°} \\ &= 1061.06 \text{ ft}\end{aligned}$$

Then add the known arc length to the station of the PC to find the station of the PT.

$$\begin{aligned}\text{PC} &= 214{+}23.61 \\ +L &= 10{+}61.06 \\ \hline \text{PT} &= 224{+}84.67\end{aligned}$$

The answer is (B).

SOLUTION 39

39.1. First, determine the central angle of the curve.

$$\text{central angle} = 90°00' - 45°58' = 44°02'$$

Since the radius is given, the arc length from the PC to the intersection can be found.

$$\begin{aligned}\text{arc length} = L &= \frac{\pi RI}{180°} \\ &= \frac{\pi(400.00 \text{ ft})(44°02')}{180°} \\ &= \frac{(17{,}613.33 \text{ ft})\pi}{180°} \\ &= 307.41 \text{ ft}\end{aligned}$$

Now the station of the intersection can be found.

$$\begin{aligned}\text{PC} &= 33{+}15.47 \\ +L &= 3{+}07.41 \\ \hline \text{intersection} &= 36{+}22.88\end{aligned}$$

The answer is (D).

39.2. The external distance is the length from the midpoint of the curve, along the prolongation of a radial line, to the tangent line. The central angle of the curve is 44°02′ and the radius is 400.00 ft. The distance from the center of the curve along the radial line through the midpoint of the curve to the tangent can be called $R + E$, that is, the radius plus the external distance. It is found using the following equation.

$$R + E = R \sec \frac{I}{2}$$

Note that the secant function is 1/cos.

$$\begin{aligned}R + E &= (400.00 \text{ ft})\left(\sec \frac{44°02'}{2}\right) \\ &= (400.00 \text{ ft})(\sec 22°01') \\ &= (400.00 \text{ ft})(1.0787) \\ &= 431.46 \text{ ft}\end{aligned}$$

To find the external distance, subtract the radius from the quantity $R + E$.

$$\begin{aligned}R + E &= 431.46 \text{ ft} \\ -R &= 400.00 \text{ ft} \\ \hline E &= 31.46 \text{ ft}\end{aligned}$$

The answer is (B).

SOLUTION 40

A reverse curve consists of two curves that share a common tangent and stand on opposite sides of that tangent. However, the curves need not have radii of different lengths or equal central angles.

The answer is (A).

SOLUTION 41

A compound curve consists of two curves on the same side of a common tangent, with different radii.

The answer is (C).

SOLUTION 42

The abbreviation PRC is used to indicate the change in the rate or direction of curvature in reverse, and PCC is used to indicate the compound curves.

The answer is (C).

SOLUTION 43

In order for the line from the PC to the PT to pass through the PRC, the central angles of the two component curves must be the same.

The answer is (B).

SOLUTION 44

44.1. This problem is most easily solved by coordinate geometry. Throughout this solution, the first coordinate given is the northing and the second is the easting.

First, assign coordinates to the center of curve 1 and use the bearings and distances given to compute the coordinates of the other two centers.

center of curve 1	N 5000.00
	E 5000.00
center of curve 2	N 5167.33
	E 4677.87
center of curve 3	N 4813.09
	E 4794.70

The only center that will move is that of curve 2. Its coordinates after a move N40°24′W of 30.00 ft will be

new center of curve 2	N 5190.18
	E 4658.45

Take the inverse between the new coordinates of the center of curve 2 and the other centers to establish the new radial lines that the reverse curves share.

center of curve 1		center of curve 2
N 5000.00	N60°53′00″ W	N 5190.18
E 5000.00	390.95 ft	E 4658.43
center of curve 3		center of curve 2
N 4813.09	N19°52′06″ W	N 5190.18
E 4794.70	400.96 ft	E 4658.43

Find the central angles by subtracting the bearings of the radial lines.

curve 1

$$\begin{array}{rr} & \text{N}19°45'00''\text{W} \\ & -\text{N}60°53'30''\text{W} \\ \hline \text{central angle} = & 41°08'30'' \end{array}$$

curve 2

$$\begin{array}{rr} & \text{S}60°53'30''\text{E} \\ & -\text{S}19°52'06''\text{E} \\ \hline \text{central angle} = & 41°01'24'' \end{array}$$

curve 3

$$\begin{array}{rr} & \text{N}70°53'00''\text{W} \\ & -\text{N}19°52'06''\text{W} \\ \hline \text{central angle} = & 51°00'54'' \end{array}$$

The answer is (A).

44.2. The statement following the problem's illustration stipulates that the radius of curve 2 remains unchanged. Therefore, following the inverse calculation illustrated in the solution to the previous problem, the remaining radii can be found by subtraction.

curve 1	
distance from the center of 1 to 2	390.95 ft
unchanged radius of 2	–173.00 ft
radius of 1	217.95 ft

The radius of curve 2 is unchanged at 173.00 ft.

curve 3	
distance from the center of 2 to 3	400.96 ft
unchanged radius of 2	–173.00 ft
radius of 3	227.96 ft

The answer is (D).

44.3. Find the central angle and the radius of each curve, then calculate the long chord of each curve.

$$\text{LC} = 2R\sin\frac{I}{2}$$

curve 1

$$\begin{aligned}\text{LC} &= (2)(217.95\ \text{ft})\left(\sin\frac{41°08'30''}{2}\right) \\ &= 153.16\ \text{ft}\end{aligned}$$

curve 2

$$\begin{aligned}\text{LC} &= (2)(173.00\ \text{ft})\left(\sin\frac{41°01'24''}{2}\right) \\ &= 121.24\ \text{ft}\end{aligned}$$

curve 3

$$\begin{aligned}\text{LC} &= (2)(227.96\ \text{ft})\left(\sin\frac{51°00'54''}{2}\right) \\ &= 196.33\ \text{ft}\end{aligned}$$

The answer is (A).

44.4. The central angle and radius are the critical elements needed to solve this problem; they were found in the solutions of Prob. 44.1 and Prob. 44.2, respectively.

$$\text{arc length} = L = \frac{\pi RI}{180°}$$

curve 1

$$\begin{aligned}L &= \frac{\pi(217.95\ \text{ft})(41°08'30'')}{180°} \\ &= \frac{28{,}170.11\ \text{ft}}{180°} \\ &= 156.50\ \text{ft}\end{aligned}$$

curve 2

$$L = \frac{\pi(173.00\ \text{ft})(41°01'24'')}{180°} = \frac{22{,}296.00\ \text{ft}}{180°} = 123.87\ \text{ft}$$

curve 3

$$L = \frac{\pi(227.96\ \text{ft})(51°00'54'')}{180°} = \frac{36{,}534.77\ \text{ft}}{180°} = 202.97\ \text{ft}$$

The answer is (D).

44.5. The bearing of the long chord of each curve can be found as follows.

curve 1

bearing of the radial line to the PC	=	N19°45′00″ W
radial line expressed as an azimuth	=	340°15′00″
		–90°00′00″
azimuth of the tangent	=	250°15′00″
one-half of the central angle	=	–20°34′15″
azimuth of the long chord	=	229°40′45″
		–180°00′00″
bearing of the long chord	=	S49°40′45″ W

curve 2

bearing of the radial line to the PC	=	S60°53′30″ E
radial line expressed as an azimuth	=	119°06′30″
		+90°00′00″
azimuth of the tangent	=	209°06′30″
one-half of the central angle	=	+20°30′42″
	=	229°37′18″
azimuth of the long chord	=	–180°00′00″
bearing of the long chord	=	S49°37′18″ W

curve 3

bearing of the radial line to the PC	=	N19°52′06″ W
radial line expressed as an azimuth	=	340°07′54″
		–90°00′00″
azimuth of the tangent	=	250°07′54″
one-half of the central angle	=	–25°30′27″
azimuth of the long chord	=	224°37′44″
		–180°00′00″
bearing of the long chord	=	S44°37′44″ W

The answer is (C).

SOLUTION 45

45.1. The first curve of the compound curve will have a tangent distance of 11.77 ft, the distance from point A to the PC. An additional element of the curve must be known to establish its radius. Since A is to be the PI, the central angle is 41°35′.

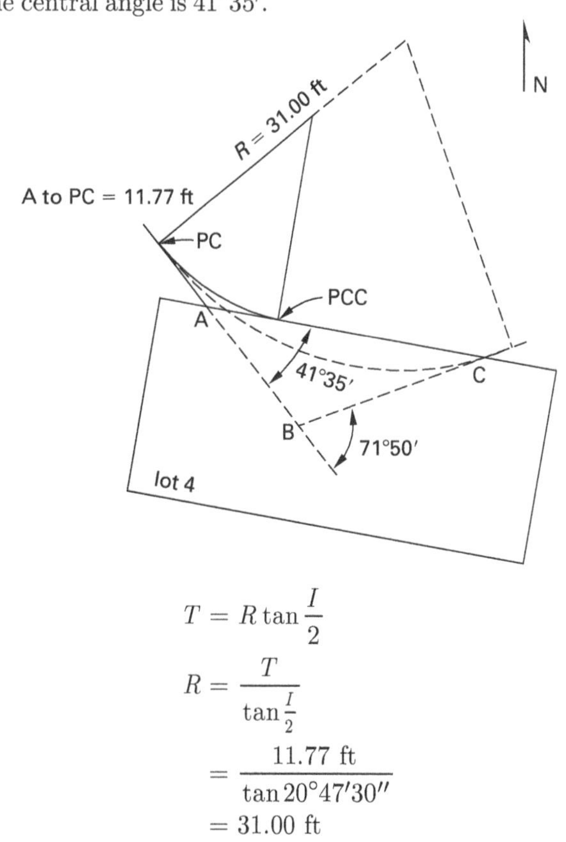

$$T = R\tan\frac{I}{2}$$

$$R = \frac{T}{\tan\frac{I}{2}} = \frac{11.77\ \text{ft}}{\tan 20°47'30''} = 31.00\ \text{ft}$$

The answer is (A).

45.2. Referring to the illustration for Prob. 45.1 through Prob. 45.3, find the tangent distance of the original single curve.

$$\begin{aligned} T &= R\tan\frac{I}{2} \\ &= (49.44\text{ ft})\left(\tan\frac{71^\circ 50'}{2}\right) \\ &= (49.44\text{ ft})(\tan 35^\circ 55') \\ &= 35.81\text{ ft} \end{aligned}$$

Subtract from this the tangent distance of the first curve of the compound curve to find the distance from A to B.

$$\begin{array}{rr} & 35.81\text{ ft} \\ & -11.77\text{ ft} \\ \hline \text{distance from A to B} = & 24.04\text{ ft} \end{array}$$

Find the angle at B.

$$\begin{array}{rr} & 180^\circ 00' \\ & -71^\circ 50' \\ \hline \text{B} = & 108^\circ 10' \end{array}$$

Then use the law of sines to find the distance from A to C.

$$\frac{24.04\text{ ft}}{\sin 30^\circ 15'} = \frac{x}{\sin 108^\circ 10'}$$

$$x = 45.34\text{ ft}$$

Next, subtract the tangent distance used for the first curve of the compound curve.

$$\begin{array}{rr} & 45.34\text{ ft} \\ \text{tangent distance for the second curve} = & -11.77\text{ ft} \\ \hline & 33.57\text{ ft} \end{array}$$

With the tangent distance and the central angle, find the radius of the curve.

$$\begin{aligned} R &= \frac{T}{\tan\frac{I}{2}} \\ &= \frac{33.57\text{ ft}}{\tan\frac{30^\circ 15'}{2}} \\ &= \frac{33.57\text{ ft}}{\tan 15^\circ 07'30''} \\ &= 124.20\text{ ft} \end{aligned}$$

The answer is (D).

45.3. The central angle and radius have been found for the two component curves in the solutions of Prob. 45.1 and Prob. 45.2. Using that information, find the long chord of each.

$$\text{LC} = 2R\sin\frac{I}{2}$$

curve 1

$$\begin{aligned} \text{LC} &= (2)(31.00\text{ ft})\left(\sin\frac{41^\circ 35'}{2}\right) \\ &= (62.00\text{ ft})(\sin 20^\circ 47'30'') \\ &= 22.01\text{ ft} \end{aligned}$$

curve 2

$$\begin{aligned} \text{LC} &= (2)(124.20\text{ ft})\left(\sin\frac{30^\circ 15'}{2}\right) \\ &= (248.20\text{ ft})(\sin 15^\circ 07'30'') \\ &= 64.81\text{ ft} \end{aligned}$$

The answer is (C).

SOLUTION 46

Three points define a unique circular curve, just as two points define a unique line. Four points usually define more than two curves.

The answer is (A).

SOLUTION 47

The perpendicular bisector of any chord that is inscribed in a circle will pass through the center of that circle. Therefore, several perpendicular bisectors of such chords will intersect at the center of the circle.

The answer is (D).

SOLUTION 48

48.1. The foundation of this solution is the principle stated in the solution to the previous problem. The perpendicular bisectors of the two lines will intersect at the center of the arc that passes through the three points.

First, establish the coordinates of points 1 and 2, then find the midpoint of each line as in *Table for Prob. 48.*

Next, use this data to find the azimuths of the perpendicular bisectors at A and B.

The line at point A that is perpendicular to line 2-3 will have an azimuth of 130°19′00″.

Table for Problem 48

point 3		point A		point 2
N5295.0298	N40°19′00″E	N5385.7649	N40°19′00″E	N5476.5001
E10,179.0676	119.00 ft	E10,256.0620 (midpoint)	119.00 ft	E10,333.0564
point 2		point B		point 1
N5476.5001	N57°43′00″E	N5542.1952	N57°43′00″E	N5607.8903
E10,333.0564	123.00	E10,437.0427 (midpoint)	123.00 ft	E10,541.0290

The line at point B that is perpendicular to line 1-2 will have an azimuth of 147°43′00″.

The following illustration shows the triangle ABC, with C at the center of the curve.

Find the side AB by inversing between the coordinates for points A and B.

point A		point B
N5385.7649	N49°09′42″E	N5542.1952
E10,256.0620	239.216 ft	E10,437.0427

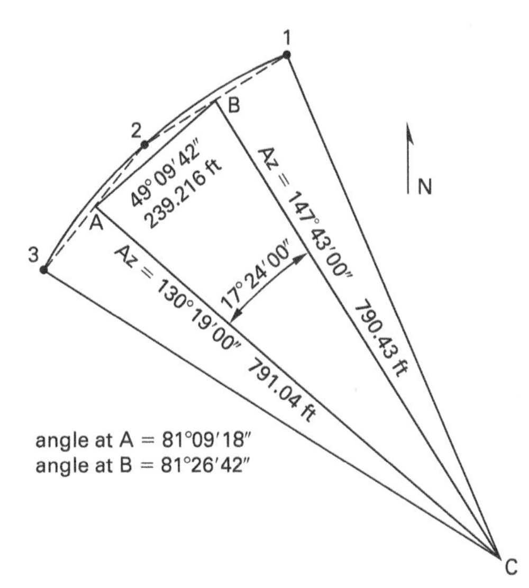

Using the law of sines, solve for the lengths of side BC and side AC.

$$\frac{AB}{\sin C} = \frac{BC}{\sin A} = \frac{AC}{\sin B}$$

$$\frac{239.216 \text{ ft}}{\sin 17°24'00''} = \frac{BC}{\sin 81°09'18''}$$

$$BC = 790.43 \text{ ft}$$

$$\frac{239.216 \text{ ft}}{\sin 17°24'00''} = \frac{AC}{\sin 81°26'42''}$$

$$AC = 791.04 \text{ ft}$$

Use the azimuths and distances from A or B to establish the coordinates of C, the center.

point A		point C
N5385.7649	Az 130°19′00″	N4873.9528
E10,256.0620	791.04 ft	E10,859.2143

With the coordinates of the center established, the radius of the curve can be found by inversing between points 1, 2, or 3.

point C		point 1
N4873.9528	Az 336°33′42″	N5607.8903
E10,859.2143	799.94 ft (radius)	E10,541.0290

The radius of the curve is 799.94 ft.

The answer is (C).

48.2. The radius is available from the solution of the previous problem, but the central angle is also needed to calculate the arc length. The central angle can be found by first finding the azimuth from point C to point 3, and then subtracting it from the azimuth from point C to point 1.

$$\begin{aligned} \text{azimuth from point C to point 1} &= 336°33'42'' \\ \text{azimuth from point C to point 3} &= \underline{-301°45'42''} \\ \text{central angle} &= 33°48'00'' \end{aligned}$$

Next, find the length of the curve.

$$\begin{aligned} L &= \frac{\pi R I}{180°} = \frac{\pi(799.94 \text{ ft})(34°48'00'')}{180°} \\ &= \frac{87{,}455.38 \text{ ft}}{180°} \\ &= 485.86 \text{ ft} \end{aligned}$$

The answer is (D).

48.3. The chord bearing and length can be found by inversing between point 1 and point 3. The chord length can also be found using the following formula.

$$\begin{aligned} \text{LC} &= 2R\sin\frac{I}{2} = (2)(799.94\ \text{ft})\left(\sin\frac{34°48'00''}{2}\right) \\ &= (1599.88\ \text{ft})(\sin 17°24') \\ &= 478.43\ \text{ft} \end{aligned}$$

The chord bearing is available immediately; it is the same as the line that connects the midpoints of the original lines, line AB. The similar triangles 123 and A2B assure that the bearings of their bases, 1-2 and AB, respectively, will be identical.

The answer is (B).

SOLUTION 49

A sector is shaped like a slice of pie. It is bounded by two radii and the included curve.

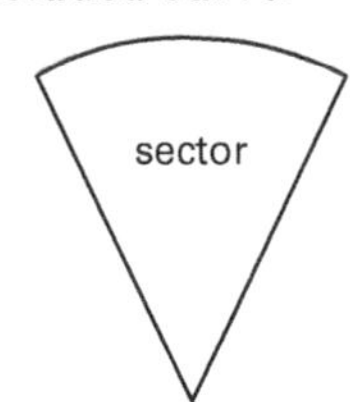

The answer is (C).

SOLUTION 50

The area within a circle is expressed as πR^2.

The answer is (C).

SOLUTION 51

The ratio of the area of a sector to the area of the circle is the same as the ratio of the central angle to 360°.

$$\frac{A}{\pi R^2} = \frac{I}{360°}$$

$$\begin{aligned} \text{area of a sector in square feet} &= \left(\frac{I}{360°}\right)\pi R^2 \\ &= \left(\frac{33°}{360°}\right)\pi(1400\ \text{ft})^2 \\ &= (0.0917)\pi(1{,}960{,}000.00\ \text{ft}^2) \\ &= 564{,}439.48\ \text{ft}^2 \end{aligned}$$

$$\begin{aligned} \text{area in acres} &= \frac{564{,}439.48\ \text{ft}^2}{43{,}560\ \frac{\text{ft}^2}{\text{ac}}} \\ &= 12.96\ \text{ac} \end{aligned}$$

The answer is (B).

SOLUTION 52

The long chord of a curve and its arc include an area known as a *segment*.

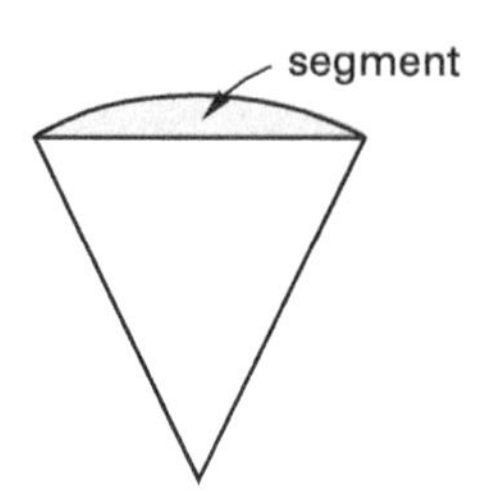

The answer is (D).

SOLUTION 53

The area within a segment is expressed as

$$\frac{LR - R^2\sin I}{2}$$

L is the length of the curve, and R is the radius. The first task is to find the length of the curve.

$$\begin{aligned} L &= \frac{\pi RI}{180°} \\ &= \frac{\pi(52.46\ \text{ft})(23°42')}{180°} \\ &= \frac{\pi(1236.19\ \text{ft})}{180°} \\ &= 21.70\ \text{ft} \end{aligned}$$

Now calculate the area of the segment.

$$\begin{aligned} \text{area of the segment} &= \frac{LR - R^2\sin I}{2} \\ &= \frac{(21.70\ \text{ft})(52.46\ \text{ft}) - (52.46\ \text{ft})^2\sin 23°42'}{2} \\ &= \frac{1138.38\ \text{ft}^2 - 1106.18\ \text{ft}^2}{2} \\ &= 16.10\ \text{ft}^2 \quad (16\ \text{ft}^2) \end{aligned}$$

The answer is (A).

SOLUTION 54

The expression RT will yield the area between the radii and the tangents of a curve. $LR/2$ will give the area of the sector. The difference between these two expressions is the area of the fillet.

$$\text{area between radii and tangent} = RT$$

$$\text{area of the sector} = \frac{LR}{2}$$

$$\begin{aligned}\text{area of the fillet} &= RT - \frac{LR}{2}\\ &= R\left(T - \frac{L}{2}\right)\end{aligned}$$

The answer is (C).

SOLUTION 55

The elements needed to solve this problem are the length of the arc, the radius, and the tangent distance. The radius, 270.00 ft, is given.

First, calculate the length of the arc.

$$\begin{aligned}L &= \frac{\pi RI}{180°}\\ &= \frac{\pi(270.00\text{ ft})(55°)}{180°}\\ &= \frac{\pi(14{,}850\text{ ft})}{180}\\ &= 259.18\text{ ft}\end{aligned}$$

Next, calculate the tangent distance.

$$\begin{aligned}T &= R\tan\frac{I}{2}\\ &= (270.00\text{ ft})\left(\tan\frac{55°}{2}\right)\\ &= 140.55\text{ ft}\end{aligned}$$

Finally, calculate the area of the fillet.

$$\begin{aligned}\text{area of the fillet} &= R\left(T - \frac{L}{2}\right)\\ &= (270.00\text{ ft})\left(140.55\text{ ft} - \frac{259.18\text{ ft}}{2}\right)\\ &= (270.00\text{ ft})(10.96\text{ ft})\\ &= 2959.20\text{ ft}^2\end{aligned}$$

The answer is (D).

SOLUTION 56

The area between the arcs is the difference between the areas of the two sectors.

$$\text{area of a sector} = \left(\frac{I}{360°}\right)(\pi R^2)$$

The larger sector has a radius of 39.73 ft.

$$\begin{aligned}\text{area of the larger sector} &= \left(\frac{53°29'}{360°}\right)\pi(39.73\text{ ft})^2\\ &= (0.14856)(4958.919\text{ ft}^2)\\ &= 736.70\text{ ft}^2\end{aligned}$$

The smaller sector has a radius of 16.50 ft less than the first sector.

$$\begin{aligned}R_1 &= 39.73\text{ ft}\\ & \underline{-16.50\text{ ft}}\\ R_2 &= 23.23\text{ ft}\end{aligned}$$

The smaller sector has a radius of 23.23 ft.

$$\begin{aligned}\text{area of the smaller sector} &= \left(\frac{53°29'}{360°}\right)\pi(23.23\text{ ft})^2\\ &= (0.14856)(1695.307\text{ ft}^2)\\ &= 251.85\text{ ft}^2\end{aligned}$$

The area between the two arcs is the difference between the area of the larger sector and the area of the smaller sector.

$$\begin{aligned}&736.70\text{ ft}^2\\ &\underline{-251.85\text{ ft}^2}\\ \text{area between the arcs} &= 484.85\text{ ft}^2\end{aligned}$$

The answer is (A).

SOLUTION 57

The curve that results is a parabola, also known as a *conic section.* Vertical curves are parabolic curves.

The answer is (D).

SOLUTION 58

In vertical curves, the tangent offsets vary as the square of the distance from the point of tangency. This relationship provides a straightforward method of calculation.

The answer is (C).

SOLUTION 59

Vertical curves, being parabolic, have the beneficial quality of having a constant rate of change in grade throughout their entire lengths. The following calculation will yield this rate of change, r, for a particular vertical curve.

$$r = \frac{g_2 - g_1}{L}$$

The answer is (C).

SOLUTION 60

The rate of change is constant and the tangent offsets vary as the square of the distance from the point of tangency. Therefore, the following expression would yield the tangent offset for any station along the curve.

$$\left(\frac{r}{2}\right)x^2$$

The answer is (A).

SOLUTION 61

At either the highest point on a summit vertical curve or the lowest point on a sag vertical curve, a tangent to the curve will be a horizontal line. That is, the slope of the tangent at that point will be zero. These points are often of particular interest when minimum clearances or drainage structures are involved.

The answer is (A).

SOLUTION 62

In a symmetrical vertical curve, the curve will cut in half the line between the midpoint of the long chord and the PVI.

The answer is (A).

SOLUTION 63

The formula $x = -g_1/r$ can be used to find the distance from the PVC to the point on the vertical curve where the slope is zero.

The answer is (A).

SOLUTION 64

64.1. The illustration for Prob. 64.1 through Prob. 64.5 indicates that the initial grade, g_1, of the tangent is +2.00%, or +0.02 ft per foot. The distance from the PVC to the PVI is 300 ft, since the curve is symmetrical and its overall length is 600 ft.

First, find the elevation of the PVI.

$$\begin{aligned} (300\text{ ft})(+0.02) &= +6.00\text{ ft} \\ \text{elevation of the PVC} &= 100.00\text{ ft} \\ & \underline{+6.00\text{ ft}} \\ \text{elevation of the PVI} &= 106.00\text{ ft} \end{aligned}$$

Then, find the elevation of the PVT.

$$\begin{aligned} (300.00\text{ ft})(-0.03\text{ ft}) &= -9.00\text{ ft} \\ \text{elevation of the PVI} &= 106.00\text{ ft} \\ & \underline{-9.00\text{ ft}} \\ \text{elevation of the PVT} &= 97.00\text{ ft} \end{aligned}$$

Finally, find the elevation of the midpoint of the long chord.

$$\begin{aligned} \text{elevation of the PVT} &= 97.00\text{ ft} \\ \text{elevation of the PVC} &= \underline{-100.00\text{ ft}} \\ & -3.00\text{ ft} \\ \frac{-3.00\text{ ft}}{2} &= -1.50\text{ ft} \\ \text{elevation of the PVC} &= 100.00\text{ ft} \\ & \underline{-1.50\text{ ft}} \\ \text{elevation of the midpoint of the chord} &= 98.50\text{ ft} \end{aligned}$$

The answer is (B).

64.2. The largest tangent offset of this curve is from the PVI to the midpoint of the curve. In a symmetrical vertical curve, the midpoint of the curve is equidistant from the midpoint of the long chord and the PVI.

$$\begin{aligned} \text{elevation of the PVI} &= 106.00\text{ ft} \\ \text{elevation of the midpoint of the chord} &= \underline{-98.50\text{ ft}} \\ & 7.50\text{ ft} \end{aligned}$$

$$\begin{array}{l} \text{the tangent offset from the PVI to} \\ \text{the midpoint of the vertical curve} \end{array} = \frac{7.50\text{ ft}}{2} = 3.75\text{ ft}$$

The answer is (C).

64.3. From the tangent line to the curve at any station, the tangent offsets vary with the square of the distance from the point of tangency.

The tangent offset at the PVI, calculated in the previous problem, is 3.75 ft.

$$\frac{\text{tangent offset at }12{+}50}{(250)^2} = \frac{\text{tangent offset at }13{+}00}{(300)^2}$$

$$\frac{\text{tangent offset at }12{+}50}{(250)^2} = \frac{3.75 \text{ ft}}{(300)^2}$$

$$\begin{aligned}\text{tangent offset at }12{+}50 &= (3.75 \text{ ft})\left(\frac{(250)^2}{(300)^2}\right)\\ &= 2.60 \text{ ft}\end{aligned}$$

The answer is (D).

64.4. The line tangent to the curve at the highest point is horizontal and the slope is zero. The expression used to find this point is

$$\text{distance (in stations) to the highest point} = \frac{-g_1}{r}$$

g_1 is the entering grade, and r is the rate of change.

First, find the rate of change. g_1 is the entering grade, and g_2 is the exiting grade.

$$\begin{aligned}r &= \frac{g_2 - g_1}{L}\\ &= \frac{-3.00 - (+2.00)}{6}\end{aligned}$$

The grades have been expressed in percentages and the length of the curve in stations.

$$\begin{aligned}r &= \frac{-5}{6}\\ &= -0.833\end{aligned}$$

The distance in stations from the PVC to the highest point on the curve can now be calculated.

$$\begin{aligned}\text{distance from PVC to highest point} &= \frac{-g_1}{r}\\ &= \frac{-(+2.00)}{-0.833}\\ &= 2.40 \text{ sta}\end{aligned}$$

$$\begin{aligned}\text{station of the PVC} &= 10{+}00\\ \text{distance to the highest point} &= \underline{+2{+}40}\\ \text{station of the highest point} &= 12{+}40\end{aligned}$$

The answer is (C).

64.5. The distance from the PVC to the highest point has been calculated in the previous problem. First, it is necessary to find the elevation at that station, 12+40, on the tangent. The grade is +2.00%.

$$\begin{aligned}(\text{distance along the tangent})(\text{grade}) &= \text{rise in elevation, PVC to }12{+}40\\ (240 \text{ ft})(0.02) &= 4.80 \text{ ft}\end{aligned}$$

$$\begin{aligned}\text{elevation of the PVC} &= 100.00 \text{ ft}\\ \text{rise to station }12{+}40 &= \underline{+4.80 \text{ ft}}\\ \text{elevation of the tangent of }12{+}40 &= 104.80 \text{ ft}\end{aligned}$$

Calculate the tangent offset at 12+40 to the vertical curve, using the expression

$$\text{tangent offset} = \left(\frac{r}{2}\right)x^2$$

r is the rate of change, and x is the distance in stations from the point of tangency.

$$\begin{aligned}\text{tangent offset} &= \left(\frac{-0.833}{2}\right)(2.4)^2\\ &= \frac{-4.80}{2}\\ &= -2.40 \text{ ft}\end{aligned}$$

Finally, find the elevation on the curve at station 12+40.

$$\begin{aligned}\text{elevation on the tangent of }12{+}40 &= 104.80 \text{ ft}\\ \text{tangent offset} &= \underline{+(-2.40 \text{ ft})}\\ \text{elevation at }12{+}40 &= 102.40 \text{ ft}\end{aligned}$$

The answer is (D).

SOLUTION 65

65.1. The length of a vertical curve can be expressed as

$$L = \frac{g_2 - g_1}{r}$$

g_1 is the entering grade, g_2 is the exiting grade, and r is the maximum allowable change in grade per station.

$$\begin{aligned}L &= \frac{-0.40 - (+0.80)}{-0.20}\\ &= 6.0 \text{ sta}\end{aligned}$$

The length of the vertical curve is 6.0, or 600 ft.

The answer is (B).

65.2. A convenient expression is

$$\text{elevation of sta on curve} = \left(\frac{r}{2}\right)x^2 + g_1x + \text{elevation of PVC}$$

r is the rate of change, and x is the distance in stations from the PVC.

$$\begin{aligned}\left(\begin{array}{c}\text{distance from the}\\ \text{PVI to the PVC}\end{array}\right) \times \left(\begin{array}{c}\text{grade from the}\\ \text{PVI to the PVC}\end{array}\right) &= \left(\begin{array}{c}\text{change in elevation,}\\ \text{PVI to PVC}\end{array}\right)\\ (300.00 \text{ ft})(-0.008) &= -2.40 \text{ ft}\end{aligned}$$

$$\begin{aligned} \text{elevation of the PVI} &= 100.00\text{ ft} \\ &\quad +(-2.40\text{ ft}) \\ \text{elevation of the PVC} &= 97.60\text{ ft} \end{aligned}$$

Next, find the distance in stations from the PVC to the station in question.

$$\begin{aligned} \text{station of the PVI} &= 90{+}00 \\ \text{distance, in stations, PVI to PVC} &= -3{+}00 \\ \text{station of the PVC} &= 87{+}00 \end{aligned}$$

$$\begin{aligned} \text{station in question} &= 92{+}00 \\ \text{station of the PVC} &= 87{+}00 \\ \text{distance in stations,} & \\ \text{PVC to station in question} &= 5{+}00 \end{aligned}$$

Now find the rate of change.

$$\begin{aligned} r &= \frac{g_2 - g_1}{L} \\ &= \frac{-0.40 - (+0.80)}{6} \\ &= -0.20 \end{aligned}$$

Note that this value is the same as the maximum allowable percentage of change in grade per station.

Finally, find the elevation of station 92+00 on the vertical curve.

$$\begin{aligned} \begin{array}{l}\text{elevation of the} \\ \text{station on curve}\end{array} &= \left(\frac{r}{2}\right)x^2 + g_1x + \text{elevation of PVC} \\ \text{elevation of } 92{+}00 &= \left(\frac{-0.20}{2}\right)(5)^2 + (+0.8)(5) \\ &\quad +97.60\text{ ft} \\ &= -2.5\text{ ft} + 4.0\text{ ft} + 97.60\text{ ft} \\ &= 99.10\text{ ft} \end{aligned}$$

The answer is (C).

65.3. The expression used to find the distance from the PVC to the point of zero slope, which is the highest point, is

$$\begin{aligned} \text{distance in stations} &= \frac{-g_1}{r} \\ &= \frac{-(+0.8)}{-0.20} \\ &= 4 \end{aligned}$$

Next, find the station of the highest point on the curve.

$$\begin{aligned} \text{station of the PVC} &= 87{+}00 \\ \text{distance to the highest point} &= +4{+}00 \\ \text{station of the highest point on the curve} &= 91{+}00 \end{aligned}$$

Finally, find the elevation of the highest point on the curve.

$$\begin{aligned} \begin{array}{l}\text{elevation of station} \\ \text{on the curve}\end{array} &= \left(\frac{r}{2}\right)x^2 + g_1x \\ &\quad +\text{elevation of PVC} \\ \text{elevation of } 91{+}00 &= \left(\frac{-0.20}{2}\right)(4)^2 \\ &\quad +(+0.80)(4) + 97.60\text{ ft} \\ &= -1.6\text{ ft} + 3.2\text{ ft} + 97.60\text{ ft} \\ &= 99.20\text{ ft} \end{aligned}$$

The answer is (B).

SOLUTION 66

66.1. An asymmetrical vertical curve can be thought of as two symmetrical vertical curves end to end. The elevation of the CVC can be found by first finding the elevations of PVI_1 and PVI_2.

PVI_1 and PVI_2 fall at the midpoints of the tangent lines on which they are located.

Find the length of the first vertical curve.

$$\begin{aligned} \text{station of the CVC} &= 18{+}00 \\ \text{station of the PVC} &= -13{+}00 \\ \text{length, in station, of the first curve} &= 5{+}00 \end{aligned}$$

Find the length of the second curve.

$$\begin{aligned} \text{station of the PVT} &= 25{+}00 \\ \text{station of the CVC} &= -18{+}00 \\ \text{length, in stations, of the second curve} &= 7{+}00 \end{aligned}$$

Since the PVIs fall at the midpoints of the tangents,

$$\begin{aligned} \text{station of PVI}_1 &= (13{+}00) + (2{+}50) = 15{+}50 \\ \text{station of PVI}_2 &= (18{+}00) + (3{+}50) = 21{+}50 \end{aligned}$$

Now find the elevations of the two PVIs.

$$\begin{aligned} &\begin{pmatrix}\text{distance PVC to PVI}_1 \\ \text{in station}\end{pmatrix} \\ &\quad\times\begin{pmatrix}\text{grade} \\ \text{PVC to PVI}_1\end{pmatrix} = \begin{pmatrix}\text{change in elevation,} \\ \text{from PVC to PVI}_1\end{pmatrix} \\ (2.50\text{ ft})(-2.00) &= -5.00\text{ ft} \\ \text{elevation of PVC} &= 627.56\text{ ft} \\ \text{change in elevation} &= -5.00\text{ ft} \\ \text{elevation of the PVI}_1 &= 622.56\text{ ft} \end{aligned}$$

$$\begin{pmatrix}\text{distance PVT to PVI}_2\\ \text{in station}\end{pmatrix} \times \begin{pmatrix}\text{grade}\\ \text{PVT to PVI}_2\end{pmatrix} = \begin{pmatrix}\text{change in elevation,}\\ \text{from PVC to PVI}_2\end{pmatrix}$$

$$(3.50\text{ ft})(-0.71) = -2.49\text{ ft}$$

$$\begin{aligned}\text{elevation of PVT} &= 622.53\text{ ft}\\ \text{change in elevation} &= \underline{-2.49\text{ ft}}\\ \text{elevation of the PVI}_2 &= 620.04\text{ ft}\end{aligned}$$

The CVC will be on the line between the PVIs.

$$\begin{aligned}\text{elevation of PVI}_2 &= 620.04\text{ ft}\\ \text{elevation of PVI}_1 &= \underline{-622.56\text{ ft}}\\ \text{change in elevation along the line PVI}_1 - \text{PVI}_2 &= -2.52\text{ ft}\end{aligned}$$

The distance between the PVIs is found by the difference between their stations.

$$\begin{aligned}\text{station of PVI}_2 &= 21{+}50\\ \text{station of PVI}_1 &= -15\\ &\quad\ \underline{+50}\\ \text{length of PVI}_1 - \text{PVI}_2\text{, in stations} &= 6{+}00\end{aligned}$$

Find the grade of the line between the PVIs.

$$\frac{\text{change in elevation along the line PVI}_1 - \text{PVI}_2}{\text{length of PVI}_1\text{–PVI}_2\text{ in stations}} = \frac{-2.52}{6}$$

The grade of the line PVI_1–PVI_2 is –0.42%.

From PVI_1 to the CVC, the grade is –0.42 and the distance is 2.5 sta.

$$(\text{stations})(\text{grade}) = \text{change in elevation PVI}_1\text{ to CVC}$$

$$(2.50)(-0.42) = -1.05\text{ ft}$$

$$\begin{aligned}\text{elevation of PVI} &= 622.56\text{ ft}\\ \text{change in elevation to CVC} &= \underline{-1.05\text{ ft}}\\ \text{elevation of CVC} &= 621.51\text{ ft}\end{aligned}$$

The answer is (B).

66.2. The elevation and station of the CVC found in the previous problem is the beginning of the second curve. The station in question is 260.00 ft beyond the CVC. In order to find the elevation of sta 20+60, first determine r, the rate of change.

$$\begin{aligned}r &= \frac{g_2 - g_1}{L}\\ &= \frac{+0.71 - (-0.42)}{7}\\ &= 0.161\end{aligned}$$

Find the elevation of sta 20+60.

$$\text{elevation of station on the curve} = \left(\frac{r}{2}\right)x^2 + g_1x + \text{elevation of CVC}$$

$$\begin{aligned}\text{elevation of 20+60} &= \left(\frac{0.161}{2}\right)(2.6)^2 + (-0.42)(2.6) + 621.51\text{ ft}\\ &= 0.54\text{ ft} + (-1.09\text{ ft}) + 621.51\text{ ft}\\ &= 620.96\text{ ft}\end{aligned}$$

The answer is (A).

SOLUTION 67

One of the most convenient aspects of using the chain as a unit of measurement is that 10 square chains is equal to 1 ac.

The answer is (B).

SOLUTION 68

Find the value, in square feet, of 3 ac.

$$(43{,}560\text{ ft}^2)(3) = 130{,}680\text{ ft}^2$$

The square root of 3 ac, in square feet, will be the length of one side of the square.

$$\sqrt{130{,}680\text{ ft}^2} = 361.50\text{ ft}$$

The answer is (D).

SOLUTION 69

The area of a trapezoid can be expressed as

$$A = \frac{(b_1 + b_2)h}{2}$$

b_1 is a base of the trapezoid, b_2 is the other base of the trapezoid, and h is the height.

The bases are shown in the illustration, but the height is not.

First, find the height of the trapezoid. The angle at the SW corner of the trapezoid is

$$\begin{array}{r} 90°00' \\ -\text{N}27°58'\text{E} \\ \hline 62°02' \end{array}$$

Next, find the height of the trapezoid.

$$(\sin 62°02')(313.40 \text{ ft}) = 276.80 \text{ ft}$$

Finally, find the area of the trapezoid.

$$\begin{aligned} A &= \frac{(b_1 + b_2)h}{2} \\ &= \frac{(751.36 \text{ ft} + 382.26 \text{ ft})(276.80 \text{ ft})}{2} \\ &= 156{,}893.0 \text{ ft}^2 \end{aligned}$$

$$\frac{156{,}893.0 \text{ ft}^2}{43{,}560.0 \ \frac{\text{ft}^2}{\text{ac}}} = 3.6 \text{ ac}$$

The answer is (C).

SOLUTION 70

The first step is to find the area, in acres, of the entire lot.

$$\begin{aligned} A = lh &= (422.00 \text{ ft})(960.00 \text{ ft}) \\ &= 405{,}120 \text{ ft}^2 \end{aligned}$$

$$\frac{405{,}120 \text{ ft}^2}{43{,}560 \ \frac{\text{ft}^2}{\text{ac}}} = 9.3 \text{ ac}$$

Since parcel 1 must contain 6 ac, the acreage of parcel 2 must be

$$\begin{array}{r} 9.3 \text{ ac} \\ -6.0 \text{ ac} \\ \hline 3.3 \text{ ac} \end{array}$$

$$(3.3 \text{ ac})\left(43{,}560 \ \frac{\text{ft}^2}{\text{ac}}\right) = 143{,}748 \text{ ft}^2$$

Parcel 2 is a right triangle. The area can be found using the expression

$$A = \tfrac{1}{2}bh$$

b is the base of the triangle, and h is its height.

$$\begin{aligned} 143{,}748 \text{ sq ft} &= \left(\frac{1}{2}\right)(422.00 \text{ ft})h \\ (2)(143{,}748 \text{ sq ft}) &= (422.00 \text{ ft})h \\ 287{,}496 \text{ sq ft} &= (422.00 \text{ ft})h \\ h &= \frac{287{,}496 \text{ ft}^2}{422.00 \text{ ft}} \\ &= 681.27 \text{ ft} \end{aligned}$$

Now that the height of the triangle is known, the length EC, which is its hypotenuse, can be found using the Pythagorean theorem.

$$\begin{aligned} \text{EC}^2 &= (681.27 \text{ ft})^2 + (422.00 \text{ ft})^2 \\ &= 464{,}128.81 \text{ ft}^2 + 178{,}084 \text{ ft}^2 \\ &= 642{,}212.81 \text{ ft}^2 \\ \text{EC} &= 801.38 \text{ ft} \end{aligned}$$

The answer is (D).

SOLUTION 71

The area of an oblique triangle can be found from the formula

$$A = \frac{ab \sin \text{C}}{2}$$

Neither triangle contains all the required values. However, the relationship between the two triangles is known, allowing the following equation to be constructed.

$$\frac{(\text{BX})(\text{AX}) \sin \text{X}}{2} = (8)\left(\frac{(\text{XZ})(\text{XY}) \sin \text{X}}{2}\right)$$

In other words, eight times the area of triangle XYZ is equal to the area of triangle XAB.

$$\frac{(288.85 \text{ ft})(308.58 \text{ ft})(\sin \text{X})}{2} = (8)\left(\frac{(142.00 \text{ ft})(\text{XY})(\sin \text{X})}{2}\right)$$

$$\begin{aligned} \frac{(89{,}133.33 \text{ ft}^2)(\sin \text{X})}{2} &= (4)(142.00 \text{ ft})(\text{XY})(\sin \text{X}) \\ (89{,}133.33 \text{ ft}^2)(\sin \text{X}) &= (8)(142.00 \text{ ft})(\text{XY})(\sin \text{X}) \\ 89{,}133.33 \text{ ft}^2 &= (1136.00 \text{ ft})(\text{XY})\left(\frac{\sin \text{X}}{\sin \text{X}}\right) \\ \text{XY} &= \frac{89{,}133.33 \text{ ft}^2}{1136.00 \text{ ft}} \\ &= 78.46 \text{ ft} \end{aligned}$$

The answer is (B).

SOLUTION 72

Drawing a sketch is a good way to begin solving this problem.

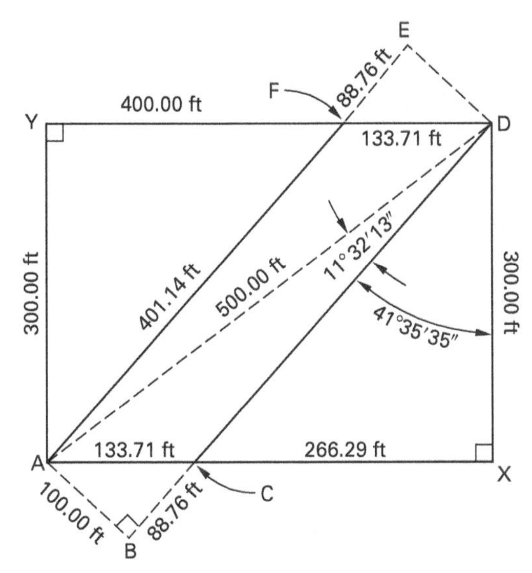

First, construct the diagonal of the rectangle from A to D and find its length using the Pythagorean theorem.

$$\begin{aligned} c^2 &= a^2 + b^2 \\ AD^2 &= (300.00 \text{ ft})^2 + (400.00 \text{ ft})^2 \\ AD &= \sqrt{250{,}000.00 \text{ ft}^2} \\ &= 500.00 \text{ ft} \end{aligned}$$

Next, construct the triangle ABD with a right angle at B, and solve for the base of the triangle, BD, using the Pythagorean theorem.

$$\begin{aligned} a^2 + b^2 &= c^2 \\ b^2 &= c^2 - a^2 \\ &= (500.00 \text{ ft})^2 - (100.00 \text{ ft})^2 \\ b &= \sqrt{240{,}000.00 \text{ ft}^2} \\ &= 489.8979 \text{ ft} \end{aligned}$$

Now find the area of the rectangle ABDE.

$$A = lh$$
$$(489.8979 \text{ ft})(100.00 \text{ ft}) = 48{,}989.79 \text{ ft}^2$$

This area must be reduced by the area within triangles ABC and DEF, which are equal in area.

Find the area in triangle ABC.

$$\begin{aligned} \text{ABD} &= \sin^{-1}\left(\frac{100}{500}\right) \\ &= 11°32'13'' \\ \text{XDA} &= \tan^{-1}\left(\frac{400}{300}\right) \\ &= 53°07'48'' \end{aligned}$$

$$\begin{array}{rr} & 53°07'48'' \\ & -11°32' \\ & 13'' \\ \hline \text{XDC} = & 41°35'35'' \end{array}$$

$$\begin{array}{rr} & 90°00'00'' \\ & -41°35' \\ & 35'' \\ \hline \text{XCD} = & 48°24'25'' \end{array}$$

$$\text{XCD} = \text{BCA}$$

$$\begin{aligned} \text{BC} &= (100.00 \text{ ft})(\cot 48°24'25'') \\ &= 88.76 \text{ ft} \\ A_{\text{ABC}} &= \tfrac{1}{2}bh \\ &= \left(\frac{1}{2}\right)(88.76 \text{ ft})(100.00 \text{ ft}) \\ &= 4438.00 \text{ ft}^2 \end{aligned}$$

Find the combined area of triangles ABC and DEF.

$$(4438.00 \text{ ft}^2)(2) = 8876.00 \text{ ft}^2$$

Finally, find the area, in acres, inside the 100.00 ft right-of-way.

$$\begin{array}{r} 48{,}990.00 \text{ ft}^2 \\ -8876.00 \text{ ft}^2 \\ \hline 40{,}114.00 \text{ ft}^2 \end{array}$$

$$\frac{40{,}114.00 \text{ ft}^2}{43{,}560.00 \ \frac{\text{ft}^2}{\text{ac}}} = 0.92 \text{ ac}$$

The answer is (C).

SOLUTION 73

There are two ways to solve this type of problem. One method is Simpson's one-third rule; the other is the trapezoidal rule.

To use Simpson's one-third rule, find the sum of the end offsets, plus twice the sum of the odd intermediate offsets and four times the sum of the even offsets. Multiply the quantity determined by one-third of the common interval between offsets. The result is the required area.

Using this rule to find the area in the problem, first find the sum of the end offsets.

$$\begin{array}{r} 16.34 \text{ ft} \\ +12.83 \text{ ft} \\ \hline 29.17 \text{ ft} \end{array}$$

Next, find the sum of the odd intermediate offsets.

$$\begin{aligned} \text{third offset} &= 28.42 \text{ ft} \\ \text{fifth offset} &= +27.64 \text{ ft} \\ \text{seventh offset} &= +23.61 \text{ ft} \\ \text{sum of the odd intermediate offsets} &= 79.67 \text{ ft} \end{aligned}$$

Double the sum of the odd intermediate offsets.

$$(79.67 \text{ ft})(2) = 159.34 \text{ ft}$$

Now find the sum of the even intermediate offsets.

$$\begin{aligned} \text{second offset} &= 31.79 \text{ ft} \\ \text{fourth offset} &= +19.55 \text{ ft} \\ \text{sixth offset} &= +30.09 \text{ ft} \\ & 81.43 \text{ ft} \end{aligned}$$

Multiply the sum of the even intermediate offsets by four.

$$(81.43 \text{ ft})(4) = 325.72 \text{ ft}$$

Next, find the sum of the required quantities.

$$\begin{aligned} \text{sum of the end offsets} &= 29.17 \text{ ft} \\ \text{twice the sum of the odd offsets} &= 159.34 \text{ ft} \\ \text{four times the sum of the even offsets} &= +325.72 \text{ ft} \\ & 514.23 \text{ ft} \end{aligned}$$

Multiply the sum above by one-third of the common interval between the offsets.

$$\left(\frac{1}{3}\right)(25.00 \text{ ft})(514.23 \text{ ft}) = 4285.25 \text{ ft}^2$$

Finally, convert square feet to acres.

$$\frac{4285 \text{ ft}^2}{43{,}560 \ \frac{\text{ft}^2}{\text{ac}}} = 0.10 \text{ ac}$$

The answer is (B).

SOLUTION 74

The trapezoidal rule can be used to solve this problem. To use this rule, add the average of the end offsets to the sum of the intermediate offsets. The product of this quantity and the common interval between the offsets is the required area.

However, in this case, the interval between the offsets is not common, so a variation of the trapezoidal rule will be helpful. Each pair of offsets and the distance between them will be treated as a right trapezoid, and the area of each of these trapezoids taken together will give the area between the fence and the bottom of the ravine.

$$A = \frac{(h_1 + h_2)d}{2}$$

This expression can be used to find the area of each trapezoid that comprises the figure. h is an offset, and d is the distance between consecutive stations.

Begin with the trapezoid between 0+00 and 1+05.

$$\begin{array}{rr} & 1{+}05 \\ & -0 \\ & +00 \\ \hline d = & 105 \text{ ft} \end{array}$$

$$h_1 = 53 \text{ ft}$$
$$h_2 = 128 \text{ ft}$$

$$\begin{aligned} A &= \frac{(h_1 + h_2)d}{2} \\ &= \frac{(53 \text{ ft} + 128 \text{ ft})(105 \text{ ft})}{2} \\ &= 9502.5 \text{ ft}^2 \end{aligned}$$

Calculate the area of the trapezoid between 1+05 and 2+61.

$$\begin{array}{r} 2{+}61 \\ -1 \\ +05 \\ \hline d = 156\ \text{ft} \end{array}$$

$$\begin{aligned} h_1 &= 128\ \text{ft} \\ h_2 &= 311\ \text{ft} \\ A &= \frac{(128\ \text{ft} + 311\ \text{ft})(156\ \text{ft})}{2} \\ &= 34{,}242.0\ \text{ft}^2 \end{aligned}$$

Calculate the area of the trapezoid between 2+61 and 3+82.

$$A = \frac{(311\ \text{ft} + 285\ \text{ft})(121\ \text{ft})}{2} = 36{,}058.0\ \text{ft}^2$$

Calculate the area of the trapezoid between 3+82 and 4+67.

$$A = \frac{(285\ \text{ft} + 227\ \text{ft})(85\ \text{ft})}{2} = 21{,}760.0\ \text{ft}^2$$

Calculate the area of the trapezoid between 4+67 and 6+38.

$$A = \frac{(227\ \text{ft} + 169\ \text{ft})(171\ \text{ft})}{2} = 33{,}858.0\ \text{ft}^2$$

Calculate the area of the trapezoid between 6+38 and 8+50.

$$A = \frac{(169\ \text{ft} + 28\ \text{ft})(212\ \text{ft})}{2} = 20{,}882.0\ \text{ft}^2$$

The area between the fence and the bottom of the ravine is the sum of the areas of the trapezoids.

$$\begin{array}{r} 9{,}502.5\ \text{ft}^2 \\ 34{,}242.0\ \text{ft}^2 \\ 36{,}058.0\ \text{ft}^2 \\ 21{,}760.0\ \text{ft}^2 \\ 33{,}858.0\ \text{ft}^2 \\ +20{,}882.0\ \text{ft}^2 \\ \hline 156{,}302.5\ \text{ft}^2 \end{array}$$

Divide the sum by 43,560 ft^2/ac to express the answer in acres.

$$\frac{156{,}302.5\ \text{ft}^2}{43{,}560.0\ \frac{\text{ft}^2}{\text{ac}}} = 3.6\ \text{ac}$$

The answer is (D).

SOLUTION 75

The cubic yard, the most common unit of measurement used in earthwork volumes, contains 27 ft^3. The cubic yard is equal to $(3\ \text{ft})^3$.

$$1\ \text{yd}^3 = (3\ \text{ft})(3\ \text{ft})(3\ \text{ft}) = 27\ \text{ft}^3$$

The answer is (D).

SOLUTION 76

A prismoid is a solid with plane sides and parallel (but unequal) ends. In the prismoidal method of calculating volumes of earthwork, these ends, or cross sections, are augmented by a section midway between, and parallel with, the two end bases. The prismoidal method is consistently more accurate than the more easily applied average end area method.

The answer is (D).

SOLUTION 77

The prismoidal method of calculating volumes is not as convenient as the average end area method, and the extra accuracy does not always justify the extra difficulty. However, when the irregularity of the ground causes large variations in the areas of successive cross sections, the difference in the accuracy between the two methods grows greater.

The answer is (D).

SOLUTION 78

The prismoidal method will generally yield a smaller volume than that found by the average end area method. An exception arises when a cross section with a large height and a narrow width is immediately followed by a cross section that is very wide but has a small height.

The answer is (B).

SOLUTION 79

The pile can be solved as a frustrum of a pyramid.

First, calculate the areas of the bases.

$$\begin{aligned}\text{area of small (top) base} &= B_1 = (50\text{ ft})(50\text{ ft})\\ &= 2500\text{ ft}^2\\ \text{area of large (bottom) base} &= B_2 = (100\text{ ft})(100\text{ ft})\\ &= 10{,}000\text{ ft}^2\end{aligned}$$

Second, using the following formula where h is the height, solve for the volume, V, of the frustrum.

$$V = \frac{h(B_1 + B_2 + \sqrt{B_1 B_2})}{3}$$

$$V = \frac{(25\text{ ft})\left(2500\text{ ft}^2 + 10{,}000\text{ ft}^2 + \sqrt{(2500\text{ ft}^2) \times (10{,}000\text{ ft}^2)}\right)}{3} = 145{,}833\text{ ft}^3$$

Finally, express the answer in cubic yards.

$$\frac{145{,}833\text{ ft}^3}{27\,\frac{\text{ft}^3}{\text{yd}^3}} = 5401\text{ yd}^3$$

The answer is (B).

SOLUTION 80

First, it is necessary to find each cross-sectional area. The area of a simple level section can be expressed as

$$A = c(b + sc)$$

c is the cross section's height, b is the length of the base, and s is the slope.

$$\begin{aligned}A = A_1 &= (3.60\text{ ft})\big(44.00\text{ ft} + (1.5)(3.60\text{ ft})\big)\\ &= (3.60\text{ ft})(44.00\text{ ft} + 5.40\text{ ft})\\ &= (3.60\text{ ft})(49.40\text{ ft})\\ &= 177.84\text{ ft}^2\\ A = A_2 &= (4.80\text{ ft})\big(44.00\text{ ft} + (1.5)(4.80\text{ ft})\big)\\ &= (4.80\text{ ft})(44.00\text{ ft} + 7.20\text{ ft})\\ &= (4.80\text{ ft})(51.20\text{ ft})\\ &= 245.76\text{ ft}^2\end{aligned}$$

Next, find the volume between the two cross sections. The volume by the average end area method can be expressed as

$$V = \left(\frac{A_1 + A_2}{2}\right)\left(\frac{L}{27}\right) \quad \text{[in cubic yards]}$$

A_1 is the area of the first cross section, A_2 is the area of the second cross section, and L is the distance between the cross sections.

$$\begin{aligned}V &= \left(\frac{177.84\text{ ft}^2 + 245.76\text{ ft}^2}{2}\right)\left(\frac{100\text{ ft}}{27}\right)\\ &= \left(\frac{423.60\text{ ft}^2}{2}\right)\left(\frac{100\text{ ft}}{27}\right)\\ &= 784.4\text{ yd}^3 \quad (780\text{ yd}^3)\end{aligned}$$

The answer is (A).

SOLUTION 81

81.1. One method that can be used to solve this problem is an adaptation of the general method of finding areas by means of coordinates. First add one-half the width of the roadbed to the cross section of 1+10, then assign algebraic signs as shown. All dimensions are in feet.

0	9.8	1.0	0.5	0
−33.00+	−62.40+	−0.00+	+34.50−	+33.00−

The rule used to find double the area of this cross section states that each upper term will be multiplied by the algebraic sum of the adjacent lower terms. The signs used to determine the algebraic sum are those facing the upper term. These products, added together, will yield double the area of the cross section.

In other words,

$$\begin{aligned}(0)\big((-33.00) + (-62.40)\big) &= 0.00\\ (9.8)\big((+33.00) + (-0.00)\big) &= 323.40\\ (1.0)\big((+62.40) + (+34.50)\big) &= 96.90\\ (0.5)\big((+0.00) + (+33.00)\big) &= 16.50\\ (0)\big((-34.50) + (-33.00)\big) &= 0.00\end{aligned}$$

The sum of the products is double the area of the cross section at 1+10.

$$\begin{array}{r} 0.00 \\ 323.40 \\ 96.90 \\ 16.50 \\ +0.00 \\ \hline 436.80 \ \text{ft}^2 \end{array}$$

Therefore, divide this sum in half to find the actual area of the cross section.

$$\frac{436.80 \text{ ft}^2}{2} = 218.40 \text{ ft}^2$$

Next, apply the same method to the cross section at station 1+20.

$$\begin{array}{ccccc} \dfrac{0}{-33.00+} & \dfrac{10.0}{-63.00+} & \dfrac{2.5}{-0.00+} & \dfrac{1.8}{+38.40-} & \dfrac{0}{+33.00-} \end{array}$$

$$(0)\big((-33.00) + (-63.00)\big) = 0.00$$
$$(10.0)\big((+33.00) + (-0.00)\big) = 330.00$$
$$(2.5)\big((+63.00) + (+38.40)\big) = 253.50$$
$$(1.8)\big((+0.00) + (+33.00)\big) = 59.40$$
$$(0)\big((-38.40) + (-33.00)\big) = 0.00$$

$$\begin{array}{r} 0.00 \\ 330.00 \\ 253.50 \\ 59.40 \\ +0.00 \\ \hline 642.90 \ \text{ft}^2 \end{array}$$

$$\frac{642.90 \text{ ft}^2}{2} = 321.45 \text{ ft}^2$$

The answer is (B).

81.2. According to the average end area method,

$$V = \left(\frac{A_1 + A_2}{2}\right)\left(\frac{L}{27}\right) \quad \text{[in cubic yards]}$$

A_1 is the area of the first cross section, A_2 is the area of the second cross section, and L is the distance between them.

$$\begin{aligned} V &= \left(\frac{218.40 \text{ ft}^2 + 321.45 \text{ ft}^2}{2}\right)\left(\frac{10 \text{ ft}}{27}\right) \\ &= \left(\frac{539.85 \text{ ft}^2}{2}\right)\left(\frac{10 \text{ ft}}{27}\right) \\ &= (269.93 \text{ ft}^2)\left(\frac{10 \text{ ft}}{27}\right) \\ &= 99.97 \text{ yd}^3 \quad (100 \text{ yd}^3) \end{aligned}$$

The answer is (B).

81.3. The prismoidal method requires the construction of a median cross section from the end sections.

$$\begin{array}{ccccc} \dfrac{0}{-33.00+} & \dfrac{9.9}{-62.70+} & \dfrac{1.75}{-0.00+} & \dfrac{1.15}{+36.45-} & \dfrac{0}{+33.00-} \end{array}$$

This middle section is found not by averaging the end areas, but by averaging the corresponding linear dimensions. In this case, the dimensions of the cross section at 1+10 were averaged with the cross section at 1+20 to arrive at the coordinates shown.

Find the area of the middle section.

$$(0)\big((-33.00) + (-62.70)\big) = 0.00$$
$$(9.9)\big((+33.00) + (-0.00)\big) = 326.70$$
$$(1.75)\big((+62.70) + (+36.45)\big) = 173.51$$
$$(1.15)\big((-0.00) + (+33.00)\big) = 37.95$$
$$(0)\big((-36.45) + (-33.00)\big) = 0.00$$

$$\begin{array}{r} 0.00 \\ 326.70 \\ 173.51 \\ 37.95 \\ +0.00 \\ \hline 538.16 \ \text{ft}^2 \end{array}$$

$$\frac{538.16 \text{ ft}^2}{2} = 269.08 \text{ ft}^2$$

The volume can now be found by the prismoidal method, which can be expressed as

$$V = \left(\frac{L}{(6)(27)}\right)(A_1 + 4A_m + A_2) \quad \text{[in cubic yards]}$$

A_1 is the area of the first cross section, A_2 is the area of the second cross section, A_m is the area of the middle section, and L is the distance between the end sections.

$$\begin{aligned} V &= \left(\frac{10 \text{ ft}}{162}\right)\left(\begin{array}{l} 218.40 \text{ ft}^2 + (4)(269.08 \text{ ft}^2) \\ \quad +321.45 \text{ ft}^2 \end{array}\right) \\ &= \left(\frac{10 \text{ ft}}{162}\right)\left(\begin{array}{l} 218.40 \text{ ft}^2 + 1076.32 \text{ ft}^2 \\ \quad +321.45 \text{ ft}^2 \end{array}\right) \\ &= \left(\frac{10 \text{ ft}}{162}\right)(1616.17 \text{ ft}^2) \\ &= 99.8 \text{ yd}^3 \end{aligned}$$

The answer is (B).

81.4. The prismoidal correction may be found by using the formula

$$C = \left(\frac{L}{(12)(27)}\right)(c_1 - c_2)(w_1 - w_2) \quad \text{[in cubic yards]}$$

c_1 and c_2 are the center heights of the end sections, w_1 and w_2 are the widths from slope intercept to slope intercept of the end sections, and L is the distance between the end sections.

$$\begin{aligned} C &= \left(\frac{10 \text{ ft}}{324}\right)(1.0 \text{ ft} - 2.5 \text{ ft})(96.90 \text{ ft} - 101.40 \text{ ft}) \\ &= \left(\frac{10 \text{ ft}}{324}\right)(-1.5 \text{ ft})(-4.5 \text{ ft}) \\ &= \left(\frac{10 \text{ ft}}{324}\right)(6.75 \text{ ft}^2) \\ &= 0.2 \text{ yd}^3 \end{aligned}$$

The answer is (C).

SOLUTION 82

The linear error of closure is found using the Pythagorean theorem.

$$\begin{array}{rr} & 5000.324 \\ & -5000.000 \\ \hline \text{difference in N} = & +0.324 \\ & 9999.871 \\ & -10{,}000.000 \\ \hline \text{difference in E} = & -0.129 \end{array}$$

$$\sqrt{(0.324 \text{ ft})^2 + (0.129 \text{ ft})^2} = \begin{array}{r} 0.349 \text{ ft linear error} \\ \text{of closure} \end{array}$$

The bearing of the direction of the linear error of closure can be found by using the tangent function.

$$\begin{aligned} \tan^{-1}\left(\frac{\text{difference in E}}{\text{difference in N}}\right) &= \tan^{-1}\left(\frac{0.129 \text{ ft}}{0.324 \text{ ft}}\right) \\ &= 21°42'36'' \end{aligned}$$

The bearing of the error of closure is N21°42′36″W.

The answer is (A).

SOLUTION 83

The compass rule uses the proportion the length of each leg bears to the entire length of the traverse as the basis of calculating the corrections in both latitude and departure applied to each leg.

The answer is (A).

SOLUTION 84

The sum of the interior angles of a closed plane figure is

$$(n-2)(180°)$$

n is the number of sides of the figure.

In the case of a closed figure of 12 sides, the calculation is

$$(12-2)(180°) = (10)(180°) = 1800°$$

The answer is (B).

SOLUTION 85

The sum of the exterior angles of a closed plane figure is

$$(n+2)(180°)$$

n is the number of sides of the figure.

In the case of a closed figure with eight sides, the calculation is

$$(8+2)(180°) = (10)(180°) = 1800°$$

The answer is (D).

SOLUTION 86

The compass rule is widely considered to be superior to the transit rule in balancing a traverse. Both assume that the errors in traversing are random; that is, that systematic errors have been eliminated. However, the transit rule assumes that the angular measurements are superior to the distance measurements, while the compass rule assumes the two types of measurements to be equally reliable.

The answer is (D).

SOLUTION 87

Like all fabricated instruments, the plate of an optical reading theodolite is not perfect. The incrementation of the plate prevents imperfections in one area of the plate from corrupting measurements of the angle. The technique cannot eliminate the effect of the systematic error in the plate; it merely attempts to lessen its effect.

The answer is (D).

SOLUTION 88

All the statements except (B) are true. No adjustment can be said to eliminate random errors from a survey.

The answer is (B).

SOLUTION 89

First, find the theoretically correct value for the sum of the interior angles of a six-sided plane figure.

$$(n-2)(180°)$$

n is the number of sides of the figure.

$$(6-2)(180°) = (4)(180°) = 720°$$

The sum of the measured angles of the figure is 719°59′00″.

Next, find the difference between the theoretically correct sum of the interior angles and the sum of the measured interior angles.

$$\begin{array}{r} 720°00'00'' \\ -719°59' \\ \underline{00''} \\ 00°01'00'' \end{array}$$

Since the measured angles are of equal precision, they each are adjusted by one-sixth of the discrepancy, that is, 10 seconds.

Therefore, the adjustment of the first angle is

$$\begin{array}{r} 118°15'58'' \\ \underline{+10''} \\ 118°16'08'' \end{array}$$

The answer is (C).

SOLUTION 90

The uncertainty of position that results from an angular error of 20″ can be found by means of the inverse of the tangent function.

$$\cot 00°00'20'' = 10{,}313.24 \quad (10{,}000)$$

The answer is (B).

SOLUTION 91

The first statement is false. Accuracy is the degree of agreement between a measurement and the true value; precision is the degree of agreement between repetitive measurements of the same value.

The answer is (D).

SOLUTION 92

If a single measurement is selected at random from a series of measurements of the same quantity, the percentage of probability is the level of certainty that the measurement will fall within the range represented by the standard deviation. For example, if the standard deviation of a series of measurements of the same quantity was ±0.2 ft, one could be 68.3% (68%) sure that any single measurement from that group of measurements would be within ±0.2 ft of the true value.

The answer is (A).

SOLUTION 93

The median is the middle value of the measurements arranged in ascending or descending order. The mode is the measurement occurring most frequently. The mean is the sum of the measurements that has been divided by the total number of observations. The mean is the best approximation of the true value.

The answer is (C).

SOLUTION 94

The compass rule states that the correction applied to the latitude (or departure) of any course is in the same proportion to the total correction in the latitude (or departure) as the length of the course is to the total length of the traverse.

The error of closure in the northing is +0.128 ft; the correction in the northing of point 100 is unknown.

The length of the course from Allen to 100 is 9718.480 ft, and the total length of the traverse is 63,954.250 ft.

$$\frac{9718.480 \text{ ft}}{63{,}954.250 \text{ ft}} = 0.1520$$

The ratio of the length of the course to the sum of the lengths of the traverse is the same as that of the correction of the northing of point 100 to the overall error in the northing.

$$(0.1520)(-0.128 \text{ ft}) = -0.019 \text{ ft}$$

The correction in the easting of point 100 must be in the same proportion.

$$(0.1520)(+0.131 \text{ ft}) = +0.020 \text{ ft}$$

Finally, add these corrections to the coordinates of point 100.

$$\begin{aligned} \text{unadjusted northing in feet} &= 2{,}132{,}478.378 \\ \text{correction} &= \underline{\qquad\quad -0.019} \\ \text{adjusted northing of point 100} &= 2{,}132{,}478.359 \\ \text{unadjusted easting in feet} &= 745{,}786.469 \\ \text{correction} &= \underline{\qquad\quad +0.020} \\ \text{adjust easting of point 100} &= 745{,}786.489 \end{aligned}$$

The answer is (B).

SOLUTION 95

First, determine the quantity of spherical excess within the figure from the area it encloses.

$$\frac{131.24 \text{ mi}^2}{75.60 \text{ mi}^2} = 1.74 \text{ sec of spherical excess}$$

The theoretical sum of the angles of a plane triangle is ordinarily 180°00′00″. With the addition of the spherical excess, the sum of the observed angles should be 180°00′01.74″.

Find the sum of the measured angles.

$$\begin{array}{r} 52°28'15'' \\ 50°15'10'' \\ \underline{+77°16'32''} \\ 179°59'57'' \end{array}$$

Find the difference between the sum of the observed angles and the theoretically correct sum.

$$\begin{array}{r} 180°00'01.74'' \\ -179°59' \\ \underline{57.00''} \\ +04.74'' \end{array}$$

The adjustment for each angle is

$$\frac{+04.74''}{3} = +1.58''$$

The answer is (D).

SOLUTION 96

Geodetic azimuths are sometimes expressed from south. The rotation of a south azimuth is normally clockwise, just as it is with a north azimuth. In this case, the bearing angle can be subtracted from 360° to find the azimuth from south.

$$\begin{aligned} & 360°00'00'' \\ \text{bearing angle} &= \underline{-10°59'01''} \\ \text{azimuth from south} &= 349°00'59'' \end{aligned}$$

The answer is (D).

SOLUTION 97

Begin by expressing the direction of line AB as an azimuth from north.

$$\begin{array}{r} 180°00'00'' \\ \underline{-89°56'15''} \\ 90°03'45'' \end{array}$$

Find the azimuth of line CD by subtracting deflections to the left and adding deflections to the right.

$$\begin{aligned} & 90°03'45'' \\ \text{deflection to the left} &= \underline{-00°04'15''} \\ & 89°59'30'' \\ \text{deflection to the right} &= \underline{+00°00'35''} \\ & 90°00'05'' \end{aligned}$$

The answer is (D).

SOLUTION 98

The azimuth of line EF is 65°14′02″, and the azimuth of line FG is

$$\begin{array}{r} 180°00'00'' \\ -75°23'10'' \\ \hline 104°36'50'' \end{array}$$

The deflection angle to the right is the difference between the two azimuths.

$$\begin{array}{r} 104°36'50'' \\ -65°14'02'' \\ \hline 39°22'48'' \end{array}$$

The answer is (B).

SOLUTION 99

The true bearing derived from the 1886 magnetic bearing is

$$\begin{aligned} \text{magnetic bearing from 1886} &= \text{N01°20′W} \\ \text{eastern declination from 1886} &= \underline{\quad 00°31'\text{E}} \\ \text{derived true bearing} &= \text{N00°49′W} \end{aligned}$$

The magnetic bearing derived from the current true bearing is

$$\begin{aligned} \text{current western declination} &= 00°12'\text{W} \\ \text{current true bearing} &= \underline{-\text{N00°09′W}} \\ \text{derived magnetic bearing} &= \text{N00°03′E} \end{aligned}$$

The differences are

$$\begin{aligned} \text{magnetic bearing from 1886} &= \text{N01°20′W} \\ \text{current magnetic bearing} &= \underline{+\text{N00°03′W}} \\ \text{difference} &= 01°23' \\ \text{current true bearing} &= \text{N00°09′W} \\ \text{true bearing from 1886} &= \underline{\text{N00°49′W}} \\ \text{difference} &= 00°40' \end{aligned}$$

The answer is (C).

SOLUTION 100

The following illustration is a general representation of the relationships involved in this problem.

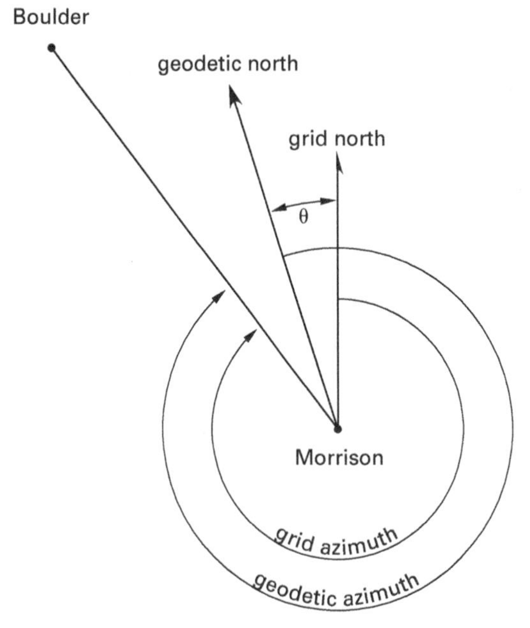

The grid azimuth is measured from grid north, which is parallel to the central meridian of the zone of a plane coordinate projection system. The geodetic azimuth is measured from geodetic north, which coincides with a true spheroidal meridian. The convergence angle, also known as the gamma angle, γ, in the Lambert projection and in the Mercator projection represents the angular difference between grid and geodetic norths.

In this case, the geodetic azimuth is equal to the grid azimuth plus the gamma angle, γ.

$$\begin{aligned} \text{geodetic azimuth} &= \text{grid azimuth} + \gamma \\ &= 348°32'24.0'' + (+00°10'53.2'') \\ &= 348°43'17.2'' \end{aligned}$$

The answer is (D).

SOLUTION 101

The problem's illustration shows the arrangement of the significant azimuths.

$$\text{geodetic azimuth} - \gamma = \text{grid azimuth}$$
$$25°17'49.11'' - (+00°34'37.61'') = 24°43'11.50''$$

The grid azimuth from Adams to Douglas is 24°43′11.50″ from south. The grid azimuth from Douglas to Adams from grid north is the same value.

The answer is (A).

SOLUTION 102

The orientation of an instrument on the earth's surface is generally considered level when its vertical axis is coincident with the direction of gravity. Therefore, such an orientation is level with respect to the surface called the *geoid.* The geoid is defined by gravity, and is the surface that is perpendicular everywhere to the direction of gravity. On the other hand, the spheroid is an imaginary model of the earth that has no such direct relation to the force of gravity.

Geodetic azimuths are measured from meridians on the spheroid; astronomic azimuths are measured by instruments oriented to the geoid. In other words, up and down are not the same on the geoid as they are on the spheroid. This difference is the deviation of the vertical. This affects the difference between geodetic and astronomic azimuths, just as using an instrument that is out of level would affect an astronomic observation.

The answer is (C).

SOLUTION 103

The limits of embankment or excavation along a route are defined by slope stakes. These stakes generally are set at the intersection of the natural ground and the designed side slope.

The answer is (B).

SOLUTION 104

The side slope indicates the ratio of the horizontal distance to the vertical distance, which is usually one. In this case, the first number of each side slope ratio is the number of feet the width of the levee's base increases for each foot of its height.

$$\left(3\ \frac{\text{ft}}{\text{ft}}\right)(15\ \text{ft}) = 45\ \text{ft}$$

$$\left(2\ \frac{\text{ft}}{\text{ft}}\right)(15\ \text{ft}) = 30\ \text{ft}$$

$$\text{width of the levee at the top} = 20\ \text{ft}$$

The sum of the values is the width of the levee at its base.

$$\begin{array}{r} 45\ \text{ft} \\ 30\ \text{ft} \\ +20\ \text{ft} \\ \hline 95\ \text{ft} \end{array}$$

The answer is (C).

SOLUTION 105

The distance, d, from the centerline of the roadway to the set slope stake can be expressed as

$$d = \frac{w}{2} + hs$$

w is the width of the roadway, h is the difference between the grade rod and the ground rod, and s is the first number in the side slope ratio.

The grade rod is found by subtracting the designed subgrade elevation of the roadway from the HI of the instrument.

$$\begin{aligned} \text{grade rod} &= 618.23\ \text{ft} - 609.45\ \text{ft} \\ &= 8.78\ \text{ft} \end{aligned}$$

This is the rod reading that would indicate that the rod was held exactly at the design elevation of the roadway subgrade.

The ground rod is the rod reading at the existing ground.

$$\begin{aligned} h &= \text{grade rod} - \text{ground rod} \\ &= 8.78\ \text{ft} - 7.3\ \text{ft} \\ &= 1.48\ \text{ft} \\ d &= \frac{w}{2} + hs \\ &= \frac{60\ \text{ft}}{2} + (1.48\ \text{ft})(1.5) \\ &= 30\ \text{ft} + 2.22\ \text{ft} \\ &= 32.22\ \text{ft} \quad (32\ \text{ft}) \end{aligned}$$

The answer is (B).

SOLUTION 106

The angle of repose varies with the material used in earthwork, but it is always an important element of design. The designed side slopes along a route frequently depend on the appropriate angle of repose for the material available.

The answer is (D).

Advanced Mathematics

QUADRATIC EQUATIONS

PROBLEM 1

A second degree equation in x is a quadratic equation and has the general form

$$ax^2 + bx + c = 0$$

a, b, and c are constants, and a is not equal to zero. A derivation of this general equation expresses x in terms of the other elements shown and is known as the *quadratic formula.* Which of the following mathematical expressions correctly represents the quadratic formula?

(A) $x = a^2 + b + \frac{c^2}{4}$

(B) $x = -a \pm \sqrt{\frac{b^2 - 4bc}{2c}}$

(C) $x = \frac{-b \pm \sqrt{b^2 - 4ac}}{2a}$

(D) $x = -c^2 + 2ab - b^2$

PROBLEM 2

Which values correctly represent the base and the height of a right triangle containing 510 ft^2 in which the base is 4 ft shorter than the height?

(A) base = 16 ft
height = 20 ft

(B) base = 25 ft
height = 29 ft

(C) base = 30 ft
height = 34 ft

(D) base = 32 ft
height = 36 ft

PROBLEM 3

Which values of x satisfy the expression $9x^2 + 30x + 20 = 0$?

(A) −2.41

(B) −0.92

(C) 0.13

(D) both A and B

PROBLEM 4

Which values of x satisfy the expression $201.93 = 6x + 5x^2$?

(A) −9.52 and +5.61

(B) −6.98 and +4.09

(C) −6.98 and +5.78

(D) +5.81 and +3.74

PROBLEM 5

The owner of a rectangular parcel that is 50 ft by 200 ft wishes to double the area of the lot. She wants to acquire a strip of land to add to the end of the lot and another strip of exactly the same width to add to the side of the lot. Which of the following widths will accomplish the goal?

(A) 35.07 ft

(B) 37.12 ft

(C) 40.00 ft

(D) 52.76 ft

SIMULTANEOUS EQUATIONS

PROBLEM 6

What is the name usually given to the following type of square array used in solving a system of linear equations?

$$\begin{vmatrix} a & b \\ c & d \end{vmatrix}$$

The equations for the form are

$$ax + by = e$$
$$cx + dy = f$$

(A) covariant

(B) cofactor

(C) determinant

(D) eigenvector

PROBLEM 7

What is the value of the expansion of the following form?

$$\begin{vmatrix} 1 & 9 \\ -6 & -4 \end{vmatrix}$$

(A) −58

(B) 33

(C) 50

(D) 216

PROBLEM 8

When a determinant is used to solve a system of simultaneous equations where there are just as many equations as there are unknowns, the system has a single solution if the determinant of the coefficients of the unknowns is not zero. The value of each unknown in the system of equations can be written as the quotient of two determinants. How are these two determinants constructed?

(A) The determinant in the denominator is different for each unknown. The denominator is identical to the numerator, with the coefficients of the desired unknown being replaced by the constants. The determinant in the numerator consists of the coefficients of the unknowns.

(B) The determinant in the denominator consists of the coefficients of the unknowns. The determinant in the numerator is different for each unknown. The numerator is identical to the denominator, with the coefficients of the desired unknown being replaced by the constants.

(C) The determinant of the denominator consists of the coefficients of unknowns, but the row or column representing the desired unknown is replaced by any numbers, which are cofactors. The determinant in the numerator is different for each unknown. The numerator is identical to the denominator, with the addition of the coefficients of the desired unknown.

(D) The determinant in the denominator is different for each unknown. The denominator is identical to the numerator, with the addition of the coefficients of the desired unknown. The determinant of the numerator consists of the coefficients of unknowns, but the row or column representing the desired unknown is replaced by any numbers, which are cofactors.

PROBLEM 9

What is the name of the following rule for solving a system of simultaneous equations by the use of determinants?

> When there are just as many equations as there are unknowns, the system has a single solution if the determinant of the coefficients of the unknowns is not zero. The value of each unknown in the system of equations can be written as the quotient of two determinants.
>
> The determinant in the denominator consists of the coefficients of the unknowns.
>
> The determinant in the numerator is different for each unknown. The numerator is identical to the denominator with the coefficients of the desired unknown being replaced by the constants.

(A) Leibniz theorem

(B) Kowa's rule

(C) Cramer's rule

(D) Pratt's axiom

PROBLEM 10

What are the values of x and y that will solve the following pair of simultaneous equations?

$$2x - 8y = 8$$
$$3x + 4y = 28$$

(A) $x = 4$
$y = 4$

(B) $x = 8$
$y = 1$

(C) $x = 10$
$y = 5$

(D) $x = 16$
$y = 3$

PROBLEM 11

What prevents the values of x and y in the following system of equations from being found by the use of determinants?

$$2x - 5y = 10$$
$$4x - 20 = 10y$$

(A) The determinant comprised of the coefficients of the unknowns is equal to zero.

(B) There is no solution for x and y in real numbers.

(C) The values of x and y are not integers.

(D) Nothing; the equations can be solved with determinants.

PROBLEM 12

What is the arrangement and order of the terms for the expansion of the following determinant?

$$\begin{vmatrix} a_1 & b_1 & c_1 \\ a_2 & b_2 & c_2 \\ a_3 & b_3 & c_3 \end{vmatrix}$$

(A) $a_3b_2c_1 + a_1b_3c_2 + a_1b_2c_3 + a_3b_1c_2 + a_2b_3c_1 - a_2b_1c_3$

(B) $a_1b_2c_3 - a_3b_1c_2 + a_2b_3c_1 - a_3b_2c_1 + a_1b_3c_2 - a_2b_1c_3$

(C) $a_2b_3c_1 + a_3b_2c_1 + a_1b_2c_3 - a_3b_1c_2 - a_1b_3c_2 - a_2b_1c_3$

(D) $a_1b_2c_3 + a_3b_1c_2 + a_2b_3c_1 - a_3b_2c_1 - a_1b_3c_2 - a_2b_1c_3$

PROBLEM 13

What is the value of the expansion of the following determinant?

$$\begin{vmatrix} 1 & 4 & 2 \\ 4 & 6 & 3 \\ 2 & -1 & 6 \end{vmatrix}$$

(A) −91

(B) −65

(C) 42

(D) 96

PROBLEM 14

What values of x, y, and z will solve the following system of simultaneous equations?

$$3x + 2y - 6z = 65$$
$$x + 3y - z = 23$$
$$x + 10y + 9z = -29$$

(A) $x = -10$
$y = -1$
$z = -1$

(B) $x = 3$
$y = 4$
$z = -8$

(C) $x = 4$
$y = 6$
$z = -7$

(D) $x = 10$
$y = 1$
$z = -10$

Advanced Mathematics

AREAS

PROBLEM 15

An existing rectangular parcel of land was measured many years ago with a Gunter's chain and was found to contain 40 ac. The original surveyor noted later that the chain used to make the measurements was one link too short. Assuming that the error was evenly distributed throughout the measuring, what was the area of the parcel?

(A) 39.20 ac

(B) 39.60 ac

(C) 40.40 ac

(D) 40.80 ac

PROBLEM 16

Two unequal square parcels adjoin one another and the perimeter of both, except for their shared side as shown in the following illustration, is enclosed by 560 ft of fencing. The area of both parcels together is 16,400 ft^2. What two ratios of the length of a side of the larger parcel divided by the length of a side of the smaller parcel will satisfy these conditions?

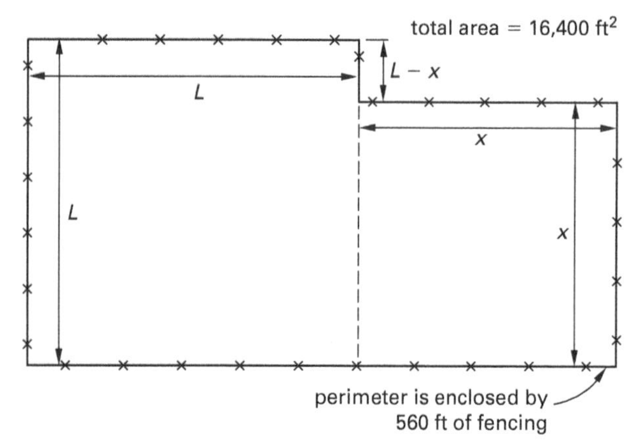

(A) 1.250, 3.875

(B) 1.472, 4.152

(C) 1.505, 3.190

(D) 2.189, 2.786

PROBLEM 17

What is the area, to the nearest square foot, of an ellipse with a major axis of 284 ft and a minor axis of 152 ft?

(A) 18,145 ft^2

(B) 33,904 ft^2

(C) 38,465 ft^2

(D) 40,746 ft^2

PROBLEM 18

Mr. Gauss sold one acre off the west side of the parcel shown. The deed stipulated that the eastern boundary of that acre be parallel with the western boundary of the parcel. Mr. Gauss is concerned that the house, which is 200 ft perpendicular from the western boundary, may not be on the acre that has been sold. What is the perpendicular distance from the western boundary of the parcel to the eastern boundary of the acre that has been sold?

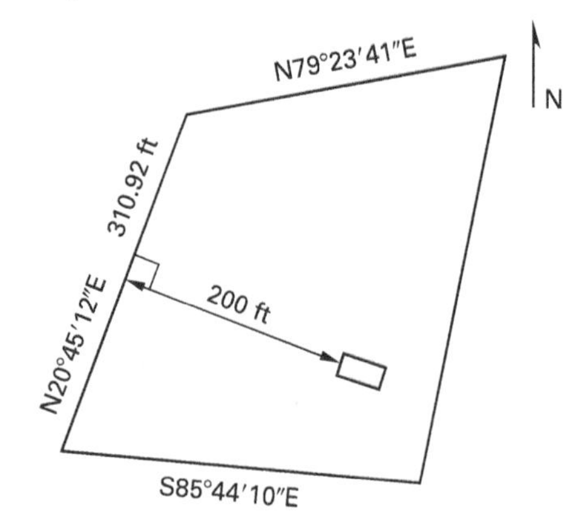

(A) 131.40 ft

(B) 135.56 ft

(C) 139.21 ft

(D) 140.10 ft

PROBLEM 19

The following two problems refer to the illustration shown.

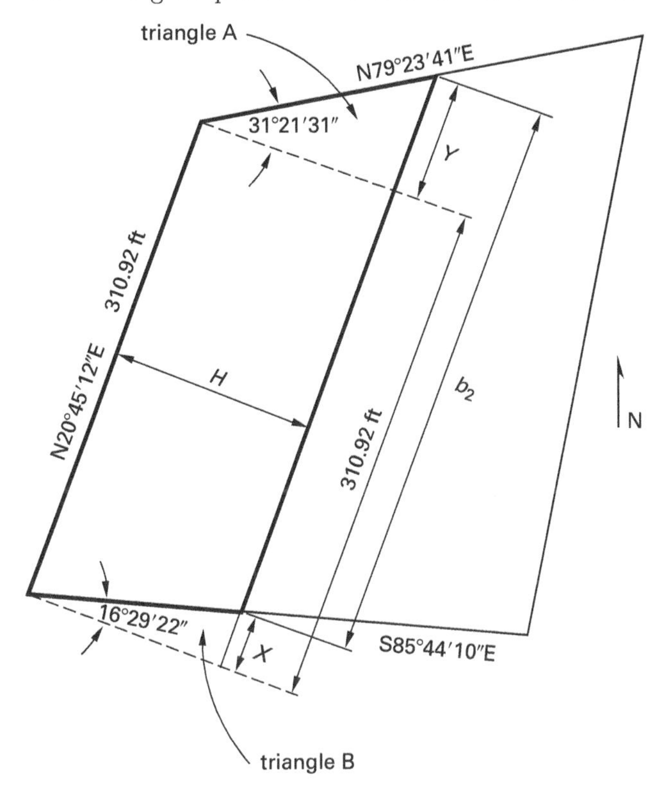

19.1. What is the length of b_2?

(A) 341.62 ft

(B) 352.10 ft

(C) 361.23 ft

(D) 389.21 ft

19.2. If H in the illustration was 200 ft, what would be the area within the heavy line in the parcel?

(A) 52,971 ft^2

(B) 63,289 ft^2

(C) 68,452 ft^2

(D) 70,104 ft^2

PROBLEM 20

The following two problems refer to the illustration shown.

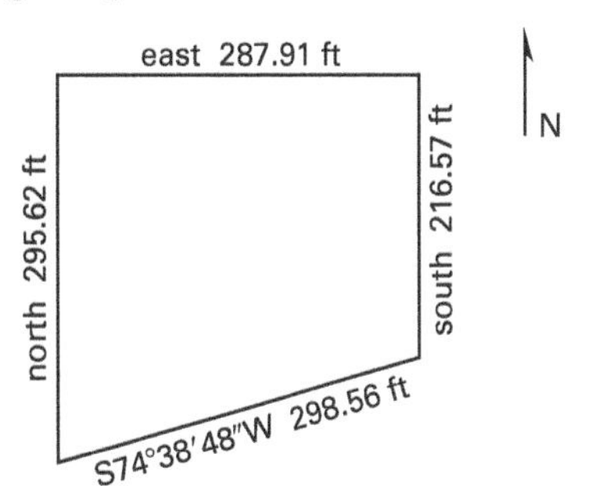

20.1. The area of the parcel shown is 73,732.31 ft^2. Its owner wishes to divide the parcel into two equal parts. The dividing line must begin at the midpoint of the northern boundary, but it may take any direction that divides the areas into halves. At what distance along the southern boundary from the southwestern corner of the parcel will the dividing line intersect?

(A) 126.24 ft

(B) 167.34 ft

(C) 186.32 ft

(D) 201.45 ft

20.2. What is the length of the line that passes through the midpoint of the northern boundary of the parcel and divides it into halves?

(A) 256.10 ft

(B) 263.13 ft

(C) 276.43 ft

(D) 287.33 ft

PROBLEM 21

You have completed the layout of the illustrated reflecting pool and its base. Both are regular polygons. The distances shown are from the midpoint of a side of the pool and the base to their shared center. You have been asked to provide the contractor with the area between the pool and the base for an estimate of the materials needed to cover it. What is the area between the seven-sided heptagonal pool and the eleven-sided undecagonal base?

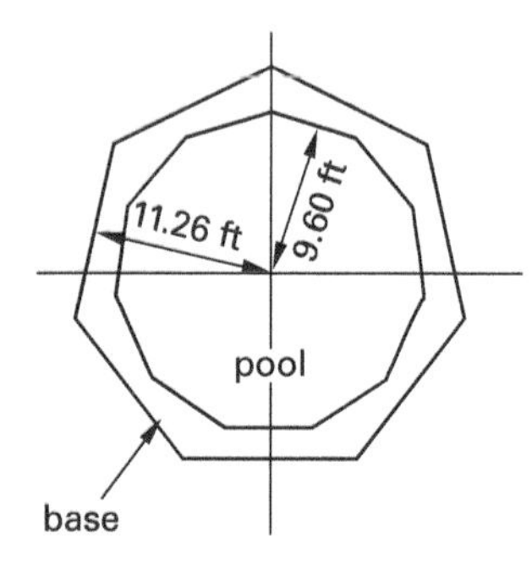

(A) 108.78 ft^2

(B) 129.81 ft^2

(C) 185.44 ft^2

(D) 259.62 ft^2

PROBLEM 22

What is the length of the second diagonal of the parcel in the following illustration?

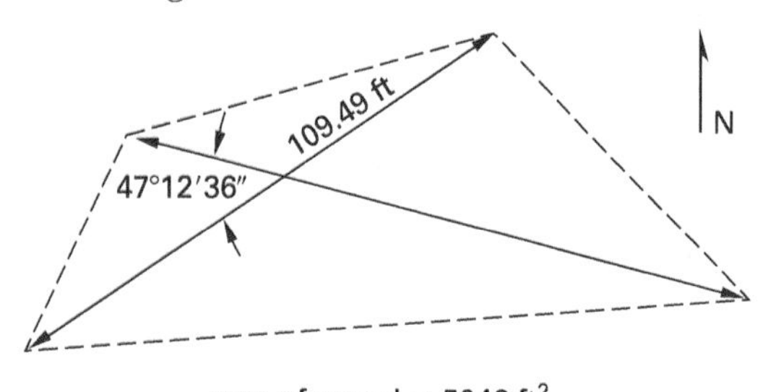

(A) 125.60 ft

(B) 142.97 ft

(C) 150.62 ft

(D) There is not sufficient information available to answer this question.

PROBLEM 23

What is the area north of the parabola in the illustration?

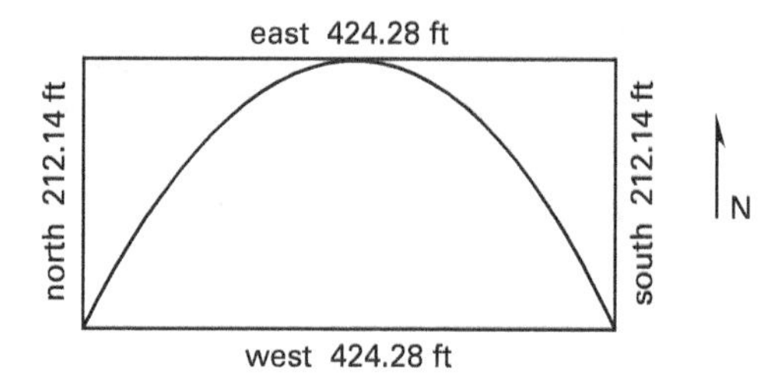

(A) 30,002.2 ft^2

(B) 45,003.4 ft^2

(C) 51,753.9 ft^2

(D) 60,004.5 ft^2

VERTICAL CURVES

PROBLEM 24

The following two problems refer to the illustration shown.

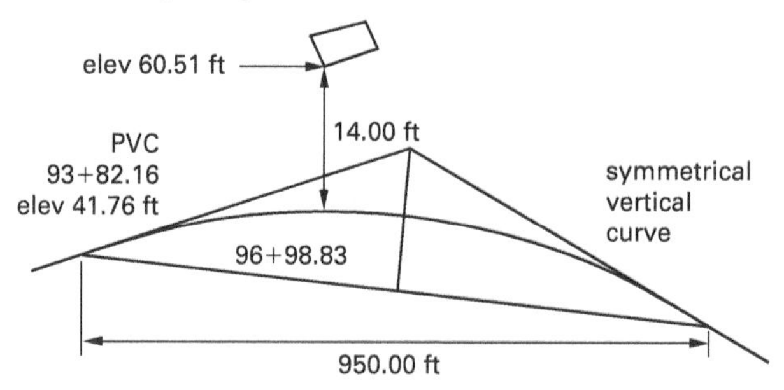

24.1. A symmetrical vertical curve with an entering grade of +3% and a length of 9.5 full stations must provide 14 ft of clearance under a bridge that passes over it at elev 60.51 ft. The PVC is at sta 93+82.16 and has an elevation of 41.76 ft. Station 96+98.83 on the vertical curve occurs directly below the bridge. What is the exiting grade that will provide the required clearance?

(A) −6%

(B) −4.2%

(C) −3.15%

(D) −3%

24.2. What is the station and elevation of the PVT in the vertical curve illustrated?

(A) 101+55.76, 33.21 ft

(B) 102+66.21, 28.64 ft

(C) 103+32.16, 27.51 ft

(D) 110+41.33, 33.47 ft

PROBLEM 25

A manhole is located in a street that has a 4 in parabolic crown. It is 16.16 ft from the centerline and the street is 42 ft wide. What is the elevation of the center of the manhole if the elevation at the street centerline is 277.18 ft?

(A) 276.74 ft

(B) 276.85 ft

(C) 276.98 ft

(D) 277.06 ft

PROBLEM 26

The following four problems refer to the illustration shown.

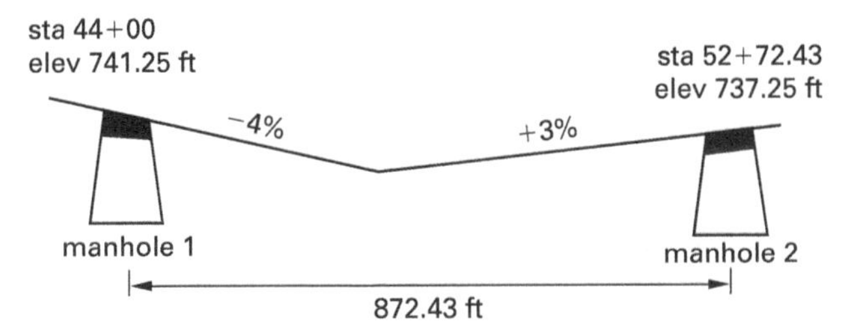

26.1. A vertical curve must begin at the center of manhole 1 (sta 44+00, elev 741.25 ft) and end at manhole 2 (sta 52+72.43, elev 737.25 ft) as illustrated. The entering grade of −4% and the exiting grade of +3% cannot be changed. At what distance from manhole 1 will the grade −4% intersect the grade +3% from manhole 2?

(A) 429 ft

(B) 431 ft

(C) 440 ft

(D) 441 ft

26.2. Using the same parameters as in Prob. 26.1, design an asymmetrical vertical curve. At what station and upon what line(s) will the CVC be found?

(A) 46+15.50 at the intersection of the entering and the exiting grade

(B) 47+23.25 at the intersection of the entering and the exiting grade

(C) 48+31.00 on the chord connecting the midpoints of the entering grade and the exiting grade

(D) 50+51.715 at the midpoint of the chord connecting the midpoints of the entering grade and the exiting grade

26.3. What are the stations and elevations of the PVIs of the vertical curves that make up the asymmetrical vertical curve illustrated?

(A) sta 41+23.25, elev 728.32 ft
sta 49+41.36, elev 727.32 ft

(B) sta 46+15.50, elev 732.63 ft
sta 50+51.715, elev 730.63 ft

(C) sta 48+31.00, elev 724.01 ft
sta 41+23.25, elev 728.32 ft

(D) sta 48+31.00, elev 724.01 ft
sta 49+41.36, elev 727.32 ft

26.4. What is the grade of the line connecting the PVI of the first vertical curve with the PVI of the second vertical curve through the asymmetrical curve?

(A) −0.46%

(B) −0.33%

(C) 3.3%

(D) 5.0%

PROBLEM 27

The following three problems refer to the illustration shown.

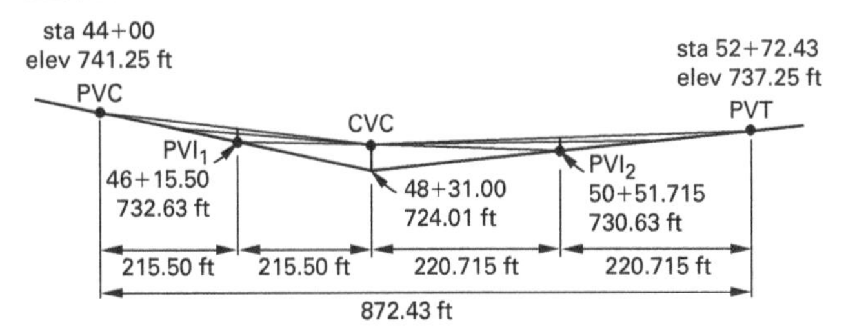

27.1. What are the rates of change of the two vertical curves?

(A) first vertical curve = −0.139
second vertical curve = 1.721

(B) first vertical curve = 0.802
second vertical curve = −0.802

(C) first vertical curve = 0.821
second vertical curve = 0.784

(D) first vertical curve = 1.624
second vertical curve = 1.586

27.2. A cross-pan drainage structure is to be built on the roadway that follows the asymmetrical vertical curve illustrated. At what station should the pan be constructed?

(A) 46+15.50

(B) 48+31.00

(C) 48+89.67

(D) 50+57.33

27.3. What is the elevation of the lowest point on the asymmetrical vertical curve illustrated?

(A) 731.50 ft

(B) 731.52 ft

(C) 731.59 ft

(D) 731.60 ft

CURVE SUPERELEVATION

PROBLEM 28

When a highway is superelevated around a curve, which statement best describes the usual treatment of the centerline, inner edge, and outer edge of that curve?

(A) The inner edge is maintained at grade, the centerline is raised by half of the superelevation, and the outer edge is raised by half of the superelevation.

(B) The inner edge is lowered by half of the superelevation, the centerline is maintained at grade, and the outer edge is raised by half of the superelevation.

(C) The inner edge is maintained at grade, the centerline is maintained at grade, and the outer edge is raised by the superelevation distance.

(D) The inner edge is lowered by the superelevation distance, the centerline is maintained at grade, and the outer edge is maintained at grade.

PROBLEM 29

What is the usual treatment of the crown of the pavement around a superelevated curve of more than 1° of curve?

(A) The usual convex crown of the pavement is continued around the curve to provide a banking effect on the inner lane.

(B) The crown is gradually changed from convex to concave in the middle of the superelevated curve.

(C) The crown is gradually changed from a convex crown to a plane-inclined surface in the middle of the curve.

(D) The crown becomes concave immediately at the beginning of the curve and continues to the end.

PROBLEM 30

Which statement best describes the meaning of the friction factor as used in the design of superelevated curves?

(A) The friction factor is a variable used to increase the extent of the transverse slope of a superelevated curve based on the centripetal force acting on the vehicle traveling around the curve.

(B) The friction factor is a quantification of the centrifugal force acting on a vehicle traveling around a superelevated curve that is used to calculate the length of the curve.

(C) The friction factor is a variable used to reduce the superelevation of a curve to prevent a vehicle traveling around the curve from skidding inward.

(D) The friction factor is a variable value based on the impedance of a vehicle traveling around a superelevated curve and is used to calculate the length of the curve.

PROBLEM 31

The following two problems refer to the illustration shown.

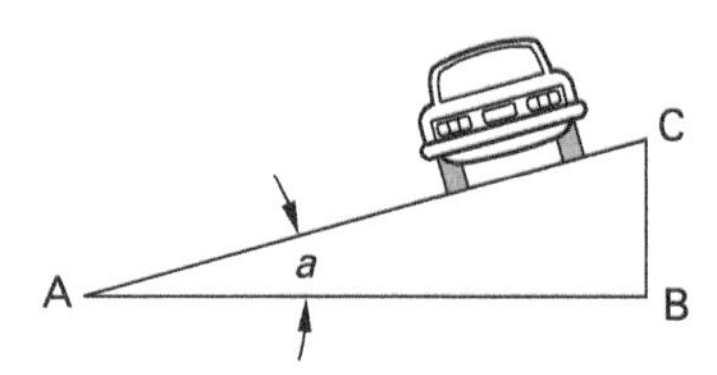

31.1. A superelevated curve has a radius of 1000 ft with a vehicle at the middle of a curve. The superelevation was designed for a speed of 70 mph and a friction factor of 0.10. What is angle a to the nearest tenth of a degree?

(A) 12.8°

(B) 21.5°

(C) 26.0°

(D) 30.6°

31.2. Assume that the design speed for a superelevated curve is 70 mph and the maximum transverse slope is tan 12.8°, or 0.2267. Under these conditions, what is the minimum length of the spiral curve through which the superelevation should occur?

(A) 1163 ft

(B) 1469 ft

(C) 1775 ft

(D) 2387 ft

PROBLEM 32

What is the approximate superelevation of the outer rail above the inner rail of a 2° railroad curve with a 70 mph design speed if the friction factor is neglected?

(A) $5\frac{1}{4}$ in

(B) $6\frac{1}{2}$ in

(C) $7\frac{1}{4}$ in

(D) $8\frac{1}{2}$ in

PROBLEM 33

What is the minimum length of a railroad spiral curve where the superelevation of the outer rail is $4\frac{1}{2}$ in above the inner rail and the design speed is 60 mph?

(A) 317 ft

(B) 340 ft

(C) 364 ft

(D) 387 ft

SPIRAL CURVES

PROBLEM 34

A spiral curve, as used in surveying, is

(A) also known as a compound circular curve

(B) a hyperbolic curve in the horizontal plane

(C) a loxodrome

(D) a curve with a uniformly varying radius

PROBLEM 35

Which statement best defines the usual purpose of an easement curve?

(A) An easement curve is commonly used in the layout of railroad and highway right-of-ways to accommodate the routes of large circular curves.

(B) An easement curve, also known as a *transition spiral curve*, is used to introduce a circular curve and the centrifugal force associated with it.

(C) An easement curve, also known as a *cubic spiral*, is most often used instead of a circular curve in railroads.

(D) An easement curve, also known as a *clothoid* or *Euler spiral*, is used in conjunction with other spiral curves in routes that must follow steep grades.

PROBLEM 36

Which of the following abbreviations are arranged in the proper order for the transition from a tangent through a spiral curve to a circular curve, followed by a return from the circular curve through a spiral curve to a tangent?

(A) SC
TS
ST
CS

(B) TS
ST
SC
CS

(C) CS
TS
ST
SC

(D) TS
SC
CS
ST

PROBLEM 37

The following ten problems refer to the illustrations shown.

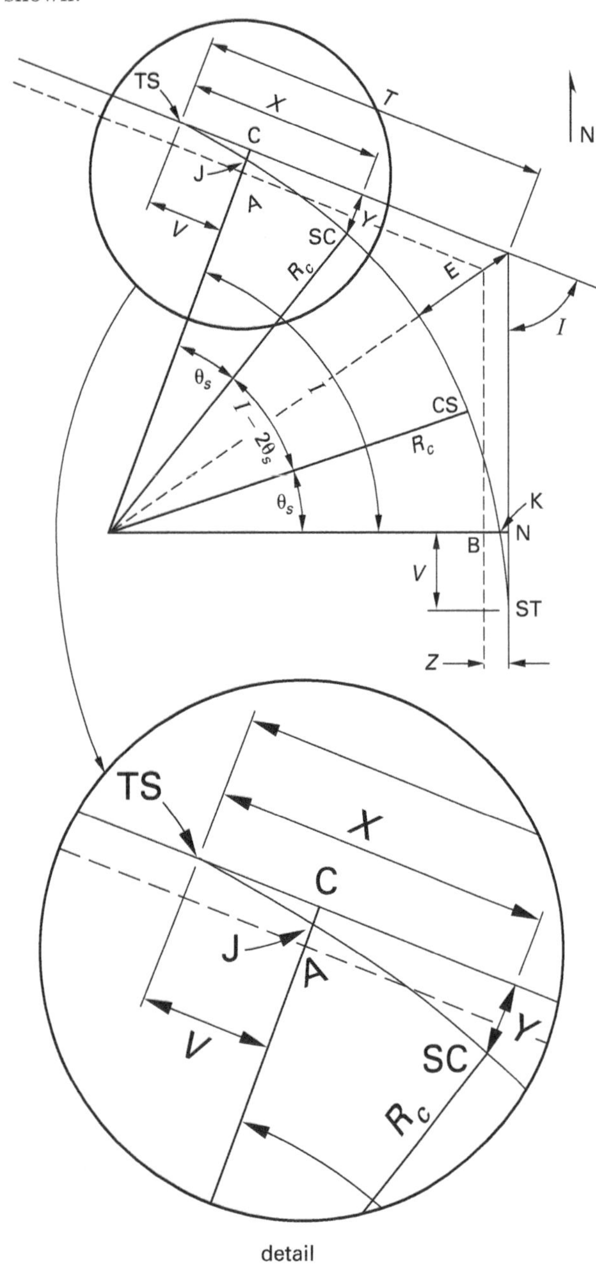

detail

37.1. Which statement best describes the distance CJ in the transition spiral?

(A) The distance CJ is equal to R_c vers θ_S.

(B) The distance CJ is equal to $R_c + Z \sec\left(\frac{1}{2}I\right) - R_c$.

(C) The distance CJ is equal to $R_c + Z \sec\left(\frac{1}{2}\theta_s\right) - R_c$.

(D) The distance CJ is approximately one-half of the distance CA.

37.2. The circular curve in the center of the illustration has been shifted to make room for the transition spirals TS to SC and CS to ST. Assume that R_c is equal to 1000.00 ft, Y is equal to 30.00 ft, and θ_s is equal to 10°. What is most nearly the length of the shift, distance CA?

(A) 6.97 ft

(B) 8.87 ft

(C) 10.9 ft

(D) 14.8 ft

37.3. Assume that R_c is equal to 955.00 ft, X is equal to 398.20 ft, and θ_s is equal to 12°. What is the length of the distance V?

(A) 199.65 ft

(B) 216.35 ft

(C) 249.62 ft

(D) 266.47 ft

37.4. Assume that R_c is equal to 955.00 ft, X is equal to 398.20 ft, and θ_s is equal to 12°. What is the length of the spiral transition curve TS to SC?

(A) 397.10 ft

(B) 399.30 ft

(C) 400.02 ft

(D) 404.21 ft

37.5. Which of the following statements concerning the detail in the illustration is correct?

(A) The length of the spiral curve TS to SC is three times longer than the circular arc of the angle θ_s with the radius R_c.

(B) The line CA very nearly bisects the spiral curve from TS to SC.

(C) The distance labeled Y is known as the throw and is equal to CA $\sec\left(\frac{1}{2}I\right)$.

(D) The spiral curve TS to SC is a cubic cycloid.

37.6. Assume that the degree of curve of the circular portion of the arc is 8° and the length of the spiral curve is 583.54 ft. Under these conditions, what is the spiral angle θ_s?

(A) 12°14′52″

(B) 18°42′15″

(C) 20°45′31″

(D) 23°20′30″

37.7. Perpendicular offsets along the spiral curve from TS to J may be constructed from the tangent line to the spiral curve for layout. These offsets may be approximately checked using which of the following relationships?

(A) The offsets should vary in proportion to the square of the distance from TS.

(B) The offsets should vary in proportion to the cube of the distance from TS.

(C) The offsets should vary inversely to the distance from TS.

(D) The offsets should decrease exponentially as the distance from TS.

37.8. Assume that the offset, distance CA, is 24.60 ft and the I angle is 20°. What is the throw of the curve—the shift inward radially from the PI?

(A) 24.41 ft

(B) 24.63 ft

(C) 24.79 ft

(D) 24.98 ft

37.9. Considering the curve illustrated, assume that R_c is equal to 955.00 ft, the I angle is 40°, and the distance CA is 6.95 ft. What is the external distance labeled E?

(A) 64.18 ft

(B) 68.69 ft

(C) 70.85 ft

(D) 72.33 ft

37.10. Considering the curve illustrated, assume that R_c is equal to 955.00 ft, the I angle is 40°, the distance CA is 6.95 ft, and the distance labeled V is 199.65 ft. What is the tangent distance labeled T?

(A) 531.87 ft

(B) 549.77 ft

(C) 553.00 ft

(D) 563.76 ft

SOLUTION 1

The quadratic formula is a convenient method to use to solve quadratic equations. It is expressed as

$$x = \frac{-b \pm \sqrt{b^2 - 4ac}}{2a}$$

The answer is (C).

SOLUTION 2

The formula for the area of a right triangle is

$$A = \frac{bh}{2}$$

b is the base of the triangle and h is its height. In this problem, the base is 4 ft less than the height.

$$\begin{aligned} b &= h - 4 \\ A &= \frac{(h-4)h}{2} \\ &= 510\ \text{ft}^2 \\ \frac{h^2 - 4h}{2} &= 510 \\ h^2 - 4h &= (510)(2) \\ h^2 - 4h &= 1020 \\ h^2 - 4h - 1020 &= 0 \end{aligned}$$

Use the quadratic formula to find the two roots of the expression.

$$\begin{aligned} h &= \frac{-b \pm \sqrt{b^2 - 4ac}}{2a} \\ &= \frac{-(-4) \pm \sqrt{(-4)^2 - (4)(1)(-1020)}}{(2)(1)} \\ &= \frac{4 \pm \sqrt{16 + 4080}}{2} \\ &= \frac{4 + 64}{2} \text{ and } \frac{4 - 64}{2} \\ &= 34\ \text{ft [height]} \\ &= 30\ \text{ft [base]} \end{aligned}$$

The answer is (C).

SOLUTION 3

The quadratic formula can be used to solve $9x^2 + 30x + 20 = 0$. The equation is arranged according to the standard form: $a = 9$, $b = 30$, and $c = 20$.

$$\begin{aligned} x &= \frac{-b \pm \sqrt{b^2 - 4ac}}{2a} \\ &= \frac{-(30) \pm \sqrt{(30)^2 - (4)(9)(20)}}{(2)(9)} \\ &= \frac{-30 \pm \sqrt{900 - 720}}{18} \\ &= \frac{-30 \pm \sqrt{180}}{18} \\ &= \frac{-30 \pm 13.416}{18} \\ &= \frac{-16.584}{18} \text{ and } \frac{-43.416}{18} \\ &= -0.92 \text{ and } -2.41 \end{aligned}$$

The answer is (D).

SOLUTION 4

First, express the equation in the standard form of a quadratic equation; then derive the values of a, b, and c.

$$\begin{aligned} 201.93 &= 6x + 5x^2 \\ 201.93 - 5x^2 &= 6x \\ 201.93 - 5x^2 - 6x &= 0 \\ -5x^2 - 6x + 201.93 &= 0 \\ a &= -5 \\ b &= -6 \\ c &= +201.93 \end{aligned}$$

Next, use the quadratic formula to solve for the values of x that will satisfy the equation.

$$\begin{aligned} x &= \frac{-b \pm \sqrt{b^2 - 4ac}}{2a} \\ &= \frac{-(-6) \pm \sqrt{(-6)^2 - (4)(-5)(201.93)}}{(2)(-5)} \\ &= \frac{+6 \pm \sqrt{36 - (-4038.60)}}{-10} \\ &= \frac{6 + 63.833}{-10} \text{ and } \frac{6 - 63.833}{-10} \\ &= -6.98 \text{ and } +5.78 \end{aligned}$$

The answer is (C).

SOLUTION 5

The area of the current lot, the width multiplied by the length, is 10,000 ft^2. The strips of equal width, or x, are to be added to the length and width to double the area.

$$\begin{aligned} (x + \text{length})(x + \text{width}) &= 20{,}000 \text{ ft}^2 \\ (x + 200 \text{ ft})(x + 50 \text{ ft}) &= 20{,}000 \text{ ft}^2 \\ x^2 + 250x + 10{,}000 &= 20{,}000 \\ x^2 + 250x - 10{,}000 &= 0 \end{aligned}$$

Use the quadratic formula to solve for x.

$$\begin{aligned} x &= \frac{-b \pm \sqrt{b^2 - 4ac}}{2a} \\ &= \frac{-250 \pm \sqrt{(250)^2 - (4)(1)(-10{,}000)}}{(2)(1)} \\ &= \frac{-250 \pm \sqrt{62{,}500 - (-40{,}000)}}{2} \\ &= \frac{-250 + 320.15}{2} \text{ and } \frac{-250 - 320.15}{2} \\ &= 35.07 \text{ ft and } -285.08 \text{ ft} \end{aligned}$$

The negative root is meaningless in this case. The positive root is correct.

The answer is (A).

SOLUTION 6

A determinant is a convenient abbreviation of the expressions used in solving systems of linear equations. Determinants are useful tools in many other applications as well. They are frequently used in processing Global Positioning System (GPS) data.

The answer is (C).

SOLUTION 7

The form for the expansion of such determinants may be represented as

$$\begin{aligned} \begin{vmatrix} a & b \\ c & d \end{vmatrix} &= ad - bc \\ &= (1)(-4) - (9)(-6) \\ &= -4 + 54 \\ &= 50 \end{aligned}$$

The answer is (C).

SOLUTION 8

The following method may be used to solve a system of simultaneous equations where there are just as many equations as there are unknowns and the determinant of the coefficients is not zero.

$$3x+4y=17 \quad \text{and} \quad 8x+12y=48$$

of the unknowns.

$$\begin{vmatrix} 3 & 4 \\ 8 & 12 \end{vmatrix}$$

The determinant in the numerator is different for each unknown. The numerator is identical to the denominator, with the coefficients of the desired unknown being replaced by the constants.

$$x = \frac{\begin{vmatrix} 17 & 4 \\ 48 & 12 \end{vmatrix}}{\begin{vmatrix} 3 & 4 \\ 8 & 12 \end{vmatrix}}$$

$$y = \frac{\begin{vmatrix} 3 & 17 \\ 8 & 48 \end{vmatrix}}{\begin{vmatrix} 3 & 4 \\ 8 & 12 \end{vmatrix}}$$

The answer is (B).

SOLUTION 9

The method was first described by the Swiss mathematician Cramer.

The answer is (C).

SOLUTION 10

The equations can be solved by substitution.

$$\begin{array}{rrrl} (2)(3x & +4y & = & 28) \\ +2x & -8y & = & 8 \\ \hline 6x & +8y & = & 56 \\ +2x & -8y & = & 8 \\ \hline 8x & & = & 64 \\ & x & = & \dfrac{64}{8} \\ & & = & 8 \end{array}$$

Substituting 8 for x in the equations and solving for y,

$$\begin{aligned} (2)(8) - 8y &= 8 \\ 16 - 8y &= 8 \\ -8y &= 8 - 16 \\ -8y &= -8 \\ \frac{-8y}{-8} &= \frac{-8}{-8} \\ y &= 1 \end{aligned}$$

Another method is to use the determinant.

$$\begin{aligned} 2x - 8y &= 8 \\ 3x + 4y &= 28 \end{aligned}$$

$$x = \frac{\begin{vmatrix} 8 & -8 \\ 28 & 4 \end{vmatrix}}{\begin{vmatrix} 2 & -8 \\ 3 & 4 \end{vmatrix}} = \frac{(8)(4) - (28)(-8)}{(2)(4) - (3)(-8)} = \frac{256}{32} = 8$$

$$y = \frac{\begin{vmatrix} 2 & 8 \\ 3 & 28 \end{vmatrix}}{\begin{vmatrix} 2 & -8 \\ 3 & 4 \end{vmatrix}} = \frac{(2)(28) - (3)(8)}{(2)(4) - (3)(-8)} = \frac{32}{32} = 1$$

The answer is (B).

SOLUTION 11

To use Cramer's rule to solve a system of simultaneous equations, the determinant of the coefficients of the unknowns must not be equal to zero. In this case, it is.

$$\begin{vmatrix} 2 & -5 \\ 4 & -10 \end{vmatrix} = (2)(-10) - (-5)(4) = 0$$

The answer is (A).

SOLUTION 12

The arrangement of the terms for the expansion of a determinant of the third order can be conveniently recalled by arranging the terms as follows.

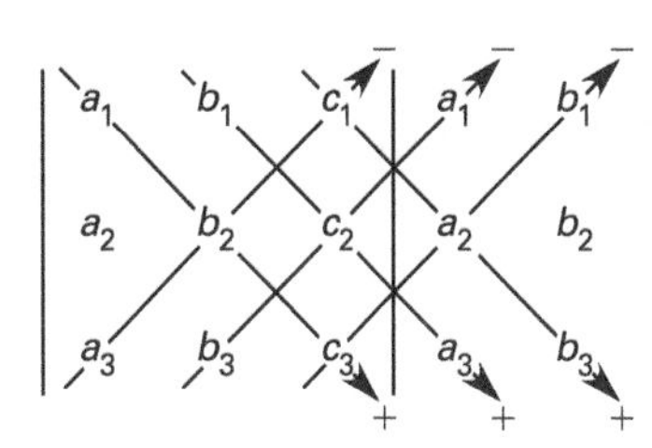

The correct form of the expansion is

$$a_1b_2c_3 + a_3b_1c_2 + a_2b_3c_1 - a_3b_2c_1 - a_1b_3c_2 - a_2b_1c_3$$

The answer is (D).

SOLUTION 13

The expansion of the determinant results in the following calculations.

$$\begin{aligned}(1)(6)(6) + (4)(3)(2) + (2)(4)(-1) - (2)(6)(2)\\ -(-1)(3)(1) - (6)(4)(4)\\ &= 36 + 24 + (-8) - 24 - (-3) - 96\\ &= 36 + 24 - 8 - 24 + 3 - 96\\ &= -65\end{aligned}$$

The answer is (B).

SOLUTION 14

The arrangement of the determinants that will solve these equations involves a system similar to that used with a pair of equations of two variables each.

$$\begin{aligned}x &= \frac{\begin{vmatrix} 65 & 2 & -6 \\ 23 & 3 & -1 \\ -29 & 10 & 9 \end{vmatrix}}{\begin{vmatrix} 3 & 2 & -6 \\ 1 & 3 & -1 \\ 1 & 10 & 9 \end{vmatrix}}\\ &= \frac{1755 + 58 + (-1380) - 522 - (-650) - 414}{81 + (-2) + (-60) - (-18) - (-30) - 18}\\ &= \frac{147}{49}\\ &= 3\end{aligned}$$

$$\begin{aligned}y &= \frac{\begin{vmatrix} 3 & 65 & -6 \\ 1 & 23 & -1 \\ 1 & -29 & 9 \end{vmatrix}}{\begin{vmatrix} 3 & 2 & -6 \\ 1 & 3 & -1 \\ 1 & 10 & 9 \end{vmatrix}}\\ &= \frac{621 + (-65) + 174 - (-138) - 87 - 585}{49}\\ &= \frac{196}{49}\\ &= 4\end{aligned}$$

$$\begin{aligned}z &= \frac{\begin{vmatrix} 3 & 2 & 65 \\ 1 & 3 & 23 \\ 1 & 10 & -29 \end{vmatrix}}{\begin{vmatrix} 3 & 2 & -6 \\ 1 & 3 & -1 \\ 1 & 10 & 9 \end{vmatrix}}\\ &= \frac{-261 + 46 + 650 - 195 - 690 - (-58)}{49}\\ &= \frac{-392}{49}\\ &= -8\end{aligned}$$

The answer is (B).

SOLUTION 15

Although the dimensions of the rectangle are not known, the error can be calculated. The original surveyor measured the length, x, and depth, y, of the rectangle and found that $xy = 40$ acres. Since the chain was missing one of its 100 links, however, the length was actually $0.99x$ and the depth $0.99y$. Therefore, the correct area is

$$\begin{aligned}A &= (0.99x)(0.99y) = 0.9801xy\\ &= (0.9801)(40)\\ &= 39.20 \text{ ac}\end{aligned}$$

The same result will be found if the parcel is presumed to be rectangular but not square.

The answer is (A).

SOLUTION 16

Three sides of both the smaller and the larger parcel are components of the 560 ft of fencing. The shared side between the parcels is not fenced.

$$\begin{aligned}560 \text{ ft} &= 3L + 3x + (L - x)\\ &= 4L + 2x\\ &= (2)(2L + x)\\ 2L + x &= \frac{560 \text{ ft}}{2}\\ &= 280 \text{ ft}\\ 2L &= 280 - x\\ L &= \frac{280 - x}{2}\\ &= 140 - 0.5x\\ x &= 280 - 2L\end{aligned}$$

Since the parcels are squares, the area may be expressed as

$$x^2 + L^2 = 16{,}400 \text{ ft}^2$$

Substituting $140 - 0.5x$ for L,

$$\begin{aligned} 16{,}400 \text{ ft}^2 &= x^2 + (140 - 0.5x)^2 \\ &= x^2 + (0.25x^2 - 140x + 19{,}600) \\ &= 1.25x^2 - 140x + 19{,}600 \\ 16{,}400 - 16{,}400 &= 1.25x^2 - 140x + 19{,}600 - 16{,}400 \\ 0 &= 1.25x^2 - 140x + 3200 \end{aligned}$$

This expression can be solved with the quadratic formula.

$$\begin{aligned} x &= \frac{-b \pm \sqrt{b^2 - 4ac}}{2a} \\ &= \frac{-(-140) \pm \sqrt{(-140)^2 - (4)(1.25)(3200)}}{(2)(1.25)} \\ &= \frac{140 \pm \sqrt{19{,}600 - 16{,}000}}{2.5} \\ &= \frac{140 \pm \sqrt{3600}}{2.5} \\ &= \frac{140 \pm 60}{2.5} \\ &= 32 \text{ ft and } 80 \text{ ft} \end{aligned}$$

Since $L = 140 - 0.5x$, the corresponding values of L are

$$\begin{aligned} L &= 140 - (0.5)(32) \\ &= 140 - 16 \\ &= 124 \text{ ft} \end{aligned}$$

and

$$\begin{aligned} L &= 140 - (0.5)(80) \\ &= 140 - 40 \\ &= 100 \text{ ft} \end{aligned}$$

The required ratios are

$$\frac{L}{x} = \frac{124 \text{ ft}}{32 \text{ ft}} = 3.875 \quad \text{or} \quad \frac{L}{x} = \frac{100 \text{ ft}}{80 \text{ ft}} = 1.250$$

The answer is (A).

SOLUTION 17

The area of an ellipse is a mean proportion between the areas of the circle described by its major axis and the circle described by its minor axis. The area of the ellipse can be found by multiplying the area of the two circles together and finding the square root of the product.

$$\begin{aligned} \text{major axis} &= 284 \text{ ft} \\ \text{radius of the circle} &= 142 \text{ ft} \\ \text{area of the circle} &= \pi r^2 \\ &= \pi(142 \text{ ft})^2 \\ &= 63{,}347.07 \text{ ft}^2 \\ \text{minor axis} &= 152 \text{ ft} \\ \text{radius of the circle} &= 76 \text{ ft} \\ \text{area of the circle} &= \pi r^2 \\ &= \pi(76 \text{ ft})^2 \\ &= 18{,}145.84 \text{ ft}^2 \\ (63{,}347.07 \text{ ft}^2)(18{,}145.84 \text{ ft}^2) &= 1{,}149{,}485{,}797 \text{ ft}^4 \\ \sqrt{1{,}149{,}485{,}797 \text{ ft}^4} &= 33{,}904 \text{ ft}^2 \end{aligned}$$

A shortcut to finding the area of an ellipse is to multiply the major axis by the minor axis and the factor 0.7854.

$$(284 \text{ ft})(152 \text{ ft})(0.7854) = 33{,}904 \text{ ft}^2$$

The answer is (B).

SOLUTION 18

Since its eastern and western boundaries are parallel, the 1 ac parcel is a trapezoid. The area of a trapezoid may be found using the formula

$$A = \frac{(b_1 + b_2)H}{2}$$

The western boundary of the parcel is 310.92 ft. The acreage of the parcel is 1 ac, or 43,560 ft^2.

$$\begin{aligned} A &= 43{,}560 \text{ ft}^2 \\ &= \frac{(310.92 \text{ ft} + b_2)H}{2} \end{aligned}$$

b_2 and H are still unknown.

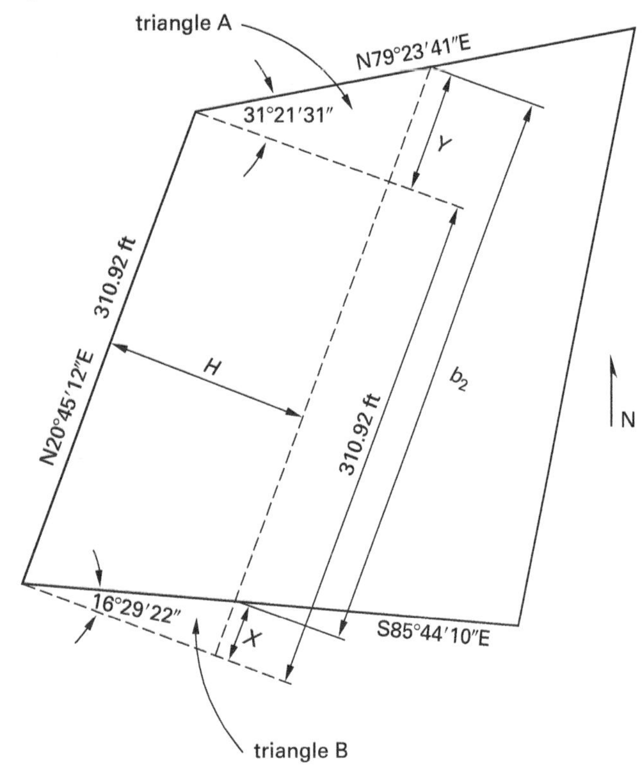

By constructing lines perpendicular to the western boundary of the parcel, two right triangles, labeled A and B, are formed with the altitudes X and Y. First, find the smallest angle of each triangle.

S69°14′48″ is a line perpendicular to the western boundary of the parcel.

$$180° - \text{N}79°23'41''\text{E} - \text{S}69°14'48''\text{E} = 31°21'31''$$

The angle at the western apex of triangle A is 31°21′31″.

$$\text{S}85°44'10''\text{E} - \text{S}69°14'48''\text{E} = 16°29'22''$$

The angle at the western apex of triangle B is 16°29′22″.

Since the base of both right triangles is the unknown H, their altitudes may be expressed as

$$\begin{aligned} Y &= \tan 31°21'31''H \\ X &= \tan 16°29'22''H \end{aligned}$$

The unknown b_2 can now be expressed as

$$\begin{aligned} b_2 &= 310.92 \text{ ft} - X + Y \\ &= 310.92 \text{ ft} - \tan 16°29'22''H + \tan 31°21'31''H \\ &= 310.92 \text{ ft} - 0.2960H + 0.6094H \\ &= 310.92 \text{ ft} + 0.3134H \end{aligned}$$

The formula for the area of the trapezoid can now be expressed with one unknown, H.

$$\begin{aligned} A &= \frac{(b_1 + b_2)H}{2} \\ 43{,}560 \text{ ft}^2 &= \frac{\big((310.92 \text{ ft}) + (310.92 \text{ ft} + 0.3134H)\big)H}{2} \\ &= \frac{(621.84 \text{ ft} + 0.3134H)H}{2} \\ &= \frac{(621.84 \text{ ft})(H) + 0.3134H^2}{2} \\ &= (310.92 \text{ ft})(H) + 0.1567H^2 \end{aligned}$$

The expression has been reduced to a quadratic equation.

$$0.1567H^2 + (310.92 \text{ ft})(H) - 43{,}560 \text{ ft}^2 = 0$$

Use the quadratic formula to solve for H.

$$\begin{aligned} H &= \frac{-b \pm \sqrt{b^2 - 4ac}}{2a} \\ &= \frac{-310.92 \text{ ft} \pm \sqrt{\begin{array}{l}(310.92 \text{ ft})^2 \\ \quad -(4)(0.1567)(-43{,}560 \text{ ft}^2)\end{array}}}{(2)(0.1567)} \\ &= \frac{-310.92 \text{ ft} \pm \sqrt{96{,}671.25 \text{ ft} - (-27{,}303.41 \text{ ft}^2)}}{0.3134} \\ &= \frac{-310.92 \text{ ft} \pm \sqrt{123{,}974.66 \text{ ft}^2}}{0.3134} \\ &= \frac{-310.92 \text{ ft} \pm 352.10 \text{ ft}}{0.3134} \\ &= 131.40 \text{ ft and } -2115.57 \text{ ft} \end{aligned}$$

H is 131.40 ft; the negative value is meaningless.

The answer is (A).

SOLUTION 19

19.1. The perpendicular distance H was found to be 131.40 ft in Prob. 18. The altitude of triangle A, symbolized by Y, can be found by the calculation

$$\begin{aligned} X &= \tan 31°21'31''H \\ &= (0.60941)(131.40 \text{ ft}) \\ &= 80.08 \text{ ft} \end{aligned}$$

The altitude of triangle B, symbolized by X, can be found by the calculation

$$\begin{aligned} X &= \tan 16°29'22''H \\ &= (0.29601)(131.40\ \text{ft}) \\ &= 38.90\ \text{ft} \\ b_2 &= 310.92\ \text{ft} - X + Y \\ &= 310.92\ \text{ft} - 38.90\ \text{ft} + 80.08\ \text{ft} \\ &= 352.10\ \text{ft} \end{aligned}$$

The answer is (B).

19.2. First, find the altitude of triangle A, symbolized by Y.

$$\begin{aligned} Y &= \tan 31°21'31''H \\ &= (0.60941)(200.00\ \text{ft}) \\ &= 121.88\ \text{ft} \end{aligned}$$

Then find the altitude of triangle B, symbolized by X.

$$\begin{aligned} X &= \tan 16°29'22''H \\ &= (0.29601)(200.00\ \text{ft}) \\ &= 59.20\ \text{ft} \end{aligned}$$

Next, find the length of the base of the trapezoid, b_2.

$$\begin{aligned} b_2 &= 310.92\ \text{ft} - X + Y \\ &= 310.92\ \text{ft} - 59.20\ \text{ft} + 121.88\ \text{ft} \\ &= 373.60\ \text{ft} \end{aligned}$$

Finally, use the formula for the area of a trapezoid to find the area of the figure.

$$\begin{aligned} A &= \frac{(b_1 + b_2)H}{2} \\ &= \frac{(310.92\ \text{ft} + 373.60\ \text{ft})(200.00\ \text{ft})}{2} \\ &= \frac{(684.52\ \text{ft})(200.00\ \text{ft})}{2} \\ &= \frac{136{,}904\ \text{ft}^2}{2} \\ &= 68{,}452\ \text{ft}^2 \end{aligned}$$

The answer is (C).

SOLUTION 20

20.1. First, draw a line from the midpoint of the northern boundary to the southwestern corner of the parcel. The length of the hypotenuse can be found as follows.

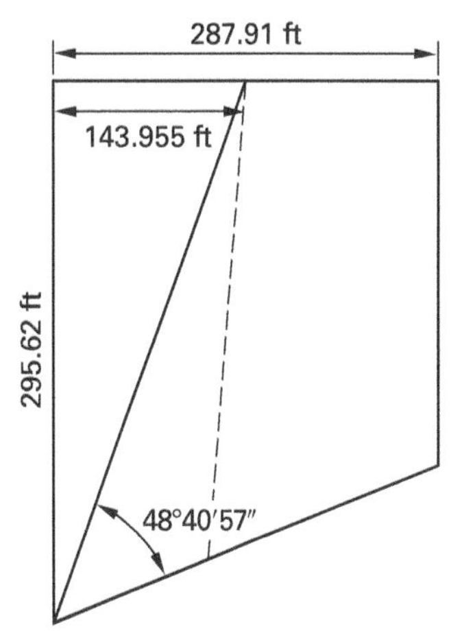

$$\frac{287.91\ \text{ft}}{2} = 143.955\ \text{ft}$$

$$\sqrt{(143.955\ \text{ft})^2 + (295.62\ \text{ft})^2} = 328.81\ \text{ft}$$

The tangent of the angle at the southerly apex of the triangle is

$$\frac{143.955\ \text{ft}}{295.62\ \text{ft}} = 0.48696$$

Therefore, the angle is

$$\arctan 0.48696 = 25°57'51''$$

The area of this triangle is

$$\begin{aligned} A &= \tfrac{1}{2}bh \\ &= \left(\frac{1}{2}\right)(143.955\ \text{ft})(295.62\ \text{ft}) \\ &= 21{,}277.99\ \text{ft}^2 \end{aligned}$$

This is not quite half of the total area of the parcel.

$$\begin{aligned} \tfrac{1}{2}\ \text{total area} &= 36{,}866.155\ \text{ft}^2 \\ \text{area of right triangle} &= \underline{-21{,}277.99\ \text{ft}^2} \\ \text{needed remaining area} &= 15{,}588.165\ \text{ft}^2 \end{aligned}$$

To find the dimensions of a triangle that will add the remaining area, use the formula

$$A = \frac{ab \sin c}{2}$$

The angle at the southwest corner of this second triangle is

$$\begin{aligned} &74°38'48'' \\ &\underline{-25°57'51''} \\ &48°40'57'' \end{aligned}$$

One side of this triangle is known to be 328.81 ft. Additionally, its area must be 15,588.165 ft^2.

$$
\begin{aligned}
15{,}588.165\ \text{ft}^2 &= \frac{(328.81\ \text{ft})b(\sin 48^\circ 40'57'')}{2}\\
&= \frac{(246.96\ \text{ft})b}{2}\\
31{,}176.33\ \text{ft}^2 &= (246.96\ \text{ft})b\\
b &= 126.24\ \text{ft}
\end{aligned}
$$

The answer is (A).

20.2. Use the law of cosines to find the length of the line.

$$
\begin{aligned}
c^2 &= a^2 + b^2 - 2ab\cos C\\
&= (126.24\ \text{ft})^2 + (328.81\ \text{ft})^2\\
&\quad -(2)(126.24\ \text{ft})(328.82\ \text{ft})(\cos 48^\circ 40'57'')\\
&= (124{,}052.55\ \text{ft}^2)\\
&\quad -(2)(41{,}510.24\ \text{ft}^2)(\cos 48^\circ 40'57'')\\
&= (124{,}052.55\ \text{ft}^2) - (2)(41{,}510.24\ \text{ft}^2)(0.66023)\\
&= 124{,}052.55\ \text{ft}^2 - 54{,}812.61\ \text{ft}^2\\
&= 69{,}239.94\ \text{ft}^2\\
c &= \sqrt{69{,}239.94\ \text{ft}^2}\\
&= 263.13\ \text{ft}
\end{aligned}
$$

The answer is (B).

SOLUTION 21

First, find the length of a side of the pool and the length of a side of the base. The length of a side of the pool can be calculated by finding the angle subtended at the center of the figure by one-half of a side of the undecagon.

$$
\begin{aligned}
\frac{360^\circ}{11} &= 32^\circ 43'38''\\
\frac{32^\circ 43'38''}{2} &= 16^\circ 21'49''
\end{aligned}
$$

Then find the length of a side of the undecagon.

$$(2)(\tan 16^\circ 21'49'')(9.60\ \text{ft}) = 5.64\ \text{ft}$$

The length of a side of the base can be calculated by finding the angle subtended at the center of the figure by one-half of a side of the heptagon.

$$
\begin{aligned}
\frac{360^\circ}{7} &= 51^\circ 25'43''\\
\frac{51^\circ 25'43''}{2} &= 25^\circ 42'51''
\end{aligned}
$$

Then find the length of a side of the undecagon.

$$L = (2)(\tan 25^\circ 42'51'')(11.26\ \text{ft}) = 10.85\ \text{ft}$$

The formula for finding the area of a regular polygon is

$$A = \frac{pLn}{2}$$

p is the length of the perpendicular from the midpoint of a side of the polygon to its center, L is the length of a side of the polygon, and n is the number of sides.

The area of the pool is

$$
\begin{aligned}
\text{area of the undecagon} &= \frac{(9.60\ \text{ft})(5.64\ \text{ft})(11)}{2}\\
&= \frac{595.58\ \text{ft}^2}{2}\\
&= 297.79\ \text{ft}^2
\end{aligned}
$$

The area of the base is

$$
\begin{aligned}
\text{area of the heptagon} &= \frac{(11.26\ \text{ft})(10.85\ \text{ft})(7)}{2}\\
&= \frac{855.20\ \text{ft}^2}{2}\\
&= 427.60\ \text{ft}^2
\end{aligned}
$$

The difference between the two figures is

$$
\begin{aligned}
\text{area of the base} - \text{area of the pool} &= 427.60\ \text{ft}^2 - 297.79\ \text{ft}^2\\
&= 129.81\ \text{ft}^2
\end{aligned}
$$

The answer is (B).

SOLUTION 22

The area of any figure enclosed by four straight lines may be found by

$$A = \frac{\sin\alpha D_1 D_2}{2}$$

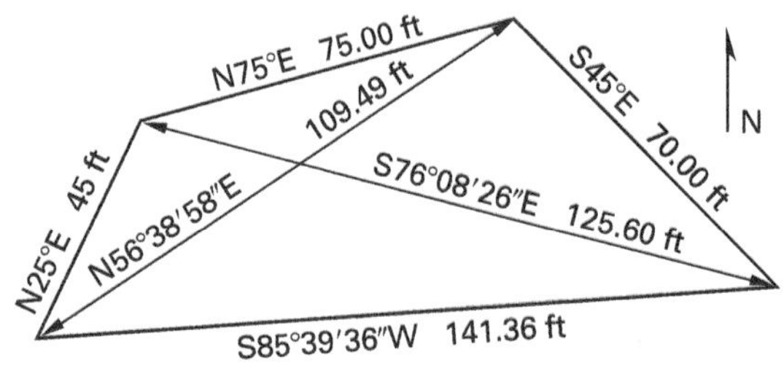

N75°E 75.00 ft
109.49 ft
S45°E 70.00 ft
N
S76°08'26"E 125.60 ft
N25°E 45 ft
N56°38'58"E
S85°39'36"W 141.36 ft

area of parcel = 5046 ft²

α is the least of the angles formed by the intersection of the diagonals, D_1 is the length of one of the diagonals, and D_2 is the length of the other diagonal. In this case, the area is known and the length of the second diagonal is the value desired. Rearranging the formula,

$$\begin{aligned} D_2 &= \frac{2A}{\sin\alpha D_1} = \frac{(2)(5046\ \text{ft}^2)}{(\sin 47°12'36'')(109.49\ \text{ft})} \\ &= \frac{10{,}092\ \text{ft}^2}{80.35\ \text{ft}} \\ &= 125.60\ \text{ft} \end{aligned}$$

The answer is (A).

SOLUTION 23

Applying Archimedes' principle as used in Simpson's one-third rule, the area south of the parabola is two-thirds of the area of the rectangle. The remaining third is north of the parabola.

$$A = \frac{(212.14\ \text{ft})(424.28\ \text{ft})}{3} = 30{,}002.2\ \text{ft}^2$$

The answer is (A).

SOLUTION 24

24.1. The elevation at sta 96+98.83 must be 14 ft below 60.51 ft, or 46.51 ft. The elevation along the entering grade at sta 96+98.83 may be found as follows.

The distance along the curve from the PVC is 3.1667 sta.

$$\begin{array}{lr} & 96{+}98.83 \\ \text{PVC} = & -93{+}82.16 \\ \hline & 3{+}16.67 \end{array}$$

The entering grade is +3%, so

$$\begin{aligned} (316.67\ \text{ft})(0.03) &= 9.50\ \text{ft} \\ \text{elevation at the PVC} &= 41.76\ \text{ft} \\ &\quad + 9.50\ \text{ft} \\ \text{elevation at 96+98.83 on the tangent} &= 51.26\ \text{ft} \end{aligned}$$

Since the elevation on the vertical curve must be 46.51 ft at sta 96+98.83 and the elevation on the tangent is 51.26 ft, the tangent offset must be

$$\begin{array}{r} 51.26\ \text{ft} \\ -46.51\ \text{ft} \\ \hline 4.75\ \text{ft} \end{array}$$

This information makes it possible to calculate the rate of change along the vertical curve, using the formula

$$\text{tangent offset} = \left(\frac{r}{2}\right)x^2$$

r is the rate of change and x is the distance along the vertical curve in stations.

$$\begin{aligned} \text{tangent offset} &= \left(\frac{r}{2}\right)x^2 \\ r &= \frac{(2)(\text{tangent offset})}{x^2} = \frac{(2)(-4.75\ \text{ft})}{(3.1667)^2} \\ &= \frac{-9.50\ \text{ft}}{10.028} \\ &= -0.9473 \end{aligned}$$

The rate of change provides the information necessary to find the exit grade, using the formula

$$r = \frac{g_2 - g_1}{L}$$

r is the rate of change, g_2 is the exiting grade, g_1 is the entering grade, and L is the length of the vertical curve in stations.

$$\begin{aligned} r &= \frac{g_2 - g_1}{L} \\ g_2 &= Lr + g_1 \\ &= (9.5\ \text{sta})(-0.9473) + (+3\%) \\ &= -9 + (+3\%) \\ &= -6\% \end{aligned}$$

The answer is (A).

24.2. The entering grade is + 3% for 475 ft, beginning at an elevation of 41.76 ft.

$$\begin{aligned} (475\ \text{ft})(0.03) &= +14.25\ \text{ft} \\ & \quad 41.76\ \text{ft} \\ & \quad +14.25\ \text{ft} \\ \text{elevation of the PVI} &= 56.01\ \text{ft} \end{aligned}$$

The exiting grade is –6% for 475 ft.

$$\begin{aligned} (475\ \text{ft})(-0.06) &= -28.50\ \text{ft} \\ & \quad 56.01\ \text{ft} \\ & \quad (-28.50\ \text{ft}) \\ \text{elevation of the PVI} &= 27.51\ \text{ft} \end{aligned}$$

The station of the PVT is 950 ft from the PVC.

$$\begin{array}{r} 93{+}82.16 \\ +\ 9{+}50.00 \\ \hline 103{+}32.16 \end{array}$$

The answer is (C).

SOLUTION 25

Begin by drawing a sketch.

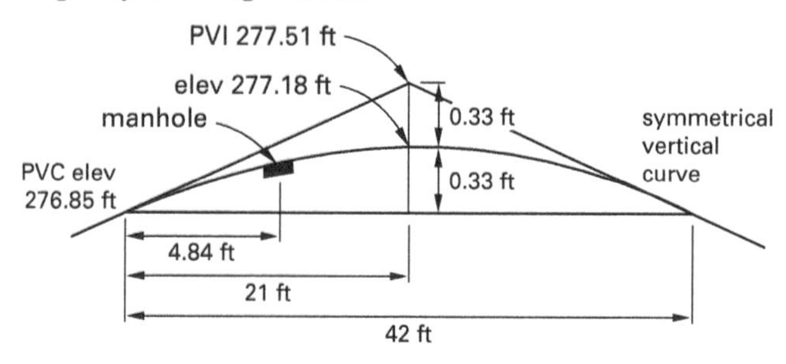

The elevation of the PVC is 4 in (0.33 ft) lower than the crown of the street at the centerline, and the PVI is 0.33 ft higher.

$$\begin{aligned} \text{elevation at the centerline} &= 277.18 \text{ ft} \\ & \underline{+0.33 \text{ ft}} \\ \text{elevation at the PVI} &= 277.51 \text{ ft} \\ \text{elevation at the centerline} &= 277.18 \text{ ft} \\ & \underline{-0.33 \text{ ft}} \\ \text{elevation at the PVC} &= 276.85 \text{ ft} \end{aligned}$$

The grade of the tangent is the difference between the elevations of the PVC and the PVI, divided by half the width of the street, or 21 ft.

$$\begin{aligned} \text{elevation at the PVI} &= 277.51 \text{ ft} \\ \text{elevation at the PVC} &= \underline{-276.85 \text{ ft}} \\ & 0.66 \text{ ft} \end{aligned}$$

$$\text{grade} = \frac{0.66 \text{ ft}}{21.00 \text{ ft}} = 0.031$$

The grade along the tangent is 3.1%. The manhole is 16.16 ft from the PVI, and it stands 4.84 ft from the PVC. The elevation on the tangent 4.84 ft from the PVC is calculated as

$$(4.84 \text{ ft})(0.031) = 0.15 \text{ ft}$$

$$\begin{array}{r} 276.85 \text{ ft} \\ \underline{+0.15 \text{ ft}} \\ 277.00 \text{ ft} \end{array}$$

The tangent offset from the tangent to the center of the top of the manhole varies as the square of its distance from the PVC.

$$\begin{aligned} \text{tangent offset} &= \left(\frac{(4.84)^2}{(21.00)^2}\right)(0.33 \text{ ft}) \\ &= \left(\frac{23.426}{441.00}\right)(0.33 \text{ ft}) \\ &= (0.0531)(0.33 \text{ ft}) \\ &= 0.0175 \text{ ft} \\ &= 0.02 \text{ ft} \quad \left[\begin{array}{c}\text{rounded to the nearest} \\ \text{hundredth of a foot}\end{array}\right] \end{aligned}$$

The elevation of the center of the manhole is the elevation of the tangent at that point less the tangent offset.

$$\begin{array}{r} 277.00 \text{ ft} \\ \underline{-0.02 \text{ ft}} \\ 276.98 \text{ ft} \end{array}$$

The answer is (C).

SOLUTION 26

26.1. The total distance between the manholes is 872.43 ft and the elevation of the PVI must be the same approached from either end of the vertical curve.

$$741.25 \text{ ft} - 0.04x = 737.25 \text{ ft} - (0.03)(872.43 \text{ ft} - x)$$

The elevation at sta 44+00 is 741.25 ft and the elevation at sta 52+72.43 is 737.25 ft. The entering grade is 0.04 and the exiting grade is 0.03. x is the distance from manhole 1 to the intersection of the grades.

$$\begin{aligned} 741.25 \text{ ft} - 0.04x &= 737.25 \text{ ft} - (0.03)(872.43 \text{ ft} - x) \\ &= 737.25 \text{ ft} - 26.17 \text{ ft} + 0.03x \\ 0.03x + 0.04x &= 741.25 \text{ ft} - 737.25 \text{ ft} + 26.17 \text{ ft} \\ 0.07x &= 30.17 \text{ ft} \\ x &= 431 \text{ ft} \end{aligned}$$

The answer is (B).

26.2. The two elements of the asymmetrical vertical curve that meet the requirements outlined are 431.00 ft and 441.43 ft long. The first begins at the PVC and ends at the CVC and sta 48+31.00.

$$\begin{aligned} \text{PVC} &= 44{+}00.00 \\ & \underline{+431.00 \text{ ft}} \\ \text{CVC} &= 48{+}31.00 \end{aligned}$$

The CVC is found on the chord connecting the midpoints of the entering grade and the exiting grade.

The answer is (C).

26.3. The station of PVI_1 is

$$\begin{aligned} \text{PVC} &= 44{+}00.00 \\ \text{half of first vertical curve} &= \underline{+215.50 \text{ ft}} \\ PVI_1 &= 46{+}15.50 \end{aligned}$$

The station of PVI_2 is

$$\begin{aligned} \text{PVT} &= 52{+}72.43 \\ \text{half of second vertical curve} &= \underline{-220.715 \text{ ft}} \\ PVI_2 &= 50{+}51.715 \end{aligned}$$

The elevation of PVI_1 can be calculated by multiplying one-half of the first vertical curve by the grade of the first tangent.

$$\begin{aligned} (215.50 \text{ ft})(-0.04) &= -8.62 \text{ ft} \\ \text{elevation of PVC} &= 741.25 \text{ ft} \\ & \quad \underline{+(-8.62 \text{ ft})} \\ \text{elevation of } PVI_1 &= 732.63 \text{ ft} \end{aligned}$$

The elevation of PVI_2 can be calculated by multiplying one-half of the second vertical curve by the grade of the first tangent from the PVT.

$$\begin{aligned} (220.715 \text{ ft})(-0.03) &= -6.62 \text{ ft} \\ \text{elevation of PVT} &= 737.25 \text{ ft} \\ & \quad \underline{+(-6.62 \text{ ft})} \\ \text{elevation of } PVI_2 &= 730.63 \text{ ft} \end{aligned}$$

The answer is (B).

26.4. The station and elevation of PVI_1 are

sta 46+15.50
elev 732.63 ft

The station and elevation of PVI_2 are

sta 50+51.715
elev 730.63 ft

The horizontal distance between the two is

$$\begin{array}{r} 50{+}51.715 \\ \underline{-46{+}15.50} \\ 436.215 \text{ ft} \end{array}$$

The vertical distance between them is

$$\begin{array}{r} 730.63 \text{ ft} \\ \underline{-732.63 \text{ ft}} \\ -2.00 \text{ ft} \end{array}$$

Therefore, the grade from PVI_1 to PVI_2 is

$$\frac{-2.00 \text{ ft}}{436.215 \text{ ft}} = -0.0046 = -0.46\%$$

The answer is (A).

SOLUTION 27

27.1. The formula for the rate of change, r, along a vertical curve is

$$\frac{g_2 - g_1}{L}$$

g_2 is the exiting grade, g_1 is the entering grade, and L is the length of the vertical curve.

For curve 1, the entering grade is −4%, the exiting grade is −0.46%, and the length is 431.00 ft

$$\begin{aligned} r &= \frac{(-0.46) - (-4.00)}{4.3100 \text{ sta}} \\ &= \frac{3.54}{4.3100 \text{ sta}} \\ &= 0.821 \end{aligned}$$

For curve 2, the entering grade is −0.46%, the exiting grade is +3.00%, and the length is 441.43 ft.

$$\begin{aligned} r &= \frac{(+3.00) - (-0.46)}{4.4143 \text{ sta}} \\ &= \frac{3.46}{4.4143 \text{ sta}} \\ &= 0.784 \end{aligned}$$

The answer is (C).

27.2. The drainage structure should be located at the lowest point of the entire asymmetrical vertical curve. Although this is possible, the lowest point may be less likely to occur on the first curve of the two since both of its grades are negative: −4% and −0.46%. Therefore, the first attempt to find the lowest point is conducted on the second of the two curves whose grades are −0.46% and +3%.

The formula for calculating the lowest point of a vertical curve is

$$\text{distance in stations} = \frac{-g_1}{r}$$

g_1 is the entering grade and r is the rate of change.

The rate of change of the second curve is 0.784 and the entering grade is −0.46%.

$$\begin{aligned} \text{distance in stations} &= \frac{-(-0.46\,\%)}{0.784} \\ &= 0.5867 \end{aligned}$$

The distance, d, from the CVC to the lowest point of the asymmetrical vertical curve is 58.67 ft.

The station may be calculated as follows.

$$\begin{aligned} \text{station of CVC} &= 48{+}31.00 \\ d &= \underline{+58.67 \text{ ft}} \\ \text{station of lowest point} &= 48{+}89.67 \end{aligned}$$

The answer is (C).

27.3. The CVC occurs at a distance of 215.50 ft from PVI_1, along a tangent whose slope is −0.46%.

$$\begin{aligned} (15.50 \text{ ft})(-0.0046) &= -0.99 \text{ ft} \\ \text{station of PVI}_1 &= 732.63 \text{ ft} \\ &\quad \underline{+(-0.99 \text{ ft})} \\ \text{elevation of CVC} &= 731.64 \text{ ft} \end{aligned}$$

The lowest point of the curve occurs at sta 48+89.67, or 58.67 ft from the CVC. Therefore, the elevation of the lowest point is calculated by

$$\text{elevation} = \left(\frac{r}{2}\right)x^2 + g_1 + \text{elevation of CVC}$$

r is the rate of change, x is the distance along the curve to the lowest point in stations, and g_1 is the entering grade.

$$\begin{aligned} \text{elevation} &= \left(\frac{0.784}{2}\right)(0.5867)^2 + (-0.46)(0.5867) \\ &\quad +731.64 \text{ ft} \\ &= \left(\frac{0.784}{2}\right)(0.3442) + (-0.2700) + 731.64 \text{ ft} \\ &= 0.1349 + (-0.2700) + 731.64 \text{ ft} \\ &= -0.1351 + 731.64 \text{ ft} \\ &= 731.50 \text{ ft} \end{aligned}$$

The answer is (A).

SOLUTION 28

Generally, the superelevation of highway curves is accomplished by lowering the inner edge of the roadway by half of the superelevation distance. The centerline is maintained at grade, and the outer edge is raised by half of the superelevation distance. On railroad work, the inner rail is maintained at grade, and the outer rail is superelevated.

The answer is (B).

SOLUTION 29

The usual convex crown of pavement would provide a slight banking effect for the inner lane of a superelevated curve, but it would have the opposite effect on the outer lane. The usual practice is to gradually change the convex crown to a plane-inclined surface in the middle of the curve and gradually return to the convex crown as the curve ends.

The answer is (C).

SOLUTION 30

The friction factor is variable and is dependent on the condition of the pavement and the speed of the vehicles traveling on a highway, among other things. It is used in the calculation of the transverse slope of a superelevated curve in order to prevent vehicles from skidding inward while traveling along the curve.

The answer is (C).

SOLUTION 31

31.1. The formula used to solve this problem is

$$\tan a = \frac{\text{v}^2}{15r} - f$$

v is the design speed in miles per hour, r is the radius of the curve in feet, f is the friction factor, and a is the angle of the transverse slope.

$$\begin{aligned} \tan a &= \frac{\text{v}^2}{15r} - f \\ &= \frac{(70)^2}{(15)(1000)} - 0.10 \\ &= \frac{4900}{15{,}000} - 0.10 \\ &= 0.3267 - 0.10 \\ &= 0.2267 \\ a &= 12.8° \end{aligned}$$

The answer is (A).

31.2. Several methods are used to calculate the minimum length of the spiral curve under the problem's conditions. One formula is

$$L = 73.3\,\text{v}S$$

L is the minimum length of the spiral curve in feet, v is the design speed of the curve in miles per hour, and S is the maximum transverse slope.

$$\begin{aligned} L &= (73.3)(70)(0.2267) \\ &= (73.3)(15.869) \\ &= 1163 \text{ ft} \end{aligned}$$

The answer is (A).

SOLUTION 32

The most commonly used formula for calculating the superelevation of railroad curves is

$$\text{superelevation of outer rail in inches} = 0.00066D\text{v}^2$$

D is the degree of curve, and v is the design speed.

$$\begin{aligned}\begin{array}{c}\text{superelevation of}\\ \text{outer rail in inches}\end{array} &= (0.00066)(2°)(70)^2\\ &= (0.00066)(9800)\\ &= 6.468\text{ in}\quad (6\,{}^1\!/_2\text{ in})\end{aligned}$$

The answer is (B).

SOLUTION 33

The usual formula used to calculate the minimum length of railroad spiral is

$$\text{minimum railroad spiral length} = 1.173ED$$

E is the superelevation of the outer rail over the inner rail in inches, and D is the design speed in miles per hour.

$$\begin{aligned}\text{minimum railroad spiral length} &= (1.173)(4\tfrac{1}{2})(60)\\ &= (1.173)(270)\\ &= 317\text{ ft}\end{aligned}$$

The answer is (A).

SOLUTION 34

A spiral curve, as used in surveying, is a curve whose radius is constantly increasing or decreasing at a uniform rate as the curve continues. There are several mathematical solutions of spiral curves.

The answer is (D).

SOLUTION 35

Transition spirals, or easement curves, are most frequently used to gradually carry vehicles from a straight to a curved path.

The answer is (B).

SOLUTION 36

The meanings of the abbreviations are

TS tangent to spiral
SC spiral to circular curve
CS circular curve to spiral
ST spiral to tangent

The answer is (D).

SOLUTION 37

37.1. A convenient approximation that may be used with transition spiral curves is that the distance indicated as CJ is approximately one-half of the distance CA.

The answer is (D).

37.2. The formula best suited to solve this problem is

$$\text{CA} = Y - R_c\text{ vers }\theta_s$$

R_c is the radius of the circular curve beginning at SC and θ_s is the spiral angle.

$$\begin{aligned}\text{CA} &= 30.00\text{ ft} - (1000.00\text{ ft})\text{vers }10°\\ &= 30.00\text{ ft} - (1000.00\text{ ft})(0.01519)\\ &= 30.00\text{ ft} - 15.19\text{ ft}\\ &= 14.81\text{ ft}\quad (14.8\text{ ft})\end{aligned}$$

The answer is (D).

37.3. The formula best suited to solve this problem is

$$V = X - R_c\sin\theta_s$$

R_c is the radius of the circular curve beginning at SC and θ_s is the spiral angle.

$$\begin{aligned}\text{CA} &= 398.20\text{ ft} - (955.00\text{ ft})\sin 12°\\ &= 398.20\text{ ft} - (955.00\text{ ft})(0.20791)\\ &= 398.20\text{ ft} - 198.55\text{ ft}\\ &= 199.65\text{ ft}\end{aligned}$$

The answer is (A).

37.4. The most convenient formula for this calculation is

$$\text{length of spiral} = 2R_c\theta_s$$

R_c is the radius of the circular curve beginning at SC, and θ_s is the spiral angle in radians.

$$\begin{aligned}\text{length of spiral} &= (2)(955.00\text{ ft})\left(\frac{12°}{57°17'44.90''}\right)\\ &= (2)(955.00\text{ ft})(0.20944)\\ &= (2)(200.01\text{ ft})\\ &= 400.02\text{ ft}\end{aligned}$$

The answer is (C).

37.5. Another convenient approximation is that the distance from TS to J along the spiral curve is very nearly equal to the distance from J to SC.

The answer is (B).

37.6. The most convenient formula for the calculation of the spiral angle, θ_s, when the degree of curve and the length of the spiral are known is

$$\begin{aligned}\theta_s &= \left(\frac{\text{spiral curve length}}{200}\right)(\text{degree of curve})\\ &= \left(\frac{583.54\ \text{ft}}{200}\right)(8^\circ)\\ &= (2.917\ \text{ft})(8^\circ)\\ &= 23^\circ 20' 30''\end{aligned}$$

The answer is (D).

37.7. A transition spiral is similar to a cubic parabola. The offsets from the tangent from TS to J are roughly proportional to the cubes of the distances from TS.

The answer is (B).

37.8. The formula that is most convenient in finding such a radial shift is

$$\begin{aligned}\text{shift} &= \text{CA}\sec\left(\tfrac{1}{2}I\right) = (24.60\ \text{ft})\sec\left(\left(\frac{1}{2}\right)(20^\circ)\right)\\ &= (24.60\ \text{ft})(1.0154)\\ &= 24.98\ \text{ft}\end{aligned}$$

The answer is (D).

37.9. The formula best suited for finding the external distance is

$$\begin{aligned}E &= (R_c + \text{CA})\sec\left(\tfrac{1}{2}I\right) - R_c\\ &= (955.00\ \text{ft} + 6.95\ \text{ft})\sec\left(\left(\frac{1}{2}\right)(40^\circ)\right) - 955.00\ \text{ft}\\ &= (961.95\ \text{ft})\sec\left(\left(\frac{1}{2}\right)(40^\circ)\right) - 955.00\ \text{ft}\\ &= (961.95\ \text{ft})\sec 20^\circ - 955.00\ \text{ft}\\ &= (961.95\ \text{ft})(1.06418) - 955.00\ \text{ft}\\ &= 1023.69\ \text{ft} - 955.00\ \text{ft}\\ &= 68.69\ \text{ft}\end{aligned}$$

The answer is (B).

37.10. The formula best suited for finding the tangent distance is

$$\begin{aligned}T &= (R_c + \text{CA})\tan\left(\tfrac{1}{2}I\right) + V\\ &= (955.00\ \text{ft} + 6.95\ \text{ft})\tan\left(\left(\frac{1}{2}\right)(40^\circ)\right) + 199.65\ \text{ft}\\ &= (961.95\ \text{ft})\tan\left(\left(\frac{1}{2}\right)(40^\circ)\right) + 199.65\ \text{ft}\\ &= (961.95\ \text{ft})\tan 20^\circ + 199.65\ \text{ft}\\ &= (961.95\ \text{ft})(0.36397) + 199.65\ \text{ft}\\ &= 350.12\ \text{ft} + 199.65\ \text{ft}\\ &= 549.77\ \text{ft}\end{aligned}$$

The answer is (B).

Land Boundary Law

BOUNDARY LAW TERMS

PROBLEM 1

A grantor and grantee are in the closing stages of the transfer of real property. A third party holds a grant that will be delivered only when all of the conditions for the transfer are complete. Which of the following terms describes the third party?

(A) a scrivener

(B) an escrow agent

(C) a disseisor

(D) a remainderman

PROBLEM 2

When the public good is involved and adequate compensation is offered, private property may be taken for public use under what principle?

(A) estoppel

(B) entitlement

(C) oyer

(D) eminent domain

PROBLEM 3

What type of evidence is usually NOT admissible in court to cure a patent ambiguity in a deed?

(A) parol evidence

(B) evidentia

(C) ewage

(D) parole evidence

PROBLEM 4

Which of the following terms includes unpaid taxes, mortgages, easements, leases, and liens?

(A) *pedis positio*

(B) escheat

(C) disseisin

(D) encumbrances

PROBLEM 5

The term *fee simple absolute* describes a type of estate in real property. Which phrase defines the idea further?

(A) Subject to the condition that if something specified comes to pass, the original grantor may reenter and retake the title.

(B) If the owner dies intestate, the title reverts to the original grantor.

(C) The title is valid for a specified period of time.

(D) The title is without limitations or conditions to the owner, his heirs, and assigns.

PROBLEM 6

Which of the following conditions is essential for the execution of a valid deed?

(A) The deed must be recorded.

(B) There must be a perambulation of the boundary.

(C) It must be sealed with a stamp specified by the laws of the state.

(D) It must have a written or printed form.

PROBLEM 7

Which of the following is a revocable right in real property?

(A) an easement

(B) a license

(C) a lien

(D) an escheat

PROBLEM 8

When an owner dies without will or heir, the state becomes the proprietor under which feudal doctrine?

(A) escheat

(B) *escobedo* rule

(C) Massachusetts rule

(D) *loquela*

PROBLEM 9

Which term applies to the land over which an appurtenant easement passes?

(A) dominant estate

(B) incorporeal tenement

(C) servient estate

(D) fee tail

PROBLEM 10

Which of the following terms applies to a written instrument that appears to convey title to real property, but actually does not?

(A) quitclaim deed

(B) indenture

(C) deed of gift

(D) color of title

PROBLEM 11

Real property is subject to the laws and jurisdiction of the state it is located in EXCEPT

(A) when concerning the governing of transactions in real property

(B) that courts of one state do not have the power to render decrees directly affecting title to land in another state

(C) where title is in the United States, or where it is specifically under the jurisdiction of the federal government

(D) in regards to the regulation of the form, execution, validity, and effect of instruments relating to land

PROBLEM 12

Real property in land is composed of a bundle of rights, and it is common for the rights of a particular parcel of land to be divided among many persons. Which of the following statements about real property rights is NOT correct?

(A) Real property includes not only the surface of the earth, but features under it and above it as well.

(B) If an owner conveys rights to the land's minerals, an implied right of entry might be created to extract the minerals, even though the owner still owns the land.

(C) Included in real property are things incidental to the use of land, such as easements and rights-of-way.

(D) The owner of the title to a tract of land cannot convey the land and retain the rights to the minerals.

PROBLEM 13

A defeasible fee simple estate is an estate that

(A) is without conditions or limitations

(B) may only be inherited by the eldest son

(C) is only measurable by the duration of a life

(D) depends on some event in the future for its continuance

PROBLEM 14

Which of the following best describes the legal doctrine of laches?

(A) It has been reenacted in most U.S. states, and therefore, any oral agreement by the party conveying land changing fixed boundaries is at best voidable.

(B) It is a legal deadline by which a plaintiff must start a lawsuit.

(C) A claim that has been for a long time undemanded cannot then be demanded.

(D) A claim will not be enforced or allowed if a long delay in asserting it has hurt the other party.

PROBLEM 15

"Fee" in the phrase "fee simple absolute," means an estate that

(A) can be inherited or devised by a will

(B) must be inherited by a specific individual

(C) has no time limitations

(D) is not subject to eminent domain, escheat, police power, or taxation

PROBLEM 16

A life estate

(A) is considered a freehold, but its duration is limited by the length of a life

(B) is limited by the duration of the life of its holder

(C) is a dower, but not a curtesy

(D) cannot be created by a deed

PROBLEM 17

What is the difference between the terms *fee simple determinable* and *fee simple to a condition subsequent*?

(A) A fee simple to a condition subsequent is the same as a fee simple absolute; a fee simple determinable is not.

(B) A fee simple to a condition subsequent has no restrictions on its inheritability; a fee simple determinable does.

(C) A fee simple to a condition subsequent is a life estate; a fee simple determinable is not.

(D) A fee simple to a condition subsequent does not automatically revert to the original grantor after an event does, or does not occur; a fee simple determinable does.

DEEDS

PROBLEM 18

There is one category of deed that does not imply that the grantor has title to the property, yet does pass any title the grantor may have. The grantor of such a deed will defend against any defects in that title that may have arisen through him or her, but not through others. What is the name of this type of deed?

(A) quitclaim deed

(B) warranty deed

(C) grant deed

(D) agreement deed

PROBLEM 19

A deed

(A) is proof of ownership of real property

(B) is the only legal method of conveying real property

(C) must be signed by the grantee to be valid

(D) must have sufficient and legal words in its description of the property

PROBLEM 20

Why is an ambiguity in a deed generally construed in favor of the grantee?

(A) The court presumes that the grantor was responsible for the language of the deed.

(B) The grantee does not sign the deed.

(C) Grantees have no title insurance.

(D) An ambiguous deed is void.

PROBLEM 21

An inconsistency in an instrument that is not on its face and generally can be cured by outside evidence is given what name?

(A) patent ambiguity

(B) dormant obscurity

(C) hearsay

(D) latent ambiguity

PROBLEM 22

A call for an adjoiner is sometimes given which of the following names?

(A) dominant tenement

(B) reservation

(C) record monument

(D) tacking

PROBLEM 23

The four corners rule is sometimes used in the interpretation of ambiguous deeds. Which statement best characterizes this rule?

(A) The granting clause may not be repugnant to the *habendum* clause.

(B) The deed description must contain clear reference to all corners of the property being conveyed.

(C) The intention of the grantor must be gathered from the language of the whole instrument rather than from separate clauses.

(D) The location of boundaries is a question of fact, but the nature of boundaries is a question of law.

PROBLEM 24

The intentions of the parties are considered to be the controlling consideration in construing the terms of a deed. Which of the following phrases limits this principle?

(A) to the extent that the intentions can be understood from the language of the instrument itself

(B) as clarified by testimony

(C) regardless of contrary words in the deed

(D) unless contradicted by survey measurements

PROBLEM 25

Which phrase is a reversion?

(A) land gradually uncovered by receding water

(B) a clause in a deed that excludes a portion of the estate that would otherwise pass

(C) land described in a deed that is not owned by the grantor

(D) an interest that will return to the grantor after some lesser estate ends

PROBLEM 26

Which statement best describes joint tenants?

(A) They can individually vest their interests in their own heirs and devisees.

(B) They have acquired their interests from the same conveyance.

(C) They can, upon the death of the other or others, acquire the interest left vacant at length to the last survivor.

(D) Both B and C are true.

PROBLEM 27

In a general warranty deed, the grantor

(A) warrants the title only against acts of the grantor's own volition

(B) limits the warranty to a five-year period

(C) must make good on the title if it is found lacking

(D) both A and B

PROBLEM 28

A surveyor who is retracing lines of a previously surveyed deed should follow what cardinal principle?

(A) distribute excesses and deficiencies with proportional measurement

(B) follow in the footsteps of the previous surveyor

(C) assign an overlap to the junior deed in a sequential conveyance

(D) record distances prevail over original monuments

PROBLEM 29

Which of the following statements about a deed is always INCORRECT?

(A) When a deed is executed and delivered, it conveys an estate in real property.

(B) A deed is the instrument by which real property is transferred from one person (the grantor) to another person (the grantee).

(C) A deed is an instrument in writing.

(D) none of the above

PROBLEM 30

The *habendum* clause in a deed usually follows which words?

(A) To have and to hold

(B) To warrant

(C) Know all men by these presents

(D) Said Grantee herein, his heirs and assigns

PROBLEM 31

If an overlap exists between sequential conveyances, what is the most likely outcome?

(A) the original grantor receives it

(B) the senior deed receives it

(C) it is divided proportionally between the senior and junior deed

(D) the junior deed receives it

PROBLEM 32

In a deed, a reservation

(A) is a pledge parties make to one another stating something is done, shall be done, or shall not be done

(B) is some interest that would otherwise be conveyed by the deed that is taken back by the grantor

(C) causes the property to revert to the state if the grantee dies without heirs

(D) indicates that some issue regarding the property has been adjudicated

UNWRITTEN RIGHTS

PROBLEM 33

Which statement correctly describes the difference in two doctrines regarding practical location of boundaries?

(A) Recognition and acquiescence differs from a parol agreement in that the former rests on an implied agreement and the latter on an actual agreement.

(B) Equitable estoppel requires that one party have knowledge of the true location of the boundary, whereas recognition and acquiescence requires that both parties be uncertain of the true location of the boundary.

(C) Establishment of a property line by a property line agreement in a location other than that described in the deed is a violation of the Statute of Frauds, but equitable estoppel is not a violation.

(D) Both A and B are true.

PROBLEM 34

Mr. Zambezi engaged a surveyor to stake the boundary of a parcel of land. He told the surveyor that he had adversely possessed the parcel for the past 11 years. The surveyor asked him if he had occupied the land under color of title. Mr. Zambezi answered yes and produced a letter that the abutting owner, Kevin Wannamaker, wrote to him 7 years ago, in which Mr. Wannamaker stated that he knew Mr. Zambezi was on his property and gave him permission to be there. What should the surveyor tell Mr. Zambezi?

(A) There is no ill will between the parties, so no adverse possession is possible.

(B) The letter does not constitute color of title, and therefore Mr. Zambezi's possession must continue another 11 years.

(C) A surveyor cannot survey land that has been stolen.

(D) Possession by permission of the record owner cannot ripen into adverse possession.

PROBLEM 35

Several doctrines, such as adverse possession and recognition and acquiescence may, through various means, become fee interests. Which statement correctly represents the priority of these unwritten rights in the order of importance of conflicting title elements?

(A) Unwritten rights overcome rights that have arisen from deeds.

(B) All claims of property through unwritten means can only be successful against property owned by a government entity.

(C) No land can be possessed by adverse means once Torrens registration has been acquired on that property.

(D) Both A and C are true.

PROBLEM 36

One necessary aspect of an adverse claim, sometimes said to be the essence of adverse possession, does not imply ill will. However, it indicates that the claimant denies all other claims on the property, including those of the actual owner. Which of the following terms describes this aspect of adverse possession?

(A) open and notorious possession

(B) color of title

(C) hostile possession

(D) continuous possession

Land Boundary Law

PROBLEM 37

One of the following terms is very similar to adverse possession. It applies to rights acquired by continuous use of another's property that is open and notorious, hostile, and for a statutory period. However, it differs from adverse possession in the extent of interest that the adverse user acquires. Generally, it is an easement right rather than a title to the land itself. Which of the following terms describes such rights?

(A) easement in gross

(B) negative easement

(C) dedication

(D) prescription

PROBLEM 38

Under which conditions is a person immune from claims of adverse possession?

(A) lack of knowledge that property is being occupied by others

(B) insanity

(C) being underage

(D) both B and C

PROBLEM 39

A person may lose property he might otherwise own if a court finds that he acquired it by implying a boundary to be correct when he knew, at the time, that it was wrong. Which of the following terms applies to this situation?

(A) quiet title

(B) estoppel

(C) laches

(D) acquiescence

PROBLEM 40

A surveyor finds that the line between his client and a neighbor appears to be defined by a picket fence constructed five years before the survey. However, the surveyor's measurements clearly indicate that the deed line is several feet over the fence. The surveyor places a prominent row of stakes along the deed line and calls the two property owners together to look at the situation. Even if the owners were inclined to accept the fence as their common line, some types of practical location cannot now be applied. Which of the following would be among them?

(A) parol agreement

(B) quitclaim deed

(C) recognition and acquiescence

(D) both A and C

PROBLEM 41

The term *tacking* is sometimes applied in cases of adverse possession. Which of the following best defines this term?

(A) neglecting to assert a right to the land in question

(B) seeking the permission of the record owner to occupy land

(C) fencing the land in question

(D) adding a claimant's time of possession together with another person's to satisfy the required statutory period

PROBLEM 42

Which of the following elements is NOT required of a claimant to title by adverse possession?

(A) hostility

(B) open and notorious possession

(C) possession for the statutory period

(D) color of title

EASEMENTS

PROBLEM 43

Gonzaga Street has not been used for 40 years. All evidence of pavement and curbs has vanished. While it still shows up on old plats of the town, the street has generally been forgotten. Essie Palisco wants her surveyor to tell her if she may extend her begonia garden into the old right-of-way without reproach. What advice might apply?

(A) An easement can be extinguished by nonuse for five years.

(B) The property within the right-of-way generally reverts to the fee owners adjoining it when the right-of-way is no longer used.

(C) The right-of-way is not extinguished unless the town has done so by an official act, even if the street has been out of use for 40 years.

(D) Mrs. Palisco's flowers should not extend beyond the old centerline, which belongs to her neighbor across the street.

PROBLEM 44

Which statement correctly defines the difference between an easement appurtenant and an easement in gross?

(A) An easement appurtenant is an interest in the land of another; an easement in gross is not.

(B) An easement in gross is attached to the person of the easement holder; an easement appurtenant is not.

(C) An easement appurtenant passes with the transfer of the land to which it is attached; an easement in gross does not.

(D) Both B and C are true.

PROBLEM 45

Which of the following will NOT extinguish an easement appurtenant that was acquired in writing?

(A) overburdening

(B) combination of the tenements in one proprietor

(C) written intent to abandon

(D) revocation by the grantor

PROBLEM 46

An implied easement arises from which of the following circumstances?

(A) A grantor mentions such an easement in the deed.

(B) Although unwritten, it is necessary for the reasonable use of the property.

(C) A claimant acquires the easement right by long and continuous use.

(D) A common-law dedication of a public street is omitted from a deed.

PROBLEM 47

Mr. A has granted Mr. B an easement appurtenant over the northerly 20 ft of his parcel, as indicated in the following illustration. A dispute over the meaning of the deed has arisen. Mr. B claims the shaded area is included in his easement; Mr. A says it is not. Which statement is most likely to apply to the situation?

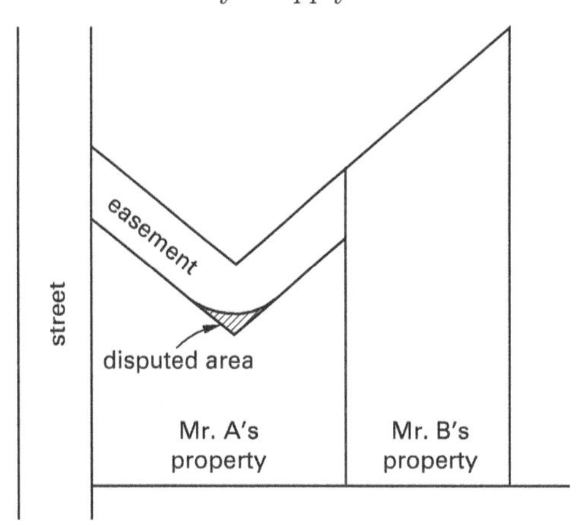

(A) The disputed area is not within the easement.

(B) The description of the easement should be construed in favor of the grantee. The disputed area is probably part of the easement.

(C) The deed is subject to two interpretations. The central question is whether the disputed area is necessary for Mr. B's use of his property.

(D) The deed is void for uncertainty.

PROBLEM 48

An easement in which the owner of the servient estate is prohibited from building any structure over 100 ft high would be called which of the following names?

(A) a negative easement

(B) a quasi-easement

(C) an easement of natural support

(D) a construction easement

PROBLEM 49

Which statement correctly defines the difference between a common-law dedication and a statutory dedication?

(A) A common-law dedication of a street can occur without writing or a map, while statutory dedications do not.

(B) A common-law dedication generally conveys an easement to the public, while statutory dedications convey a fee.

(C) A statutory dedication is always formally accepted by the government; common-law dedications are accepted by public use only.

(D) Both B and C are true.

PROBLEM 50

Which of the following describes a situation in which an easement right CANNOT arise?

(A) Easements can be created to protect a property's access to sunlight.

(B) A way of necessity may be created when property is conveyed without any access.

(C) Easements can arise by estoppel.

(D) An easement can be acquired by adverse possession.

PROBLEM 51

Mr. Parker's lot is on the boundary of Valhalla Acres as shown in the following illustration. The street north of his lot has been vacated; the underlying fee has reverted to Mr. Parker. Which of the following statements describes the extent of that reversion?

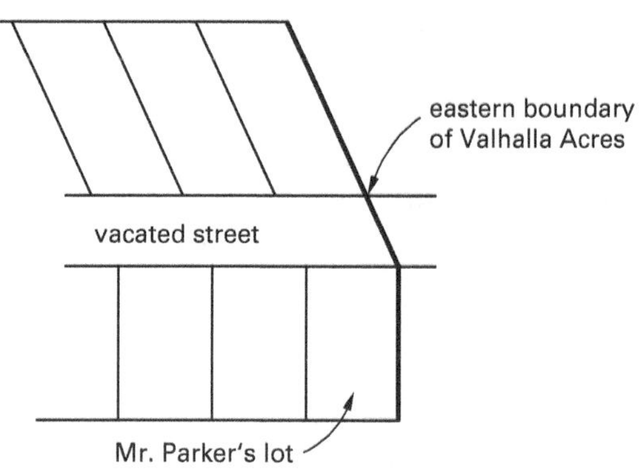

(A) The sidelines of his lot will extend to the north side of the street.

(B) The reversion rights cannot extend beyond the subdivision boundary.

(C) There is no rule for this situation.

(D) Both A and B are true.

PROBLEM 52

Which of the following statements is inconsistent with the definition of an easement?

(A) An easement is a fee interest in land.

(B) An easement is defined for specific uses.

(C) An easement is a nonpossessory interest in land.

(D) The owner of an easement appurtenant has an easement over the land of another.

PROBLEM 53

Which of the following statements regarding an easement is correct?

(A) An easement is a non-possessory right to make a specific use of a definite portion of land owned by another.

(B) The owner of the underlying ground does not have the right to make use of the land over which the easement passes, even if the use does not interfere with the exercise of the easement granted.

(C) If the tenant holds the property over which the easement passes after a term has expired, the landlord may evict the tenant.

(D) An easement provides possession of land given by the owner to another person for a specified rent and period of time.

PROPORTIONMENT

PROBLEM 54

Proportionate measurement may be described as an even distribution of a discovered excess or a deficiency of measurement to all parts of an established line. This procedure can be further defined by which of the following statements?

(A) The method should only be used in the resolution of conflicts between junior and senior property rights.

(B) The method should only be applied to the reestablishment of obliterated corners in the Public Land Surveying System.

(C) The method is best applied where the original monumentation called for on the plat of a subdivision is found at every lot corner, and the retracement survey measurements agree with the platted dimensions.

(D) The method is usually applied to parcels created simultaneously where original monumentation definitely has been lost.

PROBLEM 55

In which of the following situations would the stated conditions definitely exclude proportionate measurement from the possible solutions?

(A) The discrepancy between a retracement and the record length of a line was found to be attributable to a mistake in its measurement.

(B) A deficiency is found in a block of a platted subdivision, but all required monumentation can be established by acceptable collateral evidence.

(C) Several parcels were created at various times by a common grantor.

(D) All of the above are true.

PROBLEM 56

The exact width of streets as shown on a duly recorded subdivision plat governs their location. Except where original monuments set by the original surveyor indicate otherwise, an excess or deficiency in the measurement of adjoining blocks should not be absorbed in the public way. Which of the following statements regarding this principle is correct?

(A) The principle is seldom true. Streets and alleyways should receive a proportional part of any excess or deficiency discovered between original monuments.

(B) The principle is never true. The courts will not uphold such a procedure.

(C) The principle is correct in most circumstances.

(D) The principle can only apply when the centerlines of paved streets coincide with the centerline of the right of way.

PROBLEM 57

The following two problems refer to the illustration shown.

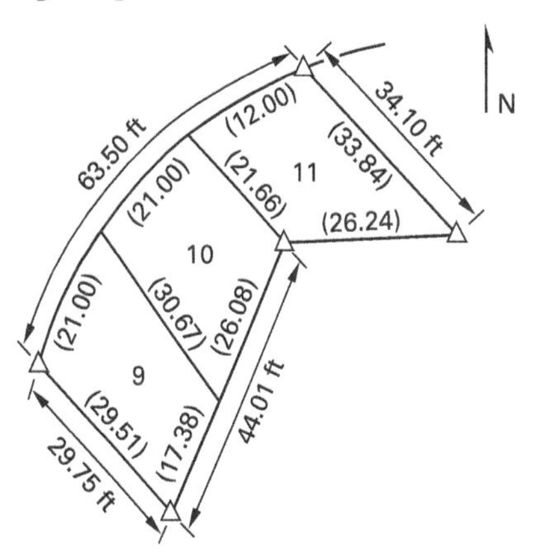

57.1. What distance should be measured from the NE corner of lot 11 along the arc to reestablish the NW corner of lot 11?

(A) 12.00 ft

(B) 18.00 ft

(C) 21.00 ft

(D) 21.17 ft

57.2. What distance should be measured from the SW corner of lot 9 to the SE corner of lot 9?

(A) 17.38 ft

(B) 17.49 ft

(C) 17.50 ft

(D) 17.60 ft

PROBLEM 58

The following three problems refer to the Fairview Tract, described as follows.

Mr. Holland owns a parcel of land known as the Fairview Tract. The Fairview Tract has been platted and recorded as a square parcel, with each side 650.00 ft long and oriented in a cardinal direction. No improvements have been made, but the parcel has been surveyed and its four corners monumented. The monuments are still in place.

58.1. Mr. Holland sold the east 325 ft of the Fairview Tract to Mr. Clark by deed dated Oct. 12, 1998. A year later, Mr. Holland sold to Mr. Brown the west 325 ft of the Fairview Tract, described as extending to the west line of Mr. Clark's property.

A new survey finds that the distance between the SW and SE corners of the Fairview Tract is 655.00 ft.

What distance should be measured from the SW corner of the Fairview Tract to establish the SE corner of Mr. Brown's property?

(A) 325.00 ft

(B) 325.50 ft

(C) 327.50 ft

(D) 330.00 ft

58.2. Mr. Holland sold the east 325 ft of the Fairview Tract to Mr. Clark by deed dated Oct. 12, 1988. Mr. Holland then sold the west 325 ft of the Fairview Tract to Mr. Brown by deed dated Jan. 29, 1990.

A new survey finds that the distance between the SW and SE corners of the Fairview Tract is 645.00 ft.

What distance should be measured from the SW corner of the Fairview Tract to establish the SE corner of Mr. Brown's property?

(A) 320.00 ft

(B) 323.00 ft

(C) 324.50 ft

(D) 330.00 ft

58.3. Mr. Holland sold the east 325 ft of the Fairview Tract to Mr. Clark and the west 325 ft to Mr. Brown by the same deed dated July 2, 1998.

The distance measured between the SW corner and the SE corner of the Fairview Tract is 658.24 ft.

What distance should be measured from the SW corner of the Fairview Tract to establish the SE corner of Mr. Brown's property?

(A) 322.23 ft

(B) 325.00 ft

(C) 329.12 ft

(D) 333.24 ft

PROBLEM 59

The following five problems refer to the statement shown and to the accompanying illustration.

The lots shown on a portion of the subdivision plat of Valhalla Estates were created simultaneously. The plat of the subdivision was recorded in 1957. The iron pipe monuments indicated were set at that time. The fence crossing blocks 3 and 4 has a verifiable age of 30 years. The owners of lots 5 and 10 in both blocks 3 and 4 have all accepted the fence as their northern boundary and have maintained it as such. Several monuments were set in a survey that was conducted in 1997; some of those monuments have been recovered as shown.

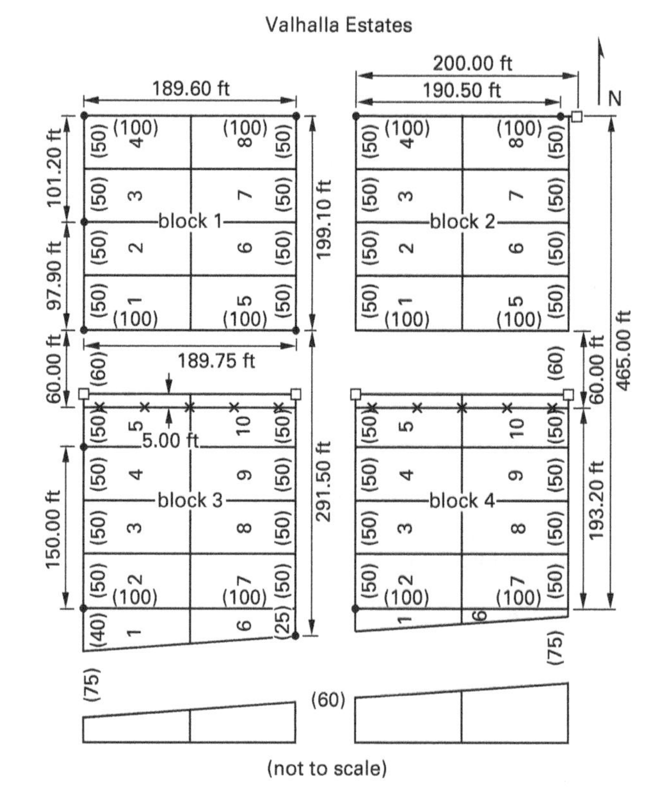

59.1. The western boundaries of lots 2 and 3 of block 1 both have lengths of 50.00 ft according to the plat of Valhalla Estates. Assuming that all evidence of the original monuments of these lots has been lost, what are the lengths of the boundaries according to the recent measurements indicated in the illustration?

(A) The west lines of both lots 2 and 3 are 50.00 ft long.

(B) The west lines of both lots 2 and 3 are 48.95 ft long.

(C) The west line of lot 2 is 49.77 ft long; the west line of lot 3 is 52.95 ft long.

(D) The west line of lot 2 is 48.95 ft long; the west line of lot 3 is 50.60 ft long.

59.2. What distance in block 2 should be measured from the NW corner of lot 4 to establish the NE corner of lot 4, block 2?

(A) 95.25 ft

(B) 96.33 ft

(C) 97.81 ft

(D) 100.00 ft

59.3. What distances should be measured south from the NW corner of block 2 along the west line to establish the lot corners?

(A) The west lines of lots 1 through 4 in block 2 are each 50.00 ft long.

(B) The west lines of lots 1 through 3 in block 2 are each 51.25 ft long, but the west line of lot 4 is 50.00 ft long.

(C) The west lines of lots 1 through 4 in block 2 are each 52.95 ft long.

(D) The west lines of lots 1 through 4 in block 2 are each 51.70 ft long.

59.4. What distance should be measured from the SE corner of block 3 to establish the NE corner of lot 6, block 3?

(A) 25.00 ft

(B) 25.67 ft

(C) 25.72 ft

(D) 26.28 ft

59.5. Lot 6 of block 4 has not been given a dimension on the plat and the original monuments of the lot have been lost. Which of the following principles might be appropriate in the reestablishment of the lost corners of lot 6?

(A) Irregular lots without dimensions on the recorded subdivision plat will take the excess or lose the deficiency in a measurement of the original lines.

(B) Lots without dimensions must be scaled from the plat, and the distances so determined control its boundaries.

(C) Irregular lots are frequently subdivisional errors and should be combined with the adjoining lot, particularly where a block shows a deficiency in measurement.

(D) The dimensions of the irregular lot should be computed area as determined from the plat.

WATER BOUNDARIES

PROBLEM 60

Reliction is NOT

(A) uncovering of land by the recession of water

(B) the sudden and permanent withdrawal of a sea, river, or lake

(C) also known as dereliction

(D) different from avulsion

PROBLEM 61

Which of the following rights is related to the bank of a stream, river, or other body of flowing water?

(A) reversionary

(B) littoral

(C) meander

(D) riparian

PROBLEM 62

Avulsion is best defined as

(A) the process that gradually builds up alluvion

(B) the rapid erosion of a shore

(C) the sudden removal of land from one owner and its attachment to the land of another owner by the action of water, by which action title to the newly attached land most often does not change

(D) both B and C

PROBLEM 63

Which of the following terms applies to the action of water gradually building up sand and soil and the subsequent formation of firm ground?

(A) revulsion

(B) dereliction

(C) erosion

(D) accretion

PROBLEM 64

Which of the following terms applies to land formed by sediment laid down by the action of water, such as that resulting from a slowing of the current of a stream or river?

(A) accretion

(B) ambit

(C) littoral

(D) alluvium

PROBLEM 65

Which of the following terms best describes the deepest or most navigable part of a river channel?

(A) mean high water mark

(B) thread

(C) batture

(D) thalweg

PROBLEM 66

Which of the following statements concerning meander lines is true?

(A) A meander line defines the property boundary of the upland owner abutting a navigable waterway, unless there are specific legal provisions to the contrary.

(B) A meander line established during a public land survey was intended to follow the mean high water mark.

(C) When the bank of a navigable river shifts, the meander line shifts with it.

(D) Nonnavigable bodies of water were not meandered during the public land surveys.

PROBLEM 67

Which of the following terms most accurately describes the rights along the shore of a pond, lake, or sea?

(A) riparian

(B) batture

(C) thweat

(D) littoral

PROBLEM 68

Title to land may be gained or lost by the action of water. What sort of change may cause a riparian owner to gain land?

(A) Land built by accretion belongs to the riparian owner.

(B) Land deposited by avulsion belongs to the riparian owner where it was deposited.

(C) Land that becomes newly inundated with water belongs to the riparian owner.

(D) Both A and B are true.

PROBLEM 69

Title to land may be gained or lost by the action of water. What sort of change may cause a riparian owner to lose land?

(A) Title to land that is cut off from a riparian owner by the sudden shift of the channel of a river is lost by that owner.

(B) Land that is gradually eroded or submerged is lost by the original riparian owner.

(C) Relicted land is lost to the original riparian owner.

(D) Both A and C are true.

PROBLEM 70

A particular navigable stream has been meandered; the meander line has been in place for 75 years. Recently, the amount of water flowing through the channel of the stream has gradually decreased from natural causes. All indications are that the reduction in the flow is permanent. The riparian owner abutting the stream on one side has asked the surveyor to state where the boundary along the stream now falls. The surveyor should tell the riparian owner that

(A) the same meander line that the owner has had for the last 75 years still holds

(B) most likely the owner holds to the thread of the stream

(C) the original mean high water line that was established before the stream shrank holds

(D) most likely the owner holds the newly relicted land

PROBLEM 71

To accurately determine mean high tide, how long must a given station be monitored?

(A) 1 wk

(B) 1 yr

(C) 6 mo

(D) 19 yr

PROBLEM 72

Coastal areas that experience one high tide and one low tide during a tidal day are said to have what type of tides?

(A) neap

(B) retrograde

(C) diurnal

(D) semidiurnal

TESTIMONY

PROBLEM 73

The testimony of a licensed land surveyor is generally held by courts to be expert. Expert testimony is afforded the distinction that it

(A) need not be given under oath

(B) normally gives the surveyor more latitude to express opinions

(C) is not permitted in trials before a jury

(D) both A and B

PROBLEM 74

A land surveyor may take testimony. In which of the following situations may a licensed land surveyor find such authority useful?

(A) the establishment of the age of a fence

(B) the drafting of a property line agreement between adjoining land owners

(C) establishing the priority between two conflicting corner monuments

(D) all of the above

PROBLEM 75

A surveyor has been hired as an expert witness in an action to quiet title and will be paid a fee. Which of the following would be inappropriate?

(A) The surveyor is asked the amount of the fee during cross-examination.

(B) The court has fixed the amount of the fee.

(C) The fee includes compensation for the surveyor's travel and lodging expenses.

(D) All of the above are appropriate.

PROBLEM 76

Which of the following statements describes something a surveyor should NOT do while testifying as an expert witness before the court?

(A) examine pertinent photographs, maps, and field notes

(B) discuss opposing viewpoints regarding the questions at issue

(C) answer hypothetical questions

(D) offer to answer the ultimate question of the case before the court

PROBLEM 77

Generally, a land surveyor appearing as an expert witness may expect the initial questions to involve which of the following?

(A) the surveyor's relationship with the defendant

(B) the name of the client and the nature of the work done

(C) opinions on the issues before the court that contradict those expressed by the surveyor

(D) the surveyor's qualifications and experience

ORDER OF CALL AND THE PRIORITY OF MONUMENTS

PROBLEM 78

Which of the following elements that can control the location of land boundaries is usually considered the most important when found to be in conflict with the others?

(A) a senior right of possession

(B) a call for an adjoiner

(C) a call for an artificial monument

(D) an unwritten right of possession

PROBLEM 79

Between two private owners of sequential conveyances, which of the following would most likely control if their deeds contain conflicting elements?

(A) call for a particular distance along a boundary

(B) call for a particular direction along a boundary

(C) call for a specific acreage

(D) senior right

PROBLEM 80

Which of the following elements that can control the location of land boundaries is usually considered the LEAST important when found to be in conflict with the others?

(A) a call for distance

(B) a call for an artificial monument

(C) a call for a survey

(D) the area of the property described

PROBLEM 81

Monuments sometimes cannot be conclusively proved or disproved. In such circumstances, a monument may be accepted by reputation. What is the term used to describe such a monument?

(A) monument of common report

(B) monument by informative evidence

(C) defective monument

(D) reversionary monument

PROBLEM 82

Which statement is most correct regarding the relative importance of measured distances as compared to measured directions when resolving conflicting title elements?

(A) Direction is superior to distance.

(B) Distance is superior to direction.

(C) There is no consistent rule regarding the relative importance of distance and direction.

(D) They are considered equally important.

PROBLEM 83

A deed describes a lot that is shown on a recorded plat showing 1 in iron pipe monuments set at each corner. However, the description in the deed makes no mention of the iron pipe monuments and describes the boundary lines only by bearings and distances. If a surveyor retracing the boundary finds some of the original 1 in iron pipe monuments in positions that do not agree with the measured bearings and distances in the deed, which evidence should yield, and why?

(A) If the surveyor is convinced that the 1 in iron pipe monuments are undisturbed, the bearings and distances should yield.

(B) The 1 in iron pipe monuments should yield since they are not called for in the deed description and were only shown on the plat.

(C) Even if the surveyor is convinced that the 1 in iron pipe monuments have not been disturbed, the bearings and distances must be considered the best evidence of the original position of the boundary, since the deed is the instrument of the conveyance, not the plat.

(D) If the surveyor is convinced that the 1 in iron pipe monuments are undisturbed and the parties to the deed acted with reference to the plat, the bearings and distances should yield.

PROBLEM 84

What is the fundamental principle behind the preference for one title element over another? For example, why are found original monuments generally preferred over a call for the measured distance between the monuments when they are in conflict?

(A) The element that is least likely to be in error is considered controlling; those that conflict with it are considered merely informational.

(B) When a well-known order of importance of conflicting title elements evolves over time in a region and is supported by a body of common law, such a ranking should be followed regardless of extrinsic evidence to the contrary.

(C) When the written elements of a deed description are superior to any other considerations as codified by the Statute of Frauds, all other conflicting evidence of title must be considered inferior.

(D) Both A and B are true.

PROBLEM 85

In retracing a particular boundary, a surveyor found three monuments that all purported to stand for the same corner. One of the monuments fell at the bearing and distance called for in the client's deed. Another was precisely the monument called for in the deed. However, both of these monuments caused his client's property to overlap the neighbor's property. Finally, the third monument found was called for in the neighbor's deed. If the neighbor's adjoining deed is senior to the client's deed, which of the following statements is correct?

(A) The monument called for in the adjoining senior deed is superior to the other two and should be controlling.

(B) The monument that is in harmony with the bearing and distance recorded in the client's deed is the correct location for the corner.

(C) The monument called for in the client's deed is superior to the other two, even if it does not fall at the bearing and distance recorded in the deed description.

(D) None of the monuments described should be controlling. The surveyor should harmonize the calls and set the monument at the best fit based on all of the evidence available.

CASES

PROBLEM 86

A condominium is a(n)

(A) multiple-unit building where there is separate ownership of individual units, and the unit owners have a right to use common elements

(B) single-family residential unit combined with other single-family units, two-family units, and multiple-family dwellings to form a neighborhood

(C) association for selling homes and lots in a residential subdivision

(D) multiple-unit building where the owner holds title to the whole premises and grants rights of occupancy to individual units through leases or similar arrangements

PROBLEM 87

What is a difference between a condominium and a cooperative?

(A) A condominium unit owner owns shares in a corporation, and a cooperative resident does not.

(B) A condominium unit owner has fee simple ownership, and a cooperative resident does not.

(C) In a cooperative, units are owned individually and appear as separate entities on tax rolls. In a condominium, the property on which all the condominium units are situated appears on tax rolls as a single real property where the condominium corporation is responsible for the taxes.

(D) A cooperative resident has fee simple ownership, and a condominium owner does not.

PROBLEM 88

Which of the following statements about common elements of a condominium is INCORRECT?

(A) General common elements are available to all condominium unit owners.

(B) General common elements are for the use of one or more, but not all, condominium unit owners.

(C) General common elements may include lobbies, elevators, and common hallways.

(D) General common elements differ from limited common elements.

PROBLEM 89

The following five problems refer to the outlined case.

An action was filed in district court by Fred Hale against Edward Pitt to quiet title to land bought by Mr. Hale from Margaret Pitt, the defendant's sister.

Both parties rely upon the prior title of Alice Pitt, the mother of Edward and Margaret Pitt, and both claims involved in the dispute stem from her ownership.

Alice Pitt was a widow with extensive land holdings. In 1950, she made out deeds to each of her 10 children. All of her land was then platted and divided on the plats, and each tract was assigned a number. The sons and daughters drew lots to determine who would receive which lot, and a deed was made out to each according to the number drawn. The deeds were not immediately delivered to the sons and daughters. They were to be held in a safe-deposit box until Alice Pitt became terminally ill.

In June 1950, Alice Pitt, then 90 years of age, became ill and was taken to the home of her daughter Lenore. During this illness, two of Mrs. Pitt's daughters were sent to the safe-deposit box to bring her the deeds. Mrs. Pitt then handed all of the deeds to Lenore and instructed her to deliver them, after Mrs. Pitt's death, to the named grantees. Lenore put the deeds in a bureau drawer in the room where her mother lay ill; they remained there until Mrs. Pitt's death. Lenore claims she then delivered the deeds as instructed.

Apparently, not all of the land that belonged to Alice Pitt was covered by the deeds. There was also some overlapping and other discrepancies that appeared in the deeds as given. Among these discrepancies, the north boundary of the land deeded to Jane Pitt was found to overlap the south boundary of the land deeded to Margaret Pitt. Furthermore, the description of the land conveyed to Jane Pitt, if followed strictly, failed to close and actually extended to the east onto property not owned by Alice Pitt.

Subsequently, Edward Pitt purchased his sister Jane's tract and 6 years later Fred Hale purchased Margaret Pitt's tract. It was at this point that the present controversy arose regarding the overlapping area, which included approximately 5 ac between the two parcels.

The district court found that the deeds were void because there had been no delivery of the deeds from Alice Pitt to her children. Consequently, the court found that neither the plaintiff nor the defendant owned the property; instead, it belonged to the estate of Alice Pitt. The plaintiff appealed the ruling.

89.1. The facts of the case indicate that the grantor, Alice Pitt, placed the deeds in the hands of a third party, Lenore Pitt, for delivery after the grantor's death. This action placed Lenore in what role with regard to all of the deeds?

(A) grantor

(B) testator

(C) trustee

(D) grantee

89.2. Prior to the appeal, which of the parties to the original action was the plaintiff?

(A) Edward Pitt

(B) Lenore Pitt

(C) Fred Hale

(D) Lenore and Jane Pitt

89.3. The court of appeal overturned the lower court's decision and held that the deeds in question were in fact delivered. The appellants next argued that the deed from the mother to Jane Pitt was void by reason of an erroneous description that failed to close. What would be the most likely response from the court?

(A) The lines of a description must be continuous, each line must begin where the other ends, and the final line must return to the point of beginning. Otherwise, the description as written does not close, and therefore does not describe a tract of land.

(B) The court will provide the missing line when doing so will close the approximate acreage called for in the description.

(C) In the construction of boundaries, the intention of the parties is the most important consideration.

(D) Both B and C are true.

89.4. Which of the following statements is most correct concerning the priority of the deeds in question?

(A) Fred Hale's deed is junior to Edward Pitt's. Mr. Hale's deed arises from the purchase of Margaret Pitt's land six years after Edward bought his sister Jane's tract.

(B) Edward Pitt's deed is junior to Fred Hale's since his title arose from within his family and he is entitled to the subsequent privity that implies.

(C) Fred Hale's deed must have priority since there is no cloud on the title he purchased from Margaret Pitt. However, Jane Pitt's title was clouded by its inclusion of land her mother did not own.

(D) There can be no preference of one title over the other between Fred Hale and Edward Pitt.

89.5. Assume that the court of appeal overturned the lower court's decision and held that the deeds in question were in fact delivered. Also assume that the deed from the mother to Jane Pitt was not void and that there is no question of seniority between the two deeds involved in the dispute. Which of the following answers is the most likely resolution of the overlap between Fred Hale and Edward Pitt?

(A) The overlap belongs to Fred Hale.

(B) They both own the overlap as a tenancy in common.

(C) The overlap remains in the estate of Alice Pitt.

(D) The overlap belongs to Edward Pitt.

PROBLEM 90

The following six problems refer to the illustration shown and the accompanying narrative. All numbers are in feet.

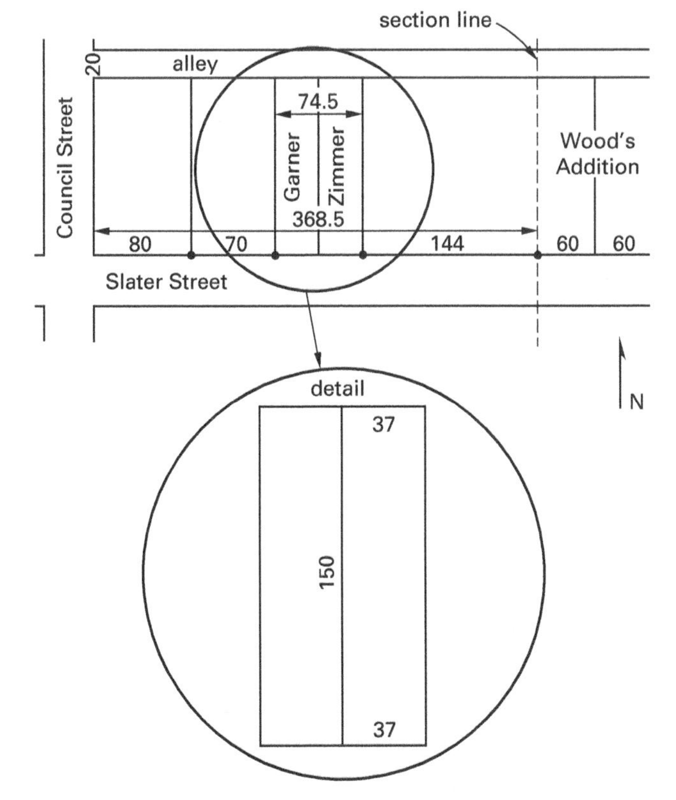

Mary Garner died intestate and the administrator of her estate, M. Thomas, filed suit to quiet title. A decree was issued in his favor and the adjoining property owners, D. Zimmer and his wife, appealed.

The action involved a boundary dispute between tracts in a portion of outlot 8 bounded on the west by Council Street, on the south by Slater Street, on the east by Wood's Addition, and on the north by a 20 ft alley as illustrated.

The city surveyor testified that the east line of the outlot is the section line but that Council Street is not parallel to the section line due to convergence. He found four monuments, as shown, and made the measurements indicated, which were not contested. He measured from the southwest corner of outlot 8, the northeast corner of the intersection of Slater and Council streets, and along the north side of Slater Street to a post at the southwest corner of Wood's Addition. He found the distance to be 368.5 ft. He found one monument 80 ft east and another 70 ft further east from the southwest corner of the outlot. The latter monument, mentioned in the deed description, marks the southwest corner of Mary Garner's lot, which is now administered by Thomas. Further measurements reveal that the monument at the southeast corner of the Zimmer parcel mentioned in the description lies 144 ft west of the post at the southwest corner of Wood's Addition and 224.5 ft east of the southwest corner of outlot 8.

The titles of the Zimmers and Mary Garner both stem from a common grantor, Rachel Garner—Mary Garner's mother. It has been suggested that she thought her property was 80 ft wide instead of 74.5 ft wide, as current measurement indicates. It has been further suggested that the error was the result of the description from Charles Schwarz to Rachel Garner that described her tract as follows.

> The SW$\frac{1}{4}$ of outlot 8, except 150 ft square in the SW corner of said outlot extending to the alley. Also commencing at the southeast corner of said SW$\frac{1}{4}$ of said outlot 8, thence north to the alley, thence east to the west line of lands sold by Jacob Baum to Mark Crandall, thence south to Slater Street, thence west to beginning, the intent being to convey all the land 80 ft wide between the lands conveyed by me to K. L. Larson and the land conveyed from Baum to Crandall.

No record was found of the lands sold by Jacob Baum to Mark Crandall.

Subsequently, Rachel Garner sold to D. Zimmer, "37 ft off the east side of the piece of ground that is described in the deed of Rachel Garner."

M. Thomas, as administrator of Mary Garner's estate, asked that the title be quieted as follows.

> Commencing 150 ft east of the southwest corner of outlot 8, thence east 43 ft to the west line of the land sold to D. Zimmer, thence north 150 ft to the south line of the alley, thence west along the south line of said alley to a point due north of the point of beginning, thence south to the beginning.

The original trial court so decreed, but reduced the recited width from 43 ft to 41 ft. The Zimmers appealed the decision.

90.1. What was the width of the lot before Rachel Garner sold the east 37 ft to Zimmer, and what is the best supporting evidence of that width (or is the width important to the question at all)?

(A) The width of the lot was 80 ft, as described in Rachel Garner's deed.

(B) No record exists of the lands sold by Jacob Baum to Mark Crandall, and the southwesterly 150 ft of outlot 8 is an ambiguous description. Therefore, the parcel was 80 ft wide as stipulated.

(C) The description of the southwesterly 150 ft is not ambiguous and is supported by the city surveyor's measurements. The fact that no deed is recorded regarding the conveyance of Jacob Baum to Mark Crandall does not alter the evidence of the current measurements. The parcel was 74.5 ft wide.

(D) The best available evidence indicates that the lot was 74.5 ft wide, but the width is not a critical consideration in resolving the boundary dispute between the two parties.

90.2. Assuming that the section line forming the western boundary of Wood's Addition is a cardinal line, what was the direction of the western boundary of Mary Garner's lot prior to the court case?

(A) There is not sufficient information available to make any assumption about the direction of the western line of Mary Garner's lot.

(B) The western line of Mary Garner's lot is parallel with the western boundary of outlot 8.

(C) The western line of Mary Garner's lot is a cardinal line.

(D) The western line of Mary Garner's lot is on an average bearing between the eastern boundary of Council Street and the section line.

90.3. What statement can be made concerning the direction of the line between the Zimmers' and Mary Garner's lots?

(A) It must be a cardinal line.

(B) It is parallel with the Zimmers' eastern boundary.

(C) It is parallel with the western boundary of outlot 8.

(D) It is parallel with the eastern line of Council Street.

90.4. During the proceedings, it was proposed that the most equitable solution to the disagreement was to proportion the 74.5 ft between the lots. Which statement best describes the merit of this proposal?

(A) The proposal is equitable. The estate of Mary Garner would receive 37.7 ft and the Zimmers would receive the remainder.

(B) The proposal is equitable. The estate of Mary Garner would receive 40.5 ft and the Zimmers would receive the remainder.

(C) The use of proportioning is restricted to simultaneous conveyances and this suit involves a sequential conveyance of property. Mary Garner's estate has senior rights in this situation and should receive the full 40 ft originally intended by Rachel Garner.

(D) The use of proportioning is restricted to simultaneous conveyances and this suit involves a sequential conveyance of property. Proportioning should not be used to resolve this dispute.

90.5. During the proceedings, the Zimmers pointed out that a fence had stood along a line 37 ft west from their eastern boundary for 15 years and that the fence was moved to this position by Mary Garner herself. The Zimmers contributed to the cost of the fence and contend that it constitutes a practical location of the boundary. M. Thomas countered by saying that the Zimmers had exercised undue influence on the elderly Mary Garner by urging her to move the fence 2.5 ft west from where it had stood for 25 years. What doctrine might M. Thomas cite to support his position on the question of the fence?

(A) recognition and acquiescence

(B) eminent domain

(C) prescription

(D) estoppel

90.6. What role, if any, does senior rights play in this case?

(A) There are no senior rights in this situation between the Zimmers and the estate of Mary Garner.

(B) The Zimmers have senior rights and are entitled to the full 37 ft in width.

(C) M. Thomas, administrator of Mary Garner's estate, is the holder of the senior rights and is entitled to a full width of 43 ft as intended when Rachel Garner conveyed the eastern 37 ft of 80 ft.

(D) There is an overlap of 2.5 ft between the two adjoining property owners and it is still vested in the estate of Rachel Garner, now held by M. Thomas.

PROBLEM 91

The following five problems refer to the illustration shown and the accompanying narrative.

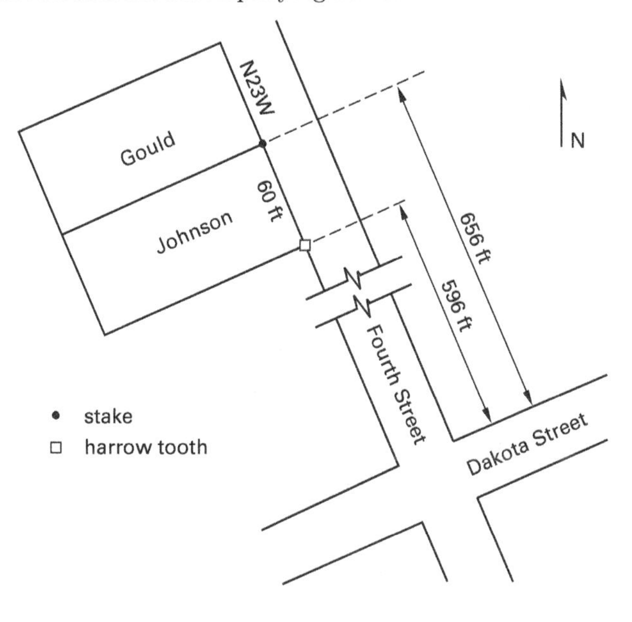

(not to scale)

May Johnson instituted a quiet title action regarding her lot on the southwest side of Fourth Street, which had been acquired from appellant George Gould by Mrs. Johnson's son Henry. The deed described the lot as follows.

> Beginning at a harrow tooth on the southwest side of Fourth Street distant N23°00′W 596.00 ft from the point of intersection of the southwest line of Fourth Street and the northwest line of Dakota Street, thence N23°00′W along said southwest line of Fourth Street 60.00 ft to a stake set at the common corner for lots 7 and 8.

Henry Johnson and his wife subsequently conveyed the lot to his mother, May Johnson, by a deed containing the same language.

The appellant, Mr. Gould, contended the description in the deed he signed was the result of a mutual mistake. He had intended to sell a lot with a frontage of 55.00 ft, not 60.00 ft. In support of his contention, the appellant presented the escrow instructions signed by both himself and Henry Johnson on the same date he conveyed the lot to Mr. Johnson. The description of the frontage in the escrow instructions were as follows.

> Beginning at a harrow tooth on the southwest side of Fourth Street distant N23°00′W 596.00 ft from the point of intersection of the southwest line of Fourth Street and the northwest line of Dakota Street, thence N23°00′W along said southwest line of Fourth Street 55.00 feet to a stake set at the common corner for lots 7 and 8.

Mr. Gould conceded that this description was inconsistent with the actual distance between the harrow tooth and the stake. He contended that the bank holding the escrow resolved the discrepancy to reflect the actual distance between the monuments without consulting the parties to the deed. He further contended that he was not aware of the discrepancy or of the alleged mistake in the deed because he could not read.

The evidence presented to the court showed that the respondent, May Johnson, lived in the house on the lot and that if the boundary Mr. Gould advocated was accepted, her house would be too near the property line to satisfy local setback ordinances. The appellant testified that the respondent had agreed to move the house at her expense.

Mr. Gould testified that he did not wish to sell the 60.00 ft because it would include part of his driveway and require him to remove several trees.

Mrs. Johnson countered that there had been no conversations regarding her house or a frontage of 55.00 ft. She further testified that she had tried to construct a fence on the boundary in question and that Mr. Gould had torn it down. She said she had not attempted to reconstruct the fence because of death threats from Mr. Gould.

In his appeal, Mr. Gould requested that the court overturn the lower court's affirmation of the deed description and reform the deed to conform with the escrow instructions.

91.1. What is the meaning of the term *mutual mistake* as used in this case?

(A) Mr. Gould contended that the escrow holder was mistaken in the presumption that Mr. Gould had read and understood the deed description, and that he was himself mistaken in trusting that the deed correctly described the property he intended to sell.

(B) The mutual mistake was the misunderstanding shared by Mr. Gould and Henry Johnson concerning the frontage of the lot. Mr. Gould believed it was 55.00 ft and Henry Johnson believed it was 60.00 ft.

(C) Mr. Gould was mistaken in believing that the actual dimension between the harrow tooth and the stake was 55.00 ft; Henry Johnson was mistaken in believing that the actual dimension of 60.00 ft was what Mr. Gould intended to convey.

(D) The mutual mistake was the misunderstanding shared by Mr. Gould and Henry Johnson concerning the frontage of the lot. Both believed the deed specified that the lot was to have a frontage of 55.00 ft.

Land Boundary Law

91.2. Could the deed between Mr. Gould and Henry Johnson have been considered void; if so, on what grounds?

(A) Yes, Mr. Gould was under a disability. One of the requirements for a valid deed is competent parties. He was not competent to convey property because he could not read.

(B) No, the fact that Mr. Gould could not read does not render him incompetent to convey property.

(C) Yes, one of the requirements for a valid deed is that it be read. Mr. Gould may be competent to convey property, but by his own testimony, he did not read the deed.

(D) Yes, the escrow agreement was changed without notification of the parties involved, which automatically voided the conveyance.

91.3. It was revealed in the testimony that Mrs. Johnson had paid the property taxes in accordance with her deed and that Mr. Gould had never attempted to pay the taxes on the disputed 5 ft strip of property. What issue may have been raised in the case that would have caused this fact to become important?

(A) the possible criminality of the alleged death threats to Mrs. Johnson attributed to Mr. Gould

(B) the possible violation of local setback ordinances that would arise from the reformation of the deed proposed by Mr. Gould

(C) the ripening of unwritten rights supporting Mr. Gould's claim on the disputed 5 ft strip

(D) possible monetary damages as compensation for the loss of Mr. Gould's trees and driveway

91.4. Assume that the 60.00 ft stipulated as the frontage of Mrs. Johnson's lot remained in the deed, but the actual distance between the monuments had been 55.00 ft. Would the distance or the monuments prevail, and why?

(A) The 60.00 ft would have been considered locative regardless of the actual distance between the monuments.

(B) The monuments would have prevailed and Mrs. Johnson's lot would have had a frontage of 55.00 ft because the deed description contained the phrase "to a stake set at the common corner of lots 7 and 8."

(C) The deed must be construed from its four corners, and such a patent ambiguity would void the deed entirely.

(D) A harrow tooth cannot be considered a proper monument for a lot corner, so the distance would have prevailed.

91.5. Mr. Gould testified and presented evidence that the deed did not reflect his intentions regarding the conveyance of the lot. What importance does this evidence have in the disposition of the case?

(A) The first and foremost rule in construing deeds is that they must reflect the intention of the parties to the conveyance. The language in the deed was not the language that the grantor intended, but Mr. Gould was not aware that that was the case because he could not read. He should not be bound by language over which he had no control. The deed should be reformed by the court.

(B) The intentions of the grantor are paramount, but only as reflected in the language in the deed. The deed should not be reformed by the court.

(C) The intention of the grantor was expressed by the monuments on the ground and the dimensions between them recited in the deed. The grantor cannot be permitted to alter such an unambiguous deed description with testimony.

(D) Both B and C are true.

PROBLEM 92

The following five problems refer to the illustration shown and the accompanying narrative.

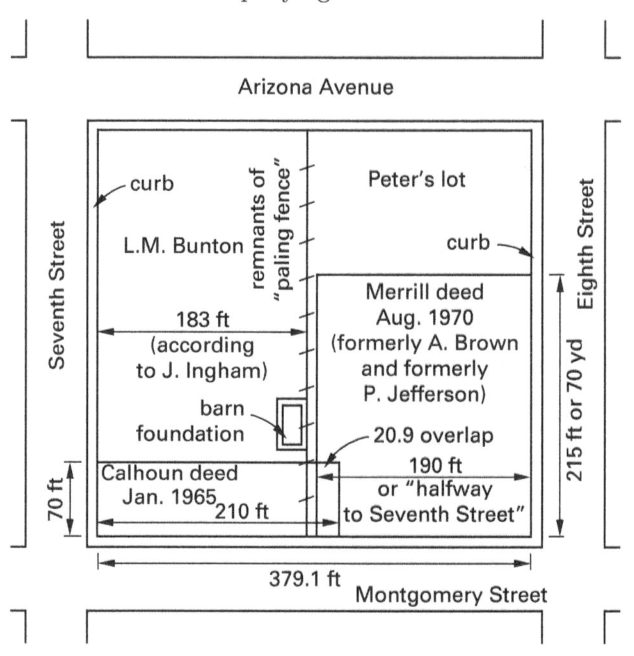

The appellant in this case, H. Calhoun, appealed a decree from a lower court establishing the boundary line between himself and R. Merrill. Both derived their titles from a common source, P. Jefferson. Mr. Jefferson conveyed directly to the appellee, R. Merrill; the appellant traces his title to Jefferson through three mesne conveyances.

Both parcels front on Montgomery Street. The measurement from the back of the curb at Seventh Street to the back of the curb at Eighth Street is 379.1 ft.

Mr. Calhoun's deed calls for 210 ft along Montgomery Street and Mr. Merrill's deed calls for 190 ft. Hence, there appears to be an overlap of 20.9 ft.

Mr. Calhoun contends that the appellee built a fence on his property and asked the court to adjudicate their boundary. Mr. Merrill countered that he had built the fence on the correct boundary line, west of where the appellant asserted it to be. Mr. Merrill contended that the west line of his property was the west line of the A. Brown lot, which was well defined on the ground.

Mr. Calhoun's deed, dated January 1965, describes his property as follows.

> Beginning at the southwest corner of the lot owned by L. M. Bunton; thence along the east side of Seventh Street 70 ft to Montgomery Street; thence east along the north side of Montgomery Street 210 ft; thence north along the west boundary of the lot formerly owned by A. Brown, now owned by P. Jefferson, 70 ft; thence west 210 ft parallel with Montgomery Street to the starting point. Said lot containing one-third acre.

Mr. Merrill's deed, dated August 1970, describes his property as follows.

> Commencing at the crossing of Eighth Street and Montgomery Street and thence halfway to Seventh Street; thence north 70 yd to a lot now known as Peter's lot; thence east to Eighth Street; thence south back to the starting point, containing 1 ac more or less.
>
> Said lot more particularly described as follows: Commencing at the northwest corner of Eighth Street and Montgomery Street, thence northward along the west boundary of Eighth Street 215 ft to Peter's lot; thence westward along the line of Peter's lot and parallel with Montgomery Street 190 ft; thence southward parallel with Eighth Street along the west boundary of the lot formerly owned by A. Brown 215 ft to the north boundary line of said Montgomery Street; thence eastward along the north boundary line of Montgomery Street 190 ft to the starting point.

The lower court found in favor of the appellee, Mr. Merrill. Mr. Calhoun appealed the decision.

The appellant contended that the court issuing the decree under appeal had erred by using a fictitious monument in fixing the boundary. This contention stemmed from the following paragraph in the decree.

> It is decreed by this court that the true boundary between the Calhoun lot and the Merrill lot begins at the northern extremity of the old brick foundation of a barn once located thereon near the southeast corner of the L. M. Bunton lot. Said boundary then extends south along said brick foundation. The boundary subsequently follows the old rusted wire fence and rotten or decayed fence posts above and below the surface of the ground to Montgomery Street.

L. M. Bunton was called as a witness for the appellee and testified that he had purchased the lot north of the Calhoun lot in 1937 and had lived on it for several years. He said he had built a cypress picket fence from the southeast corner of his lot north to Arizona Avenue. Additionally, he had built a brick foundation for the barn that stood at his southeast corner, and there was a fence that had extended from the barn to the north side of Montgomery Street. He testified that he did not know if the fences were still up or not, but that the barn foundation was still in place.

B. Cooper, a witness for the appellee, testified that he had for many years cultivated the land up to the fence that extended both north and south from the barn mentioned by Mr. Bunton.

The appellee testified that he had erected his fence along the brick foundation of the barn south to Montgomery Street, following the old fence posts that had rotted to the ground.

Mr. Calhoun testified that he had possessed the 210 ft called for in his deed and that no one had pointed out the boundary lines when he purchased the property. He further testified that he had no knowledge of the A. Brown lot, except that it was mentioned in his deed and that there had been no fence extending south to Montgomery Street when he bought the lot.

J. Ingham, a surveyor, testified that he found no evidence of a fence on the line claimed by the appellants, but that he did find evidence of a boundary line 183 ft east of the back of the curb on Seventh Street. This boundary evidence was an extension of the line of a paling fence running south from Arizona Avenue on the Bunton property. Mr. Ingham further testified that he could find no information as to the width of Seventh Street or Eighth Street before the curbs were installed.

92.1. In this case, *mesne conveyances*

(A) refer to unrecorded deeds

(B) are intervening instruments

(C) are deeds that have been voided

(D) refer to instruments that give color of title

Land Boundary Law

92.2. The description of Mr. Merrill's property contains several contradictions. For example, the call for 70 yd contradicts the call for 215 ft. How should this be resolved?

(A) The call for 70 yd occurs earlier in the description than does the call for 215 ft and should, therefore, have priority.

(B) The more particular description should prevail over the more general one. The 215 ft occurs in the more specific section of the description and should be used over the more general call for 70 yd.

(C) Neither the 70 yd nor the 215 ft need be of critical importance as long as the parcel known as Peter's lot can be located with certainty.

(D) Both B and C are true.

92.3. What role, if any, does senior rights play in the location of the boundary between Mr. Calhoun and Mr. Merrill?

(A) The deeds of the two property owners both stem from the same grantor. Since Mr. Calhoun's deed predates Mr. Merrill's, the first deed need not contribute to any discovered shortage.

(B) The deeds of the two property owners both stem from the same grantor and Mr. Calhoun's deed predates Mr. Merrill's. However, while the deeds may furnish the means of locating the particular boundary between them, the actual location of that boundary is an independent inquiry.

(C) The description in Mr. Calhoun's deed calls for the west boundary of the adjoining A. Brown lot. The A. Brown lot must have priority over Mr. Calhoun's parcel.

(D) Both A and C are true.

92.4. What significance, if any, do the widths of Seventh and Eighth streets have in the resolution of the boundary dispute between Merrill and Calhoun?

(A) Seventh Street is a record monument called for in Mr. Calhoun's deed. If the eastern boundary of Seventh Street has moved since his deed was written, then the eastern boundary of his lot has moved with it.

(B) Seventh and Eighth streets are both record monuments called for in Mr. Merrill's deed. If the streets have been widened since his deed was written, this may have caused his western boundary to overlap Mr. Calhoun's property.

(C) Even if Seventh and Eighth streets have been widened since the two deed descriptions in question were written, the boundaries near the middle of the block would not have been altered.

(D) Both A and B are true.

92.5. Mr. Calhoun contends that he is in possession of the senior deed and that he should be entitled to the entire 210 ft of frontage along Montgomery Street called for in his deed description. Is he right or wrong, and what is the principle underlying the answer?

(A) Mr. Calhoun is right. He possesses the senior right and even though a call for a record monument (the west line of the A. Brown lot) is included in his deed, his senior right is superior to such a call.

(B) Mr. Calhoun is wrong. While his senior right can overcome the call for 190 ft in Merrill's deed, it cannot prevail over the remains of the fence extending south from the old barn foundation. Testimony and the apparent acceptance of the fence for 28 years before his deed was executed is the best evidence of the position of the west line of A. Brown's lot, which is the east boundary of his property.

(C) Mr. Calhoun is right. P. Jefferson owned both of the parcels in question. He first conveyed the lot owned by Calhoun and created a boundary 210 ft east from the east line of Seventh Street. Mr. Calhoun is entitled to rely on that boundary.

(D) Mr. Calhoun is wrong. The call for the west line of the A. Brown lot is not merely a call for that line alone, but for the entire A. Brown tract. The determination of the position of that tract must depend on the deed description included in the Merrill deed. The Merrill deed contains several patent ambiguities and should be considered void. Therefore, Mr. Calhoun's deed is void for uncertainty as well.

PROBLEM 93

The following three problems refer to the narrative given.

This is an appeal from a judgment in favor of the plaintiff in an action brought to quiet title arising from conflicting interpretations placed on a warranty deed.

The grantors and defendants, the Andersons, allege that the warranty deed reserved to them a one-half interest in the oil and gas estate, plus the fee interest in all other minerals located on or beneath the surface. The grantees and plaintiffs, the Footes, claim title to the fee interest in the surface and mineral estates, except for a one-half interest in the oil and gas estate reserved by the defendants.

At the trial, the parties stipulated that the only evidence admissible would be the deed on which the interest rests. Therefore, the only question is whether the trial court was correct in the construction of the deed.

The deed, dated May 1954, makes the following recitation.

> Grantor L. Anderson and her husband E. R. Anderson ... convey and warrant to H. Foote and K. Foote, his wife, as joint tenants and not as tenants in common with full rights of survivorship ... the following described real estate, to wit.
>
> The SW $\frac{1}{4}$ of the SW $\frac{1}{4}$ of Section 5, Township 5 South, Range 7 West of the Bolton Meridian and containing 40 ac more or less. Subject to an easement for a roadway 20 ft wide along the north side of said lands.
>
> Reserving unto the grantors the undivided one-half interest to all oil and gas in and under and that may be produced from the above-described lands, together with the right of ingress and egress at all times for the purpose of mining, drilling, exploring, operating, and developing said lands for oil, gas, and storing, handling, transporting, and marketing the same therefrom.
>
> It is the intention of the grantors to convey an undivided one-half interest in and to all of the oil and gas that may be produced from the above-described lands, also the surface of the described lands.

The court that heard the original action found for the plaintiffs; the defendants appealed the decision.

93.1. Interest in the surface estate is conveyed by the granting clause of the deed description quoted so that both grantees hold the surface estate

(A) together as an undivided interest and should one of them die, the other would automatically acquire the entire surface estate

(B) together as an undivided interest and should one of them die, that person's interest would automatically pass to their heirs

(C) but should one of them die, that person's interest would automatically revert to the grantor

(D) but should one of them die, the entire surface estate would automatically revert to the grantor

93.2. Interest in the mineral estate is conveyed by the granting clause of the deed description quoted so that the mineral estate

(A) is not specifically mentioned and therefore is reserved to the grantor. It does not pass to the grantee

(B) is not specifically mentioned and therefore the minerals are conveyed to the grantee, except in and under the easement across the northerly 20 ft of the tract

(C) is not specifically mentioned and therefore the minerals are conveyed to the grantee

(D) cannot be conveyed by deed. It must be conveyed by a mineral lease

93.3. How does the last clause beginning "It is the intention of the grantors ..." affect the conveyance to the grantees?

(A) It does not alter the effect of the preceding paragraphs.

(B) It prevents the passing of the mineral estate to the grantees.

(C) It expresses the intent of the grantors, which is the primary consideration in construing the meaning of a deed.

(D) It voids the entire deed.

PROBLEM 94

The following five problems refer to the illustration and the accompanying narrative.

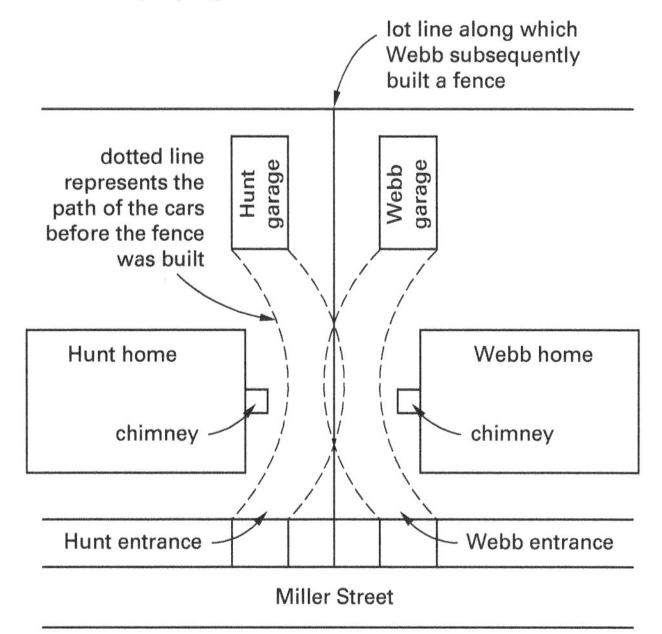

This case involves a dispute between two neighbors, the Hunts and the Webbs, over a driveway. The plaintiffs, the Hunts, claimed an easement over a strip of land 6 ft wide on the Webbs' property. After an equity action, the court found against the plaintiffs and denied the easement. The Hunts appealed the decision.

The difficulties between the neighbors began when the Webbs erected a fence along their west line. There is no dispute between the parties regarding the location of the lot line. However, with only $6\frac{1}{2}$ ft between the chimney on the east side of the Hunts' house and the Webbs' fence, the Hunts did not have enough room to comfortably drive their car through to their garage. Nevertheless, the Hunts did not commence the action until five years after the fence was constructed.

The driveway entrances from the street are separate and distinct. The only place the neighbors found it necessary to drive onto the others' property was where they veered to clear the chimneys that extended from the sides of each of their houses, as illustrated.

The Hunts claim an easement by prescription over the Webbs' property. The state in which this claim was made requires that prescriptive easements be acquired by use that is continuous for a statutory period of 10 years under claim of right, and that the use be open, notorious, and hostile. Additionally, the state requires that the claimant formally notify the party against whom the claim is made prior to the running of the statutory period.

The Hunts called several witnesses who had lived in the house they now occupy and established that the use of the driveway had been as illustrated for 38 years. These witnesses further testified that they had no conversations with their neighbors about the use of the driveway since the use was never questioned. The Webbs testified that they had no communication with the Hunts regarding their driving on the Webbs' property prior to the construction of the fence.

The Webbs widened their driveway by removing their chimney and then built a fence along the property line. When Mrs. Hunt saw the post holes the Webbs had dug, she filled them and had her attorney call the Webbs to inform them that she objected to the fence. Nevertheless, it was constructed, and five years elapsed before the Hunts issued a written notice to the Webbs demanding a quitclaim deed granting them an easement. The Webbs refused and the Hunts immediately filed their action in court.

94.1. What is the meaning of the phrase "claim of right" as applied to this case?

(A) the Hunts' initial filing of a suit to acquire the easement rights

(B) the Webbs' construction of a fence on the property line

(C) the Hunts' demand for a quitclaim deed from the Webbs

(D) the Hunts' belief that they were entitled to the easement right they exercised for many years across the Webbs' property

94.2. What is the meaning of the phrase "open and notorious" as applied to this case?

(A) The Hunts must have used the easement across the Webbs' property in such a way that the Webbs could reasonably be expected to have known that their property was being used in a manner inconsistent with their ownership. This would have provided the Webbs with the opportunity to take measures to prevent the Hunts from using the property improperly.

(B) In the state where the case occurred, the notorious aspect must include formal notification of the Webbs, not merely constructive notice alone.

(C) The Hunts should have constructed a fence of their own 6 ft across the property line on the Webbs' parcel in order to establish their claim in an open and notorious manner. Their use of the easement alone is not sufficient.

(D) Both A and B are true.

94.3. There are doctrines other than prescription by which an unwritten easement right may be acquired. Which of the following methods of procuring an unwritten easement might the Hunts have used?

(A) an implied easement

(B) a way of necessity

(C) estoppel easement

(D) common-law dedication

94.4. The Hunts alleged that the use of the driveway across the Webbs' property was continuous for 38 years. Assuming that this evidence was true, did it satisfy the requirement of continuous use for the statutory period?

(A) Yes, the statutory period in the state where the case was heard was 10 years.

(B) No, the use was not continuous; the Webbs used the same area of the driveway as well.

(C) Yes, the use of the driveway was intermittent, but since that was consistent with the very nature of such an easement right, the continuity was not disturbed.

(D) No, the statutory period did not run at all during the entire 38 years since the state law where the case was heard required express formal notification of the parties against whom the claim was made.

94.5. Which of the following statements is correct regarding this case?

(A) The burden of proof was on the Webbs.

(B) The formal notice required by the state law occurred when Mrs. Hunt had her lawyer call the Webbs.

(C) The Hunts would have had a better case if they had claimed an easement by necessity rather than a prescriptive easement.

(D) The burden of proof was on the Hunts.

PROBLEM 95

The following five problems refer to the narrative given.

This is an action to quiet title in a roadway on land formed by accretion along the Manterbrod River in Lamp City. The accretion land was originally platted as part of Lamp City between block 75 and the Manterbrod River. The plaintiff, Mr. Clark, acquired block 75 and the accretion land by a warranty deed, but the warranty was limited to "that part of said real estate that is not accretion land." The defendant in this action is the city of Lamp City.

The accretion land east of a high bank of the river was unenclosed, uncultivated land covered with brush, timber, and willows. A road ran north and south along the river on the most easterly portion of the accretion land. The evidence offered to the court indicated that the road had existed for 25 years before the action to quiet title was brought by Mr. Clark. The earliest use was by hunters, fishermen, and sightseers; however, 20 years before Mr. Clark's action, stabilization work was done and the road began to be used by rock trucks. During the 11 years immediately prior to Mr. Clark's action, the road was used continuously by the public and was maintained by Lamp City.

The requirements for acquisition of an easement by prescription, as well as fee title by adverse possession in the state where Mr. Clark's action was brought, include adverse use that is open, notorious, hostile, and uninterrupted for a statutory period of 10 years. Claim of right is a rebuttable presumption in the ripening of adverse possession or prescriptive easement rights.

The lower court stated, "Lamp City and the public have maintained and used said road continuously and adversely for more than 10 years," and quieted and confirmed a fee simple absolute title to the road in Lamp City. Mr. Clark appealed the decision.

Mr. Clark contended that Lamp City had the power of eminent domain and was thereby barred from obtaining real property by adverse possession without just compensation. He also contended that Lamp City had accepted his payment of taxes on the land and was thereby barred from acquiring it by adverse possession.

Mr. Clark further contended that the fee simple absolute title awarded to Lamp City by the lower court was improper. The decree did not define a specific use for the road, and Mr. Clark argued that it is not necessary that a road acquired by prescriptive use be equal in all respects to the title that would result from a formal dedication.

95.1. Which of the following statements is most correct concerning Mr. Clark's contention that Lamp City was barred from acquiring the road by adverse possession when it could have exercised its power of eminent domain?

(A) Mr. Clark is correct. If property is to be taken for public use, compensation must be paid to the private land owner.

(B) The public acquired the right to the river road by continuous use. The individual members of the public did not have the power of eminent domain but, by their concerted action, created their right to the road now held by the city for the benefit of the public.

(C) At any time during the running of the statutory period of 10 years, Mr. Clark could have exercised his right to the roadway. His inaction during the statutory period contributed to the ripening of the public's right to the road.

(D) A and B may each be true depending on the state in which the action is taken. C is true in all jurisdictions.

Land Boundary Law

95.2. Which of the following statements is most correct concerning Mr. Clark's contention that the acceptance of his payment of taxes on the property covered by the road prevents the city from acquiring the road by adverse possession?

(A) In most jurisdictions, the payment of taxes is an element that may be considered in establishing ownership in cases of adverse possession, but barring a statutory provision, it is not conclusive.

(B) The payment of taxes by Mr. Clark confirmed his dominion over the lands in question and prevented the running of the statute against him.

(C) The payment of taxes is not a consideration in cases of adverse possession.

(D) It was the acceptance of Mr. Clark's payment of taxes by the municipality rather than his attempt to pay them that confirmed his dominion over the lands in question and prevented Lamp City from having any claim of adverse possession.

95.3. Assume that the court of appeals ruled that the lower court erred in granting a fee interest rather than an easement right in the road to Lamp City. Which of the following statements would be a justification for the higher court's ruling?

(A) The fee title to the road remains with Mr. Clark since the municipality is not entitled to a grant of greater interest than is essential for public use.

(B) All of the streets and roadways in Lamp City are held in trust for the purposes for which the street was dedicated as an easement right and not as a fee simple absolute.

(C) The easement was acquired by prescription rather than by adverse possession.

(D) All of the above

95.4. The foundation of Mr. Clark's ownership of the land in question was a warranty deed whose warranty did not cover the accreted land along the Manterbrod River. Why was the accreted land not covered by the warranty?

(A) Accreted land is unstable and may be easily eroded away.

(B) Accreted land is generally not covered by those insuring titles because it is virgin property without a chain of title.

(C) Accreted land generally belongs to the state.

(D) Accreted land is not considered real property before the law.

95.5. If Mr. Clark had been able to establish that the road in question had not always followed the same route over the 10 years during the running of the statute, might the court have found in his favor?

(A) Yes, the road must have a definite and consistent route for a prescriptive right to ripen.

(B) No, since the public merely acquired the right to travel across the accretion land, the exact path is not critical.

(C) Yes, the roadway must have a definite and consistent route that is maintained at public expense and improved to a level equal with the other streets and roads in the municipality for the prescriptive right to ripen.

(D) No, the municipality could have rerouted the roadway any number of times. As long as it was maintained at public expense, the statute would have continued to run.

PROBLEM 96

The following five problems refer to the illustration shown and the accompanying narrative.

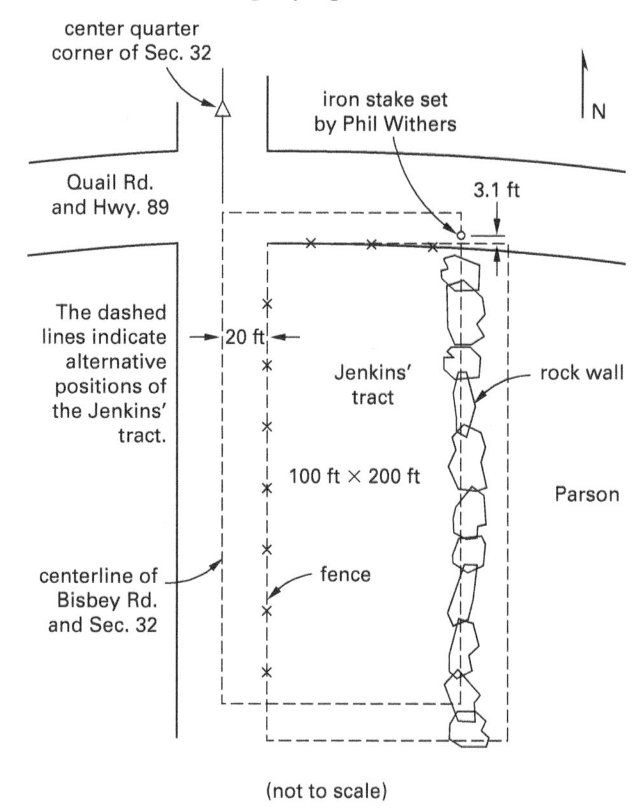

(not to scale)

The appellants in this action are the Jenkins; the appellees are the Parsons.

This appeal involves a decree that determined the boundaries of a 100 ft by 200 ft tract of land conveyed to the appellants' grantors, the Kings, by the appellees, the

Parsons. The Parsons sold the tract in question to Jack and Jean King in June 1966. The Kings conveyed the same land to the Jenkins by a deed dated May 1968. The particular description in both deeds is as follows.

> Beginning at a point on Quail Road (now State Highway 89) 9 rods south of the center of section 32, T19S, R34W; thence south along Bisbey Road 200 ft; thence east 100 ft; thence north 200 ft parallel with the centerline of said section 32 to said Quail Road, also known as Highway 89; thence west along the right-of-way of said Quail Road 100 ft to the place of beginning.

The Parsons retained the land to the east and south of the described tract.

The appellants contend that the issue on appeal is the correct location of the lot described. The question hinges on whether the lot includes any part of the roads. Mrs. Parson instituted the original action, contending that about 20 ft of the lot lies in Bisbey Road. The appellants assert that it was the intention of all the parties involved to convey a lot 200 ft long by 100 ft wide without easements or encroachments.

After hearing the arguments, the court found that the northeast corner of the appellants' lot was correctly marked by an iron stake set by Phil Withers in 1966. The iron stake was described in the decree as being on the south right-of-way line of Quail Road, approximately 100 ft east from a point on the centerline of Bisbey Road. The northwest corner of the property was described in the decree as being in Bisbey Road, approximately 348.5 ft south of the center of section 32. The decree further stated that the east boundary of the appellants' lot ran south from the iron stake 200 ft along a rock wall, grapevines, and other vegetation, parallel with the centerline of section 32. The south line of the appellants' lot was described in the decree as proceeding west from a point 200 ft south of the iron stake to the centerline of the section, which is also the centerline of Bisbey Road.

The Jenkins appealed the decision. The appellants asserted that the court's decree was incorrect because it did not give effect to the Parsons' intention in their original conveyance of the parcel, because it was based on hearsay evidence, because certain distances were described by the court as being approximate, and because it did not describe the boundary with such certainty that it could be located.

In her testimony, Mrs. Parson steadfastly maintained that her husband had caused a survey to be made by Phil Withers, now deceased, as the basis of their conveyance of the lot in question to the Kings. The survey was not recorded. She identified the iron stake that was 100 ft east of the centerline of section 32 and 3.1 ft north of the south right-of-way line of Highway 89 as one placed by Withers. Under cross-examination, she admitted she had been in the hospital at the time Withers made the survey.

The appellants offered testimony from another surveyor, Mr. Landers. The point of disagreement was the location of the west line of the Kings', now the Jenkins', lot. Mrs. Parson testified that Withers located it in Bisbey Road, so that about 20 ft of the road would be included in the Jenkins' lot. Landers, the Jenkins' surveyor, assumed that the west boundary of the Jenkins' lot was marked by the fence on the east border of Bisbey Road. Landers also presumed that the north boundary of the Jenkins' lot was marked by the southerly right-of-way fence of Highway 89, also known as Quail Road.

Landers testified that in order to connect the northwest corner of the Jenkins' lot (as he construed it) to the center of section 32, one would have to proceed on a bearing of N07°40′W for a distance of 149.9 ft. Landers also testified that the iron stake was approximately 20 ft west and 3.1 ft north of the northeast corner of the Jenkins' lot as he located it. Landers admitted that if he had proceeded due south from the center of section 32, he would have gone down the center of Bisbey Road. The reason he varied from this course was that he interpreted the word *along* in the deed to mean that the line, in both cases, ran on the edges of the roads.

Mr. Jenkin testified that he had built the rock wall as a retaining wall to fill his yard, not to mark the boundary of his property. Jenkin said he had no clear idea of the proper location of the eastern boundary of his property at the time. He admitted that the iron stake was in line with the rock wall, but testified that he had no idea why it was there.

Ruth Riesen, who had lived in sight of the Jenkins' lot since 1936, testified that the Jenkins had replaced a fence of long standing with their rock wall. Edith Stander, who lived slightly more than a quarter mile from the Jenkins' property, testified that the Parsons had built a fence on the east line of the lot at the time they conveyed the property to the Kings.

The appellants placed great reliance on the location of the fence on the eastern edge of Bisbey Road and on the fence along the south right-of-way of Highway 89 when they bought the property. No one told them the fences marked the west and north lines of the lot, but the Jenkins presumed this to be the case.

96.1. The appellants, the Jenkins, asserted that one of the errors made by the original trial court was its acceptance of hearsay testimony regarding the location of the boundaries of their lot. What testimony would be considered hearsay?

(A) Mrs. Parson's testimony regarding the Withers survey

(B) Mrs. Riesen's testimony regarding the replacement of a fence with the Jenkins' rock wall

(C) Mr. Jenkin's testimony about the building of the rock wall

(D) Landers' testimony that he found an iron stake 20 ft west of his determination of the east line of the lot

96.2. There appears to be a discrepancy concerning the distance from the center of section 32 along the centerline of the section to the northwest corner of the Jenkins' lot. In which of the pairs of distances are contradictory values given for this dimension?

(A) 9 rods and 148.5 ft

(B) 9 rods and 149.9 ft

(C) 148.5 ft and 149.9 ft

(D) 348.5 ft and 9 rods

96.3. The appellants contended in their appeal that the original court's decree did not describe their boundary with such certainty that it could be located. What are some of the elements of that decree that might have reasonably been considered to contribute to uncertainty of location?

(A) The iron stake was described in the decree as being on the south right-of-way line of Quail Road.

(B) The iron stake was described in the decree as being approximately 100 ft east from a point on the centerline of Bisbey Road.

(C) The northwest corner of the property was described in the decree as being in Bisbey Road, approximately 348.5 ft south of the center of section 32.

(D) Both A and C are true.

96.4. Which of the following questions is pertinent in considering whether the iron stake stands at the northeast corner of the Jenkins' lot?

(A) Does the state have an easement or a fee interest in the right-of-way of Highway 89?

(B) Does the word *along* indicate the centerline of the roads in the description?

(C) Did Withers set the iron stake during his survey of the property in preparation for its conveyance from the Parsons to the Kings?

(D) All of the above are pertinent.

96.5. What is the proper interpretation of the word *along* as used in the deed description of the Jenkins' lot?

(A) Along means with the centerlines of the roads.

(B) Along means along the edge or boundary of the roads.

(C) Along most often is understood to mean along the centerline of a road; however, this is a rebuttable presumption.

(D) Along most often is understood to mean along the edge or boundary of a road; however, this is a rebuttable presumption.

SOLUTION 1

The grantor places the deed with an escrow agent to be held until all conditions of the sale contract are fulfilled.

The answer is (B).

SOLUTION 2

Eminent domain is considered to be the assumption of dominion over private property by the state. It is one of the few limitations on a fee simple absolute.

The answer is (D).

SOLUTION 3

Spoken evidence, or parol evidence, is not usually admissible in court to cure a patent ambiguity in a deed because it would violate the Statute of Frauds.

The answer is (A).

SOLUTION 4

In real property, an encumbrance is a right or interest in the land of another that diminishes its value, but not to the extent that the land cannot be sold.

The answer is (D).

SOLUTION 5

For all practical purposes, the fee simple absolute is the greatest available interest in real property.

The answer is (D).

SOLUTION 6

The only condition that is essential is the written or printed form. The others may occur, but are not necessary in all cases. For example, a valid consideration must be exchanged between the grantee and the grantor, but that consideration need not be money.

The answer is (D).

SOLUTION 7

A license, such as a privilege to enter, is nonpossessory and revocable. One of the salient differences between a license and an easement is the license's revocability.

The answer is (B).

SOLUTION 8

When no one is available to inherit the estate of a property owner who has died intestate, escheat is exercised. The property reverts to the state.

The answer is (A).

SOLUTION 9

The servient estate is subject to rights such as ingress and egress to the dominant estate. The dominant estate is served by the servient estate.

The answer is (C).

SOLUTION 10

An instrument that, despite its appearance to the contrary, conveys no title, is said to show color of title. It cannot be so defective, however, that a person of ordinary capacity would be misled by it. Color of title has an effect in claims of adverse possession in most states. It can shorten the length of the required statute of repose when it is possessed by the claimant.

The answer is (D).

SOLUTION 11

Real property is exclusively subject to the laws and jurisdiction of the state it is located in, and transactions in real property are governed by the laws of that state, *except* where title is in the name of the United States, or where it is specifically under the jurisdiction of the federal government.

State law controls the form, execution, validity, and effect of instruments relating to land in that state. For example, state laws determine inheritability. Personal property, on the other hand, is usually regarded as situated at the home of its owner regardless of its actual location; it is governed by the law of the owner's home.

The answer is (C).

SOLUTION 12

Land is composed of a bundle of rights, and commonly, those rights are divided among many persons. However, it is important to note that rights and title are not the same thing.

Real property rights include rights that are incidental to the use of land, such as easements and rights-of-way, as well as objects both under and above the surface of the earth. For example, land includes ores, metals, coal, and other minerals in, or on, the land. While located in the land, these minerals are real property and the rights to them can be divided and subdivided in various ways.

For example, the owner of the title to a tract of land might convey the land and retain personal rights to the minerals. Or, the owner could convey the rights to the minerals to another person and retain ownership of the land. However, if the owner did convey the rights to the minerals to another person, an implied right of entry might automatically be created for the mineral's extraction, even though the owner still owned the land. To avoid such an implied right of entry, the owner might specifically exclude right of entry from the land's surface, and instead convey just the rights to the minerals 500 ft or more below the ground. Doing so would mean the minerals would need to be removed from neighboring areas.

The answer is (D).

SOLUTION 13

A defeasible fee simple estate arises when title is conditional on a future event. For example, something must be done within a certain time, or alternatively, that something must not be done.

The answer is (D).

SOLUTION 14

The word laches is derived from the French *lecher*, meaning negligence. As a legal doctrine, laches states that a claim will not be enforced or allowed if a long delay in asserting it has hurt the other party. For example, if one party knew the correct property line, yet watched a neighbor complete a house over the line and then sued, it would be a legal ambush, which laches prevents.

The answer is (D).

SOLUTION 15

The word "fee" in "fee simple absolute" means an estate can be inherited or devised by a will.

The answer is (A).

SOLUTION 16

The duration of a life estate is measured by the life of some person. However, this person does not need to be the holder of the estate, as a life estate can be conveyed to a third party. It is a freehold.

The answer is (A).

SOLUTION 17

A *fee simple to a condition subsequent* is similar to a *fee simple determinable.* However, with a fee simple to a condition subsequent, the reversion to the grantor when some event does or does not happen is not automatic. With a fee simple determinable, the reversion is automatic.

The answer is (D).

SOLUTION 18

The quitclaim deed acts like a release. It conveys any interest the grantor may have, though the grantor may, in fact, have none at all.

The answer is (A).

SOLUTION 19

A deed is evidence of, but not proof of, ownership. A deed is only one of several legal methods by which title to real property may be conveyed. It need not be signed by the grantee or the county clerk, but it must be signed by the grantor to be valid. A deed must have sufficient and legal words.

The answer is (D).

SOLUTION 20

Occasionally, courts hold an ambiguous deed against the grantee—in cases involving oil leases, for example. Generally, the grantor is presumed to have been responsible for the wording of the instrument. It is therefore considered equitable to hold any patent ambiguity against the grantor.

The answer is (A).

SOLUTION 21

Unlike a patent ambiguity, which appears on the face of a deed, a latent ambiguity appears only in the light of information outside the deed itself. A court will generally allow a latent ambiguity to be cured by testimony, unlike a patent ambiguity.

The answer is (D).

SOLUTION 22

An adjoining property, whether marked by physical monuments or not, becomes a record monument when called for in a description of property. It is important to note that the entire adjoining property, and not merely the coincident line, constitutes the record monument.

The answer is (C).

SOLUTION 23

The four corners rule has supplanted, to a large degree, the repugnancy doctrine. The four corners rule relies upon the entire instrument for the construction of its meaning. Formerly, some jurisdictions relied upon the idea that earlier language in a deed could not be contradicted by later language in the same instrument.

The answer is (C).

SOLUTION 24

If the language of a deed is unambiguous under the rules of construction brought to bear by the court, evidence of a contrary intent is often considered immaterial. The intent that prevails is the intent expressed in the written language of the deed.

The answer is (A).

SOLUTION 25

A reversion is the residue of an estate that remains with the grantor. This type of future interest usually relies on the occurrence or nonoccurrence of some specified event. For example, a life estate may revert to the grantor upon the death of the grantee.

The answer is (D).

SOLUTION 26

Unlike tenants in common, joint tenants do not hold an undivided interest in the property. They cannot devise their part to their heirs; rather, upon the death of a joint tenant, the survivors acquire the interest.

The answer is (D).

SOLUTION 27

Under a general warranty deed, the grantor is liable for any defects in the conveyed title, no matter when they occurred, and must be made good on the title if it is found lacking.

The answer is (C).

SOLUTION 28

It is a well established principle that a surveyor who is retracing the lines of a previously surveyed deed should strive to follow the footsteps of the previous surveyor.

The answer is (B).

SOLUTION 29

A deed is an instrument in writing, which when executed and delivered, conveys an estate in real property. It is the instrument by which the title to real property is transferred from one person (the grantor) to another person (the grantee). There are various categories of deeds.

The answer is (D).

SOLUTION 30

The *habendum* clause in a deed begins with "To have and to hold."

The answer is (A).

SOLUTION 31

Sequential conveyances most often arise because there is a period of time between successive conveyances. If there happens to be an overlap, the senior deed receives all that is described, and the second conveyance receives the remainder.

The answer is (B).

SOLUTION 32

In a deed, a reservation is some interest that would otherwise be conveyed, but is taken back by the grantor. It creates an encumbrance and a right, or privilege, for the grantor in the land described.

The answer is (B).

SOLUTION 33

None of the doctrines mentioned is a violation of the Statute of Frauds. The practical location of boundaries is not seen as changing established property lines by unwritten means, but rather as a clarification of their location when it is unknown to the parties involved, in most cases.

The answer is (D).

SOLUTION 34

Adverse possession is possession that is hostile to the interests of the record owner. Permission given by the owner to Mr. Zambezi not only calls into question the hostility of the latter's claim, but also shows that Mr. Zambezi recognizes Mr. Wannamaker's proprietorship.

The answer is (D).

SOLUTION 35

Choice (B) is definitely untrue because government and quasi-government entities are generally held immune to any claims by unwritten means. However, rights in property acquired by deed are inferior to successful claims through unwritten means. Registration in the Torrens title system is one way a property owner can, by deed, protect his property from future claims by such means.

The answer is (D).

SOLUTION 36

Hostility in cases of adverse possession is generally manifested by such things as the erection of fences. The intent to claim and defend ownership against all others constitutes hostility in such circumstances.

The answer is (C).

SOLUTION 37

While it frequently has a larger meaning, prescription is the term most often applied to nonownership rights. Adverse possession, however, applies to the title to lands.

The answer is (D).

SOLUTION 38

The premise of open and notorious possession is that the record owner should know that the property is being occupied by others. Insanity and being underage are recognized as disabilities (along with conviction to prison, in most jurisdictions).

The answer is (D).

SOLUTION 39

If a person becomes injured by relying on a boundary because of the assurances of someone he had reason to believe spoke with authority, the doctrine of estoppel may become involved. The relied-upon person may be estopped from any gain he enjoyed if he had knowledge of the truth and did not reveal it.

The answer is (B).

SOLUTION 40

Recognition and acquiescence is not applicable since the period of repose, 20 years by common law, has not run. A parol agreement may have been possible before the row of stakes was set, but with the stakes in place, the owners cannot agree without knowledge of the deed line's location. An exchange of quitclaim deeds is still an option.

The answer is (D).

SOLUTION 41

Tacking, the adding together the time of occupation of more than one claimant, is sometimes permitted in order to complete the statutory period.

The answer is (D).

SOLUTION 42

While color of title may shorten the length of the statutory period, it is not generally held as a requirement for a successful claim of adverse possession.

The answer is (D).

SOLUTION 43

A right-of-way is not extinguished by mere nonuse. The town must officially declare it abandoned or discontinued for the right-of-way to be truly extinguished.

The answer is (C).

SOLUTION 44

An easement appurtenant is attached to the land. An easement in gross, on the other hand, is attached to the person or corporation that holds it.

The answer is (D).

SOLUTION 45

All the choices except (D) will extinguish an easement because a written easement is not a revocable right in property.

The answer is (D).

SOLUTION 46

This type of unwritten easement most frequently arises when a grantor sells a portion of his property without mentioning an easement that is clearly required for the reasonable use of the parcel sold. Even though the deed makes no reference to the easement, it is presumed to exist by implication.

The answer is (B).

SOLUTION 47

When more than one interpretation of the language in a deed is possible, the one that gives greatest advantage to the grantee will prevail. The grantor is presumed to have been responsible for the language of the deed and is therefore subject to reasonable consequences of that language.

The answer is (B).

SOLUTION 48

A negative easement has the effect of preventing an owner of land from doing something that he or she would otherwise be entitled to do.

The answer is (A).

SOLUTION 49

Generally, a statutory dedication is accomplished with the filing of a map that shows the areas dedicated for public use. A common-law dedication requires no map or writing and grows from the intention of the owner, indicated by words or acts, to dedicate the land for public use. Both types of dedications may be formally accepted. Common-law dedications generally convey an easement right to the public; statutory dedications may or may not convey a fee.

The answer is (A).

SOLUTION 50

All the statements may produce easement rights except the one regarding adverse possession. Adverse possession, where successfully claimed, produces a fee simple absolute, not an easement, in most jurisdictions.

The answer is (D).

SOLUTION 51

The original subdivider of Valhalla Acres had no interest in the land beyond the boundary of the subdivision. Since his title is through that subdivider, Mr. Parker has no reversionary right beyond that boundary once the street is vacated.

The answer is (B).

SOLUTION 52

An easement is a nonpossessory interest in the land of another, and as such cannot be considered a fee interest in land.

The answer is (A).

SOLUTION 53

An easement is a non-possessory right to make a specific use of a definite portion of land owned by another. The owner of the underlying ground has the right to also make use of the land to the extent that the use does not interfere with the exercise of the easement granted. The holder of the easement cannot use the land for purposes other than intended by the easement.

The answer is (A).

SOLUTION 54

Proportionate measurement is a last resort in the restoration of lost monumentation where no parcel has priority over any of the others under consideration.

The answer is (D).

SOLUTION 55

Proportionment is not applicable to sequential conveyances. It should not be used to distribute excesses or deficiencies that are attributable to mistakes, and should always be a rule of last resort.

The answer is (D).

SOLUTION 56

The weight of most common law supports the idea that proportionate measurement should not be used to determine the width of rights-of-way. Infrequently, index corrections are used in reestablishing streets, but these are exceptions to the rule.

The answer is (C).

SOLUTION 57

57.1. Proportionate measurement should not be used to distribute the effect of a mistake. The 12.00 ft shown on the plat as the frontage of lot 11 is clearly a mistake on the part of the draftsman. The intended dimension was 21.00 ft. Proportionment of the frontage indicates that each of the lots should receive 21.17 ft.

$$\frac{21.00\text{ ft}}{63.00\text{ ft}} = \frac{x}{63.50\text{ ft}}$$

$$1333.50\text{ ft}^2 = (63.00\text{ ft})x$$

$$x = \frac{1333.50\text{ ft}^2}{63.00\text{ ft}} = 21.17\text{ ft}$$

The answer is (D).

Land Boundary Law

57.2. Single proportionate measurement is applicable to this situation.

$$\frac{17.38 \text{ ft}}{43.46 \text{ ft}} = \frac{x}{44.01 \text{ ft}}$$
$$764.89 \text{ ft}^2 = (43.46 \text{ ft})x$$
$$x = \frac{764.89 \text{ ft}^2}{43.46 \text{ ft}} = 17.60 \text{ ft}$$

The answer is (D).

SOLUTION 58

58.1. Proportionate measurement has no place in the solution of this situation. Senior and junior rights are the deciding factors. Mr. Clark's property is senior to Mr. Brown's. If no lines of possession are involved, Mr. Clark would receive the east 325 ft. Mr. Brown's deed calls for Mr. Clark's property as an adjoiner, so the weight of court opinion would hold that Mr. Brown would acquire the remaining 330.00 ft of the Fairview Tract.

The answer is (D).

58.2. Mr. Clark has senior rights and receives the east 325 ft. Mr. Brown receives the remaining 320.00 ft. Proportionate measure is not applicable in such circumstances.

The answer is (A).

58.3. Proportionate measurement is appropriate in this situation. The critical factor is the simultaneous execution of the deeds to Mr. Brown and Mr. Clark. They each will receive half of the measured frontage: 329.12 ft.

The answer is (C).

SOLUTION 59

59.1. Since the lots in Valhalla Estates were created simultaneously, there are no senior rights involved between the two lots. The discrepancy between the plat distances and the found original monuments may be resolved by single proportioning. The measurement between the SW corner of lot 1 and the NW corner of lot 2 is 97.90 ft; the record indicates 100.00 ft for the same line.

$$\frac{50.00 \text{ ft}}{100.00 \text{ ft}} = \frac{x}{97.90 \text{ ft}}$$
$$4895.00 \text{ ft}^2 = (100.00 \text{ ft})x$$
$$x = 48.95 \text{ ft}$$

The length of the west line of lot 2 is 48.95 ft.

The single proportionment of the measurement along the west line of lots 3 and 4 can be proportioned in the same way.

$$\frac{50.00 \text{ ft}}{100.00 \text{ ft}} = \frac{x}{101.20 \text{ ft}}$$
$$5060.00 \text{ ft}^2 = (100.00 \text{ ft})x$$
$$x = 50.60 \text{ ft}$$

The answer is (D).

59.2. Any excess or deficiency of a new measurement when compared to a record measurement should not be distributed beyond a found verified original monument. Despite evidence that a subsequent surveyor set a NE corner of block 2 beyond the original monumentation, the proportionment should not be influenced by the new monumentation. Therefore, the distance measured in block 2 is

$$\frac{190.50 \text{ ft}}{2} = 95.25 \text{ ft}$$

The answer is (A).

59.3. First, establish the length of the complete western line of block 2. The next original found monument available is at the SW corner of lot 2, block 4. However, there is a fence that has been accepted by the adjoining lot owners as their northern boundary. If the intersection of the fence with the western boundary of block 4 is accepted as its NW corner, then the length of block 2 and the intervening street is

$$\begin{array}{r} 465.00 \text{ ft} \\ -\,193.20 \text{ ft} \\ \hline 271.80 \text{ ft} \end{array}$$

The dedicated right-of-way is recorded as 60.00 ft and should remain that wide. Proportionate measurement should not be used to change the width of a street.

$$\begin{array}{r} 271.80 \text{ ft} \\ -\,60.00 \text{ ft} \\ \hline 211.80 \text{ ft} \end{array}$$

Therefore, the west line of block 2 is 211.80 ft long. The record indicates that the four lots will share that frontage equally.

$$\frac{211.80 \text{ ft}}{4} = 52.95 \text{ ft}$$

The answer is (C).

59.4. The measured distance from the SE corner of block 3 to the SE corner of block 2 is 291.50 ft. The plat indicates that this distance is 285.00 ft. However, the street must receive its full 60.00 ft before any excess of the measurement is distributed among the lot frontages.

$$\begin{array}{r} 291.50 \text{ ft} \\ -60.00 \text{ ft} \\ \hline 231.50 \text{ ft} \end{array}$$

$$\begin{aligned} \frac{25.00 \text{ ft}}{225.00 \text{ ft}} &= \frac{x}{231.50 \text{ ft}} \\ 5787.50 \text{ ft}^2 &= (225.00 \text{ ft})x \\ x &= \frac{5787.50 \text{ ft}^2}{225.00 \text{ ft}} \\ &= 25.72 \text{ ft} \end{aligned}$$

The answer is (C).

59.5. When a lot is shown without dimensions on a subdivision plat, particularly at the end of a block or adjoining the subdivision boundary, any excess or deficiency is frequently given to that lot. The presumption is that the original proprietor intended that the irregular lot absorb the discrepancies of the measurements.

The answer is (A).

SOLUTION 60

Reliction is a slow and imperceptible recession of water and the attendant uncovering of the land once submerged. The recession of the water must be permanent to be considered reliction.

The answer is (B).

SOLUTION 61

Along streams and rivers, abutting owners are entitled to riparian rights, which originated in common law. The precise nature of these rights varies from jurisdiction to jurisdiction, but is generally related to the use of the water, as well as the banks and bed, of a stream.

The answer is (D).

SOLUTION 62

Avulsion is a sudden, perceptible loss or addition of land caused by the action of water. An example of avulsion is when a river suddenly breaks its banks and forms a new channel, thereby cutting off a large portion of land.

The answer is (D).

SOLUTION 63

Accretion is the gradual and imperceptible process of building up new land by the action of water.

The answer is (D).

SOLUTION 64

The action that causes the building of land along a water boundary is known as *accretion*. The land built by such action is known as *alluvium*.

The answer is (D).

SOLUTION 65

The line connecting the lowest part of a valley and the deepest part of a river is known as the *thalweg*. The term is also used to denote the middle of the main navigable channel.

The answer is (D).

SOLUTION 66

The meander lines established by public land surveys were intended to follow the mean high water line and were not intended to establish the legal boundaries of the abutting upland owners.

The answer is (B).

SOLUTION 67

Littoral is the word most often used to describe rights associated with water that is not flowing. The word usually pertains to the seashore, but is also used regarding lakes and ponds.

The answer is (D).

SOLUTION 68

Even if the state is considered to own the bed of a stream, in most jurisdictions the riparian owner may nevertheless acquire land by accretion. The alluvium that is gradually deposited by the natural action of water is gained by the owner of the land to which it attaches.

The answer is (A).

SOLUTION 69

A riparian owner may lose title to land that gradually becomes eroded or permanently inundated with water. If the change is sudden, however, the title will not change.

The answer is (B).

SOLUTION 70

In the majority of jurisdictions the riparian owner holds the relicted land to the new mean high water line. He does not hold to the old or a new meander line. Meander lines are not intended to be property boundaries.

The answer is (D).

SOLUTION 71

A full 18.6 years is required for the relative positions of the sun, moon, and earth to be repeated. This tidal cycle is often rounded off to 19 years. To obtain the correct mean high tide at a given station, that station should be monitored for the full period. However, such monitoring is normally not practical.

The answer is (D).

SOLUTION 72

A diurnal tide is one that recurs daily. Along the Atlantic coast, the two daily high tides and two daily low tides are much alike and are called a *semidiurnal tide.*

The answer is (C).

SOLUTION 73

The expression of opinion is generally more restricted in testimony that is not considered expert.

The answer is (B).

SOLUTION 74

It would not be useful to attempt to resolve a patent ambiguity in the language of a deed with testimony. Such a contradiction cannot normally be cured by testimony.

The answer is (D).

SOLUTION 75

All of the aspects of the fee listed are entirely proper. It would be improper for the amount of the surveyor's fee to depend on the final decision of the court. The attorney performing the cross-examination has the right to ask the surveyor about the fee.

The answer is (D).

SOLUTION 76

A decision on the central question of the case is the exclusive purview of the court, not the expert witness.

The answer is (D).

SOLUTION 77

In order to establish the credibility of the testimony, the initial questions commonly pertain to the surveyor's experience and qualifications.

The answer is (D).

SOLUTION 78

Generally, an unwritten right of occupancy that has duly ripened into a legal right is considered superior to all others, including senior rights and writings. However, it is difficult for a right of occupancy to so ripen.

The answer is (D).

SOLUTION 79

The junior deed will generally yield to the senior deed where there are no unwritten rights of possession. The seniority of one deed over another will overcome all other conflicts between the two deeds in most situations.

The answer is (D).

SOLUTION 80

The area described in a deed is usually considered the least specific element of the title and therefore the least locative.

The answer is (D).

SOLUTION 81

A monument of common report is one that is supported by hearsay evidence. The acceptance of such a monument can only be considered reasonable when the monument cannot be conclusively proved or disproved by any other means. In other words, it should be a rule of last resort.

The answer is (A).

SOLUTION 82

There is no consistent rule from jurisdiction to jurisdiction concerning whether distance should yield to direction or vice versa. In sectionalized lands, distances have generally been held as superior to direction, but in Texas and Kentucky, the opposite has generally been the case.

The answer is (C).

SOLUTION 83

The 1 in iron pipe monuments should be held as controlling if the parties to the deed acted with specific reference to the plat. If the deed mentions the plat in the writings, the plat becomes part of the deed, and the monuments may then be said to have been called for. If they are undisturbed, called for, and identifiable, they are superior to the informative bearings and distances.

The answer is (D).

SOLUTION 84

While common law does tend to support a particular ranking of conflicting title elements, such an order of importance should not be followed slavishly. Extrinsic evidence to the contrary should be considered. The written elements of a deed are important, but properly ripened unwritten rights may defeat them.

The answer is (A).

SOLUTION 85

The senior right of the adjoining deed should not be violated. The monument called for in the senior deed is most likely the correct position.

The answer is (A).

SOLUTION 86

The formation of a condominium is typically governed by specific state statutes that provide for individual ownership of units within a multiple-unit building. Usually, unit owners have a right to use common elements of the condominium's property, such as driveways, parking areas, mailboxes, and so on.

The answer is (A).

SOLUTION 87

The owner of a condominium unit has fee simple ownership of the unit, but a cooperative resident does not. Cooperative residents own shares in a corporation but do not own the real estate in which they live. Since a condominium unit is owned by the resident, the condominium unit appears on tax rolls as a separate entity. The property on which all the units of a cooperative are situated appears on tax rolls as a single property where the cooperative corporation is responsible for the taxes.

The answer is (B).

SOLUTION 88

In a condominium, there are general common elements and limited common elements. General common elements are available to all condominium unit owners. Limited common elements are exclusively reserved for the use of one or more, but not all, condominium unit owners.

The answer is (B).

SOLUTION 89

89.1. Regarding the deeds, Lenore was placed in the position of trustee by her mother. An administrator is appointed by a court to handle the assets and liabilities of a decedent.

The answer is (C).

89.2. Fred Hale brought the action before the court. He was the plaintiff.

The answer is (C).

89.3. In general, courts provide closure for descriptions that are deficit in that regard. There has been a strong desire to construe deeds so that they will be given effect and pass title, rather than be declared void. However, a description cannot be so vague that the parcel cannot be located at all.

The answer is (D).

89.4. The two deeds were created simultaneously by Alice Pitt. Neither is senior to the other; they both have equal standing.

The answer is (D).

89.5. The court would most likely hold that Hale and Edward Pitt hold title to the overlap as tenants in common. Each of them has an equal claim to the property and there is no priority of one deed over the other.

The answer is (B).

SOLUTION 90

90.1. Rachel Garner's deed described a remainder after previous tracts were excepted. The width of the parcel was mentioned as 80 ft, but that call is merely informative as it is dependent on senior conveyances. However, the actual width of the parcel is not a central question in the resolution of the boundary dispute between the estate of Mary Garner and the Zimmers.

The answer is (D).

90.2. The language in Rachel Garner's deed describing the 150 ft square implies that the eastern boundary of that square is parallel with the western boundary of outlot 8. If the section line is cardinal, the western boundary of Mary Garner's lot cannot also be cardinal since the city surveyor testified the line and the boundary were not parallel due to convergence.

The answer is (B).

90.3. The boundary in question was created in Rachel Garner's deed to the Zimmers, which calls for 37 ft off the east side. The implication is that the lines bounding the 37 ft on the east and west are parallel.

The answer is (B).

90.4. The use of proportionate measurement is not appropriate to sequential conveyances of the type described in this case.

The answer is (D).

90.5. M. Thomas might wish to cite the doctrine of estoppel. The Zimmers may have influenced Mary Garner to move her fence to her detriment by alleging the boundary that had stood for 25 years was incorrect. Their claim of practical location might then be estopped.

The answer is (D).

90.6. When two parties derive their titles to adjoining tracts from a common grantor and it is revealed that there is a shortage, the party claiming under the grantor's first deed need not contribute to the shortage. The Zimmers hold the first deed executed by Rachel Garner and are entitled to the full 37 ft stipulated in their deed. The estate of Mary Garner receives the remainder.

The answer is (B).

SOLUTION 91

91.1. A mutual mistake occurs when both parties to a deed share the same misconception. When the deed in its written form does not reflect the intentions of the parties to the deed and the two parties are in agreement as to what it should have said, there is a mutual mistake.

The answer is (D).

91.2. A disability in legal terms is an incapacity such as insanity, being underage, or a felony conviction that prevents the enjoyment of normal rights. An inability to read is not such a legal disability.

The answer is (B).

91.3. The payment of taxes is often cited in cases where the question of adverse possession is raised. Mr. Gould may have contemplated such a claim based on the location of his trees and driveway.

The answer is (C).

91.4. Monuments called for in the deed and positively identified are controlling; the distances cited are considered merely descriptive.

The answer is (B).

91.5. The intention of the grantor as expressed in the deed's language is paramount; an unambiguous description cannot be changed by subsequent testimony.

The answer is (D).

SOLUTION 92

92.1. The word *mesne* is used to describe intermediate or intervening profits, assignments, and other processes. In this problem, the term *mesne conveyances* means that intervening deeds were executed between P. Jefferson's ownership and Calhoun's ownership of the lot.

The answer is (B).

92.2. The more specific description will most often prevail. The measurements of 70 yd and 215 ft may be considered of comparable precision, but the latter occurs in a section of the description that begins, "more particularly described as follows ..." However, as long as the location of Peter's lot can be ascertained, neither measurement is critical.

The answer is (D).

92.3. The Calhoun deed was issued before the Merrill deed and they both call for the boundary of the A. Brown lot. However, this fact by itself does not settle the question of the location of the eastern boundary of the Calhoun property. A call for an adjoiner does not always imply that the record monument will prevail.

The answer is (B).

92.4. It is clear from the conveyances from P. Jefferson to Calhoun and Merrill that Mr. Jefferson or his predecessors believed the block from Seventh to Eighth Street was 400 ft long. Should those streets have been widened and that 400 ft thereby reduced to 379.1 ft, the apparent overlap may be explained by Calhoun's and Merrill's reliance on the distances recited in their deeds. However, the boundary shared by Calhoun and Merrill cannot have been moved by such street widening. If the overlap did not exist before the streets were widened, it cannot exist after such a change. The distances in both deeds are informative only since their common boundary is described as the west boundary of the lot formerly owned by A. Brown.

The answer is (C).

92.5. The critical consideration here is the position of the west boundary of the A. Brown lot. The parcels now owned by Calhoun and Merrill existed as separate entities long before P. Jefferson owned them. When

Mr. Jefferson conveyed them to the present proprietors, he described their common boundary as the west line of the A. Brown lot. The best evidence of the position of that line is the old barn foundation and the remains of the old fence.

The answer is (B).

SOLUTION 93

93.1. The Footes hold the surface estate as joint tenants. As such, the death of one of them would cause that person's interest to automatically become vested in the other person.

The answer is (A).

93.2. The mineral estate is conveyed to the grantee unless it is specifically excepted.

The answer is (C).

93.3. The last clause expresses the intention of the grantors to convey one-half of the oil and gas and the entire surface estate to the grantees. This has already been accomplished by the granting and reservation clauses. The absence of any mention of the mineral estate in the last clause does not prevent the estate from passing to the grantee.

The answer is (A).

SOLUTION 94

94.1. Claim of right rests on the intention of the claimant in a matter of prescription. An example of this is the Hunts' belief that they had a full right to use the driveway that extended onto the Webbs' property as their own.

The answer is (D).

94.2. The Hunts did not need to fence the property involved to satisfy the requirement of open and notorious use. They did not intend to acquire a fee interest in the 6 ft strip of land, but merely an easement right across the land. However, they were required to formally notify the Webbs in the state where the action was filed. While open and notorious use may be sufficient for constructive notice in many states, here constructive notice alone was not adequate.

The answer is (D).

94.3. Both an implied easement and a way of necessity require a unity of title of the two parcels involved before their separation and the subsequent creation of the easement. A common-law dedication involves appropriation for use by the public. These conditions did not exist in this case. However, an easement by estoppel may have had some application in this case. The Hunts believed that the servitude imposed on the Webbs' property (the driveway) was a permanent condition. The Webbs did not object and allowed the Hunts to use the driveway for many years, as did their predecessors in title. The Hunts may have been able to claim that an easement by estoppel existed.

The answer is (C).

94.4. The Hunts did not formally inform the Webbs of their intention to possess the easement right across the driveway until five years after the Webbs constructed their fence. The testimony of the Hunts' predecessors verified the absence of any formal notification between the neighbors for 38 years. Under the law of the state where the action was filed, the statutory period did not run at all.

The answer is (D).

94.5. The burden of proof of a prescriptive easement was on the Hunts. Mrs. Hunt merely told the Webbs through her lawyer that she did not approve of their fence, not that the Hunts intended to claim an easement right. The payment of taxes is of no significance in an action to acquire an incorporeal right in property.

The answer is (D).

SOLUTION 95

95.1. The right of eminent domain may be exercised in the public interest; however, in this case, the acquisition of the road by other means is not barred by that fact. Some states do bar governmental entities with the power of eminent domain from acquiring property by adverse possession.

The answer is (D).

95.2. In most jurisdictions (but not all), the payment of taxes and color of title are not absolute requirements in claims of adverse possession, but they are elements that the court may consider.

The answer is (A).

95.3. All of the reasons given would point to the acquisition of an easement right by the municipality rather than a fee interest.

The answer is (D).

95.4. Aside from the difficulty of establishing the ownership of accretion land, title insurance companies consistently except such land from their policies because the history of the ownership of such virgin land is usually not yet a matter of record.

The answer is (B).

95.5. If Mr. Clark had successfully shown that the roadway took different routes at different times during the running of the statute, the municipality would likely have lost its case. The location of a roadway must remain consistent for the prescriptive right to ripen.

The answer is (A).

SOLUTION 96

96.1. Hearsay evidence is a statement that is made by someone other than the person testifying, but is recounted by a witness as evidence to prove the truth of an assertion. In the absence of a recorded survey, Mrs. Parson's account of the placement of the iron stake could be considered hearsay since she was in the hospital at the time.

The answer is (A).

96.2. The distance of 9 rods, or 148.5 ft, is given in the deed description of the property and is contradicted by the distance of 348.5 ft stipulated in the original court's decree. The distance of 149.9 ft was not measured along the centerline of section 32.

The answer is (D).

96.3. The iron stake was not found to be on the south right-of-way line of Quail Road, but 3.1 ft north of it. The distance of 348.5 ft is more nearly the distance from the center of section 32 to the southwest corner of the Jenkins' lot. However, the iron stake can be said to be approximately 100 ft east of the centerline of Bisbey Road.

The answer is (D).

96.4. The questions each raise significant aspects of the location of the northeast corner of the Jenkins' property. If the state holds a fee interest in Highway 89, the Parsons could not sell what they did not own. The location of the northeast corner is dependent on whether the western boundary of the lot is along the centerline of Bisbey Road. The placement of the iron stake by Withers as the basis for the conveyance would give it great significance, even though the stake is not called for in the deed description.

The answer is (D).

96.5. The usual meaning of the word *along*, as used in deed descriptions, carries the boundary to the centerline of the strip of land. However, the context cannot be ignored. Here, the evidence of the iron stake and the fence, which was reconstructed as a rock wall along the eastern boundary of the Jenkins' lot, indicates a different meaning of the word.

The phrase "thence west along the right-of-way of said Quail Road 100 ft to the place of beginning" means that the boundary is coincident with the southern right-of-way line of Quail Road, not its centerline. This meaning is supported by Withers' iron stake and the fact that the place of beginning is on the southern right-of-way line of Quail Road.

On the other hand, the phrase "thence south along Bisbey Road 200 ft" means that the boundary is the centerline of Bisbey Road. This meaning is supported by the fence at the eastern boundary of the Jenkins' lot that was rebuilt as a rock wall and is further supported by the iron stake set by Withers.

The answer is (C).

Surveying Astronomy

ASTRONOMICAL COORDINATE SYSTEMS

PROBLEM 1

The spherical triangle known as the PZS triangle has sides that are segments of three great circles. Which two of those great circles intersect at the star?

(A) the prime vertical and the celestial equator

(B) the hour circle and the horizon

(C) the equinoctial colure and the celestial equator

(D) the hour circle and the vertical circle

PROBLEM 2

Which astronomical coordinates are constantly changing and can readily be measured by a theodolite?

(A) right ascension and declination

(B) hour angle and declination

(C) hour angle and parallactic angle

(D) altitude and azimuth

PROBLEM 3

The angle between an observer's zenith and the celestial equator will always be equal to which of the following?

(A) 90°

(B) the observer's longitude

(C) the right ascension

(D) the observer's latitude

PROBLEM 4

One sidereal hour after a star has set exactly in the west, the First Point of Aries stands precisely at upper transit on an observer's meridian. At that instant, what is the right ascension of the star?

(A) 105°30′

(B) 180°00′

(C) 255°00′

(D) 330°15′

PROBLEM 5

If your latitude was 39°47′30″N, what would be the altitude of the sun at apparent noon on the day of the summer solstice?

(A) $23°26\frac{1}{2}'$

(B) $39°47\frac{1}{2}'$

(C) $50°17\frac{1}{2}'$

(D) 73°39′

PROBLEM 6

What is the latitude of the place where the maximum altitude of the sun above the horizon achieved at any time of the year is $23°26\frac{1}{2}'$?

(A) 23°26′30″N

(B) 90°00′00″N

(C) 90°00′00″S

(D) both B and C

PROBLEM 7

What is the declination of a star if that star passes through your zenith where an observer's latitude is 35°N?

(A) 35°N

(B) 35°S

(C) 55°N

(D) 90°N

PROBLEM 8

Standing at 40°N latitude, you observe the sun rise precisely in the east. What is the declination of the sun and what is the approximate date?

(A) declination 0°, June 22 or Dec. 22

(B) declination 0°, Mar. 21 or Sept. 23

(C) declination $23°26\frac{1}{2}'$, June 22 or Dec. 22

(D) declination $23°26\frac{1}{2}'$, Mar. 21 or Sept. 23

PROBLEM 9

Concerning the PZS triangle, colatitude is to latitude as polar distance is to which of the following?

(A) declination

(B) altitude

(C) hour angle

(D) zenith distance

PROBLEM 10

Which value corresponds to the sum of a star's hour angle and its right ascension?

(A) the azimuth of the star

(B) the longitude of the observer

(C) the zenith distance of the star

(D) the sidereal time at the observer's position

GEOGRAPHICAL COORDINATES AND THEIR ELEMENTS

PROBLEM 11

Which of the following terms describes a great circle on which every point is equidistant from the north and the south pole?

(A) the prime meridian

(B) the hour circle

(C) the Tropic of Capricorn

(D) the terrestrial equator

PROBLEM 12

Which of the following statements correctly describes properties of the geographical coordinate known as *latitude*?

(A) Latitude is measured in degrees, minutes, and seconds north and south of the equator.

(B) Every position on the earth has a unique latitude, which is unlike the latitude of any other position on the earth.

(C) A parallel of latitude is a great circle on the surface of the earth.

(D) All of the above are true.

PROBLEM 13

A second of latitude near the equator bears what relationship to a second of latitude near one of the poles?

(A) A second of latitude near the poles is approximately a foot longer than a second of latitude near the equator.

(B) A second of latitude has the same length in feet regardless of its distance from the equator.

(C) A second of latitude near the poles is approximately a foot shorter than a second of latitude near the equator.

(D) Each second of latitude north of the equator grows progressively shorter until finally it reaches a point at the pole.

PROBLEM 14

Which of the following statements correctly describes properties of the geographical coordinate known as *longitude*?

(A) Each meridian of longitude lies in the plane of a great circle.

(B) The distance along a parallel of latitude through a degree of longitude grows smaller as the latitude approaches 90°.

(C) 15° of longitude equals one mean solar hour.

(D) All of the above are true.

PROBLEM 15

The longitude of a place is 100°15′32″. What would this longitude be if it were expressed in mean solar hours, minutes, and seconds from Greenwich?

(A) 6 hr 01 min 02 sec

(B) 6 hr 40 min 37 sec

(C) 6 hr 41 min 02 sec

(D) 6 hr 55 min 45 sec

PROBLEM 16

Which of the following Greek letters is the standard symbol for longitude?

(A) β

(B) ς

(C) ϕ

(D) λ

PROBLEM 17

Which of the following Greek letters is the standard symbol for latitude?

(A) θ

(B) ϕ

(C) ψ

(D) μ

THE PZS TRIANGLE

PROBLEM 18

Which of the following correctly identifies the points of intersection of the three great circles that form the spherical triangle called the *PZS triangle*?

(A) P: parallactic
Z: zone
S: south

(B) P: pole
Z: zero
S: station

(C) P: prime meridian
Z: zenith
S: Sirius

(D) P: pole
Z: zenith
S: star

PROBLEM 19

The angle between the great circles that form two of the sides of the PZS triangle is the azimuth of the star being observed. Which are those two circles?

(A) the prime vertical and the hour circle through the star

(B) the observer's meridian and the hour circle through the star

(C) the vertical circle through the star and the observer's meridian

(D) the vertical circle through the star and its hour circle

PROBLEM 20

Which of the following arcs of the great circles, both complements of the sides of the PZS triangle, are used to find the azimuth angle of a star using the hour angle method?

(A) latitude and declination

(B) zenith distance and declination

(C) coaltitude and latitude

(D) polar distance and zenith distance

PROBLEM 21

Which element of the PZS triangle is the least used in surveying astronomy?

(A) zenith angle

(B) hour angle

(C) parallactic angle

(D) colatitude

PROBLEM 22

When a star is at upper transit on the observer's meridian, which statement concerning its PZS triangle is correct?

(A) The hour angle is 90°.

(B) The hour circle is coincident with the celestial equator.

(C) The colatitude is equal to the coaltitude.

(D) It is coincident with the observer's meridian and is a straight line.

THE EARTH'S ORBIT AND THE ECLIPTIC

PROBLEM 23

Which of the following defines the ecliptic?

(A) the sun's apparent yearly path

(B) the intersection of the earth's orbital plane with the celestial sphere

(C) the name of the point in the earth's orbit where it is farthest from the sun

(D) both A and B

PROBLEM 24

What are the names given to the two points of intersection between the celestial equator and the ecliptic?

(A) summer solstice and the winter solstice

(B) vernal equinox and the autumnal equinox

(C) solstitial points

(D) Tropic of Cancer and the Tropic of Capricorn

PROBLEM 25

Which of the following symbols is used to indicate the vernal equinox?

(A) ♈

(B) ζ

(C) Ψ

(D) α

PROBLEM 26

Which of the following angles most closely indicates the value known as the obliquity of the ecliptic?

(A) 16°56′45″

(B) 23°26′30″

(C) 59°43′12″

(D) 74°12′30″

PROBLEM 27

Moving along its yearly path, the sun crosses the celestial equator in an apparent motion from north to south in September. This position of the sun has more than one name. Which of the following names apply to the position?

(A) vernal equinox and the First Point of Aries

(B) summer solstice and the First Point of Taurus

(C) autumnal equinox and the First Point of Libra

(D) winter solstice and the First Point of Scorpio

PROBLEM 28

The orbit of the earth around the sun follows a path that quite closely resembles the shape of what geometric figure defined in Kepler's laws of planetary motion?

(A) cycloid

(B) hyperbolic paraboloid

(C) cardioid

(D) ellipse

PROBLEM 29

A planet's positions at its closest approach and farthest distance from the sun have been given what names?

(A) apogee: the position closest to the sun
perigee: the position farthest from the sun

(B) aphelion: the position farthest from the sun
perihelion: the position closest to the sun

(C) apogee: the position farthest from the sun
perigee: the position closest to the sun

(D) aphelion: the position closest to the sun
perihelion: the position farthest from the sun

PROBLEM 30

When is the earth nearest to and farthest from the sun?

(A) The earth is nearest to the sun about Jan. 4 of each year.

(B) The earth is farthest from the sun about July 6 of each year.

(C) The earth is nearest to the sun about Mar. 21 of each year.

(D) Both A and B are true.

PROBLEM 31

Approximately how many months separate the summer solstice from the autumnal equinox?

(A) 1 mo

(B) 3 mo

(C) 6 mo

(D) 7 mo

PROBLEM 32

The following three problems refer to the illustration of the yearly path of the sun along the ecliptic. Note that the position labeled 1 is the First Point of Aries.

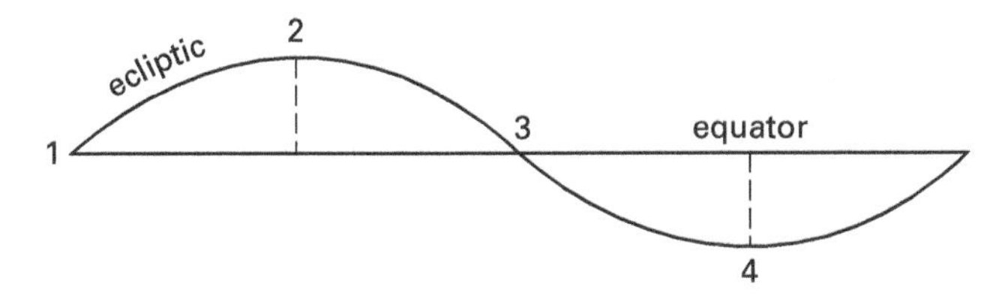

32.1. Which of the labeled positions would be found on the Tropic of Cancer and the Tropic of Capricorn, respectively?

(A) Tropic of Cancer: 2
Tropic of Capricorn: 4

(B) Tropic of Cancer: 4
Tropic of Capricorn: 2

(C) Tropic of Cancer: 1
Tropic of Capricorn: 3

(D) Tropic of Cancer: 2
Tropic of Capricorn: 3

32.2. What angular values of right ascension and declination would be correct for the sun at the moment of the autumnal equinox?

(A) $\alpha = 0°$
$\delta = 90°$

(B) $\alpha = 0°$
$\delta = 180°$

(C) $\alpha = 90°$
$\delta = 180°$

(D) $\alpha = 180°$
$\delta = 0°$

32.3. Which number indicates the position of the sun at the winter solstice?

(A) 1

(B) 2

(C) 3

(D) 4

PROBLEM 33

Which of the following best describes the meaning of the term *geoid*?

(A) the figure generated by revolving an ellipse about its minor axis

(B) a surface that is everywhere perpendicular to the direction of gravity

(C) a gently undulating surface that departs from a true spheroidal surface by amounts up to about 300 ft

(D) both B and C

SOLAR OBSERVATION FOR AZIMUTH

PROBLEM 34

When working with an observation to the sun's center that was made to one of its limbs, a correction based on its semidiameter is used. Which statement correctly characterizes the variation in that correction?

(A) It will grow larger as the altitude of the sun increases.

(B) It will grow smaller as the altitude of the sun increases.

(C) It will frequently exceed $00°16\frac{1}{2}'$.

(D) Both A and C are true.

PROBLEM 35

The true sun does not travel across the sky at a constant rate, but the mean sun does. Which statement is true of the relationship between the two?

(A) Coordinated Universal Time (UTC), or Zulu time, more nearly corresponds to the movement of the mean sun than that of the true sun.

(B) The true sun moves along the ecliptic, while the mean sun is imagined to move along the celestial equator.

(C) The difference between the true sun's and the mean sun's positions can be found using the equation of time.

(D) All of the above are true.

PROBLEM 36

The local hour angle of the sun can be found using which formula?

(A) GHA − west longitude

(B) SHA + local sidereal time

(C) GHA + east longitude

(D) both A and C

PROBLEM 37

Which small circles on the earth's surface have latitudes that correspond to the maximum declination of the sun?

(A) the Tropics of Cancer and Capricorn

(B) the Arctic and Antarctic circles

(C) the parallel of declination of Sirius

(D) the equinoctial colures

PROBLEM 38

During an observation to determine the altitude of the sun, which of the following variables must be corrected for?

(A) the temperature of the air

(B) the pressure of the air

(C) the azimuth of the sun

(D) both A and B

PROBLEM 39

What significance does the following formula have in preparing a solar observation?

$$\text{UTC} + \text{DUT1} = \text{UT1}$$

(A) The calculation has no significance to a solar observation.

(B) The calculation corrects UTC for significance to tenths of a second.

(C) The calculation applies to sidereal time rather than solar time.

(D) The calculation is the correction of Zulu time to reflect time intervals measured by the true sun rather than the mean sun.

PROBLEM 40

What would be the right ascension of the sun at the moment of the vernal equinox?

(A) 0°

(B) 90°

(C) 180°

(D) 270°

PROBLEM 41

Which of the following correctly identifies the symbols in the hour angle formula?

$$\tan z = \frac{\sin t}{\cos \phi \tan \delta \cos t}$$

(A) $z =$ azimuth
$\phi =$ longitude
$\delta =$ latitude
$t =$ GHA

(B) $z =$ zenith angle
$\phi =$ latitude
$\delta =$ declination
$t =$ LHA

(C) $z =$ zenith distance
$\phi =$ declination
$\delta =$ longitude
$t =$ right ascension

(D) $z =$ hour angle
$\phi =$ latitude
$\delta =$ longitude
$t =$ LHA

PROBLEM 42

The right ascension of so-called fixed stars is a virtually constant value for each star. The sun is not fixed, and its right ascension is constantly changing. Which other astronomical coordinate is virtually constant for fixed stars but not for the sun?

(A) azimuth

(B) altitude

(C) hour angle

(D) declination

PROBLEM 43

Which of the following defines *obliquity of the ecliptic*?

(A) the precession and nutation of the earth's rotation

(B) the tidal force exerted on the earth by the moon

(C) the difference between the sidereal day and the solar day

(D) the angle of the earth's rotational axis with respect to the orbital plane of the earth around the sun

PARALLAX, REFRACTION, AND THE SEMIDIAMETER OF THE SUN

PROBLEM 44

The following five problems refer to the illustration shown.

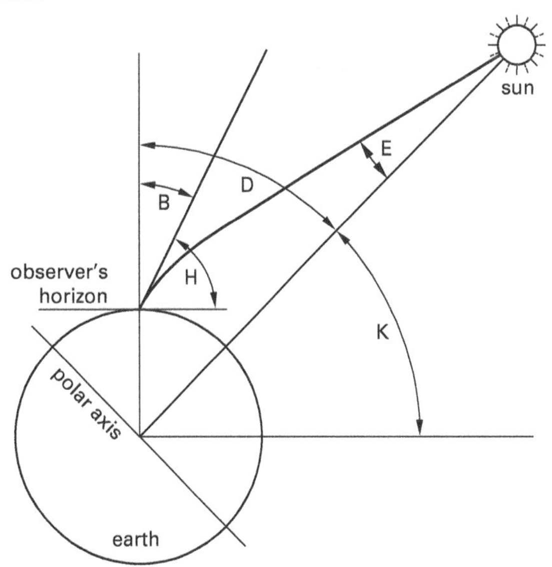

44.1. Which of the following terms best defines the effect indicated by the letter E in the illustration?

(A) parallax

(B) refraction

(C) halation

(D) diffusion

44.2. Which of the following terms best defines the value indicated by the letter H in the illustration?

(A) refraction

(B) true altitude of the sun

(C) zenith distance of the sun

(D) observed altitude of the sun

44.3. Which of the following terms best defines the effect indicated by the letter B in the illustration?

(A) refraction

(B) halation

(C) diffusion

(D) parallactic angle

44.4. Which of the following terms best defines the value indicated by the letter D in the illustration?

(A) polar distance of the sun

(B) zenith distance of the sun

(C) codeclination of the sun

(D) declination of the sun

44.5. Which of the following terms best defines the value indicated by the letter K in the illustration?

(A) declination of the sun

(B) coaltitude of the sun

(C) true altitude of the sun

(D) codeclination of the sun

PROBLEM 45

The sun's parallax, which is always less than 10 sec of arc, is sometimes applied to an observed altitude that has been corrected for the effect of refraction. How is this correction applied?

(A) It is always added.

(B) It is always subtracted.

(C) It is sometimes added and sometimes subtracted.

(D) It is multiplied by the observed altitude.

PROBLEM 46

It is sometimes said that parallax has no effect on the observed altitudes of fixed stars. Which of the following statements most correctly explains this idea?

(A) Observations of the altitude of the sun, moon, or planets in the solar system are subject to distortion as light passes through the earth's atmosphere. Similar observations of fixed stars are not subject to such distortion and therefore require no correction for parallax.

(B) Parallax is the difference between the direction of a celestial body as seen from the earth's center and as seen from an observer's point of view on the earth's surface. Strictly speaking, this difference must exist in all astronomical observations, but in the case of fixed stars, the effect is so small as to be immeasurable.

(C) The distance from the earth to the celestial body under observation affects parallax.

(D) Both B and C are true.

PROBLEM 47

At what position of the sun would an observer find the effect of parallax to be zero?

(A) at the observer's horizon

(B) at the observer's zenith

(C) at an altitude of 45°

(D) at an altitude of 23°26′30″

PROBLEM 48

The effect of refraction causes a celestial body to appear higher in the sky than its true altitude. Which of the following statements correctly describes consequences of this fact?

(A) The sun can appear to be just above the observer's horizon when it is actually entirely below it.

(B) A fixed star that is apparently just above an observer's horizon may, in fact, be half a degree below it.

(C) Refraction is zero for a celestial body at the observer's zenith.

(D) All of the above are true.

PROBLEM 49

Which of the following statements correctly describes the effect of refraction on the observation of the altitude of celestial bodies?

(A) Observed altitudes of the sun, moon, and planets must be corrected for the effect of refraction in order to find their true altitudes, but observed altitudes of fixed stars need not be corrected.

(B) As with the sun, moon and planets, the observed altitudes of fixed stars must be corrected for the effect of refraction in order to find their true altitudes.

(C) The corrections applied to an observed altitude to compensate for the effects of parallax and refraction are equal.

(D) Refraction decreases with the density of the atmosphere.

PROBLEM 50

Which of the following statements correctly describes the relationship between variations in temperature and barometric pressure and variations in the effect of refraction?

(A) As the air temperature and barometric pressure increase, the effect of refraction increases.

(B) As the air temperature increases and the barometric pressure decreases, the effect of refraction increases.

(C) As the air temperature decreases and the barometric pressure increases, the effect of refraction increases.

(D) As the air temperature and barometric pressure decrease, the effect of refraction increases.

PROBLEM 51

An observed altitude of the sun is 35°15′42″, the refraction correction is 00°01′24″, and the parallax correction is 00°00′07″. What is the true altitude of the sun?

(A) 35°14′11″

(B) 35°14′25″

(C) 35°16′59″

(D) 35°17′13″

PROBLEM 52

In preparation for an observation on the star Sirius, a true altitude is calculated for the star at the moment of observation. The calculated true altitude of the star is 19°23′11.9″. The calculated refraction correction for that altitude is 00°02′24.3″. What altitude will be measured when the star is observed if these calculated values are correct?

(A) 19°20′47.6″

(B) 19°22′15.5″

(C) 19°23′11.9″

(D) 19°25′36.2″

PROBLEM 53

An observation for azimuth was taken on the trailing limb of the sun. The angle right from the azimuth mark to that limb was 324°01′22″. The semidiameter of the sun at the time of the observation was 00°15′59.5″, and the true altitude of the sun was 42°19′06.1″. If the azimuth of the sun at the moment of observation was 146°01′47.6″, what is the azimuth of the line to the azimuth mark?

(A) 178°15′33.9″

(B) 178°21′12″

(C) 181°38′48″

(D) 181°44′26.1″

PROBLEM 54

The semidiameter of the sun gradually changes from approximately 00°16.3′ in the winter months to approximately 00°15.8′ in the summer months. Which of the following statements best describes the reason for this variation?

(A) The earth reaches aphelion in the winter and perihelion in the summer.

(B) Refraction is greater in the summer than it is in the winter.

(C) Refraction is less in the summer than it is in the winter.

(D) The earth is closer to the sun in the winter than it is in the summer.

OBSERVATIONS OF POLARIS

PROBLEM 55

Polaris, the pole star, is found in which of the following constellations?

(A) Ursa Major

(B) Sagittarius

(C) Canis Major

(D) Ursa Minor

PROBLEM 56

The polar distance of Polaris never exceeds which of the following values?

(A) 00°01′

(B) 00°30′

(C) 00°45′

(D) 01°00′

PROBLEM 57

The maximum and minimum altitudes of circumpolar stars, including Polaris, are sometimes described by which names?

(A) upper and lower elongation

(B) inner and outer elliptic

(C) upper and lower zenith

(D) upper and lower culmination

PROBLEM 58

Which of the stars listed is closest to the south celestial pole?

(A) Mensa

(B) α Hydrus

(C) β Chamaeleon

(D) σ Octantis

PROBLEM 59

The positions of Polaris at which it appears to reach the maximum east and west extents of its path around the northern celestial pole are called by which names?

(A) eastern and western extension

(B) upper and lower elongation

(C) eastern and western parallactics

(D) eastern and western elongation

PROBLEM 60

The altitude of Polaris corrected for refraction provides a good approximation of which of the following?

(A) right ascension

(B) declination

(C) longitude of the observer

(D) latitude of the observer

PROBLEM 61

Which of the following statements describes advantages of observing Polaris for azimuth at elongation?

(A) The calculation of the azimuth of the star is simplified.

(B) The effect of refraction is minimized.

(C) The star moves less than 5 sec in azimuth within 3 min of elongation.

(D) Both A and C are true.

PROBLEM 62

Polaris is seen at its eastern elongation by an observer at the position

$$\begin{aligned} \text{N}\phi &= 39°49'49'' \\ \text{W}\lambda &= 104°59'12'' \end{aligned}$$

The declination of Polaris is 89°13′16″. What is the azimuth of the star?

(A) 00°48′19″

(B) 00°59′56″

(C) 01°00′45″

(D) 01°00′51″

PROBLEM 63

At the moment of the eastern or western elongation of Polaris, which of the angles of the PZS triangle is 90°?

(A) hour angle

(B) parallactic angle

(C) codeclination

(D) coaltitude

PROBLEM 64

The following three problems refer to the data given.

An observer occupied a station with the following latitude and longitude.

$$N\phi = 39°49'46''$$
$$W\lambda = 104°59'10''$$

The eastern local hour angle of Polaris at the moment of observation was 118°46′35″, and its declination was +89°12′37″. The observer backsighted a reference mark and measured a horizontal angle right of 258°19′58″ to Polaris.

64.1. What is the azimuth of Polaris at the moment of observation?

(A) 00°53′12″

(B) 00°53′47″

(C) 00°55′42″

(D) 00°56′30″

64.2. What is the azimuth of the line from the observer to the reference mark?

(A) 90°56′12″

(B) 100°46′15″

(C) 102°33′49″

(D) 102°33′55″

64.3. What was the altitude of Polaris at the moment of observation?

(A) 37°14′52″

(B) 38°50′21″

(C) 39°26′45″

(D) 39°49′46″

OBSERVATION FOR LATITUDE

PROBLEM 65

The lower limb of the sun was observed at the moment of transit on the observer's meridian, which is at apparent noon. The declination of the sun at the moment of observation was +5°41′07″. The refraction correction was 00°01′05″ and the parallax correction was 00°00′06″. The semidiameter was 00°15′40″. The measured altitude of the sun was 45°15′42″. What is the latitude of the observer?

(A) 40°17′03″

(B) 49°42′57″

(C) 50°10′44″

(D) 51°42′11″

THE EQUATION OF TIME

PROBLEM 66

Which of the following expressions correctly describes the difference between mean solar time and true solar time at any particular instant?

(A) the right ascension of the mean sun

(B) sidereal time

(C) the right ascension of the true sun

(D) the equation of time

PROBLEM 67

Which of the following values is never exceeded by the equation of time?

(A) 00 hr 01 min

(B) 00 hr 05 min

(C) 00 hr $07\frac{1}{4}$ min

(D) 00 hr $16\frac{1}{2}$ min

PROBLEM 68

An hour measured by the apparent motion of the true sun, which is sundial time, varies with the time of year. An hour measured by the imaginary mean sun is constant throughout the year. Which of the following reasons account for this effect?

(A) The orbit of the earth around the sun is elliptical. Therefore, the apparent motion of the true sun is more rapid when the earth is closer to the sun and slower when the earth is farther away.

(B) The apparent motion of the true sun is along the ecliptic and the motion of the imagined mean sun is along the celestial equator. Therefore, the mean sun moves steadily in terms of right ascension, while the true sun does not.

(C) The tidal forces of the moon and other planets of the solar system cause the rate of the rotation of the earth on its axis to vary significantly during the year. This variation causes the apparent motion of the true sun to vary.

(D) Both A and B are true.

ZONE TIME

PROBLEM 69

Which of the following phrases correctly defines the meaning of the abbreviation LMT?

(A) local meridian time

(B) lunar median time

(C) longitudinal meridian time

(D) local mean time

PROBLEM 70

If the following local hour angles are expressed in terms of hours, minutes, and seconds, which formula correctly represents local mean time?

(A) LMT = LHA (of the true sun) + equation of time

(B) LMT = LHA (of the true sun) + 6 hr

(C) LMT = LHA (of the mean sun) + 12 hr

(D) LMT = LHA (of the mean sun)

PROBLEM 71

Which of the following would reckon noon from the upper transit of the mean sun on the Greenwich meridian?

(A) Greenwich sidereal time

(B) 7th zone time

(C) international atomic time

(D) universal time

PROBLEM 72

There are four time zones across the contiguous 48 states of the United States. Which of the following correctly represents the central meridians of those time zones?

(A) 60°, 70°, 80°, and 90°

(B) 75°, $87\frac{1}{2}$°, 100°, and $112\frac{1}{2}$°

(C) 75°, 90°, 105°, and 120°

(D) 80°, 85°, 90°, and 95°

PROBLEM 73

A ship's clock set to universal time reads 20 hr 00 min 00 sec as the ship reaches port. If the west longitude of the port is 150°24′15″, what is the LMT at the moment of the ship's arrival?

(A) 09 hr 34 min 56 sec

(B) 09 hr 58 min 23 sec

(C) 10 hr 01 min 37 sec

(D) 10 hr 12 min 15 sec

PROBLEM 74

The date is June 20 and the LMT at west longitude 103°15′00″ is 17 hr 08 min 01 sec. What is the corresponding universal time?

(A) 00 hr 01 min 01 sec on June 21

(B) 06 hr 53 min 00 sec on June 21

(C) 10 hr 15 min 01 sec on June 20

(D) 12 hr 01 min 01 sec on June 21

PROBLEM 75

Pacific Standard Time has 120°Wλ as its central meridian. Eastern Standard Time has 75°Wλ as its central meridian. Given this information, if someone in New York City phones a friend in Los Angeles at 9 hr 15 min 24 sec Eastern Standard Time, what time is it in Los Angeles?

(A) 03 hr 15 min 24 sec

(B) 06 hr 15 min 24 sec

(C) 10 hr 15 min 24 sec

(D) 12 hr 15 min 24 sec

PROBLEM 76

An aircraft left Chicago at 17 hr 00 min Central Standard Time (CST) and arrived in Tucson at 17 hr 23 min Mountain Standard Time (MST). The central meridian for CST is 90°Wλ. The central meridian for MST is 105°Wλ. To the nearest minute, what was the elapsed flight time of the aircraft?

(A) 00 hr 23 min

(B) 01 hr 10 min

(C) 01 hr 23 min

(D) 02 hr 15 min

PROBLEM 77

An aircraft left Boston at 14 hr 06 min EST and arrived in Des Moines at 16 hr 37 min CST. The central meridian for EST is 75°Wλ. The central meridian for CST is 90°Wλ. To the nearest minute, what was the elapsed flight time of the aircraft?

(A) 01 hr 12 min

(B) 02 hr 16 min

(C) 03 hr 10 min

(D) 03 hr 31 min

PROBLEM 78

A solar observation was taken at 105°00′00″Wλ on July 10. The local hour angle to the true sun at the moment of the observation was 46°12′42″ and the equation of time was −00 hr 06 min 22 sec. The mean sun is ahead of the true sun. To the nearest minute, what was the UT at the moment of the observation?

(A) 10 hr 11 min

(B) 10 hr 31 min

(C) 22 hr 11 min

(D) 22 hr 31 min

SIDEREAL TIME

PROBLEM 79

A local sidereal day is the interval of time between successive upper transits of the First Point of Aries on the observer's meridian. Which of the following transits also occurs at intervals of 24 sidereal hours?

(A) successive upper transits of the true sun

(B) successive upper transits of the mean sun

(C) successive lower transits of Saturn

(D) successive upper transits of Aldebaran

PROBLEM 80

An observation was made at the moment of upper transit of the star Antares. If the right ascension of the star was 247°10′45″, what was the sidereal time at the moment of the observation?

(A) 04 hr 28 min 43 sec

(B) 07 hr 31 min 17 sec

(C) 16 hr 28 min 43 sec

(D) 19 hr 31 min 17 sec

PROBLEM 81

There are approximately $365\frac{1}{4}$ solar days in a year. What is the number of sidereal days in a year?

(A) $304\frac{1}{4}$ sidereal days

(B) $364\frac{1}{4}$ sidereal days

(C) 365 sidereal days

(D) $366\frac{1}{4}$ sidereal days

PROBLEM 82

Which of the following intervals of mean solar time corresponds to the interval between the sidereal times of 02 hr 12 min 15 sec and 06 hr 45 min 32 sec?

(A) 04 hr 31 min 47 sec

(B) 04 hr 32 min 32 sec

(C) 04 hr 33 min 17 sec

(D) 04 hr 34 min 02 sec

PROBLEM 83

Which of the following intervals of sidereal time corresponds to the mean solar time interval of 24 hr 00 min 00 sec?

(A) 23 hr 56 min 03 sec

(B) 24 hr 00 min 00 sec

(C) 24 hr 03 min 57 sec

(D) 24 hr 05 min 27 sec

SOLUTION 1

The acronym PZS stands for three points on the celestial sphere: the pole, the zenith, and the star. Each pair of points is included in a great circle. The great circle through the pole and the star is known as the *hour circle.* The great circle through the zenith and the star is known as the *vertical circle.* The great circle through the pole and the zenith is known as the *observer's meridian.*

The answer is (D).

SOLUTION 2

The altitude and azimuth of a celestial body can be measured with respect to the observer's horizon and astronomic north. An instrument such as a theodolite can be successfully oriented to these references with relative ease. The references of the other astronomic coordinates include the First Point of Aries and the celestial equator, among others. It would be considerably more difficult to orient a theodolite to such references.

The answer is (D).

SOLUTION 3

This problem involves one of the principles of the imaginary system known as the *celestial sphere.* Terrestrial positions are transferred to the celestial sphere by lines prolonged from the center of the earth to the surface of the celestial sphere. Therefore, the angular distance from the terrestrial equator to an observer's position on the earth, which is the observer's latitude, remains unchanged when the points between which the latitude is reckoned are transferred to the surface of the celestial sphere.

The answer is (D).

SOLUTION 4

Sidereal time, also known as *star time*, is the system used to measure the movement of the stars. While this system of time is not synchronous with solar time, there are nevertheless 24 sidereal hours in a sidereal day. Zero hour in local sidereal time occurs when the First Point of Aries, a point on the celestial equator, is at upper transit on an observer's meridian. Twenty-four sidereal hours later, the First Point of Aries once again stands at the same meridian. Since it must travel 360° in those 24 sidereal hours, it must travel 15° per sidereal hour.

The star that sets precisely in the west must also be on the celestial equator. As with all stars, its right ascension is measured from the First Point of Aries in the direction opposite the apparent rotation of the celestial sphere. Had the star in question set at the same moment that the First Point of Aries reached upper transit, its right ascension would be 270° (remember, right ascension is measured opposite the direction of apparent rotation). However, the star set one sidereal hour, or 15°, before the First Point of Aries reached upper transit. Therefore, the star's right ascension is 270° less 15°, or 255°.

The answer is (C).

SOLUTION 5

The sun is on an observer's meridian at upper transit each day at what is known as apparent noon for that particular location. The sun's declination, that is, its angular distance from the celestial equator, changes throughout the year. Its declination is at its maximum in the northern hemisphere on the summer solstice, which occurs around June 22 each year. The sun's declination on the summer solstice is about +23°26′30″.

The angular distance from the celestial equator to the observer's zenith is 39°47′30″ (the observer's latitude).

The angular distance from the observer's horizon to the observer's zenith is 90°00′00″.

The angular distance from the observer's horizon to the celestial equator is 90°00′00″ – 39°47′30″ = 50°12′30″.

The declination of the sun, its angular distance from the celestial equator, on the summer solstice is 23°26′30″.

The altitude of the sun at apparent noon on the summer solstice for an observer at latitude 39°47′30″N can be found by addition.

angular distance from the celestial equator to the observer's horizon	=	50°12′30″
angular distance from the celestial equator to the sun on the summer solstice	=	+23°26′30″
altitude of the sun on the summer solstice for an observer at latitude 39° 37′30″N	=	73°39′00″

The answer is (D).

SOLUTION 6

Since the maximum declination of the sun is $23°26\frac{1}{2}'$ north and south of the celestial equator, any latitude less than 90° must witness an altitude greater than $23°26\frac{1}{2}'$ at some time. The only place where the maximum altitude of the sun is the same as its maximum declination is where the celestial equator corresponds with the horizon —at one of the earth's poles.

The answer is (D).

SOLUTION 7

The angle between the celestial equator and the zenith of the observer is the latitude of the observer (in this problem, the latitude is 35°N). Since declination of stars is also measured from the celestial equator along the stars' hour circles, the star in question must have a declination equal to that of the observer's latitude. Any star that passes through an observer's zenith must have a declination equal to the observer's latitude. This principle is the foundation of one technique for determining the latitude of a position.

The answer is (A).

SOLUTION 8

In order to rise exactly in the east, the sun must lie directly on the celestial equator. This occurs twice a year, on Mar. 21 (the vernal equinox) and Sept. 23 (the autumnal equinox). These dates vary somewhat from year to year because the exact times of the equinoxes vary.

The answer is (B).

SOLUTION 9

The latitude of an observer is the complement of the side of the PZS triangle called the *colatitude*. The complement of the polar distance is the declination of a star.

The answer is (A).

SOLUTION 10

The right ascension of a star is measured in the opposite direction from the apparent rotation of the celestial sphere. Hour angles are measured in the same direction as the apparent rotation of the celestial sphere. Both values are measured to the hour circle of the star, but from different origins. The right ascension is from the First Point of Aries and the local hour angle is from the observer's meridian. Their sum is equal to the local sidereal time at the observer's position, since zero hour sidereal would have occurred when the First Point of Aries was last at the upper transit on the observer's meridian.

The answer is (D).

SOLUTION 11

The terrestrial equator lies on a plane that passes through the earth's center and is perpendicular to the polar axis of the earth. These two properties assure that every point on the terrestrial equator must be equidistant from the two poles of the earth.

The answer is (D).

SOLUTION 12

A parallel of latitude is a small circle on the surface of the earth. Every point along a parallel of latitude has the same latitude, so it cannot be said that any position, except perhaps the north and south poles, has a latitude unique to itself. Latitude is measured in degrees, minutes, and seconds north and south of the equator.

The answer is (A).

SOLUTION 13

A second of latitude near the poles is approximately a foot longer than a second of latitude near the equator. A degree of latitude near the equator is approximately 68.7 mi, while a degree of latitude near the poles is approximately 69.4 mi. The length of a second can be approximated by dividing these quantities by 3600—the number of seconds in a degree—and multiplying the answer by 5280 ft.

Near the equator

$$\frac{68.7 \text{ mi}}{3600 \text{ sec}} = 0.0191$$

$$(0.0191)(5280 \text{ ft}) = 100.76 \text{ ft}$$

Near the poles

$$\frac{69.4 \text{ mi}}{3600 \text{ sec}} = 0.0192$$

$$(0.0192)(5280 \text{ ft}) = 101.79 \text{ ft}$$

The debate over the length of a degree of latitude measured along a meridian became so intense in 1735 that the French Academy of Sciences sent geodetic surveying expeditions to settle the issue. Expeditions went to Peru and Lapland to take their measurements. The expeditions discovered that the earth is flattened at the poles and determined that a degree of latitude near the poles is longer. The earth is an oblate spheroid.

The answer is (A).

SOLUTION 14

The plane of each meridian of longitude passes through the center of the earth. Therefore, each meridian is an arc of a great circle. The distance between meridians along a parallel of latitude grows quite short near the poles.

Since there are 24 hours in a mean solar day and 360° of longitude around the earth, 15° equals one mean solar hour.

$$\frac{360^\circ}{24 \text{ hr}} = 15^\circ$$

The answer is (D).

SOLUTION 15

Greenwich has been the prime, or zero, meridian since 1884. Each 15° of longitude is one mean solar hour. Since Greenwich's longitude is zero, the answer may be found by dividing the given longitude by 15.

$$\frac{100^\circ 15' 32''}{15} = 6 \text{ hr } 41 \text{ min } 02 \text{ sec}$$

The answer is (C).

SOLUTION 16

The standard notation for longitude is the lowercase Greek letter lambda, or λ. The abbreviation Wλ is used to indicate west longitude and Eλ is used for east longitude.

The answer is (D).

SOLUTION 17

The standard notation for latitude is the lowercase Greek letter phi, or ϕ. Nϕ indicates north latitude and Sϕ indicates south latitude. North latitude is sometimes called *positive* (+), and south latitude is called *negative* (−).

The answer is (B).

SOLUTION 18

The PZS triangle is formed by the arcs of three great circles that intersect at the pole, the observer's zenith, and the star. The pole may be either the north or south, and the star may be any celestial body.

The answer is (D).

SOLUTION 19

The PZS triangle has six elements, one of which is called the *zenith angle*. The zenith angle is the angle between the vertical circle through the star and the observer's meridian. Both the vertical circle through the star and the observer's meridian are thought of as great circles on the celestial sphere.

The answer is (C).

SOLUTION 20

The three sides of the PZS triangle are segments of three great circles. These segments are known as the *colatitude*, the *polar distance*, and the *zenith distance*. The polar distance is also known as the *codeclination* and the zenith distance is also known as the *coaltitude*.

The two sides of the PZS triangle whose complements are critical to solving the azimuth of a star by the hour angle method are the colatitude and the polar distance. The complements are the latitude of the observer and the declination of the star.

The formula for the hour angle is

$$\tan z = \frac{\sin t}{\cos \phi \tan \delta \cos t}$$

The latitude (complement of the colatitude) and the declination (complement of the polar distance) are both essential to the formula's solution.

The answer is (A).

SOLUTION 21

The parallactic angle, formed by the intersection of the vertical circle and the hour circle, is the least-used element of the PZS triangle. While it certainly can be calculated, the angle is not nearly as useful as the hour angle or the zenith angle, the other two angles in the PZS triangle.

The answer is (C).

SOLUTION 22

Two of the three constituent points of the PZS triangle, the observer's zenith and the pole, always lie on the observer's meridian. Most of the time the third point, the star, does not. However, when the star reaches upper transit on the observer's meridian, the PZS triangle, from the observer's point of view, is reduced to a straight line that is coincident with the observer's meridian.

The answer is (D).

SOLUTION 23

The earth's yearly orbit around the sun creates the apparent movement of the sun through the sky. The apparent path of the sun traced out on the celestial sphere is known as the *ecliptic.*

The answer is (D).

SOLUTION 24

The two equinoctial points, where the ecliptic intersects the celestial equator, indicate the position of the sun when the hours of daylight and the hours of darkness are approximately equal.

The answer is (B).

SOLUTION 25

The vernal equinox is also known as the First Point of Aries. It is symbolized by the uppercase Greek upsilon (or the ram's horn), ϒ.

The answer is (A).

SOLUTION 26

The axis of rotation of the earth is not perpendicular to the plane of its orbit around the sun. The axis is tilted with respect to the orbital plane by an angle of 66°33′30″. Therefore, the acute angle between the ecliptic and the celestial equator, known as the *obliquity of the ecliptic*, is the complement of 66°33′30″, or 23°26′30″.

The answer is (B).

SOLUTION 27

The autumnal equinox occurs about Sept. 23 each year when the sun's apparent motion carries it from the northern to the southern hemisphere. The position on the celestial equator where this crossing occurs is called the *First Point of Libra.*

The answer is (C).

SOLUTION 28

The first of Kepler's laws states that every planet orbits the sun in an ellipse, with the sun itself at one focus of that ellipse. Observation has since proven this assertion fundamentally correct. The path of the earth around the sun, though nearly a circle, is elliptical.

The answer is (D).

SOLUTION 29

The terms *apogee* and *perigee* refer to the farthest and closest positions, respectively, of celestial objects relative to the earth. The terms *aphelion* and *perihelion* apply to the farthest and closest positions, respectively, of objects relative to the sun.

The answer is (B).

SOLUTION 30

Unlikely as it seems, the earth is closest to the sun in January and farthest from it in July. The difference in the warmth of these months in the northern hemisphere is explained by the tilt of the earth's axis with respect to its orbital plane, rather than by its distance from the sun.

The answer is (D).

SOLUTION 31

The summer solstice occurs about June 22 each year; the autumnal equinox occurs about Sept. 23. The period between the autumnal equinox and the summer solstice is approximately three months.

The answer is (B).

SOLUTION 32

32.1. The Tropic of Cancer is the parallel of latitude that is defined by the most northerly extent of the sun's yearly path; the Tropic of Capricorn is similarly defined by its most southerly extent.

The answer is (A).

32.2. The autumnal equinox occurs at the position labeled 3. The equinox occurs when the sun is on the equator and the declination is 0°. Since declination is measured from the equator, it must be zero. Right ascension is measured from the First Point of Aries. It is measured along the equator, opposite the direction of apparent rotation of the celestial sphere. The autumnal equinox occurs opposite or 180° from the vernal equinox, which occurs at the position labeled 1.

The answer is (D).

32.3. The position labeled 4 indicates the most southerly position of the sun on its yearly path. The winter solstice occurs when the sun is in that position, about Dec. 22 of each year.

The answer is (D).

SOLUTION 33

The geoid is the reference surface for astronomical observations. Its surface is called *equipotential* due to its property of being always perpendicular to the direction of gravity.

The answer is (D).

SOLUTION 34

The angular distance from the center of the sun to one of its limbs is known as its semidiameter. From our point of view on earth, the sun is never more than $00°16\frac{1}{2}'$ from center to limb, but the angular correction under consideration is nearly always greater.

The mathematical expression used to calculate the angular correction, dH, applied to an observation of one of the sun's limbs to correct it to the center is

$$dH = \frac{\text{the sunís semidiameter}}{\cos h}$$

h is the altitude of the sun. Two calculations illustrate the effect that altitude plays in the size of the correction.

$$\begin{aligned} dH &= \frac{00°16'30''}{\cos 45°} \\ &= 00°23'20'' \\ dH &= \frac{00°16'30''}{\cos 50°} \\ &= 00°25'40'' \end{aligned}$$

As the altitude of the sun increases, the dH correction increases.

The answer is (D).

SOLUTION 35

The mean sun is a product of human imagination. Its constant motion along the celestial equator is in contrast to the true sun's irregular motion along the ecliptic, making the mean sun a much better time keeper than the true sun. The relationship between the two suns is known as the *equation of time*, which varies from 0 to 16 min.

The answer is (D).

SOLUTION 36

The Greenwich hour angle (GHA) is measured from the prime meridian in the direction of the apparent rotation of the celestial sphere, which is east to west. This corresponds to the direction of measurement of west longitude. However, the measurement of the GHA counters the direction of the measurement of east longitude; therefore, east longitude is added to the GHA to produce the local hour angle (LHA).

The answer is (D).

SOLUTION 37

Due to the tilt of the earth's axis, the maximum declination reached by the sun each year is 23°26′30″. In other words, at some time during the year the sun is directly overhead between the latitudes of 23°26′30″N and 23°26′30″S. These lines of latitude have been given the names Tropic of Cancer and Tropic of Capricorn, respectively.

The answer is (A).

SOLUTION 38

The apparent altitude of a celestial body is greater than its true altitude due to the refraction of light as it passes through the earth's atmosphere. In an attempt to quantify that refraction, adjustments are made to the observed altitude. These adjustments are based on an approximation of the density of the air at the time of the observation made by measuring the air temperature and pressure.

The answer is (D).

SOLUTION 39

DUT1 is a correction expressed in tenths of a second that can be determined from a series of double ticks broadcast on radio station WWV by the National Bureau of Standards. This correction is applied to UTC to significantly improve the accuracy of solar calculations.

The answer is (B).

SOLUTION 40

The right ascension of a celestial body is measured from the First Point of Aries, which is an arbitrary point on the celestial equator. The sun would coincide with that position at the moment of the vernal equinox.

The answer is (A).

SOLUTION 41

The use of Greek letters in surveying astronomy is fairly consistent. Ordinarily, ϕ stands for latitude, δ stands for declination, and so on.

The answer is (B).

SOLUTION 42

The declination of a fixed star, like right ascension, remains virtually constant throughout the year. Because it is so close to the earth, the same values for the sun change constantly.

The answer is (D).

SOLUTION 43

Among the consequences of the obliquity of the ecliptic are the seasons of the year and the Tropics of Cancer and Capricorn. The earth's rotational axis differs by approximately $23°26\frac{1}{2}$ from a perpendicular to the orbital plane created by the earth and sun.

The answer is (D).

SOLUTION 44

44.1. Parallax is generally the shift in the apparent position of an observed object due to the shift of the position of the observer. As used in surveying astronomy, parallax refers to the difference between the apparent position of a celestial body as it would be seen from the earth's center and the body as seen from the observer's location.

The answer is (A).

44.2. The altitude of a celestial body is measured from the observer's horizon to the apparent position of the body. However, the body consistently appears to be higher in the sky than it actually is due to the density of the atmosphere through which the observation must be made. Therefore, there is a difference between the observed altitude of a celestial body and its true altitude.

The answer is (D).

44.3. When light from a celestial body passes through the atmosphere of the earth, it bends toward the surface of the earth. This refraction causes the body to appear higher in the sky than it actually is.

The answer is (A).

44.4. The angular distance from the zenith of the observer to the celestial body is the zenith distance.

The answer is (B).

44.5. The center of the earth is assumed to be the center of the imaginary celestial sphere. The true altitude of celestial bodies is the altitude that would be measured from the center of the earth without the effects of refraction or parallax, if such a thing were possible.

The answer is (C).

SOLUTION 45

The correction for parallax is always added to an observed altitude. The effect of parallax is to reduce the altitude that would be measured at the earth's center. Therefore, when calculating the true altitude, which is the observed altitude at the earth's center, parallax must be added.

The answer is (A).

SOLUTION 46

Parallax is the difference in the direction of a celestial body as seen from the earth's center and from an observer's location on earth. Bodies in our own solar system are much closer to earth than even the nearest fixed stars; as a consequence, the effect of parallax is much larger when taking observations on the sun, moon, and planets than on fixed stars. For all practical purposes, the altitude of even the nearest fixed star is exactly the same no matter where on earth the observation is made.

The answer is (D).

SOLUTION 47

As the altitude of the sun increases, the effect of parallax diminishes. If the sun reaches the observer's zenith, the effect of parallax is zero.

The answer is (B).

SOLUTION 48

The effect of refraction is greatest near the horizon, where the observer must see the celestial body through the thickest layer of atmosphere. Refraction near the horizon reaches nearly 00°33′. The effect of refraction diminishes as the altitude of a celestial body increases, and it reaches zero at the observer's zenith.

The answer is (D).

SOLUTION 49

Light from a fixed star is refracted before reaching an observer on earth just as light from the sun is. The denser the atmosphere, the larger the correction that must be applied to the observed altitude to find the true altitude.

The answer is (B).

SOLUTION 50

As the atmosphere through which an observer sees a celestial body becomes denser, the effect of refraction increases. A decrease in air temperature causes the air to become more dense, as does an increase in barometric pressure.

The answer is (C).

SOLUTION 51

The true altitude of the sun can be derived from the observed altitude using the following expression.

$$\begin{aligned}\text{true altitude} &= \text{measured altitude} - \text{refraction} \\ &\quad + \text{parallax} \\ &= 35°15'42'' - 00°01'24'' + 00°00'07'' \\ &= 35°14'25''\end{aligned}$$

The answer is (B).

SOLUTION 52

Parallax is so small as to be undetectable for observations on fixed stars. In cases where the observed altitude is known and the true altitude is calculated, the refraction correction is subtracted.

$$\begin{matrix}\text{true} \\ \text{altitude}\end{matrix} = \begin{matrix}\text{observed} \\ \text{altitude}\end{matrix} - \begin{matrix}\text{refraction} \\ \text{correction}\end{matrix}$$

However, in this case it is the true altitude that is known and the observed altitude that must be calculated. Therefore, the refraction correction is added.

$$\begin{aligned}\begin{matrix}\text{observed} \\ \text{altitude}\end{matrix} &= \begin{matrix}\text{true} \\ \text{altitude}\end{matrix} - \begin{matrix}\text{refraction} \\ \text{correction}\end{matrix} \\ &= 19°23'11.9'' + 00°02'24.3'' \\ &= 19°25'36.2''\end{aligned}$$

The answer is (D).

SOLUTION 53

The first step in solving this problem is to find the correction in azimuth for the semidiameter, sometimes known as dH.

$$dH = \frac{\text{SD}}{\cos h}$$

SD is the semidiameter, and h is the true altitude.

$$\begin{aligned}dH &= \frac{00°15'59.5''}{\cos 42°19'06.1''} \\ &= 00°21'37.6''\end{aligned}$$

Next, the angle right is increased by the semidiameter correction since the observation was made to the trailing limb.

$$324°01'22 + 00°21'37.6'' = 324°22'59.6''$$

Finally, find the azimuth of the line to the azimuth mark.

$$\begin{aligned}\text{azimuth of the sun} &= \quad 146°01'47.6'' \\ \text{corrected angle right} &= \underline{-324°22'59.6''} \\ &\quad -178°21'12'' \\ &\quad -178°21'12'' \\ &\quad \underline{+360°00'00''} \\ \text{azimuth of the line} &= \quad 181°38'48''\end{aligned}$$

The answer is (C).

SOLUTION 54

The earth reaches perihelion, which is its closest point to the sun, in January, and aphelion in July. Therefore, the sun's semidiameter appears to be largest about Jan. 3 of each year and then slowly decreases until approximately July 4.

The answer is (D).

SOLUTION 55

Polaris is the α star in the constellation Ursa Minor—the brightest star in the Little Bear. The constellation is also known as the Little Dipper.

The answer is (D).

SOLUTION 56

Polaris is known as a circumpolar star because it never dips below the horizon in the northern hemisphere. It is very close to the north celestial pole, always within 1°, and appears to rotate counterclockwise in a circular path.

The answer is (D).

SOLUTION 57

Both the maximum and minimum altitudes of Polaris occur on the meridian of the observer and are called the *upper* and *lower transit* of the star. These positions of Polaris have been given the names *upper* and *lower culmination.*

The answer is (D).

SOLUTION 58

σ Octantis is far from the brightest star in its constellation, but it appears to rotate in a circle of less than 1° around the southern celestial pole. In this way, σ Octantis is similar to Polaris in the northern hemisphere.

The answer is (D).

SOLUTION 59

Eastern and western elongation are the names given to the farthest east and west positions of the path of Polaris around the north celestial pole. However, it is not correct to say that the altitude of Polaris when it reaches elongation is equal to that of the north celestial poles. Polaris at elongation is slightly higher than the elevation of the north celestial pole from an observer's point of view.

The answer is (D).

SOLUTION 60

Since the path of Polaris never departs from the north celestial pole by more than 1°, its true altitude provides a fairly good approximation of the latitude of the observer.

The answer is (D).

SOLUTION 61

Polaris appears to move almost vertically for approximately 3 min before and after elongation, making the observation of the star convenient. The calculation of the azimuth at elongation is also much more convenient than at other positions. If a disadvantage exists for observations of Polaris at elongation, it is that the position may be reached at inconvenient times.

The answer is (D).

SOLUTION 62

The calculation of the azimuth of Polaris at elongation is simpler than it is in any other position and may be found using the expression

$$\sin \text{Az (azimuth of Polaris at elongation)} = \frac{\cos \delta}{\cos \phi}$$

δ is the declination of Polaris and ϕ is the latitude of the observer.

$$\begin{aligned} \sin \text{Az} &= \frac{\cos 89°13'16''}{\cos 39°49'49''} \\ &= 0.0177015 \\ \text{Az} &= 01°00'51'' \end{aligned}$$

The answer is (D).

SOLUTION 63

The eastern and western elongations of Polaris occur when the parallactic angle is 90°. The hour circle and the vertical circle through the star are perpendicular to one another.

The answer is (B).

Surveying Astronomy

SOLUTION 64

64.1. The hour angle formula is convenient for this calculation.

$$\tan \text{Az} = \frac{\sin t}{\cos\phi \tan\delta - \sin\phi \cos t}$$

ϕ is the latitude, δ is the declination, and t is the hour angle.

$$\begin{aligned}\tan \text{Az} &= \frac{\sin 118°46'35''}{(\cos 39°49'46'')(\tan +89°12'37'') - (\sin 39°49'46'')(\cos 118°46'35'')} \\ &= \frac{0.876505}{(0.767954)(72.54722) - (0.640504)(-0.481392)} \\ &= \frac{0.876505}{55.7129 - (-0.308334)} \\ &= \frac{0.876505}{56.021295} \\ &= 0.015646 \\ \text{Az} &= 00°53'47''\end{aligned}$$

The answer is (B).

64.2. The angle right from the reference mark to Polaris was 258°19′58″, and the azimuth of Polaris was 00°53′47″ at the moment of observation.

$$258°19'58'' - 00°53'47'' = 257°26'11''$$

The angle to the left from north is 257°26′11″. Therefore, the azimuth will be this angle less 360°.

$$\begin{array}{rr} & 360°00'00' \\ & -257°26'11'' \\ \hline \text{azimuth from observer to reference mark} = & 102°33'49'' \end{array}$$

The answer is (C).

64.3. The formula used to find the altitude of a star is

$$\sin h = \sin\phi \sin\delta + \cos\phi \cos\delta \cos t$$

h is the altitude, ϕ is the latitude, δ is the declination, and t is the local hour angle.

$$\begin{aligned}\sin h &= (\sin 39°49'46'')(\sin 89°12'37'') \\ &\quad + (\cos 39°49'46'')(\cos 89°12'37'')(\cos 118°46'35'') \\ &= 0.640444 + (-0.005095) \\ &= 0.635648 \\ h &= 39°26'45''\end{aligned}$$

The answer is (C).

SOLUTION 65

The angular distance from the celestial equator to the observer's zenith is equal to the observer's latitude.

measured altitude of sun at transit	=	45°15′42″
correction for parallax	=	+00°00′06″
		45°15′48″
correction for refraction	=	−00°01′05″
		45°14′43″
correction for semidiameter	=	+00°15′40″
true altitude of sun's center	=	45°30′23″
altitude of sun	=	45°30′23″
sun's declination	=	−05°41′07″
height of celestial equator	=	39°49′16″
angular distance from horizon to zenith	=	90°00′00″
height of celestial equator	=	−39°49′16″
observer's latitude	=	50°10′44″

The answer is (C).

SOLUTION 66

The length of a true solar day—the time required for the earth to make a complete revolution on its axis with respect to the true sun—varies throughout the year. The length of a mean solar day—the time required for a complete revolution with respect to the imaginary mean sun—is constant. The difference between the time at any instant as measured by the mean sun and as measured by the true sun is called the *equation of time.*

The answer is (D).

SOLUTION 67

The equation of time reaches its maximum negative value in February and its maximum positive value in November, but neither value exceeds 00 hr $16\frac{1}{2}$ min.

The answer is (D).

SOLUTION 68

According to Kepler's law of planetary motion, the earth moves more rapidly along its orbital path at perihelion and more slowly at aphelion. While this accounts for some variation in the motion of the true sun through the sky, it is not the largest factor. The true sun appears to move along the ecliptic due to the inclination of the earth's axis with respect to its orbital plane with the sun. Therefore, it does not move in right ascension at a constant rate as does the mean sun. The mean sun is imagined to move steadily in right ascension along the celestial equator.

The answer is (D).

SOLUTION 69

The term *local mean time* describes the time reckoned by the mean sun at a particular meridian. The local mean time will be different for any two places on the earth that are not precisely north or south of one another.

The answer is (D).

SOLUTION 70

Noon, or 12 hr, is determined in LMT to be the moment of the upper transit of the mean sun on the meridian of a particular place. Since the local hour angle of the mean sun would be considered zero at the same moment, 12 hr should be added to the local hour angle of the mean sun to find the local mean time. Similarly, the addition of 12 hr to the local hour angle of the mean sun at any moment would yield the local mean time.

The answer is (C).

SOLUTION 71

The Greenwich, or zero, meridian remains the foundation for timekeeping around the world, but the term *Greenwich mean time* (GMT) has been largely supplanted by the term *universal time.* Midnight on the Greenwich meridian is considered 0 hr universal time.

The answer is (D).

SOLUTION 72

Beginning with the 75th meridian of west longitude, the central meridians of the time zones across the 48 states increase by 15°. This incrementation of 15° is due to the rate of the movement of the mean sun, which is 15° of longitude per hour.

The answer is (C).

SOLUTION 73

The division of the longitude of the port by 15 will yield the hours, minutes, and seconds by which the LMT is earlier than UT.

$$\frac{150°24'15''}{15} = 10 \text{ hr } 01 \text{ min } 37 \text{ sec earlier than UT}$$

$$\begin{array}{rr} \text{UT} = & 20 \text{ hr } 00 \text{ min } 00 \text{ sec} \\ & -10 \text{ hr } 01 \text{ min } 37 \text{ sec} \\ \hline \text{LMT at the port} = & 09 \text{ hr } 58 \text{ min } 23 \text{ sec} \end{array}$$

The answer is (B).

SOLUTION 74

The LMT is specific to the meridian 103°15′00″. Division of the longitude by 15 will give the number of hours, minutes, and seconds that must be added to 17 hr 08 min 01 sec to find the universal time.

$$\frac{103°15'00''}{15} = 06 \text{ hr } 53 \text{ min } 00 \text{ sec}$$

$$\begin{array}{rr} \text{LMT} = & 17 \text{ hr } 08 \text{ min } 01 \text{ sec} \\ & +06 \text{ hr } 53 \text{ min } 00 \text{ sec} \\ \hline \text{UT} = & 24 \text{ hr } 01 \text{ min } 01 \text{ sec} \end{array}$$

Since the UT exceeds 24 hr, the following day has begun at the Greenwich meridian.

$$\begin{array}{rl} 24 \text{ hr } 01 \text{ min } 01 \text{ sec} & \\ -24 \text{ hr } 00 \text{ min } 00 \text{ sec} & \\ \hline 00 \text{ hr } 01 \text{ min } 01 \text{ sec} & \text{on June 21} \end{array}$$

The answer is (A).

SOLUTION 75

There is 45° of longitude between the central meridians of the two time zones involved, creating a difference of 3 hr. The time in Pacific Standard Time (PST) is earlier than that in Eastern Standard Time (EST).

$$\begin{array}{rr} \text{time in EST} = & 9 \text{ hr } 15 \text{ min } 24 \text{ sec} \\ & -3 \text{ hr } 00 \text{ min } 00 \text{ sec} \\ \text{time in PST} = & 06 \text{ hr } 15 \text{ min } 24 \text{ sec} \end{array}$$

The answer is (B).

Surveying Astronomy

SOLUTION 76

One way to solve this problem is to convert the arrival time of the aircraft to the zone time of the zone from which it departed.

$$\begin{aligned} \text{MST} &= 17 \text{ hr } 23 \text{ min} \\ & +01 \text{ hr } 00 \text{ min} \\ \text{CST} &= 18 \text{ hr } 23 \text{ min} \end{aligned}$$

The difference between the departure and arrival times is found by subtraction.

$$\begin{aligned} \text{CST of the arrival} &= 18 \text{ hr } 23 \text{ min} \\ \text{CST of the departure} &= -17 \text{ hr } 00 \text{ min} \\ \text{elapsed flight time} &= 01 \text{ hr } 23 \text{ min} \end{aligned}$$

The answer is (C).

SOLUTION 77

Converting the arrival time to its equivalent in EST makes the computation of the elapsed flight time more convenient.

$$\begin{aligned} \text{CST} &= 16 \text{ hr } 37 \text{ min} \\ & +01 \text{ hr } 00 \text{ min} \\ \text{EST} &= 17 \text{ hr } 37 \text{ min} \end{aligned}$$

$$\begin{aligned} \text{arrival time in EST} &= 17 \text{ hr } 37 \text{ min} \\ \text{departure time in EST} &= -14 \text{ hr } 06 \text{ min} \\ \text{elapsed flight time} &= 03 \text{ hr } 31 \text{ min} \end{aligned}$$

The answer is (D).

SOLUTION 78

First, express the hour angle of the true sun in hours, minutes, and seconds.

$$\frac{46°12'42''}{15} = 03 \text{ hr } 04 \text{ min } 51 \text{ sec}$$

Add 12 hr to this since the LHA is measured from the upper transit of the sun.

$$\begin{aligned} &03 \text{ hr } 04 \text{ min } 51 \text{ sec} \\ &+12 \text{ hr } 00 \text{ min } 00 \text{ sec} \\ &15 \text{ hr } 04 \text{ min } 51 \text{ sec} \end{aligned}$$

Since the equation of time is negative, the mean sun is ahead of the true sun by 00 hr 06 min 22 sec.

$$\begin{aligned} &15 \text{ hr } 04 \text{ min } 51 \text{ sec} \\ &+00 \text{ hr } 06 \text{ min } 22 \text{ sec} \\ \text{MST of the observation} = &15 \text{ hr } 11 \text{ min } 13 \text{ sec} \end{aligned}$$

The result is in MST since the observation was taken on the 105th meridian. The UT is found by adding 7 hr.

$$\frac{105° \text{W} \lambda}{15} = 7 \text{ hr}$$

$$\begin{aligned} &15 \text{ hr } 11 \text{ min } 13 \text{ sec} \\ &+07 \text{ hr } 00 \text{ min } 00 \text{ sec} \\ &22 \text{ hr } 11 \text{ min } 13 \text{ sec} \end{aligned}$$

To the nearest minute, the UT was 22 hr 11 min.

The answer is (C).

SOLUTION 79

All of the bodies listed except the star Aldebaran have variable right ascensions. Aldebaran's angular distance from the First Point of Aries, which is its right ascension, is virtually constant. This is the case with all of the so-called fixed stars. Twenty-four sidereal hours pass between successive transits of Aldebaran.

The answer is (D).

SOLUTION 80

The right ascension of a star is measured from the First Point of Aries opposite the direction of the apparent rotation of the celestial sphere. The right ascension of Antares is more than 180°, so the sidereal time must be greater than 12 hr when the star reaches upper transit. The exact sidereal time can be found by dividing the right ascension of the star by 15, since the First Point of Aries is considered to move 15° every sidereal hour.

$$\frac{247°10'45''}{15} = 16 \text{ hr } 28 \text{ min } 43 \text{ sec}$$

The answer is (C).

SOLUTION 81

The time required for the earth to orbit the sun is precisely 365.24222 solar days. The same interval of time in sidereal days requires exactly one more revolution of the earth upon its axis, or 366.24222 sidereal days, or $366\,\frac{1}{4}$ sidereal days.

The answer is (D).

SOLUTION 82

First, determine the sidereal interval between the two sidereal times given.

$$\begin{array}{r} 06 \text{ hr } 45 \text{ min } 32 \text{ sec} \\ -\ 02 \text{ hr } 12 \text{ min } 15 \text{ sec} \\ \hline 04 \text{ hr } 33 \text{ min } 17 \text{ sec} \end{array}$$

Next, the ratio between the number of mean solar days in the year and the number of sidereal days in the year provides a convenient factor for finding the equivalent solar interval.

$$\frac{365.24222}{366.24222} = 0.997269567$$

Multiply this value by the sidereal interval to find the mean solar interval.

$$\begin{aligned} &(04 \text{ hr } 33 \text{ min } 17 \text{ sec})(0.997269567) \\ &\quad = 04 \text{ hr } 32 \text{ min } 32 \text{ sec} \end{aligned}$$

The answer is (B).

SOLUTION 83

Use the ratio developed in the solution to Prob. 82 to convert the mean solar interval to sidereal interval.

$$\begin{aligned} &\left(\frac{365.24222}{366.24222}\right)(24 \text{ hr } 00 \text{ min } 00 \text{ sec}) \\ &\quad = 24 \text{ hr } 03 \text{ min } 57 \text{ sec} \end{aligned}$$

The answer is (C).

Surveying Astronomy

5 Public Land Surveying System

HISTORY OF PUBLIC LANDS

PROBLEM 1

The first surveys of public lands in the United States were conducted under the provisions of which of the following laws?

(A) the Greenville Treaty

(B) the Act of February 11, 1805

(C) the Virginia Charter

(D) the Land Ordinance Act

PROBLEM 2

Which of the following authorities supervised the first surveys of public lands?

(A) the General Land Office

(B) the Department of the Interior

(C) the Bureau of Land Management

(D) the Geographer of the United States

PROBLEM 3

Thomas Hutchins was the surveyor in charge of the work now known as the Seven Ranges in Ohio, the first public land surveys done under the auspices of the Land Ordinance Act of 1785. He personally ran the line extending east and west from the Ohio River that is known as the Geographer's Line. Lines that serve the same function as the Geographer's Line are now found throughout the Public Land Surveying System. What are such lines called?

(A) sectional guide meridians

(B) guide meridians

(C) sectional correction lines

(D) baselines

PROBLEM 4

Rufus Putnam, the Surveyor General in 1797, established the contract system, which continued until 1910. Which statement best characterizes this system?

(A) The contract system provided for credit purchases of public lands by the highest bidders.

(B) The contract system allowed those wishing to purchase land within a township to make a deposit with the government to cover the surveying costs.

(C) The contract system abolished the credit system of purchase of public lands. Under the contract system, buyers were required to pay with hard currency.

(D) The contract system allowed the Surveyors General to arrange contracts with deputy surveyors for the execution of public land surveys.

PROBLEM 5

In 1807, Surveyor General Jared Mansfield ordered that another baseline be established 24 mi from the first in the Vincennes District to compensate for distortions in the survey. In 1818, the same idea was used to compensate for the effect of convergence. Which of the following terms is currently used to refer to such lines?

(A) correction lines

(B) guide meridians

(C) standard parallels

(D) both A and C

PROBLEM 6

In 1849, the General Land Office became part of the Department of the Interior. What department of government had jurisdiction over public land surveys before then?

(A) Department of Commerce

(B) Department of the Treasury

(C) Department of Agriculture

(D) Department of Defense

PROBLEM 7

The Act of May 10, 1872, sometimes called the Mining Act, established the legal size of a lode claim in the public lands. Which of the following sizes correctly represents this provision of the Mining Act?

(A) 200.00 ft^2

(B) 600.00 ft by 1500.00 ft

(C) 1320.00 ft by 600.00 ft

(D) 40 chains by 40 chains

PROBLEM 8

In 1847, Richard Young, commissioner of the General Land Office, instructed Charles Morse in the sectional subdivision of a township that had a particularly crooked eastern boundary. Young's solution was unpopular with Morse, but it limited the alignment problem to only one row of sections by the use of a procedure still used in similar circumstances. What is the procedure?

(A) the establishment of a sectional guide meridian

(B) the establishment of a sectional correction line

(C) subdivision by protraction

(D) a completion survey

GUNTER'S CHAIN

PROBLEM 9

The chain has been the standard unit of length in the Public Land Surveying System since its beginning and is still the standard prescribed by law. One of the reasons for its durability is its convenient relationship with the description of land in acres. Which of the following equations correctly describes this relationship?

(A) 1 sq chain = 1 ac

(B) 10 sq chains = 1 ac

(C) 80 sq chains = 40 ac

(D) 100 sq chains = 1 ac

PROBLEM 10

Which of the following lengths is equal to 0.28 chains?

(A) 3 links

(B) 12.00 ft

(C) 19.50 ft

(D) 28 links

PROBLEM 11

Early field notes from public land surveys sometimes refer to a two-pole chain. What is a two-pole chain?

(A) A two-pole chain is a chain that is equipped with two long wooden handles.

(B) The phrase two-pole chain is simply another term used to describe a standard-length Gunter's chain.

(C) A two-pole chain is 132.00 ft long.

(D) A two-pole chain is one-half of a standard Gunter's chain, or 33.00 ft long.

PROBLEM 12

Gunter's chain was invented by Edmund Gunter in seventeenth century England. The instrument was used in public land surveys throughout the nineteenth century. These chains tended to increase in length with extended use. Why?

(A) the constant repairs required by frequent breaks

(B) the elasticity of the materials from which the chains were made

(C) the wearing down of the links and the rings that joined them together

(D) all of the above

PROBLEM 13

Which of the following lengths is equivalent to 79.52 chains?

(A) 320.00 rods

(B) 1600.00 m

(C) 1920.41 varas

(D) 5248.32 ft

PLSS

THE SOLAR COMPASS

PROBLEM 14

William Burt invented his True Meridian Finding Instrument in 1833. Which of the following statements about Burt's invention is correct?

(A) Burt found that the iron ore deposits in Michigan made surveying with a magnetic needle nearly impossible. This led him to invent a solar compass.

(B) Burt's invention provided an instant mechanical solution of the astronomic triangle for the establishment of direction.

(C) Using Burt's invention, the magnetic declination of a compass can be determined.

(D) All of the above are true.

MANUAL OF INSTRUCTIONS

PROBLEM 15

The *Manual of Surveying Instructions* is prepared and published by the

(A) Public Land Survey System Foundation

(B) U.S. Department of the Interior

(C) National Society of Professional Surveyors

(D) Interagency Cadastral Coordination Council

PROBLEM 16

Which of the following statements correctly describes the applicability of the *Manual of Surveying Instructions* with regard to lands that have passed from the public domain into private ownership?

(A) All lands once in the public domain are under the jurisdiction of the Bureau of Land Management.

(B) Lands that have passed from the public domain into private ownership are under the jurisdiction of state law.

(C) Many states have adopted the principles and rules laid down in the *Manual of Surveying Instructions* through state court decisions and statute laws.

(D) Both B and C are true.

PROBLEM 17

Which of the following is NOT covered in the *Manual of Surveying Instructions*?

(A) a summary of the federal statute laws governing the surveying of lands in the public domain

(B) instructions on the adjustment and use of a solar attachment for a transit

(C) rules and procedures for the meandering of swamp and overflowed lands

(D) procedures for the reestablishment of lost corners

PROBLEM 18

According to the *2009 Manual of Surveying Instructions*, the period from 1804 to 2009 saw a shift in the dominant federal policy regarding public lands. What was it?

(A) from the regulation of their occupancy and use to the preservation of the land and its resources from destruction or unnecessary injury

(B) from favoring disposal and settling of unreserved public lands to favoring retention, administration, and control

(C) from a policy of cessions to one of graduation or distribution

(D) from a policy of privatization of public lands to promotion of outdoor recreation

PROBLEM 19

Which of the following correctly states a difference between the 1973 *Manual of Surveying Instructions* and the 2009 *Manual of Surveying Instructions*?

(A) The 1973 manual includes instructions for mineral resurveys and mineral segregation surveys, but the 2009 manual does not.

(B) The 1973 manual does not assert that coordinates may provide the best evidence of a corner position, but the 2009 manual does.

(C) The 1973 manual explains that "substantial evidence" is adequate to determine whether a corner is existent, obliterated, or lost, but the 2009 manual insists that such determination must be "beyond reasonable doubt."

(D) The 1973 manual does not address water boundaries, but the 2009 manual does.

LIMITS AND ADJUSTMENTS

PROBLEM 20

According to the *Manual of Surveying Instructions*, what is a survey's maximum allowable error of closure for the perimeter of either a regular or irregular township, in either latitude or departure?

(A) 1/1280

(B) 1/2000

(C) 1/4000

(D) 1/10,000

PROBLEM 21

According to the *Manual of Surveying Instructions*, a section may be considered regular if its boundaries do not depart from cardinal directions by more than what prescribed limit?

(A) 00°01′

(B) 00°10′

(C) 00°12′

(D) 00°21′

PROBLEM 22

The *Manual of Surveying Instructions* establishes a limit regarding the maximum acceptable adjustment in measurement between regular corners. Which of the following limits must NOT be exceeded if a section is to be considered regular?

(A) 5 links in 80 chains

(B) 10 links in 20 chains

(C) 15 links in 40 chains

(D) 25 links in 40 chains

THE QUADRANGLE

PROBLEM 23

Which of the following statements is INCORRECT regarding the use of the term *initial point* in the Public Land Surveying System?

(A) Principal meridians and baselines both originate at an initial point.

(B) The position of an initial point is given in geographical coordinates in the *Manual of Surveying Instructions.*

(C) Some public lands states have more than one initial point.

(D) There is an initial point in each of the public lands states.

PROBLEM 24

Under current instructions of the Public Land Survey System, which of the following types of corners are not to be monumented during the original survey of the principal meridians and baselines?

(A) section corners

(B) quarter-section corners

(C) meander corners

(D) closing corners

PROBLEM 25

The standard lines of the Public Land Surveying System include principal meridians, baselines, standard parallels, and guide meridians. Which of the following statements is NOT true regarding the general plan for the establishment of these standard lines?

(A) Guide meridians and principal meridians are intended to follow astronomical meridians within 00′50″.

(B) Standard parallels and baselines are intended to follow parallels of latitude within 00′50″.

(C) The difference between two sets of measurements of a standard line should not differ by more than 25 links in 40 chains.

(D) Double corners will eventually stand on standard parallels and baselines.

PROBLEM 26

The following six problems refer to the illustration shown.

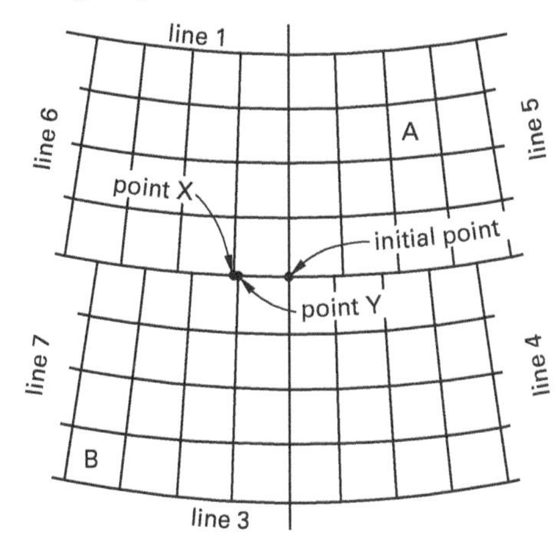

26.1. Which of the following labels correctly identify lines 4 and 5?

(A) Line 4 is the First Guide Meridian East. Line 5 is the Second Guide Meridian East.

(B) Line 4 is the First Correction Line East. Line 5 is the Second Correction Line East.

(C) Line 4 is the First Standard Parallel. Line 5 is the Second Standard Parallel.

(D) Lines 4 and 5 are portions of the First Guide Meridian East.

26.2. Which of the following labels correctly identify lines 1 and 3?

(A) Line 1 is the First Standard Parallel North. Line 3 is the First Standard Parallel South.

(B) Line 1 is the First Correction Line North. Line 3 is the First Correction Line South.

(C) Line 1 is the First Latitudinal Section Line North. Line 3 is the First Latitudinal Section Line South.

(D) Both A and B are true.

26.3. Which of the following labels correctly identify points X and Y?

(A) Point X is a standard corner. Point Y is a closing corner.

(B) Point X is a meander corner. Point Y is a witness corner.

(C) Point X is a standard corner. Point Y is a witness corner.

(D) Point X is a closing corner. Point Y is a standard corner.

26.4. Which of the following labels would correctly identify the township labeled A?

(A) T2N R3E

(B) T3N R3W

(C) T3S R2W

(D) T3N R3E

26.5. Which of the following statements is NOT true concerning the boundaries of township B under current instructions?

(A) The eastern township boundary should be run parallel with the principal meridian.

(B) The northern township boundary should be run on a random line and returned on a true line.

(C) As the township boundaries are run, quarter-section and section corners should be set alternatively at intervals of 40 chains.

(D) The eastern and western township boundaries should be run from the south to the north.

26.6. Township corners are normally established at intervals of 480 chains. However, current instructions specify a somewhat different procedure regarding the original closing of meridional township boundaries on standard parallels. Which of the following statements correctly characterizes those instructions?

(A) The excess or deficiency of measurement must be placed in the last quarter mile, between the last set quarter-quarter corner monument and the closing corner.

(B) A closing township corner must be established 1920 chains from the standard parallel to the south.

(C) The meridional township boundary must be ended at its intersection with the standard parallel, and the excess or deficiency of measurement placed in the last half mile.

(D) Meridional Township boundaries run from north to south.

PROBLEM 27

In the past, guide meridians and standard parallels were not always established at intervals of 24 mi. Which other interval would be most likely?

(A) 21 mi

(B) 23 mi

(C) 33 mi

(D) 36 mi

PROBLEM 28

Which of the following would be laid out east and west from an initial point?

(A) baseline

(B) principal meridian

(C) sectional guide meridian

(D) guide meridian

PROBLEM 29

The interval between corner monuments that were set during the layout of a principal meridian is how many chains?

(A) 20

(B) 40

(C) 60

(D) 80

PROBLEM 30

Which of the following is intended to be bounded by meridians of longitude and parallels of latitude by current instructions in the Public Land Surveying System?

(A) section

(B) fractional lot

(C) quarter section

(D) quadrangle

PROBLEM 31

How many townships are usually created from a single quadrangle in the Public Land Surveying System by current instructions?

(A) 8

(B) 10

(C) 12

(D) 16

THE TOWNSHIP

PROBLEM 32

The following three problems refer to the illustration shown.

6	5	4	3	2	1
7	8	9	10	11	12
18	17	16	15	14	13
19	20	21	22	23	24
30	29	28	27	26	25
31	32	33	34	35	36

N

T4N R3E

32.1. Under normal conditions and by current instructions, which section's boundaries are established first in the subdivision of a regular township?

(A) section 1

(B) section 6

(C) section 31

(D) section 36

32.2. Under normal conditions and by current instructions, which of the following statements correctly describes the alignment of the eastern and northern boundaries of section 22?

(A) The eastern boundary is intended to follow a true astronomical meridian of longitude. The northern boundary is intended to follow a parallel of latitude.

(B) The eastern boundary is intended to be parallel with the eastern boundary of the township. The northern boundary is intended to be parallel with the southern boundary of the township.

(C) The eastern boundary is intended to be parallel with the western boundary of the township. The northern boundary is intended to be parallel with the northern boundary of the township.

(D) The eastern boundary is intended to follow a geodetic meridian of longitude. The northern boundary is intended to follow a geodetic parallel of latitude.

PLSS

32.3. Referring to the illustration of T4N R3E, which of the following statements is correct regarding the typical placement of closing corners in such a township?

(A) Closing corners would normally occur along its eastern boundary.

(B) Closing corners would normally occur along its northern boundary.

(C) There would normally be no closing corners in such a township.

(D) Closing corners would normally occur along its western boundary.

PROBLEM 33

Under normal conditions and by current instructions, which of the following statements is NOT true regarding the establishment of the quarter-section corners during the subdivisional survey of a township?

(A) Quarter-section corners are normally established on the line halfway between section corners along the section lines of the 25 regular sections in a township.

(B) Quarter-section corners are usually established alternatively with section corners at intervals of 40 chains along the meridional section lines of the 25 regular sections in a township.

(C) Temporary quarter-section corners are established during the running of a random line, and are corrected to their proper position during the running of the true line along each section line.

(D) Quarter-section corners are established along the northern tier of sections, 40 chains north of the standard section corner, leaving the excess or deficiency of measurement in the last half mile.

PROBLEM 34

Which of the following correctly defines the term *fractional township*?

(A) a township that has been subdivided by protraction

(B) a township that has not been fully subdivided

(C) a township that contains significantly fewer than 36 sections, usually due to an invasion by a segregated body of water

(D) a township that contains fractional sections

PROBLEM 35

North-south township and section lines in the Public Land Surveying System are also known as which of the following?

(A) latitudinal

(B) random

(C) meridional

(D) meander

PROBLEM 36

Which of the following corner monuments were in place on the township boundary when the normal subdivision of a quadrangle into townships was completed according to plan?

(A) township corners, but not section corners or quarter corners

(B) township corners and section corners, but not quarter corners

(C) township, section, and quarter corners

(D) neither township, section, or quarter corners

PROBLEM 37

When the normal subdivision of a regular township into sections is completed according to plan, section boundaries are intended to be

(A) aligned with astronomically determined latitude and longitude

(B) parallel to the east and south boundaries of the township

(C) parallel to the west and north boundaries of the township

(D) aligned with magnetically determined directions

PROBLEM 38

Which of the following statements concerning township plats in the Public Land Surveying System is NOT correct?

(A) They represent the township included in the survey and show the direction and length of each line and their relation to the adjoining surveys.

(B) They include some indication of the relief and the boundaries.

(C) They show the areas of the section subdivisions.

(D) They only show lines actually surveyed on the ground during the original survey.

THE SECTION

PROBLEM 39

The following seven problems refer to the illustration shown.

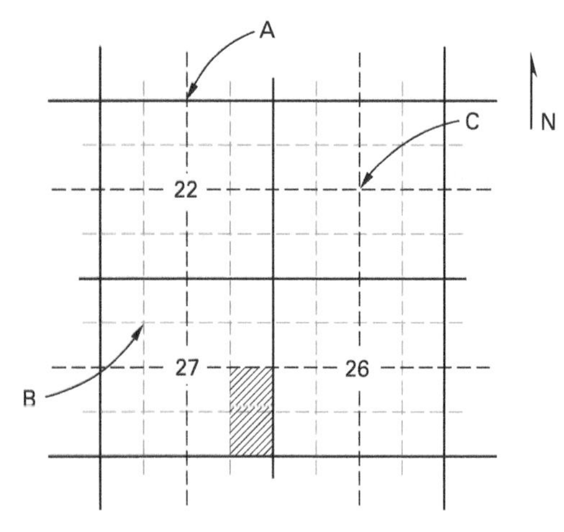

39.1. Which name is correct for the corner labeled A?

(A) an aliquot corner

(B) a center sixteenth corner

(C) the $S\frac{1}{4}$ corner of section 15

(D) both A and C

39.2. Which name is correct for the corner labeled B?

(A) the $NW\frac{1}{16}$ corner of section 27

(B) the NW corner of the $SE\frac{1}{4}$ of the $NW\frac{1}{4}$ of section 27

(C) the NE corner $SW\frac{1}{4}NW\frac{1}{4}$ of section 27

(D) all of the above

39.3. Which name is correct for the corner labeled C?

(A) the $C\frac{1}{4}$ of section 21

(B) the NE corner of the $SW\frac{1}{4}SW\frac{1}{4}$ of section 23

(C) the SW corner of the $SE\frac{1}{4}NE\frac{1}{4}$ of section 23

(D) none of the above

39.4. Which of the labeled corners is located by the intersection of lines rather than by measurement along one line?

(A) corner A

(B) corners A and C

(C) corners C and B

(D) corners A and B

39.5. Under current instructions, which of the labeled corners would usually be monumented during the subdivisional survey of a township?

(A) corner A

(B) corner B

(C) corner C

(D) all of the above

39.6. If section 27 is a regular section containing 640 ac, what is the acreage of its crosshatched aliquot part?

(A) 10 ac

(B) 40 ac

(C) 60 ac

(D) 80 ac

39.7. Which of the following phrases best describes the crosshatched aliquot part of section 27?

(A) the $E\frac{1}{4}$ of the $S\frac{1}{2}$ of section 27

(B) the $SE\frac{1}{4}$ of the $E\frac{1}{2}$ of section 27

(C) the $E\frac{1}{2}$ of the $SE\frac{1}{4}$ of section 27

(D) the $E\frac{1}{4}$ of the $SE\frac{1}{2}$ of section 27

PROBLEM 40

The following three problems refer to the illustration shown.

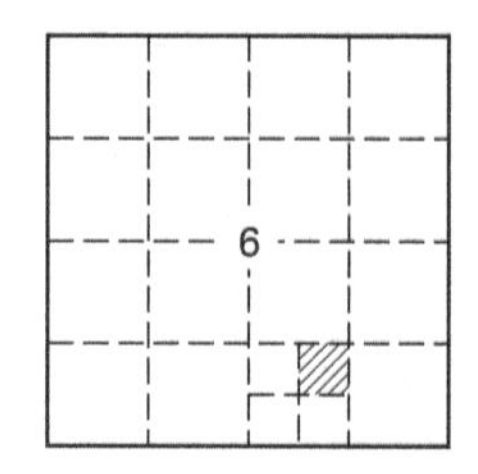

40.1. If the SE$\frac{1}{4}$ of section 6 is regular and contains 160 ac, what is the acreage of the crosshatched aliquot part?

(A) 5 ac

(B) 10 ac

(C) 20 ac

(D) 40 ac

40.2. Which of the following phrases best describes the crosshatched aliquot part in section 6?

(A) the NE$\frac{1}{4}$ of the SW$\frac{1}{4}$ of the SE$\frac{1}{4}$ of section 6

(B) the NE$\frac{1}{4}$ of the SE$\frac{1}{4}$ of the SW$\frac{1}{4}$ of section 6

(C) the NE$\frac{1}{4}$ of the SE$\frac{1}{2}$ of the SE$\frac{1}{4}$ of section 6

(D) the NE$\frac{1}{4}$ of the NE$\frac{1}{4}$ of the SE$\frac{1}{4}$ of section 6

40.3. Which of the following names would correctly apply to the crosshatched aliquot part in section 6?

(A) quarter-quarter

(B) sixteenth

(C) two hundred fifty-sixth

(D) sixty-fourth

PROBLEM 41

Referring to the illustration and coordinates given (which are based on field measurements taken between the found corner monuments), what are the coordinates of the center quarter corner of section 15?

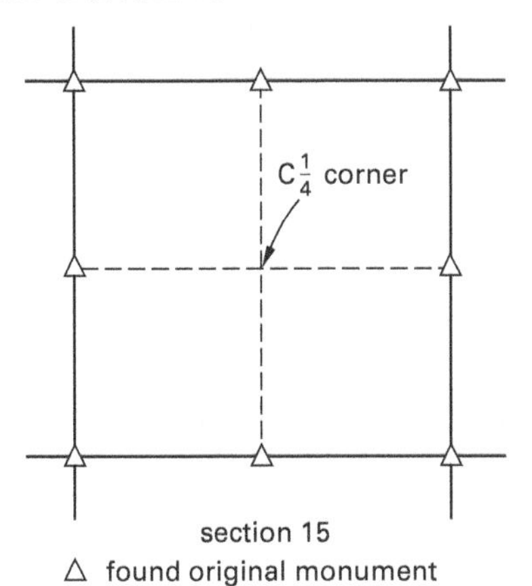

section 15
△ found original monument

N$\frac{1}{4}$ corner	N10,280.3212 E7618.1717	NE corner	N10,287.9181 E10,229.7806
E$\frac{1}{4}$ corner	N7642.6827 E10,241.1944	SE corner	N4997.4476 E10,252.6029
S$\frac{1}{4}$ corner	N4998.7326 E7614.0897	SW corner	N5000.0000 E5000.0000
W$\frac{1}{4}$ corner	N7636.0259 E5004.6007	NW corner	N10,272.7320 E5009.2027

(A) N7634.9800
E7622.7431

(B) N7639.3431
E7616.1252

(C) N7639.3543
E7622.8976

(D) N7639.5269
E7616.0965

PROBLEM 42

Which of the following statements correctly describes the position of the CS$\frac{1}{16}$ corner in the illustration shown?

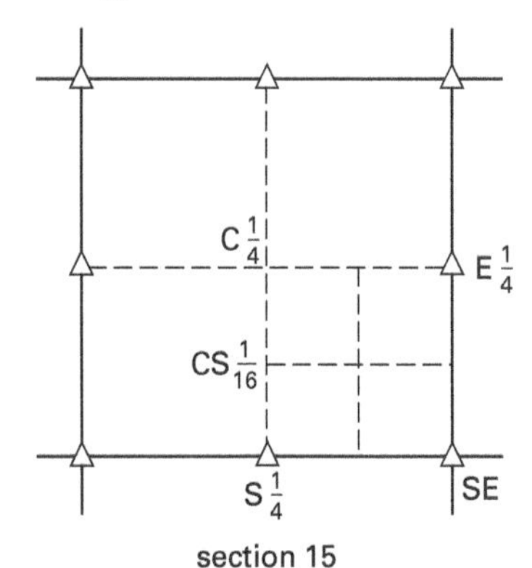

section 15
△ found original monument

(A) The quarter-quarter corner labeled CS$\frac{1}{16}$ should be placed 20 chains north from the S$\frac{1}{4}$ corner.

(B) The $\frac{1}{16}$ corner labeled CS$\frac{1}{16}$ should be placed 20 chains south from the center quarter corner.

(C) The CS$\frac{1}{16}$ corner should be placed at the intersection of the first sixteenth line north of the south section line with the meridional quarter line.

(D) The CS$\frac{1}{16}$ corner should be placed at the midpoint of the line between the C$\frac{1}{4}$ and the S$\frac{1}{4}$ of the section.

PLSS

PROBLEM 43

Which of the following statements is true regarding the center quarter corner of a regular section under normal circumstances?

(A) The center quarter corner does not exist until it is monumented.

(B) The center quarter corner stands at the midpoint of the north-south (meridional) centerline of a section.

(C) The center quarter corner stands at the midpoint of the east-west (latitudinal) centerline of a section.

(D) The center quarter corner almost certainly does not stand at the midpoint of any aliquot line.

PROBLEM 44

How many public land corners are indicated in the illustration shown?

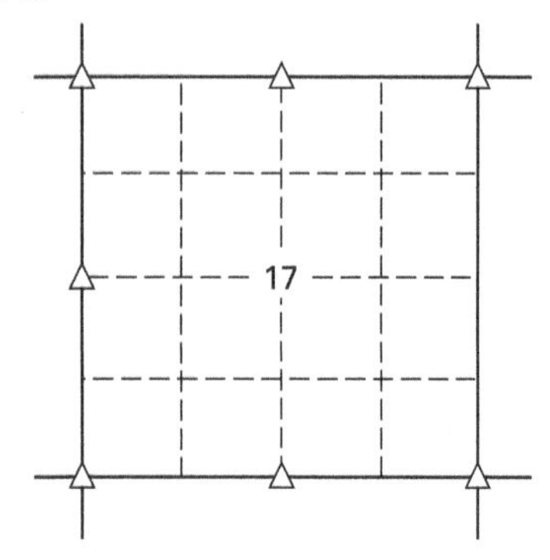

(A) 7

(B) 8

(C) 9

(D) 25

PROBLEM 45

The following five problems refer to the illustration shown.

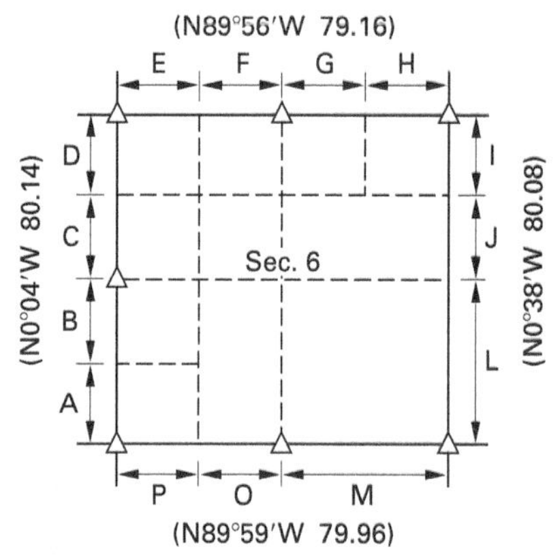

Section 6 is part of a township where the normal closing distances occur on the northern and western boundaries of the township.

45.1. The broken lines shown in the illustration are typical of those shown in closing sections on official township plats. Which of the following statements is correct regarding the broken lines?

(A) They are lines that were normally measured during the original survey of the township.

(B) The broken lines in section 6 indicate surveyed lines dividing the section into two half-quarter sections, one quarter section, and eight quarter-quarter sections.

(C) The broken lines are created by protraction.

(D) The broken lines divide section 6 into two half-quarter sections, one quarter section, and eight lots.

45.2. What is the measurement implied by the information given in the illustration for the distance labeled E?

(A) 19.16 chains

(B) 19.96 chains

(C) 20.08 chains

(D) 20.11 chains

45.3. What is the measurement implied by the information given in the illustration for the distance labeled L?

(A) 39.97 chains

(B) 40.00 chains

(C) 40.04 chains

(D) 40.07 chains

45.4. Which of the pairs contains the label of a distance that is NOT intended to be 20.00 chains on the official plat?

(A) H and I

(B) C and B

(C) B and F

(D) G and O

45.5. A quarter-quarter corner is also known as which of the following?

(A) a sixteenth corner

(B) a sixty-fourth corner

(C) a two hundred fifty-sixth corner

(D) a meander corner

PROBLEM 46

Which of the following aliquot parts would NOT be found in section 6?

(A) one quarter section

(B) two half-quarter sections

(C) one quarter-quarter section

(D) eight quarter-quarter sections

CLOSING CORNERS

PROBLEM 47

The following three problems refer to the illustration shown, which shows a standard parallel under usual conditions surveyed by current instructions.

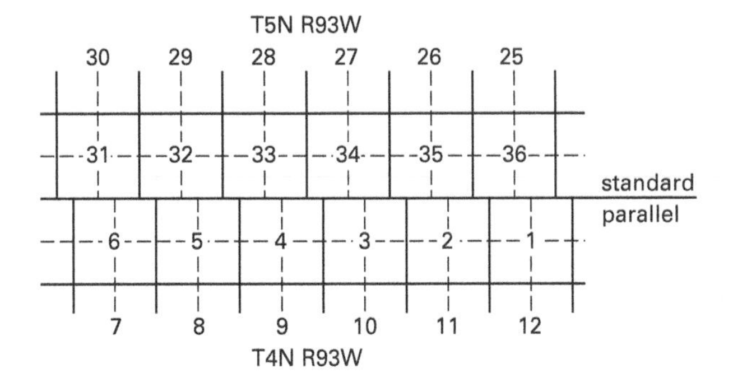

47.1. Which of the corners along the standard parallel would be called closing corners, and which would be called standard corners?

(A) All of the corners would be known as closing corners.

(B) All of the corners would be known as standard corners.

(C) The corners of sections 31 through 36 would be known as standard corners, and the corners of sections 1 through 6 would be known as closing corners.

(D) The corners of sections 31 through 36 would be known as closing corners, and the corners of sections 1 through 6 would be known as standard corners.

47.2. In what order were the corners shown established by the original surveyors?

(A) First: the township, section, and quarter-section corners for T5N R93W

Second: the township corners for T4N R93W

Third: the closing section and quarter-section corners for T4N R93W

(B) First: the township corners for both T5N R93W and T4N R93W

Second: the section and quarter-section corners for T5N R93W

Third: the closing section and quarter-section corners for T4N R93W

(C) First: the township, section, and quarter-section corners for T5N R93W

Second: the closing section and quarter-section corners for T4N R93W

Third: the township corners for T4N R93W

(D) First: the township corners for T4N R93W

Second: the township, section, and quarter-section corners for T5N R93W

Third: the closing section and quarter-section corners for T4N R93W

47.3. A monument said to be a closing corner is found near the intersection of the west line of section 1 and the standard parallel. The monument appears to have been set by the original surveyor of T4N R93W. However, it is 12.00 ft north of the standard parallel. Which of the

PLSS

following statements describes the best course of action under these circumstances?

(A) If set by the original surveyor, the monument is indeed the closing corner.

(B) The monument should be destroyed and the proper closing corner set at the intersection of the west line of section 1 and the standard parallel.

(C) If set by the original surveyor, the monument should be used to govern the direction of the west line of section 1, but not the line's termination.

(D) Both A and C are best.

PROBLEM 48

Which of the following terms would correctly complete this excerpt from the original field notes?

> 74.84 Intersected Base Line at a point 21 links East of Standard, at which point I set a post in mound with a charred stick for a ________________ to sections 4 and 5.

(A) township corner

(B) meander corner

(C) witness corner

(D) closing corner

PROBLEM 49

Closing corners are sometimes set where a surveyed line intersects a previously established boundary. Which of the following situations would not cause an original surveyor of the public domain to set a closing corner?

(A) the intersection of a section line with the boundary of a previously established Spanish land grant

(B) the intersection of a township line with a meander line previously surveyed around a navigable river

(C) in a resurvey, the intersection of a section line with a sectional guide meridian

(D) the intersection of a township boundary with the boundary of a state

PROBLEM 50

Which of the following markings would a surveyor not expect to find on the brass tablet of a monument at a closing corner along the boundary between two states?

(A) abbreviation CC

(B) year the monument was set

(C) names of the adjoining states

(D) names of the adjoining townships, ranges, and sections

PROBLEM 51

During the original subdivision of a township into sections, under current instructions, which of the following statements is correct concerning closing corners?

(A) The northern township boundary must always include closing corners.

(B) The section lines that close upon the northern and western township boundaries are always surveyed by the random and true line procedure.

(C) Following the establishment of a closing corner upon a standard parallel, the surveyor should retrace the line closed upon to the east and west of the intersection.

(D) Current instructions prohibit the establishment of closing corners on the southern and eastern boundaries of townships.

CORNERS, MONUMENTS, AND ACCESSORIES

PROBLEM 52

The principle behind the use of double proportionate measurement in the restoration of lost corners in the public lands system is best defined by which of the following statements?

(A) Monuments and corners are the same.

(B) Distance is superior to direction.

(C) Original measurements are nearly always incorrect.

(D) Modern surveying instruments can correct mistakes of the past.

PROBLEM 53

Corner accessories include which of the following?

(A) line trees

(B) bearing trees and bearing objects

(C) pits and memorials

(D) both B and C

PROBLEM 54

Which of the following best describes a corner for which no evidence of any kind can be found?

(A) obliterated corner

(B) lost corner

(C) extant corner

(D) existent corner

PROBLEM 55

Which of the following statements best defines the difference between a corner and a monument?

(A) Corner and monument are synonymous terms.

(B) A monument that can be verified as having been set by the original surveyor in a public lands survey is exactly at the position of the corner it purports to be, except off-line closing corner monuments.

(C) A corner is a point where boundaries meet, while a monument is physical evidence of a corner's location on the ground.

(D) Both B and C are true.

PROBLEM 56

Should the monument for an obliterated corner be reestablished by double or single proportionate measurement?

(A) double

(B) single

(C) neither—an obliterated corner is not lost

(D) single proportionment if it is on a township boundary; double if it is an interior corner

PROBLEM 57

A found closing corner monument, stamped AM, is found off the line upon which the closing was intended. Which statement best describes the likely circumstances?

(A) The found monument, being off of the intended line, was stamped AM, meaning *amended monument.*

(B) The found monument will help describe the closing line, but not its terminus.

(C) A new monument should have been established at the intersection of the closing line and the line intended to have been closed upon.

(D) A, B, and C are true.

PROBLEM 58

What is the function of a location monument, sometimes designated by the abbreviation USLM?

(A) Where no public land corners are available within 2 mi, a USLM is established.

(B) The function of a USLM is to provide control for isolated special surveys or mineral surveys.

(C) A USLM is usually located on a prominent topographic feature.

(D) A, B, and C are true.

PROBLEM 59

A meander corner is established at which of the following points?

(A) every point where a standard, township, or section line intersects the mean high water mark at the bank of a navigable body of water

(B) every point where an aliquot line intersects the mean high water mark at the bank of a meanderable body of water

(C) every point where a township line intersects the mean high water mark at the bank of a meanderable body of water

(D) only on banks of lakes 50 ac or larger

PROBLEM 60

Which statement best defines the difference between a bearing tree and a line tree?

(A) A bearing tree is a corner accessory and a line tree is not.

(B) A bearing tree is blazed and a line tree is hacked.

(C) A line tree is intersected by a boundary line, while a bearing tree is scribed with the description of the section in which it stands.

(D) A, B, and C are true.

PROBLEM 61

Original monuments in the Public Land Surveying System are held in high regard. Which statement correctly defines their standing?

(A) Verified undisturbed original monuments in the public land system are, by law, coincident with the corners they were established to represent, except off-line closing corners.

(B) Verified undisturbed original monuments in the public land system are to remain fixed unless technical errors are discovered in the survey by which they were established. If such technical errors are discovered, the monuments may be moved to their correct location by any licensed land surveyor.

(C) The bearings and distances shown on official township plats are considered inferior evidence of the position of corners in the public land system when contradicted by verified original corner monumentation.

(D) Both A and C are true.

PROBLEM 62

Under current instructions, what is a regulation monument for public lands surveys established by the Bureau of Land Management?

(A) a no. 5 reinforcing bar with a plastic cap

(B) 4 in × 4 in × 36 in wooden post

(C) $3\frac{1}{2}$ in outer diameter alloyed iron pipe, 28 in to 30 in long, with a brass cap

(D) $2\frac{1}{2}$ in outer diameter alloyed iron pipe, 28 in to 30 in long, with a brass cap

PROBLEM 63

An original stone monument from a public lands survey of 1891 is found to have two notches cut into its eastern edge and three notches cut into its southern edge. What corner has been found?

(A) the SE corner of section 10

(B) the SW corner of section 14

(C) the NW corner of section 3

(D) the NE corner of section 28

PROBLEM 64

Under current instructions, which statement does NOT correctly define a difference between a bearing tree and a line tree?

(A) Bearing trees are corner accessories; line trees are not.

(B) Line trees are marked with horizontal hacks; bearing trees are scribed.

(C) Bearing trees are recorded in the field notes; line trees are not.

(D) Line trees are sometimes known also as *sight trees* or *station trees;* bearing trees are not.

PROBLEM 65

A surveyor is retracing a section line that was blazed through a forested area following current instructions. The surveyor finds a tree with two blazes that are quite close together. The blazes are on the side of the tree facing the surveyor. Where is the section line in relation to the surveyor as he stands facing the blazes on the tree?

(A) The section line lies in front of the surveyor, approximately 3 chains distant.

(B) The section line lies in front of the surveyor, approximately 50 links distant.

(C) The section line lies behind the surveyor, a foot or two distant.

(D) The section line lies behind the surveyor, approximately 50 links distant.

PROBLEM 66

If found on a regulation BLM brass cap monument, the stamping shown in the illustration would be evidence of which of the following corners?

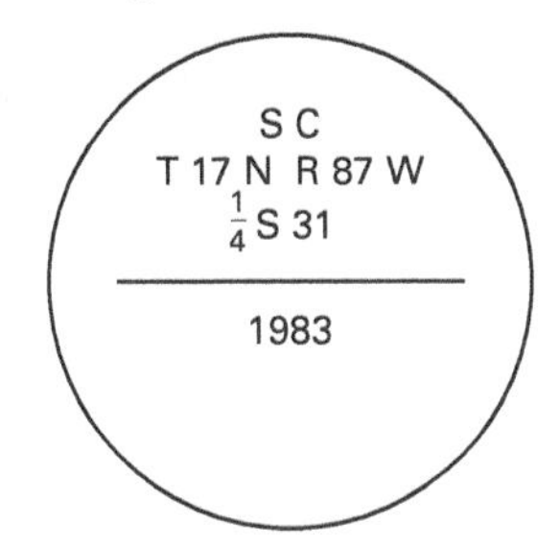

(A) the SW corner of section 17 on a standard parallel

(B) the SW corner of section 31, T17N R87W

(C) the S$\frac{1}{4}$ corner of section 31, T17N R87W, found on a standard parallel

(D) the S$\frac{1}{4}$ corner of section 17

PROBLEM 67

If found on a regulation BLM brass cap monument, the stamping shown in the illustration would be evidence of which of the following corners?

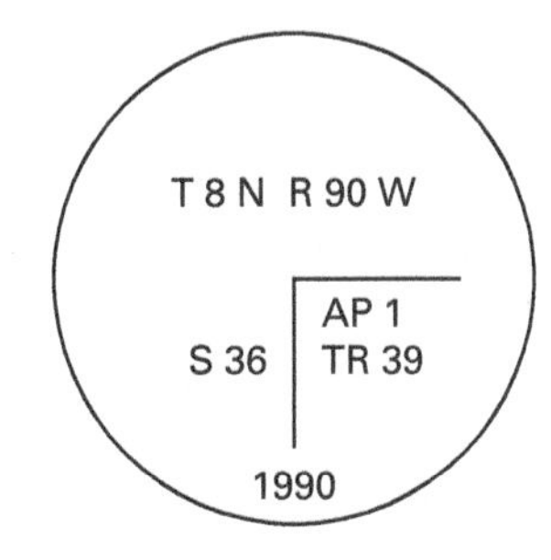

(A) a witness point for section 36, T8N, R90W

(B) an auxiliary point for a land transfer in section 36, T8N R90W

(C) the angle point number 1 for Tract 39 in section 36, T8N R90W

(D) a witness corner for a desert land entry corner in section 36, T8N R90W

PROBLEM 68

If found on a regulation BLM brass cap monument, the stamping shown in the illustration would be evidence of which of the following corners?

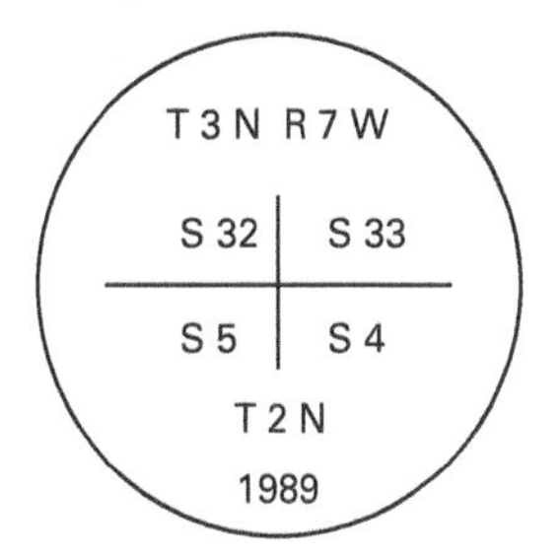

(A) a section corner common to four sections on a standard parallel

(B) a section corner common to four sections that is not on a standard parallel

(C) a section corner on a township line

(D) both B and C

PROBLEM 69

Which statement does NOT correctly represent a criterion for the placement of a witness corner monument?

(A) The preferred location of a witness corner is on one of the surveyed lines that leads to the true corner.

(B) A witness corner monument is established near a corner when the true corner falls in a position that makes monumentation and subsequent use impractical.

(C) The preferred distance between a witness corner and the true corner is 10 chains or less when the witness corner can be placed on a surveyed line that leads to the true corner.

(D) The *Manual of Surveying Instructions* calls for two witness corners to be placed near inaccessible true corners.

PROBLEM 70

Which of the following abbreviations might be found on a corner accessory that was established near a section corner that fell in an unimproved roadway where neither bearing trees nor nearby bearing objects were available?

(A) AMC

(B) RM

(C) BT

(D) TR

PROBLEM 71

A stone monument 20 in wide and 6 in thick is found projecting approximately 15 in from the ground. It has an "X" chiseled into the top and "USLM 34" is chiseled into the side. The monument was set in approximately 1906. Nearby rocks also have Xs and the letters "BR" chiseled into them. What sort of point does the monument represent and what is its purpose?

(A) It serves the same purpose as the U.S. Mineral Monuments formerly did.

(B) No corners of the public land system were available within 2 mi of the vicinity of a mineral survey when this USLM was established as a reference for the mineral surveys of the area.

(C) The monument was set by a U.S. Mineral Surveyor.

(D) All of the above are true.

PROBLEM 72

Which monument would NOT be evidenced by an accessory under normal circumstances?

(A) witness corner

(B) closing corner

(C) reference monument

(D) meander corner

PROBLEM 73

An old set of original field notes describes a cairn as a corner accessory. What is a cairn?

(A) charred wooden post

(B) tree species

(C) mound of stone

(D) pit

CORNER RECORDS

PROBLEM 74

Several of the states covered by the Public Land Surveying System have enacted corner record legislation. Which of the following correctly defines a common objective of such legislation?

(A) The filing of corner records provides a mechanism by which original public lands survey corner monuments may be replaced with monuments established at the correct positions determined by private land surveyors.

(B) The filing of corner records helps to ensure the perpetuation of found original public lands survey corner monuments, and makes the evidence by which lost or obliterated monuments are reestablished a matter of public record.

(C) The filing of corner records provides a mechanism by which a private land surveyor may make monumentation of the aliquot corner of a section a matter of public record.

(D) Both B and C are true.

PROBLEM 75

Most states require professional surveyors to file monument records under which of the following specific circumstances?

(A) A professional surveyor utilizes a monument representing a Public Land Survey System (PLSS) corner to control a survey.

(B) A professional surveyor establishes, restores, or rehabilitates any monument representing a property corner.

(C) A professional surveyor utilizes a monument representing a PLSS corner to control a survey, and the monument and its accessories are not substantially described in an existing monument record filing.

(D) A professional surveyor utilizes a monument representing a PLSS corner to control a survey, and the monument is unlikely to be destroyed by construction.

INTERPRETATION OF FIELD NOTES

PROBLEM 76

The following illustration is representative of a page found in most field notes of the public lands system. What is the significance of the illustration and the numbers shown on the township and section lines?

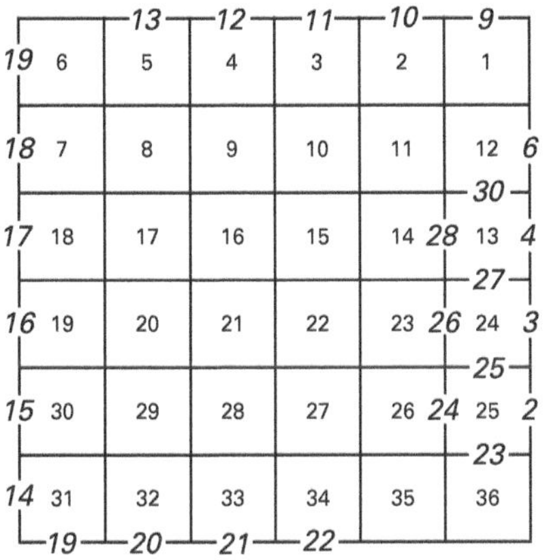

		13	12	11	10	9	
19	6	5	4	3	2	1	
18	7	8	9	10	11	12	6
						30	
17	18	17	16	15	14 28	13	4
						27	
16	19	20	21	22	23 26	24	3
						25	
15	30	29	28	27	26 24	25	2
						23	
14	31	32	33	34	35	36	
	19	20	21	22			

T1N R32W

(A) It is an index, and the numbers refer to the consecutively numbered special instructions that were issued by the Bureau of Land Management to govern the survey.

(B) It is a diary, and the numbers refer to the days of the month upon which the section lines were surveyed.

(C) It is an index, and the numbers match the consecutive labels attached to the random lines recorded in the notes themselves.

(D) It is an index, and the numbers indicate the page of the notebook on which the measurement of the line is recorded.

PROBLEM 77

Which of the following governmental entities is the official custodian of the original field notes and plats of the completed public land surveys?

(A) All of the official public lands records are under the exclusive jurisdiction of the individual states.

(B) All official public lands records are under the exclusive jurisdiction of the director of the Bureau of Land Management in Washington, D.C.

(C) The official public lands records for all of the states east of the Mississippi River have been transferred to the states. All other records remain under the jurisdiction of the Bureau of Land Management.

(D) Where public lands surveys have been substantially completed, the official public lands records have been transferred to the states, though duplicates are on file in Washington, D.C. The remaining records are under the jurisdiction of the Bureau of Land Management.

PROBLEM 78

Which of the following statements is NOT true of the official field note records of the public lands surveys that are prepared under the instructions in the *Manual of Surveying Instructions*?

(A) Field notes are typewritten transcriptions of the information recorded in field tablets.

(B) Random lines with fallings are recorded in the field notes.

(C) A full description of all of the monuments, both found and set, must be included in the field notes of a new survey.

(D) A general description of the soil, land, and forest along surveyed lines must be summarized at the conclusion of each mile completed.

THE ESTABLISHMENT OF ALIQUOT CORNERS

PROBLEM 79

The term *aliquot* is part of a section in the Public Land Surveying System that is

(A) a half or a quarter of a previously larger subdivision

(B) any portion of a previously larger subdivision

(C) any portion of a previously larger subdivision that includes a meander line

(D) a fractional lot

PROBLEM 80

Which of the following statements is NOT correct concerning the establishment of a monument at the center of a section in the Public Land Surveying System?

(A) A center of section monument should be established at the intersection of two lines—one straight from the east quarter corner to the west quarter corner, and the other from the north quarter corner to the south quarter corner.

(B) The government is responsible for setting a monument at the center of a section.

(C) A center of section is also known as the center quarter corner of a section.

(D) The establishment of center of section monuments was, and is left, to local surveyors.

PROBLEM 81

A private surveyor finds that his remeasurement along a section line between found original quarter corner and section corner monuments differs from that of the official record. The surveyor's objective is to establish an initial sixteenth corner monument along the line between the two found original corners. Which of the following techniques is the surveyor likely to find helpful in resolving the difference between the measurements?

(A) protraction

(B) double centering

(C) triangulation

(D) proportionate measurement

PROBLEM 82

The following three problems refer to the illustration shown.

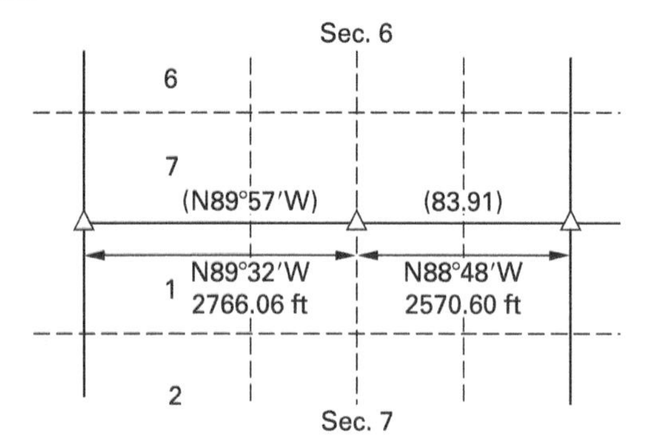

82.1. The NE, NW, and N$\frac{1}{4}$ corners of section 7 are found original corner monuments. The official township plat indicates that the record bearing and distance across the north line of section 7 is N89°57′W 83.91 chains. A private land surveyor intends to set the NE corner of lot 1. The surveyor has measured N89°32′W 2766.06 ft between the N$\frac{1}{4}$ and the NW corners of section 7. What distance should the surveyor measure west along the section line from the N$\frac{1}{4}$ corner to set the NE corner of lot 1?

(A) 1259.88 ft

(B) 1320.00 ft

(C) 1383.03 ft

(D) 1506.18 ft

82.2. The NE, NW, and N$\frac{1}{4}$ corners of section 7 are found original corner monuments. The official township plat indicates that the record bearing and distance across the north line of section 7 is N89°57′W 83.91 chains. A private land surveyor intends to set the NE corner of NW$\frac{1}{4}$NE$\frac{1}{4}$ of section 7. The surveyor has measured N88°48′W 2570.60 ft between the NE and the N$\frac{1}{4}$ corners of section 7. What distance should the surveyor measure west along the section line from the NE corner to set the NE corner of NW$\frac{1}{4}$NE$\frac{1}{4}$?

(A) 1250.98 ft

(B) 1285.30 ft

(C) 1320.00 ft

(D) 1349.20 ft

82.3. Under normal circumstances and by current instructions, would the original surveyor of section 7 have established a monument at the SW corner of lot 1?

(A) Yes, the lot corners along a township boundary are set during the course of the original survey of the township boundary.

(B) The SW corner of lot 1 would have been monumented during the course of the original survey only if the western boundary of section 7 was a guide meridian.

(C) No, neither the quarter-quarter corners nor the lot corner are monumented during the course of the original survey of a township, unless mandated by special instructions.

(D) The SW corner of lot 1 would have been monumented during the course of the original survey only if it was a closing corner.

PROBLEM 83

The following two problems refer to the illustration shown.

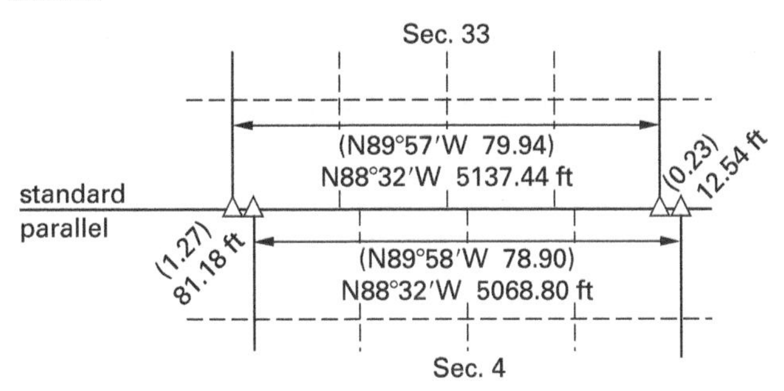

12.54 ft as measured
(0.23) record (in chains)
△ found original monument

83.1. Retracing a standard parallel, a surveyor discovers that the quarter corners were not established by the original surveyor. Using the record and measured information given, what distance should be measured along the standard parallel from the eastern standard corner to the northern quarter corner of section 4?

(A) 2521.86 ft

(B) 2534.40 ft

(C) 2603.70 ft

(D) 2638.02 ft

83.2. What distance would be measured between the N$\frac{1}{4}$ corner of section 4 and the S$\frac{1}{4}$ corner of section 33?

(A) 46.86 ft

(B) 49.50 ft

(C) 55.12 ft

(D) 68.64 ft

PROBLEM 84

The following two problems refer to the illustration shown.

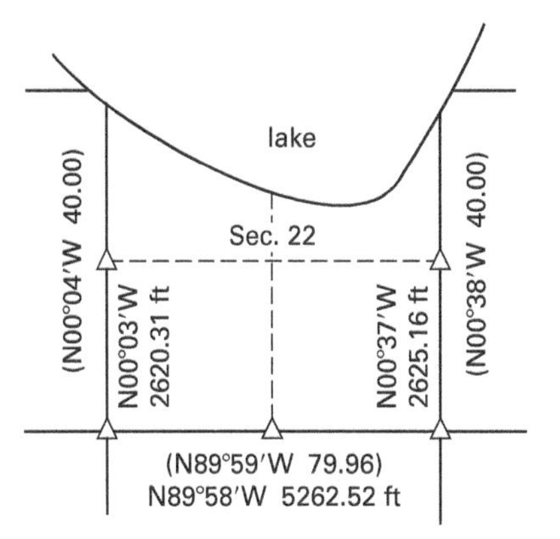

N89°58′W 5262.52 ft new measurement
(N89°59′W 79.96) record (in chains)
△ found original monument

84.1. The illustration indicates the record dimensions and the measurements of a retracement of a fractional section 22. The surveyor retracing the section wishes to monument the center of the section. What bearing should the surveyor use along the meridional centerline of the section?

(A) N00°00′00″W

(B) N00°19′24″W

(C) N00°20′00″W

(D) N00°20′12″W

84.2. The retracement surveyor wishes to set the NE corner of the SE$\frac{1}{4}$SE$\frac{1}{4}$, also known as the ES$\frac{1}{16}$. Which bearing and distance would represent the correct measurement from the SE corner of section 22 to this quarter-quarter corner?

(A) N00°37′W, 1312.58 ft

(B) N00°37′W, 1320.00 ft

(C) N00°38′W, 1312.58 ft

(D) N00°38′W, 1320.00 ft

PLSS

THE REESTABLISHMENT OF CORNERS

PROBLEM 85

Which statement best defines the meaning of the term *obliterated corner* as used in the Public Land Surveying System?

(A) The monument of an obliterated corner remains in place and can be verified by other record evidence.

(B) The position of an obliterated corner cannot be recovered beyond a reasonable doubt.

(C) There are no traces of the monument at an obliterated corner, but its position can be recovered beyond a reasonable doubt.

(D) The position of an obliterated corner can only be restored by reference to one or more interdependent corners.

PROBLEM 86

Which statement best defines the meaning of the term *existent corner* as used in the Public Land Surveying System?

(A) A corner should not be declared existent until every possible means has been unsuccessful in locating its true original position.

(B) No traces of the original monument or its accessories are found at the true position of an existent corner, nor can they be located by record evidence.

(C) It is a corner whose position is identifiable by the monument, accessories, or other physical or record evidence.

(D) It is a corner whose position can only be proven by reference to one or more other interdependent corners.

PROBLEM 87

The meaning of the term *lost corner* as it is used in the Public Land Surveying System is best defined as a corner whose position is indeterminable, EXCEPT by

(A) reference to one or more interdependent corners

(B) testimony of one or more witnesses who have dependable knowledge of its original location

(C) reference to its accessories

(D) reference to an acceptable supplemental survey record, physical evidence, or testimony

PROBLEM 88

Which of the following statements is NOT true of the use of proportionate measurement for the restoration of corners in the Public Land Surveying System?

(A) It should be considered a procedure of last resort.

(B) The new survey of a line must be adjusted so that it will be in proportionate agreement with the original reported length. The original reported length is considered the true length of the line.

(C) The extension of any proportionate measurement beyond an original monument is incorrect. Proportionate measurement should only be used between original monuments.

(D) Proportionate measurement cannot be applied to restore lost closing corners along a standard parallel.

PROBLEM 89

A double proportionate measurement would restore a lost

(A) quarter corner

(B) corner on a township boundary that is common to four sections

(C) standard corner on a correction line

(D) corner common to four townships

PROBLEM 90

Which of the following lost corners would NOT be restored by single proportionate measurement?

(A) a lost quarter section corner on a township boundary

(B) a lost quarter section corner on a section line within a township

(C) a lost meander corner

(D) none of the above

PLSS

PROBLEM 91

The following three problems refer to the illustration shown.

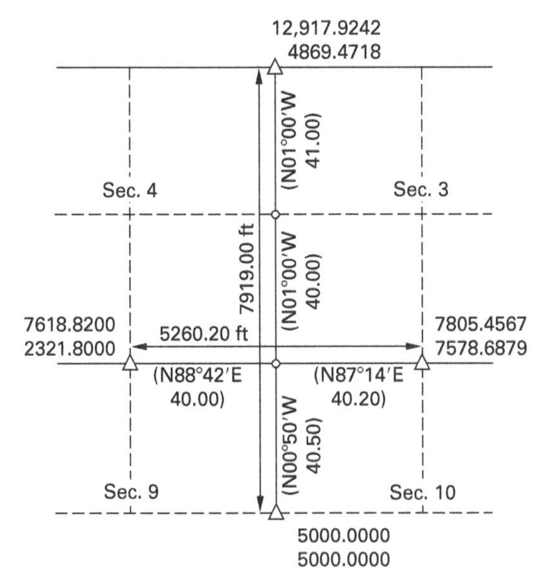

91.1. The surveyor who retraced the section lines indicated in the illustration wishes to reestablish both of the lost corners on the meridional line between sections 3 and 4. Which statement correctly describes the proper sequence of the work?

(A) First, the lost quarter section corner should be restored by single proportionment.

(B) First, the lost section corner should be restored by double proportionment.

(C) It makes no difference which of the two lost corners is restored first.

(D) First, the lost quarter section corner should be restored by double proportionment.

91.2. What are the coordinates of the restored section corner common to sections 3, 4, 9, and 10? (In the following work, the first coordinate is the northing and the second is the easting.)

(A) 7639.3081
4943.6892

(B) 7675.6068
4950.0899

(C) 7703.9512
4932.7405

(D) 7711.9056
4956.4906

91.3. Referring to the illustration, what are the coordinates of the restored quarter-section corner common to sections 3 and 4?

(A) 10,246.0321
4907.0386

(B) 10,263.1875
4943.6892

(C) 10,296.8641
4910.5398

(D) 10,307.6741
4982.2100

PROBLEM 92

In the following illustration, the NE corner of section 25 is lost. Which bearings and distances from the adjacent found original corners to the position of the restored NE corner of section 25 are correct?

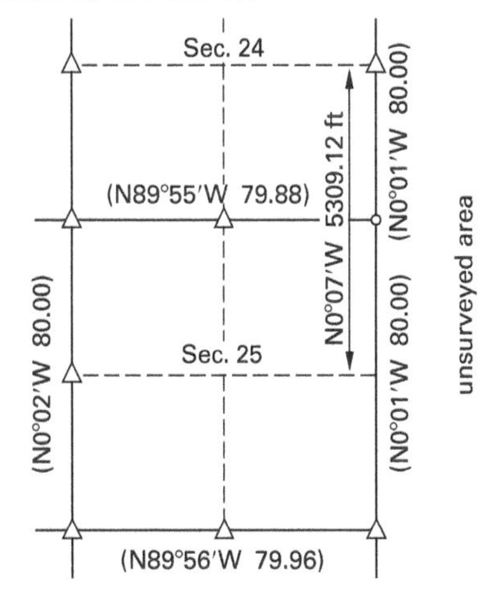

(A) from the E$\frac{1}{4}$ corner of section 25, N0°01'W 2640.00 ft to the restored NE corner of section 25

(B) from the N$\frac{1}{4}$ corner of section 25, N89°55'W 2636.04 ft to the restored NE corner of section 25

(C) from the E$\frac{1}{4}$ corner of section 25, N0°07'W 2654.56 ft to the restored NE corner of section 25

(D) both A and B

PLSS

PROBLEM 93

The following two problems refer to the illustation shown.

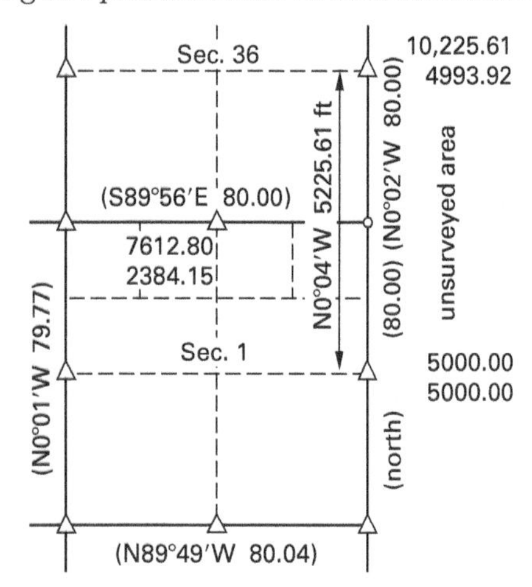

93.1. Which of the following procedures would be used to restore the lost NE corner of section 1?

(A) single proportionate measurement

(B) double proportionate measurement

(C) one-point control

(D) three-point control

93.2. Which of the following coordinates represent the position of the properly restored corner at the NE corner of section 1?

(A) 7609.73
5024.15

(B) 7612.80
4996.96

(C) 7612.80
5024.15

(D) 7702.71
5001.43

PROBLEM 94

The following two problems refer to the illustration shown.

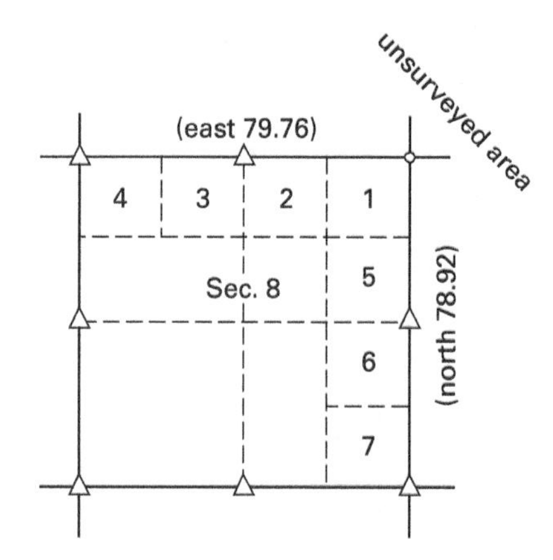

94.1. Which of the following procedures would be used to restore the lost NE corner of section 8?

(A) one-point control

(B) two-point control

(C) single proportionate measurement

(D) double proportionate measurement

94.2. To the nearest tenth of a chain, what would be the distances from the $E\frac{1}{4}$ and $N\frac{1}{4}$ corners of section 8 to the restored NE corner of the section?

(A) The distance from the $N\frac{1}{4}$ corner would be 39.9 chains. The distance from the $E\frac{1}{4}$ corner would be 39.5 chains.

(B) The distance from the $N\frac{1}{4}$ corner would be 39.9 chains. The distance from the $E\frac{1}{4}$ corner would be 38.9 chains.

(C) The distance from the $N\frac{1}{4}$ corner would be 39.8 chains. The distance from the $E\frac{1}{4}$ corner would be 39.5 chains.

(D) The distance from the $N\frac{1}{4}$ corner would be 39.8 chains. The distance from the $E\frac{1}{4}$ corner would be 38.9 chains.

PROBLEM 95

The following two problems refer to the illustration shown.

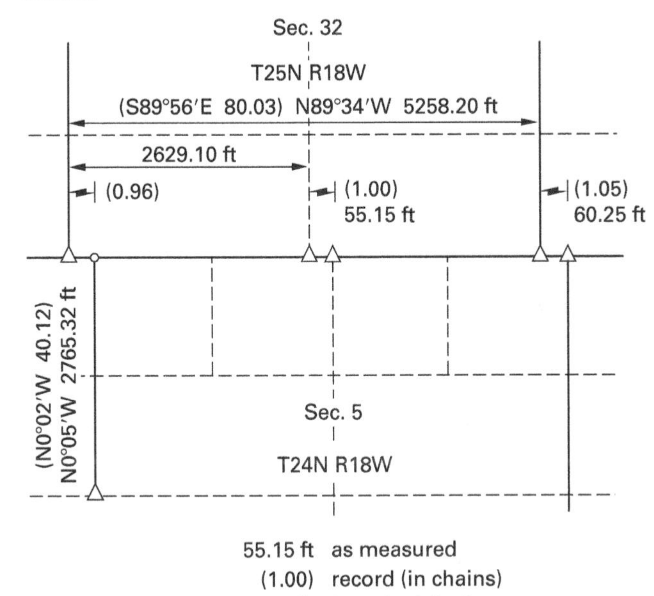

95.1. Which of the following procedures would be used to restore the lost NW corner of section 5?

(A) one-point control

(B) two-point control

(C) single proportionate measurement

(D) double proportionate measurement

95.2. To the nearest tenth of a foot, what distance would be measured from the properly restored closing corner at the NW corner of section 5 to the SW corner of section 32?

(A) 56.7 ft

(B) 57.7 ft

(C) 63.1 ft

(D) 63.4 ft

RESURVEYS

PROBLEM 96

Which of the following statements best defines the meaning of the term *dependent resurvey*?

(A) It is a resurvey that is based upon existing corners and acceptable control points from the original survey. It includes the restoration of lost corners by proportionate measurement.

(B) It is a resurvey that overrides the original survey, establishes new section lines, and also protects bona fide private land holdings in segregated tracts.

(C) It is a resurvey that is made to find the direction and distance between previously established corners, and to rehabilitate recovered corners. It does not include the restoration of lost corners.

(D) It is a resurvey that is made with the intention of finding the lengths of the lines previously established.

PROBLEM 97

Which of the following terms has the same meaning as dependent resurvey in the Public Land Surveying System?

(A) retracement

(B) remeasurement

(C) independent resurvey

(D) relocation

PROBLEM 98

Which of the following statements is NOT true of an independent resurvey in the Public Land Surveying System?

(A) It is a resurvey that supersedes the original survey of a portion of the public lands.

(B) Private bona fide rights are protected in the creation of tracts by metes and bounds surveys.

(C) The outboundary, which is the limit of the land to be independently surveyed, is established using the rules of a dependent resurvey.

(D) The lines established in an independent resurvey close upon the corners established by the original survey.

PROBLEM 99

The numbering of tracts that are to be segregated by a metes and bounds survey within an independent resurvey always begins with what number?

(A) 1

(B) 25

(C) 37

(D) 100

PROBLEM 100

Which of the following abbreviations is likely to appear on a monument at the corner of a tract in an independently resurveyed township?

(A) LM

(B) WP

(C) BO

(D) AP

PROBLEM 101

Which of the following statements is NOT true of a dependent resurvey in the Public Land Surveying System?

(A) Local points of control may be accepted, and such points are afforded the same authority as an identified original corner.

(B) The records kept of a dependent resurvey must conform to the specifications for original surveys.

(C) Monuments found remaining in place from the original survey are destroyed unless the point is needed to control a bona fide private claim.

(D) Both A and C are untrue.

LOTTING

PROBLEM 102

Which of the following best describes the meaning of the word *lot* as it is used in the Public Land Surveying System?

(A) a quarter-quarter section that adjoins the northern or western boundary of a township and bears a lot number

(B) any nonaliquot portion of a section

(C) a quarter-quarter section that contains less than 5 ac

(D) subdivisions of a section that are not described as aliquot parts but bear a lot number

PROBLEM 103

Which of the following conditions would likely NOT result in the creation of lots in a section?

(A) the sections along the exterior of a township absorbing the excess or deficiency of measurement accumulated within the township

(B) sections adjoining a meandered lake, pond, or river

(C) the closure of lines of the rectangular survey upon the boundaries of a reservation or national park

(D) a retracement of the north line of a section that previously contained no lots reveals a discrepancy with the original measurement of 3 min in direction and 4 links in length

PROBLEM 104

Which of the following statements is true of an elongated section?

(A) It contains no quarter-quarter sections.

(B) It contains more than two tiers of lots.

(C) It exceeds 85 chains in its length, width, or both.

(D) Both B and C are true.

PROBLEM 105

Which of the following statements correctly describes the system of lotting used in a section that has been the subject of an independent resurvey?

(A) The designation of the parts of such a section as lots is determined in the same way as in any other section that is subject to lotting.

(B) A portion of a resurveyed section may be given a lot number if it would otherwise have the same aliquot description it had before the resurvey.

(C) Lot numbers begin with the next number higher than the highest lot number previously used in the section.

(D) Both B and C are true.

PROBLEM 106

The following two problems refer to the illustration shown.

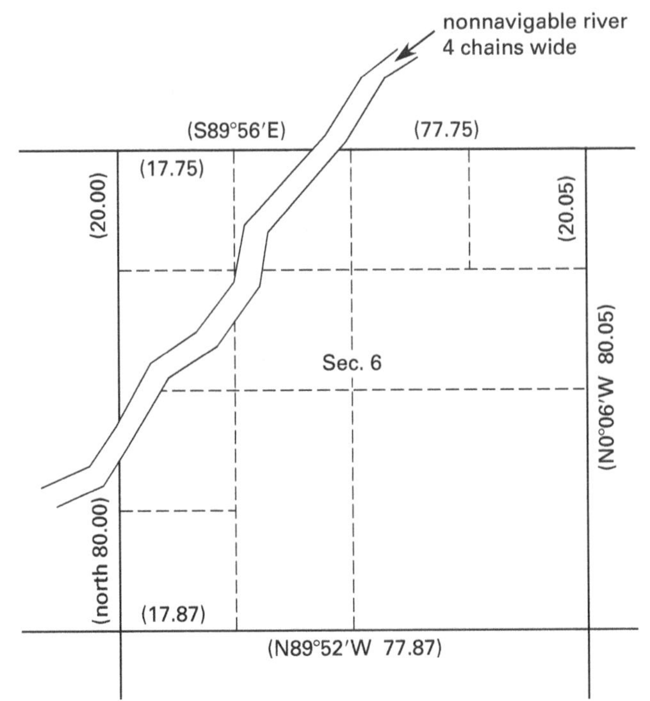

106.1. What number would be assigned to the most northwesterly lot in the section?

(A) 1

(B) 2

(C) 4

(D) 5

106.2. How many lots are there in the section?

(A) 6

(B) 7

(C) 8

(D) 10

PROBLEM 107

The following four problems refer to the illustration shown.

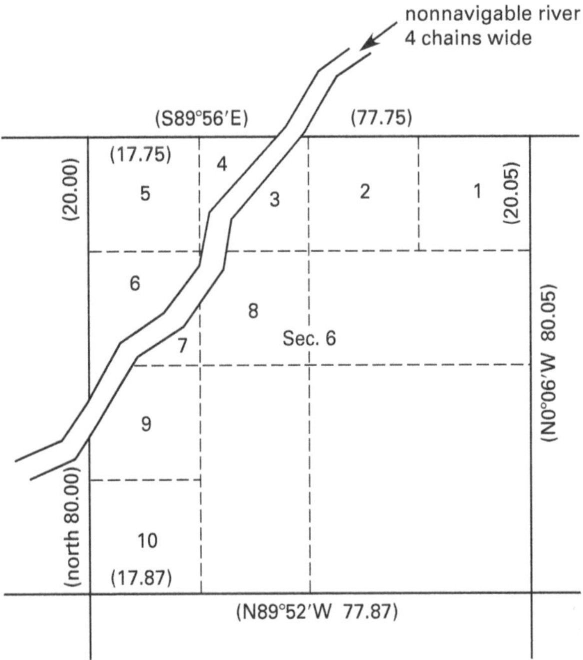

107.1. Which of the following statements gives the most likely reason that the southern boundary of lot 6 is NOT a portion of the latitudinal centerline of the section?

(A) An error was made by the draftsman who protracted the section.

(B) The southern boundary of lot 6 is a portion of the centerline of the section, regardless of what appears on the plat.

(C) The portion of land created between the meander line, the centerline, and the western boundary of the section was considered too small to be a lot.

(D) The area labeled as lot 6 is actually not a lot.

PLSS

107.2. What is the implied record distance from the southeastern corner of lot 7 to the W$\frac{1}{4}$ corner of section 6?

(A) 17.75 chains

(B) 17.77 chains

(C) 17.81 chains

(D) 20.00 chains

107.3. What are the implied record lengths of the eastern and southern boundaries of lot 5?

(A) eastern = 20.00 chains
southern = 17.75 chains

(B) eastern = 20.01 chains
southern = 17.78 chains

(C) eastern = 20.0125 chains
southern = 17.78 chains

(D) eastern = 20.10 chains
southern = 17.90 chains

107.4. On the west side, the ownership of the patent holder of lot 3 most likely ends at which of the following lines?

(A) meander line

(B) mean high water line

(C) low water line

(D) thread of the stream

PROBLEM 108

In modern practice, the creation of poorly shaped or odd-sized lots is avoided. Which of the following statements correctly defines the desirable limits on the areas of lots?

(A) A lot with a width of 20 chains should contain between 10 ac and 50 ac.

(B) A lot along an irregular boundary should contain between 5 ac and 45 ac.

(C) A lot along an irregular boundary should contain between 10 ac and 60 ac.

(D) Both A and B are true.

PROBLEM 109

The following two problems refer to the illustration shown.

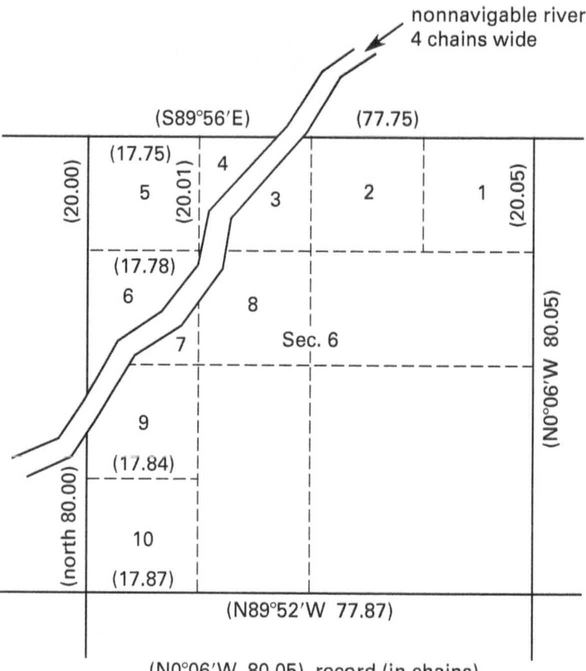

109.1. What is the record area of lot 10?

(A) 35.60 ac

(B) 35.71 ac

(C) 36.21 ac

(D) 40.00 ac

109.2. What is the record area of lot 5?

(A) 35.54 ac

(B) 35.59 ac

(C) 35.65 ac

(D) 35.71 ac

PROBLEM 110

A lot with regular boundaries has a northern and southern boundary of 20.00 chains. The record area of the lot is 39.72 ac. The eastern boundary of the lot has a record length of 19.84 chains. What is the record length of the western boundary of the lot?

(A) 19.84 chains

(B) 19.88 chains

(C) 19.90 chains

(D) 20.00 chains

EXCEPTIONS AND DISCREPANCIES IN THE PUBLIC RECORD

PROBLEM 111

Which of the following statements is NOT correct regarding corners and lines that have been accepted as the official record by the U.S. Secretary of the Interior?

(A) They are established as the proper corners and boundary lines, and their positions and lengths cannot be changed by any private land survey.

(B) They are considered to enclose the exact quantity of land expressed on the official township plat.

(C) The lines of subdivision that have not actually been run by the official government survey should be established by running straight lines between opposite corresponding corners.

(D) The Bureau of Land Management has the authority to conduct an official resurvey of any township.

PROBLEM 112

In general, meander lines are not legal boundaries. Which statement correctly describes a circumstance that could cause a meander line to become a legal boundary?

(A) Lotting within a section is shown abutting a meander line on the official township plat.

(B) The meander line occurs along a river that is part of the boundary between two states.

(C) The meander line occurs on an island.

(D) The meander line was established where no meanderable body of water ever existed.

PROBLEM 113

Much of the early surveying in the Public Land Surveying System was done with a solar compass and a linked chain. The discrepancies between the design and the reality of the monumented corners may be larger than would be expected if the work were done today with modern equipment. Regarding this, which of the following statements is correct?

(A) The monuments set during the execution of the original survey, and the boundary lines they describe, are usually incorrect.

(B) Subsequent retracements take precedence over the monuments set during the execution of the original survey and the boundary lines they describe

(C) Subsequent retracements do not take precedence over the monuments set during the execution of the original survey, but they do override the boundary lines they describe.

(D) The monuments set during the execution of the original survey and the boundary lines they describe are correct and inviolable by law.

RIGHTS OF POSSESSION IN PUBLIC LANDS

PROBLEM 114

In general, unwritten property rights will not ripen against the federal government, but exceptions can arise. Which of the following statements correctly represents an EXCEPTION to the general rule?

(A) The Bureau of Land Management is authorized to issue a patent to individuals who claim adverse possession of public domain lands under the Color of Title Act.

(B) In some cases, land may be acquired by adverse possession from federal agencies when such land has been previously patented into private ownership and has since been reacquired by the federal government.

(C) In some cases, land held by federal agencies never was part of the public domain; some of these lands may be acquired by adverse possession.

(D) All of the above are true.

MEANDERING

PROBLEM 115

Which of the following best defines the meaning of the term *meander line*?

(A) a traverse run along the mean low water line of a natural body of water

(B) a line run to indicate the sinuosities of the bank or shore line of a permanent body of water

(C) a line used to determine the approximate area of land in fractional sections after the segregation of the water area

(D) both B and C

PROBLEM 116

Which of the following best defines the meaning of the term *meander corner*?

(A) a corner established at each deflection point of a meander line

(B) a corner established at the intersection of each aliquot line with a meander line

(C) a randomly placed corner used to control a meander line

(D) a corner established at the intersection of a meander line with a standard, township, or section line

PROBLEM 117

Which of the following intersections would be properly monumented by a special meander corner?

(A) where a surveyed aliquot line intersects a meander line

(B) where a calculated section centerline intersects a meander line

(C) where a meander line encloses swamp or overflow lands

(D) both A and B

PROBLEM 118

Which of the following conditions would be accommodated by the establishment of an auxiliary meander corner?

(A) a meandered lake entirely enclosed within a quarter section

(B) a meanderable body of water in an unsurveyed area of a township

(C) an island too small to subdivide that lies entirely in one section

(D) both A and C

PROBLEM 119

Under rules outlined in the *Manual of Surveying Instructions*, which of the following should be meandered?

(A) rivers and streams having a width of 3 chains or more, whether navigable or not

(B) lakes covering 40 ac or more

(C) swamp lands

(D) islands above the mean high water line that have arisen after the state in which they stand was admitted into the United States

MINERAL SURVEYS

PROBLEM 120

Which of the following acts contains provisions that are still in effect regarding the regulation of mining claims on public land?

(A) Act of July 25, 1866

(B) Act of March 3, 1853

(C) Act of February 5, 1805

(D) Act of May 10, 1872

PROBLEM 121

Which type of surveyor is authorized to survey a mineral claim on public land in preparation for a patent?

(A) licensed land surveyor

(B) Bureau of Land Management cadastral surveyor

(C) U.S. deputy mineral land surveyor

(D) licensed geologist

PROBLEM 122

In surveying a mineral lode claim for patent, a U.S. deputy land surveyor must make sure that the claim conforms to which of the following specifications?

(A) an aliquot part of a section

(B) 20 chains by 40 chains

(C) 600 ft by 1500 ft

(D) 800 ft by 2000 ft

PROBLEM 123

A claimant may wish to make the end lines of a mineral lode claim parallel to one another. Why is such a configuration advantageous to the claimant?

(A) It guarantees that the claim contains the maximum possible area on the surface.

(B) It is required by the 1872 Mining Act.

(C) It allows for the maximum extralateral rights.

(D) The claimant may not otherwise amend the lines when the lode claim is surveyed for patent.

PROBLEM 124

What sort of mineral deposits are appropriate for inclusion in a placer claim?

(A) alluvial deposits

(B) veins in place

(C) glacial deposits

(D) both A and C

PROBLEM 125

Currently governing the configuration of a placer claim is the provision that it must

(A) be no more than 600 ft by 1500 ft

(B) include no more than 300 ft on each side of the thread of a stream

(C) enclose no more than 160 ac

(D) conform as nearly as is practical to the rectangular subdivisions of the Public Land Surveying System

PROBLEM 126

What is the maximum area of public land an individual may hold as a placer claim under current provisions?

(A) 10 ac

(B) 20 ac

(C) 40 ac

(D) 80 ac

PROBLEM 127

What is the maximum area of public land an individual may claim as a mill site?

(A) $2\frac{1}{2}$ ac

(B) 5 ac

(C) 20 ac

(D) 40 ac

PROBLEM 128

Which of the following provisions is NOT required of a mineral survey for the patent of a lode claim?

(A) The lengths of the lines are to be returned in feet rather than chains.

(B) A true astronomic bearing must be determined on one of the lines of the survey.

(C) The mineral surveyor may move the claim stakes in, but not out, to bring it within statutory limits.

(D) The mineral survey must show the patented lode claim in the same position as recorded in the original location certificate.

SOLUTION 1

Passed by Congress under the Articles of Confederation in 1785, the Land Ordinance Act established many elements of the Public Land Surveying System that remain virtually unchanged today. The Act created 640 ac sections (which were called *lots*) and 6 mi townships, and required that public lands be surveyed before they were sold.

The answer is (D).

SOLUTION 2

The General Land Office was proposed by Alexander Hamilton as early as 1790, but was not actually created until 1812. The office of the Surveyor General was first mentioned in the Act of 1796; the Bureau of Land Management did not begin until 1946. The office of the Geographer of the United States was created in the Land Ordinance of 1785. Thomas Hutchins was the first and only person to hold the office, and he conducted the first survey of the public lands in the United States.

The answer is (D).

SOLUTION 3

The baselines in the Public Land Surveying System are lines that follow a parallel of latitude from an initial point.

The answer is (D).

SOLUTION 4

From 1797 to 1910, public land surveys and the preparation of field notes were done by surveyors under contract with the General Land Office. These private contractors were generally paid per mile of line surveyed. The contract system was replaced by the direct system, which required that the surveys and resurveys be performed by government employees.

The answer is (D).

SOLUTION 5

From the beginning, the contradictory rules that townships be both 6 mi square and bounded by cardinal lines plagued the public land surveys. Compensation for convergence of meridians finally was addressed under instructions from Edward Tiffin for the survey in Indiana. The standard parallel, or correction line, was used.

The answer is (D).

SOLUTION 6

One of the initial purposes of the sale of public lands was to raise revenue. After the Revolutionary War, the United States had a great deal of land but little money. The sale of public lands provided a method for paying the huge foreign debt, among other things. Therefore, it was natural that the Department of the Treasury was in charge of the early public land surveys.

The answer is (B).

SOLUTION 7

Under the 1872 Mining Act, a patented lode claim is 300 ft on each side of the vein for a maximum length of 1500 ft.

The answer is (B).

SOLUTION 8

When the alignment of the governing east boundary of a township is irregular, the first meridional line west from the defective boundary is extended north to its intersection with the north boundary of the township. This line is called a *sectional guide meridian*. It allows the subdivision of the sections to its west in accordance with standard practice. East of the sectional guide meridian, fractional portions of the latitudinal section lines are placed in the east half-mile.

The answer is (A).

SOLUTION 9

One of the reasons for the eventual rejection of Thomas Jefferson's proposal that the nautical mile be used in the public land surveys was that it was not evenly divisible into chains. The statute mile, equal to 80 chains, was preferred. Under this system, 10 sq chains is equal to 1 ac.

The answer is (B).

SOLUTION 10

Gunter's chain is composed of 100 links. One hundredth of a chain is 1 link; therefore, 0.28 chain is equal to 28 links.

The answer is (D).

SOLUTION 11

A pole, also known as a *rod* or a *perch*, has a length of 16.5 ft. Therefore, a two-pole chain is a chain that is 33.00 ft in length.

The answer is (D).

SOLUTION 12

The earliest Gunter's chains were constructed of iron (later, steel) links joined together by rings. The deterioration of the metal-on-metal joints caused the chains to gradually increase in length.

The answer is (C).

SOLUTION 13

A chain is 66.00 ft.

$$(79.52 \text{ chains})\left(66.00 \ \frac{\text{ft}}{\text{chain}}\right) = 5248.32 \text{ ft}$$

The answer is (D).

SOLUTION 14

In 1844, the townships of Saginaw Bay were resurveyed under special appropriations from Congress, and the solar compass was the instrument specified for the work. The instrument's faculty for avoiding the hazards of local magnetic attraction made it a valuable tool in areas with high concentrations of mineral deposits.

The 1890 *Manual of Surveying Instructions* outlawed the use of the magnetic needle except in subdivision and meandering; the 1894 *Manual* required that all lines refer to the true meridian.

The answer is (D).

SOLUTION 15

The U.S. Department of the Interior Bureau of Land Management is responsible for preparing and publishing the *Manual of Surveying Instructions.* The Public Land Survey System Foundation and the National Society of Professional Surveyors work in conjunction with the U.S. Department of the Interior Bureau of Land Management to sell and distribute the *Manual,* but they do not prepare or publish it.

The answer is (B).

SOLUTION 16

The *Manual of Surveying Instructions* provides the guidelines used by the Bureau of Land Management to survey public lands. Original surveys of public lands are presumed to have been in accordance with the *Manual* or special procedures enforced at the time. However, such instructions are only advisory to land surveyors in private practice retracing privately held land, unless the instructions have been adopted by a state statute or state court. Many states have adopted the instructions published by the Bureau of Land Management, making these procedures mandatory in the retracement of public lands within their boundaries.

The answer is (D).

SOLUTION 17

Meander lines are not established in swamp or overflowed lands; therefore, such procedures are not covered in the *Manual of Surveying Instructions.*

The answer is (C).

SOLUTION 18

Per the 2009 *Manual of Surveying Instructions,* the shift in dominant federal policy regarding public lands is from one favoring disposal and settling of unreserved public lands to one favoring retention, administration, and control. The Federal Land Policy Management Act of 1976 (FLPMA) halted privatization of public lands and formally recognized the federal government's then long-established policy of retaining public lands. Under the FLPMA, public lands are not to be disposed of unless their disposal would "serve the national interest." The Taylor Grazing Act of 1934 (TGA) directed the Secretary of the Interior to manage the public domain "pending ... final disposal" so as to "regulate their occupancy and use, to preserve the land and its resources from destruction or unnecessary injury," rather than there being a shift from one policy to another. Cession had to do with the early 19th century view that states were entitled to be given title to public lands within their borders. Newer states favored reducing the price of public lands offered for sale and giving any leftover lands to the states (graduation), and older states favored policies ensuring the older states derived some direct value from public lands (distribution). Cession referred to both these options. The Land and Water Conservation Fund, established in 1964, provided substantial federal funding for a series of outdoor recreation plans to achieve the objectives of federal land management agencies and had nothing to do with opposing privatization of public lands.

The answer is (B).

SOLUTION 19

The 1973 *Manual of Surveying Instructions* does not assert that coordinates may provide the best evidence of a corner position, but the 2009 manual does. To some degree, this is a recognition of the advancements in technology since 1973 and the fact that repeatable coordinates from satellite positioning technology (GNSS) enable corner positions to be "witnessed" by them. In theory, a subsequent surveyor can determine the same point with an acceptable degree of confidence by following the previous surveyor's computational footsteps. In other words, coordinates may provide the best available evidence of a corner position in some circumstances. Both the 1973 manual and the 2009 manual include information on mineral surveys. However, the subject has been expanded in the 2009 edition with incorporation of information from the Bureau of Land Management's (BLM) *Mineral Survey Procedures Guide* and instructions

PLSS

on the resurvey of mineral lands. The 2009 manual explains "substantial evidence" is adequate to determine whether a corner is existent, obliterated, or lost, whereas the 1973 manual insists that such a determination must be "beyond reasonable doubt." Both the 1973 manual and the 2009 manual include information on water boundaries, but the 2009 manual includes more details about submerged lands, navigability, and un-surveyed islands in meandered, non-navigable streams.

The answer is (B).

SOLUTION 20

Upon completing a survey of one or more exteriors of either a regular or irregular township, the maximum allowable error of closure is 1/4000 of the perimeter in either latitude or departure.

The answer is (C).

SOLUTION 21

A section is considered regular if its boundaries do not depart from cardinal directions by more than 00°21′. This prescribed limit, among others, is considered in determining whether lines within a survey are defective.

The answer is (D).

SOLUTION 22

The distance between regular corners in a public lands survey cannot exceed 25 links in 40 chains if the section is to be considered regular.

The answer is (D).

SOLUTION 23

The initial point is the origin of the principal meridian and baseline that control the public lands surveys of an area. However, some public lands states have no initial point within their borders (Minnesota, for example), and others (Ohio, for example) have more than one.

The answer is (D).

SOLUTION 24

The procedure for running the principal meridians and baselines of the Public Land Surveying System calls for the monumentation of all quarter-section, section, and township corners along the lines. Meander corners are to be monumented as well when these standard lines intersect meanderable bodies of water. Closing corners are set during subsequent surveys.

The answer is (D).

SOLUTION 25

The maximum allowable discrepancy in measurement of rectangular limits in subdivisional surveys is 25 links in 40 chains, but the survey of the standard lines must be run to a higher precision. The allowable discrepancy between the two sets of measurements along a standard line is 2 links per 80 chains.

The answer is (C).

SOLUTION 26

26.1. The guide meridians are numbered consecutively east and west from the principal meridian. They follow true astronomical meridians, so they converge.

The answer is (D).

26.2. The standard parallels are numbered consecutively north and south from the baseline. Standard parallels are also known as *correction lines*.

The answer is (D).

26.3. Point X is a standard corner and would be set 6 mi west from the initial point in an idealized quadrangle. Due to the convergence of meridians, point Y would be set less than 6 mi from the initial point. It is known as a *closing corner*.

The answer is (A).

26.4. Ranges are numbered consecutively east and west from the principal meridian. Townships are numbered consecutively north and south from the baseline. This system continues throughout the area covered by a particular principal meridian and baseline.

The answer is (D).

26.5. The meridional township boundaries are intended to follow true astronomical meridians, not to be parallel with the principal meridian.

The answer is (A).

26.6. The closing of a meridional township boundary on a standard parallel is similar to the closing of a guide meridian on a standard parallel. In both cases, a closing corner is set at the intersection of the meridian with the parallel. The meridional township boundaries are begun at the standard corners set at intervals of 480 chains along the standard parallel on the southern boundary of a quadrangle. The boundaries end at the closing corners, which, due to the convergence of the meridians, will be closer together than 480 chains along the standard parallel, which is the northern boundary of a quadrangle.

The answer is (C).

SOLUTION 27

Townships are intended to be 6 mi square, so the interval between successive guide meridians or standard parallels would most likely be divisible by six. Of the answer choices given, only 36 is divisible by six.

The answer is (D).

SOLUTION 28

A baseline was laid out east and west from the initial point. It was intended to follow an astronomically determined parallel of latitude.

The answer is (A).

SOLUTION 29

A principal meridian was extended north and south from the initial point following an astronomically determined meridian of longitude on which corner monuments were set every 40 chains, or every half mile.

The answer is (B).

SOLUTION 30

In the Public Land Surveying System, quandrangles are large rectangular areas bounded by meridians of longitude and parallels of latitude.

The answer is (D).

SOLUTION 31

Townships are the Public Land Surveying System's unit of survey. The quadrangle is the framework enclosing the land from which 16 townships could be created.

The answer is (D).

SOLUTION 32

32.1. The most easterly range of sections is the first to be surveyed in the subdivision of a regular township under current instructions. The subdivisional survey begins at the SW corner of section 36. The western boundary of section 36 is established first, then the survey proceeds across its northern boundary and continues along the western boundary of section 25.

The answer is (D).

32.2. The eastern and southern boundaries of a township are often known as the *governing boundaries*. The meridional section lines in a township are intended to be parallel with the eastern boundary of the township, while the latitudinal boundaries are intended to be parallel with the southern boundary of the township.

The answer is (B).

32.3. The northern boundary of T4N R3E would normally be a standard parallel. The meridional section lines along the northern tier would, under typical conditions, terminate in closing corners.

The answer is (B).

SOLUTION 33

Temporary quarter-section corners are not set on the meridional section lines since these lines are not established using the random and true line method, except when closing on a standard parallel. Most meridional section lines are run on the true line, unlike the latitudinal section lines, which are most often established using a random and true line method.

The answer is (C).

SOLUTION 34

A township that cannot be normally subdivided into its full complement of 36 sections due to the presence of water—which cannot be properly included within the township or closings—is considered a fractional township.

Closings on state boundaries or conditions that cause the creation of half-ranges or half-townships are typical causes of fractional townships.

The answer is (C).

SOLUTION 35

North-south township and section lines are known as meridional in the Public Land Surveying System.

The answer is (C).

SOLUTION 36

The township, section, and quarter corner monuments were in place on the township boundary when the normal subdivision of a quadrangle into townships was completed according to plan. Most of this work was carried out every year from 1785 to the end of the 19th century.

According to the design of the Public Land Surveying System, when the quadrangle was subdivided into four tiers and four ranges of townships, 16 townships were created with boundaries very nearly meridians of longitude and parallels of latitude. The inevitable errors in the east-west distance measurements were taken up in the last half mile on the west side of each township. The errors in the north-south measurements were thrown into the last half mile on the north side of the northern tier of townships, just south of the north boundary of the quadrangle. All four boundaries of all 16 townships had corner monuments in place approximately every 40 chains.

The answer is (C).

SOLUTION 37

Section lines are intended to be parallel to the governing boundaries of the township. The governing boundaries are the east and the south township bound aries, unless those township boundaries are defective.

The answer is (B).

SOLUTION 38

When the notes of the original surveys of the townships in the Public Lands were returned to the General Land Office (GLO), and later to the Bureau of Land Management (BLM), plats were developed from the work. These plats represented the township included in the survey. They showed the direction and length of each line, and their relation to the adjoining surveys. The township plats included some indication of the relief, boundaries, descriptions, and areas of each subdivision of the sections.

The government did not, and will not, convey lands in the public domain to others until a survey is completed and the official plat has been filed and approved. This is true in part because the deed, or the patent, that eventually grants the ownership of the land, does so by direct reference to the township plat. It is an integral part of the transaction and binds the parties to the specific dimensions of the land as shown on the township plat. Therefore, the sections on the township plat's orientation and size are significant, as are the creation of the lines on the township plat by protraction. Protraction in this context refers to drawing lines on the township plat that were not actually run or monumented on the ground during the official subdivision of the land.

The answer is (D).

SOLUTION 39

39.1. An aliquot corner in the Public Land Surveying System is a corner of an aliquot part. Aliquot parts are established by the legal subdivision of sections into halves and fourths. These halves and fourths may then be subdivided into further aliquot parts, ad infinitum.

The corner labeled A is an aliquot corner. It is also the S$\frac{1}{4}$ section corner of section 15. Section 15 is directly north of section 22 in a regular township.

The answer is (D).

39.2. The corner labeled B is described by all of the answers given. Other names would also be correct, including the SW corner of the NE$\frac{1}{4}$NW$\frac{1}{4}$ of section 27.

The answer is (D).

39.3. The corner labeled C in section 23 is not correctly described by any of the names given. One name that would correctly describe it is the center quarter (C$\frac{1}{4}$) of section 23.

The answer is (D).

39.4. The center quarter corners and the center sixteenth corners of sections are located by intersection of quarter lines and sixteenth lines, respectively. The N$\frac{1}{4}$ corner of section 22, like an aliquot corner along a sectional boundary, is located by measurement along the section line.

The answer is (C).

39.5. Quarter-section corners that occur on the section lines within a township are usually monumented during the subdivision of a township into sections. The sixteenth and center quarter-section corners may occasionally be set under special instructions.

The answer is (A).

39.6. The crosshatched area is approximately 40 chains by 20 chains. The calculation of the acreage of the crosshatched area is made more convenient by remembering that 10 sq chains are equal to 1 ac.

$$(40 \text{ chains})(20 \text{ chains}) = 800 \text{ sq chains}$$

$$(800 \text{ sq chains})\left(\frac{1 \text{ ac}}{10 \text{ sq chains}}\right) = 80 \text{ ac}$$

The answer is (D).

39.7. The crosshatched area is in the SE$\frac{1}{4}$ of section 27, and is its E$\frac{1}{2}$.

The answer is (C).

SOLUTION 40

40.1. The crosshatched area is approximately 10 chains by 10 chains.

$$(10 \text{ chains})(10 \text{ chains}) = 100 \text{ sq chains}$$

$$(100 \text{ sq chains})\left(\frac{1 \text{ ac}}{10 \text{ sq chains}}\right) = 10 \text{ ac}$$

The answer is (B).

40.2. The crosshatched aliquot part in section 6 is in the SE$\frac{1}{4}$ of the section. It is in the SW$\frac{1}{4}$ of the SE$\frac{1}{4}$ of the section. Finally, it is the NE$\frac{1}{4}$ of the SW$\frac{1}{4}$ of the SE$\frac{1}{4}$ of the section.

The answer is (A).

40.3. The names given to aliquot parts of a section are based on the ratio of their area to the area of the full section. For example, a quarter section would be approximately one-quarter of the area of the full section, or 160 ac.

The crosshatched portion of section 6 contains approximately 10 ac.

$$\frac{\text{approximate area of the aliquot part}}{\text{approximate area of the full section}} = \frac{10 \text{ ac}}{640 \text{ ac}} = \frac{1}{64}$$

The aliquot part crosshatched in section 6 is a sixty-fourth part of the area of the full section. It is called a sixty-fourth.

The answer is (D).

SOLUTION 41

The straight lines that connect the quarter-section corners on opposite sides of the section intersect at the center quarter corner. The point of intersection can be readily found by using a programmable calculator, or by solving the triangle created by the E$\frac{1}{4}$, S$\frac{1}{4}$, and C$\frac{1}{4}$ corners.

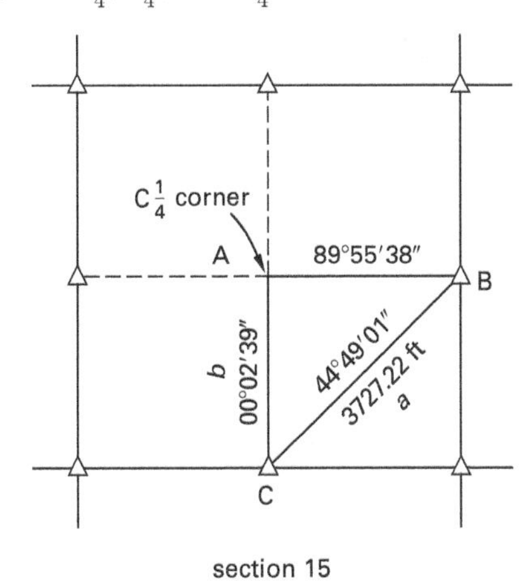

section 15

△ found original monument

The azimuths and distance shown are found by inversing between the given coordinates. The triangle may be solved by the law of sines.

$$\begin{aligned} A &= 180° - 89°55'38'' + 0°02'39'' \\ &= 90°07'01'' \\ B &= 89°55'38'' - 44°49'01'' = 45°06'37'' \\ C &= 44°49'01'' - 0°02'39'' = 44°46'22'' \end{aligned}$$

$$\frac{a}{\sin A} = \frac{b}{\sin B} = \frac{c}{\sin C}$$

$$\frac{3727.22 \text{ ft}}{\sin 90°07'01''} = \frac{b}{\sin 45°06'37''}$$

$$b = 2640.62 \text{ ft}$$

$$\frac{3727.22 \text{ ft}}{\sin 90°07'01''} = \frac{c}{\sin 44°46'22''}$$

$$c = 2625.08 \text{ ft}$$

These distances and the azimuths found by inversing make it possible to find the coordinates of the C$\frac{1}{4}$ corner.

C$\frac{1}{4}$ corner N7639.3431
E7616.1252

The answer is (B).

SOLUTION 42

The quarter-quarter corners, or sixteenth corners, of a section are normally placed at the midpoint of the lines between established corners. The exceptions to such placement usually involve closing corners, which normally occur in the sections along the northern and western boundaries of a township.

The answer is (D).

SOLUTION 43

The center quarter corner is found by intersecting the centerlines through a regular section. The intersection of the two lines almost certainly does not bisect either line.

The answer is (D).

SOLUTION 44

Each of the quarter-quarter corners, the quarter corners, and the section corners is considered a public land corner. Even though all 25 corners are not monumented, they are nevertheless corners.

The answer is (D).

SOLUTION 45

45.1. The broken lines found on official township plats subdivide the sections by protraction. These lines exist only on paper and were not normally surveyed in the field.

The answer is (C).

45.2. The excess or deficiency of the original subdivision along the latitudinal boundaries of the township is absorbed by the last quarter mile. The last half mile is not subdivided at its midpoint. The sixteenth corner is set 20 chains west from the last regular quarter section corner, throwing the excess or deficiency into the last quarter mile. Therefore, 79.16 chains − 60 chains = 19.16 chains.

The answer is (A).

45.3. The measurement implied for the distance labeled L is 40 chains. One of the principles of the subdivision of sections within a township is to provide as many regular aliquot parts as possible. Throwing the excess or deficiency into the northern and western sides of the township allows the remainder of the subdivisions to be regular. This principle was first established by the Act of May 10, 1800.

The answer is (B).

45.4. The only distance that closes on the north or west boundary of the township is I. The distance labeled I should be 20.08 chains rather than 20.00 chains.

The answer is (A).

45.5. In the Public Land Surveying System, the term *aliquot* means the division of a section or other aliquot part by one-half or one-quarter. This division is not precise in terms of area, but is based upon the bisection and intersection of the lines bounding other aliquot parts. A section contains approximately 640 ac, a quarter section contains approximately 160 ac, and a quarter-quarter ($1/16$) of a section contains approximately 40 ac. Therefore, the corners of the quarter-quarter of a section are sometimes called sixteenth corners.

The answer is (A).

SOLUTION 46

Since they absorb any excess or deficiency, the portions of section 6 protracted along the north and west boundaries are irregular. They are called *lots*, not aliquot parts.

The answer is (D).

SOLUTION 47

47.1. Township and section lines are closed on standard parallels, also known as *correction lines.* Closing corners are normally established at the intersection of these closing lines and the correction lines. Therefore, the corners of the sections that are south of the correction line are known as *closing corners*, and the corners of the sections that are north of the correction line are known as *standard corners.*

The answer is (C).

47.2. Under normal circumstances, the first corners established on a standard parallel are the standard township, section, and quarter-section corners. These corners are set during the initial running of the standard parallel. When the quadrangle defined by the standard parallels and guide meridians is divided into townships, the township lines from south of the standard parallel are closed upon it, and closing corners are set at the intersections. Finally, when the township south of the standard parallel is subdivided into sections, the section lines are closed upon the standard parallel and closing corners are set at the intersections.

The answer is (A).

47.3. Original monuments should not be destroyed. A closing corner monument found off the standard parallel does determine the direction of the closing line, in this case the west line of section 1. However, while the monument may not stand on the standard parallel, the closing corner must. The monument controls the direction of the closing line, but not its terminus. The closing corner is at the intersection.

The answer is (C).

SOLUTION 48

The baseline referred to in the field notes is a standard parallel or correction line; standard refers to a standard corner that had been previously established when the baseline was surveyed. The line between sections 4 and 5 would be a closing line in such a situation, so the missing words are "closing corner."

The answer is (D).

SOLUTION 49

Once a boundary is established and bona fide rights have arisen, those rights cannot be violated by subsequent surveys of the public lands. This principle guides the establishment of closing corners in situations where previously settled boundaries would otherwise be disturbed.

In the case of the intersection of a standard, section, or township line with a meander line, no closing corner is set. A meander corner is established, but such a corner is distinctly different from a closing corner. A meander line is not intended to represent a legal boundary, but rather to indicate the segregation of a body of water to ascertain the quantity that remains. Therefore, the meander corner, unlike a closing corner, is not intended to be a legal boundary monument.

The answer is (B).

SOLUTION 50

Corners defining boundaries between states are not set under general authority of the Bureau of Land Management. They were occasionally set under specific authorization of Congress by the General Land Office, and sometimes are resurveyed under the direction of the Supreme Court or the states, with Congress' consent. Closing corners set during the normal course of a survey of the public domain are not stamped with the names of the states. The absence of such stamping is intended to prevent the closing corners from being misconstrued as definitions of the state boundaries.

The answer is (C).

SOLUTION 51

Careful measurement of the distance and direction from a closing corner monument to the nearest standard to the east and west corners is suggested as a check on the closing corner's location. This procedure also facilitates the reestablishment of a lost closing corner.

The answer is (C).

SOLUTION 52

Directions are ignored in double proportionate measurement in the Public Land Surveying System. This method of reestablishment of lost monuments is generally applied to lost corners common to four townships and four interior sections.

The answer is (B).

SOLUTION 53

Corner accessories are considered part of the monumentation of the corner itself. They are physical objects whose distances and/or directions from the corner are known.

The answer is (D).

SOLUTION 54

In the Public Land Surveying System, three terms are used to describe the condition of monumentation: existent, obliterated, and lost. An existent corner can be found by its monumentation, accessories, field notes, or some other acceptable supplemental evidence or testimony. An obliterated corner cannot be located by evidence of its monumentation or accessories, but may be located beyond a reasonable doubt by testimony or some acceptable record evidence. A lost corner cannot be located by testimony, physical, or record evidence. A lost corner can only be reestablished by reference to other known corners.

The answer is (B).

SOLUTION 55

A corner is a point on the earth's surface where two lines meet, and a monument is a physical object that marks the position of a corner. While the terms are frequently used synonymously, strictly speaking they are not the same. In the Public Land Surveying System, a monument set by an original surveyor to represent a particular corner is understood to exactly locate the position of that corner, except off-line closing corner monuments. However, subsequent monuments may or may not stand precisely in the position of the corners they are intended to represent.

The answer is (D).

SOLUTION 56

Proportionate measurement is a technique of last resort, used to reestablish the monumentation of lost corners. An obliterated monument is not considered lost; it can be located beyond a reasonable doubt by acceptable evidence.

The answer is (C).

SOLUTION 57

Amended monuments are not only used to rectify off-line closing corners, but are also frequently found in such situations. The off-line monument, after being stamped AM, is connected by course and distance to the newly established monument at the true intersection of the closing line and the line closed upon.

The answer is (D).

SOLUTION 58

Location monuments were frequently established in areas where mineral claims or other bona fide private claims existed before the rectangular survey of the public lands in an area. A location monument in such a situation provided a common reference to which all the claims in an area could be tied. Later, the public lands survey would tie into the USLM and locate all the claims.

The answer is (D).

SOLUTION 59

Meander lines are not established as legal boundaries, but rather as a method of setting off certain bodies of water from public lands. Meander lines are intended to follow the mean high water line along meanderable bodies of water. Meander corners are only established at the points where the meander line is intersected by standard, township, and section lines.

The answer is (A).

SOLUTION 60

Trees have been marked since the inception of the Public Land Surveying System. Today, a line tree is used to perpetuate a surveyed line, while a bearing tree is considered a corner accessory. In both cases, the trees are marked to preserve healthy growth, in an attempt to ensure their future integrity.

The answer is (D).

SOLUTION 61

Once the returns from an original public lands survey have been officially accepted by the Bureau of Land Management, the monumentation set during that survey is inviolable.

The answer is (D).

SOLUTION 62

Current Bureau of Land Management instructions specify a zinc-coated $2\frac{1}{2}$ in iron alloy pipe, split at its base, with a brass cap. A $3\frac{1}{4}$ in diameter brass tablet with a $3\frac{1}{2}$ in stem is also an official monument for use in rock and concrete. When authorized, durable native stone of an appropriate shape and a volume of at least 1000 in^3 may be used in lieu of an iron post.

The answer is (D).

SOLUTION 63

The instructions for early public land surveys specified that the edges of stone monuments face the cardinal directions. These edges were then notched to indicate the number of miles that the monument stood from the governing eastern and southern township boundaries.

The answer is (B).

SOLUTION 64

The marking of trees has been part of the procedure of public lands surveying since its beginning. However, the nature and significance of the marks has changed over the years. Under the current instructions, trees that are intersected by a surveyed line are marked with two parallel horizontal hacks on the sides facing the line and are known as *line trees.* Bearing trees are corner accessories. They are generally within 3 chains of the corner. Normally, one bearing tree is selected in each section when trees of sufficient size and maturity are available. Each bearing tree is scribed with the letters BT and identification appropriate to the section in which it stands.

Both line trees and bearing trees should be recorded in the field notes of a survey.

The answer is (C).

SOLUTION 65

Trees are blazed within 50 links of a section line. Two blazes are made facing the line; the closer the blazes stand together on the tree, the farther the tree stands from the line.

The answer is (D).

SOLUTION 66

The abbreviation SC indicates that the monument stands at a standard corner, which must lie on a standard parallel.

The answer is (C).

SOLUTION 67

The abbreviation AP stands for *angle point.* The abbreviation TR stands for *tract.* The tract is inside section 36, T8N R90W.

The answer is (C).

SOLUTION 68

The absence of the abbreviations SC, for *standard corner*, or CC, for *closing corner*, indicates that the corner is probably not on a correction line. Also, Township 2 north is not likely to have a standard parallel as its northern boundary.

The answer is (D).

SOLUTION 69

The *Manual of Surveying Instructions* calls for one witness corner near inaccessible true corners, not two. The bearing and distance from the witness corner to the true corner are recorded in the field notes, and the witness corner is shown on the plat.

The answer is (D).

SOLUTION 70

When more common corner accessories are not available and the monument at a corner must be buried or is otherwise likely to be destroyed, a reference monument is set. A true corner that falls within a roadway may have two or four reference monuments. Reference monuments are described in field notes, but are not usually indicated on the official plat of the township. The abbreviation for a reference monument is RM.

The answer is (B).

SOLUTION 71

The U.S. Location Monuments, formerly known as the U.S. Mineral Monuments, were set in mining districts as permanent references for mineral surveys before the Public Land Surveying System corners were available. The monuments were established by the U.S. Mineral Surveyors, and their latitudes and longitudes recorded.

The answer is (D).

SOLUTION 72

Accessories found in the vicinity of a reference monument refer to the position of the true corner rather than the reference monument, which is itself a corner accessory. It is not likely that other corner accessories will be found near a reference monument, since it is normally established in an area where other accessories are unavailable.

The answer is (C).

SOLUTION 73

A cairn is a mound of stone. A mound of stone may still be used as a corner accessory per current instructions.

The answer is (C).

SOLUTION 74

After public lands have been properly monumented, the official plats and field notes have been prepared, and the lands have passed into private ownership, the work of the BLM and its surveyors has been completed. The privately owned sectionalized lands then fall within the venue of the private land surveyors and under the jurisdiction of the state laws. These laws recognize the inviolability of the original public lands survey and its corner monumentation, almost without exception.

However, the original corner monuments may become obliterated or lost over time; many of the aliquot corners may not have been monumented at all by the original surveyors. These difficulties have been addressed in many states by corner recordation laws, which help to perpetuate privately set aliquot corner monuments, found original controlling monumentation, and lost or obliterated original corner monumentation.

The answer is (D).

SOLUTION 75

Monument, or corner, recordation laws are designed to make corner evidence a matter of public record. Most monument recordation laws apply to monuments representing PLSS corners, though these laws typically do not require a new monument record to be filed in cases where the monument is already substantially described in or unchanged from its existing monument record filing.

The answer is (C).

SOLUTION 76

An index diagram appears near the front of each field book. It is a convenient reference by which the notes of a particular survey line may be quickly found.

The answer is (D).

SOLUTION 77

As provided by federal law, the public lands survey records of states in which the survey is substantially complete have been transferred to the states' authority. These include all of the states east of the Mississippi, plus North Dakota, South Dakota, Nebraska, Kansas, Arkansas, and Louisiana.

The answer is (D).

SOLUTION 78

The random lines and fallings are not recorded in the final transcribed field notes of a public lands survey. These trial lines, whose fallings are the basis of the calculation of true lines, were once included in the field notes of the public lands surveys, but are no longer.

The answer is (B).

SOLUTION 79

A division of previously larger subdivision is said to be aliquot when the divisor is an integer and there is no remainder. However, in practical terms, an aliquot is part of a section in the Public Land Surveying System that is a half or a quarter of a previously larger subdivision. These are a section's legal subdivisions.

The answer is (A).

SOLUTION 80

The government is not responsible for setting a monument at the center of a section. Center of section monuments are protracted because even though the quarter corner monuments were set, the official government surveyors never ran the quarter lines themselves. Rules were drafted to guide the local surveyors who would run the lines and stipulated that the monuments must be straight from quarter corner to quarter corner, and their intersection would be the center of the section.

The answer is (B).

SOLUTION 81

Proportionate measurement, also known as proration, is the technique of accommodating the principle that says that original record measurements cannot be disturbed. An even distribution of the discovered excess or deficiency is the objective of a proportionate measurement. The new values given to several parts of the line will bear the same relation to the record lengths as the new measurement of the whole line bears to the record of the whole line.

The answer is (D).

SOLUTION 82

82.1. The first step in solving this problem is to find the record distance from the N$\frac{1}{4}$ to the NE corner of lot 1. Since section 7 is a closing section, the record distance 83.91 is not divided in half at the N$\frac{1}{4}$ corner. The record distance from the NE corner of section 7 to its N$\frac{1}{4}$ corner is 40.00 chains. The remainder, 43.91 chains, is the record distance from the N$\frac{1}{4}$ corner of section 7 to its NW corner. This 43.91 chains is not divided in half at the NE corner of lot 1. The record distance from the N$\frac{1}{4}$ corner to the NE corner of lot 1 is 20.00 chains; the remainder, 23.91 chains, is the record distance along the northern boundary of lot 1.

The proportion of the record distance between the N$\frac{1}{4}$ corner of section 7 and its NW corner, 43.91 chains, to the record distance from the N$\frac{1}{4}$ to the NE corner of lot 1, 20.00 chains, must be preserved.

$$20.00 \text{ chains} = 1320.00 \text{ ft}$$
$$43.91 \text{ chains} = 2898.06 \text{ ft}$$

$$\begin{aligned}\frac{1320.00 \text{ ft}}{2898.06 \text{ ft}} &= \frac{x}{2766.06 \text{ ft}}\\ 3{,}651{,}199.20 \text{ ft} &= 2898.06x\\ x &= \frac{3{,}651{,}199.20 \text{ ft}}{2898.06}\\ &= 1259.88 \text{ ft}\end{aligned}$$

The answer is (A).

82.2. The proportion of the record distance between the NE corner of section 7 and its N$\frac{1}{4}$ corner, 40.00 chains, to the record distance from the NE corner of section 7 to the NE corner of NW$\frac{1}{4}$NE$\frac{1}{4}$, 20.00 chains, must be preserved.

$$20.00 \text{ chains} = 1320.00 \text{ ft}$$
$$40.00 \text{ chains} = 2640.00 \text{ ft}$$

$$\begin{aligned}\frac{1320.00 \text{ ft}}{2640.00 \text{ ft}} &= \frac{x}{2570.60 \text{ ft}}\\ 3{,}393{,}192.00 \text{ ft} &= 2640.00x\\ x &= \frac{3{,}393{,}192.00 \text{ ft}}{2640.00}\\ &= 1285.30 \text{ ft}\end{aligned}$$

The answer is (B).

82.3. Unless otherwise instructed, the original surveyor does not monument the quarter-quarter corners or lot corners of a township.

The answer is (C).

SOLUTION 83

83.1. The record distance between the closing corners at the NE and NW corners of section 4 is 78.90 chains, or 5207.40 ft. The measured distance between the same two corners is 5068.80 ft. The record distance between the NE corner of section 4 and the N$\frac{1}{4}$ corner is half the record distance between the NE and NW corners, or 2603.70 ft. The quarter corner monument should be placed on the standard parallel and halfway between the NE and NW corners of section 4 by proportionate measure.

PLSS

$$\frac{2603.70 \text{ ft}}{5207.40 \text{ ft}} = \frac{x}{5068.80 \text{ ft}}$$
$$13{,}197{,}634.56 \text{ ft} = 5207.40x$$
$$x = \frac{13{,}197{,}634.56 \text{ ft}}{5207.40} = 2534.40 \text{ ft}$$

The distance from the NE corner to the N$\frac{1}{4}$ corner of section 4 is 2534.40 ft. However, the eastern standard corner is the SE corner of section 33, which lies 12.54 ft west of the NE corner of section 4.

$$\begin{array}{r} 2534.40 \text{ ft} \\ -12.54 \text{ ft} \\ \hline 2521.86 \text{ ft} \end{array}$$

The answer is (A).

83.2. The record distance from the SE corner to the SW corner of section 33 is 79.94 chains, or 5276.04 ft. The record distance from the SE corner to the S$\frac{1}{4}$ corner of the section is half that, or 2638.02 ft. The measured distance along the south line of section 33 is 5137.44 ft; the S$\frac{1}{4}$ corner must be half that distance from the SE corner.

$$\left(\frac{1}{2}\right)(5137.44 \text{ ft}) = 2568.72 \text{ ft}$$

Similarly, the measured distance from the NE corner of section 4 to its N$\frac{1}{4}$ corner is half the measured distance between the NE and NW corners.

$$\left(\frac{1}{2}\right)(5068.80 \text{ ft}) = 2534.40 \text{ ft}$$

The distance from the NE corner of section 4 to its N$\frac{1}{4}$ corner must be reduced by 12.54 ft to find the distance from the SE corner of section 33 to the N$\frac{1}{4}$ corner of section 4.

$$\begin{array}{r} 2534.40 \text{ ft} \\ -12.54 \text{ ft} \\ \hline 2521.86 \text{ ft} \end{array}$$

The distance between the N$\frac{1}{4}$ corner of section 4 and the S$\frac{1}{4}$ corner of section 33 can now be found.

$$\begin{array}{r} 2568.72 \text{ ft} \\ -2521.86 \text{ ft} \\ \hline 46.86 \text{ ft} \end{array}$$

The answer is (A).

SOLUTION 84

84.1. Where opposite corresponding corners cannot be set in a fractional section, it is proper to run from the established corner as nearly parallel to the section lines as possible.

The procedure most often used to satisfy the spirit of the rule is to adopt a mean, or average, course from the established corner. In this case, the retracement surveyor has found the eastern section line to have a bearing of N00°37′W and the western section line to have a bearing of N00°03′W.

$$\frac{\text{N00°37′W} + \text{N00°03′W}}{2} = \text{N00°20′00″W}$$

The answer is (C).

84.2. The subdivision of the SE$\frac{1}{4}$ of section 22 includes the division of its boundaries by half. The ES$\frac{1}{16}$ corner is halfway between the SE corner and E$\frac{1}{4}$ corners of the section. Therefore, the record indicates a bearing of N00°38′W and a distance of 20.00 chains from the SE corner to the ES$\frac{1}{16}$. However, a retracement of the line indicates N00°37′W and 1312.58 ft.

$$\frac{1320.00 \text{ ft}}{2640.00 \text{ ft}} = \frac{x}{2625.16 \text{ ft}}$$
$$3{,}465{,}211.20 \text{ ft} = 2640.00x$$
$$x = \frac{3{,}465{,}211.20 \text{ ft}}{2640.00} = 1312.58 \text{ ft}$$

The answer is (A).

SOLUTION 85

The monument is missing from an obliterated corner, as well as its accessories. However, the location has been perpetuated and can be located beyond a reasonable doubt without relying completely on reference to other, interdependent corners.

The answer is (C).

SOLUTION 86

The evidence of the monument or its accessories may be present at an existent corner. An existent corner can sometimes be identified by reference to the description in field notes or by an acceptable supplemental survey record, physical evidence, or testimony.

The answer is (C).

SOLUTION 87

A corner is not lost until all efforts to locate its position by original marks, acceptable record evidence, and testimony are unsuccessful. A corner whose position can only be restored by reference to one or more interdependent corners is considered lost.

The answer is (A).

SOLUTION 88

Double proportionate measurement is not applicable to the restoration of corners along a standard parallel, since the alignment of the standard line is more important than the section lines that close upon the parallel.

The answer is (D).

SOLUTION 89

Double proportionate measurement is used to restore a lost corner between four known corners on lines of equal importance. A lost corner common to four interior sections or a lost corner common to four townships would be subject to double proportionate measurement.

The answer is (D).

SOLUTION 90

Single proportionate measurement is applied to lost corners where one line has precedence over another.

The answer is (D).

SOLUTION 91

91.1. The lost section corner must be restored by double proportionate measurement before the lost quarter section corner is restored, using single proportionate measurement.

The answer is (B).

91.2. The lost section corner must be reestablished by double proportionate measurement. The first step is the establishment of a temporary stake at the proper proportional distance from the nearest found original corner to the north and the south on the meridional section line. The record indicates that the distance from the $W\frac{1}{4}$ corner of section 10 to the NW corner of section 10 is 40.50 chains, or 2673.00 ft.

$$(40.50 \text{ chains})\left(66.00 \ \frac{\text{ft}}{\text{chains}}\right) = 2673.00 \text{ ft}$$

The record also indicates that the distance from the SW corner of section 3 to the NW corner of section 3 is 81.00 chains, or 5346.00 ft.

$$\begin{array}{r} 40.00 \text{ chains} \\ +41.00 \text{ chains} \\ \hline 81.00 \text{ chains} \end{array}$$

$$(81.00 \text{ chains})\left(66.00 \ \frac{\text{ft}}{\text{chain}}\right) = 5346.00 \text{ ft}$$

The measured distance from the $W\frac{1}{4}$ corner of section 10 to the NW corner of section 3 is 7919.00 ft. The total record distance for the same line is 8019.00 ft.

$$\begin{array}{r} 2673.00 \text{ ft} \\ +5346.00 \text{ ft} \\ \hline 8019.00 \text{ ft} \end{array}$$

Apply single proportionate measurement to find the distance from the $W\frac{1}{4}$ corner of section 10 to the temporary stake on the meridional line.

$$\begin{aligned} \text{record} \quad & \quad \text{measured} \\ \frac{2673.00 \text{ ft}}{8019.00 \text{ ft}} &= \frac{x}{7919.00 \text{ ft}} \\ 21{,}267{,}487.00 \text{ ft}^2 &= (8019.00 \text{ ft})x \\ x &= \frac{21{,}267{,}487.00 \text{ ft}^2}{8019.00 \text{ ft}} \\ &= 2639.67 \text{ ft} \end{aligned}$$

Check the calculation by finding the proportional distance from the NW corner of section 3 to the temporary stake on the meridional line.

$$\begin{aligned} \text{record} \quad & \quad \text{measured} \\ \frac{5346.00 \text{ ft}}{8019.00 \text{ ft}} &= \frac{x}{7919.00 \text{ ft}} \\ 42{,}334{,}974.00 \text{ ft}^2 &= (8019.00 \text{ ft})x \\ x &= \frac{42{,}334{,}974.00 \text{ ft}^2}{8019.00 \text{ ft}} \\ &= 5279.33 \text{ ft} \end{aligned}$$

The sum of the two proportional distances calculated should equal the total measured distance.

$$\begin{array}{r} 5279.33 \text{ ft} \\ +2639.67 \text{ ft} \\ \hline 7919.00 \text{ ft} \end{array}$$

The azimuth of the line from the $W\frac{1}{4}$ corner of section 10 to the NW corner of section 3 is found by inversing the coordinates.

PLSS

$W\frac{1}{4}$ corner of section 10		NW corner of section 3
5000.0000	359°03′20″	12,917.9242
5000.0000	7919.00 ft	4869.4718

Using this azimuth and the proportionate distance calculated, find the coordinates of the temporary stake on the meridional line.

$W\frac{1}{4}$ corner of section 10		temporary stake on the meridional line
5000.0000	359°03′20″	7639.3081
5000.0000	2639.67 ft	4956.4906

Next, find the coordinates of the temporary stake on the latitudinal line.

The record distance from the $N\frac{1}{4}$ corner of section 9 to the lost NE corner of section 9 is 40.00 chains, or 2640.00 ft. The record distance from the $N\frac{1}{4}$ corner of section 9 to the $N\frac{1}{4}$ corner of section 10 is 80.20 chains, or 5293.20 ft. The new measurement of the same distance is 5260.20 ft.

$$\begin{aligned} \text{record} \quad & \quad \text{measured} \\ \frac{2640.00 \text{ ft}}{5293.20 \text{ ft}} &= \frac{x}{5260.20 \text{ ft}} \\ 13{,}886{,}928.00 \text{ ft}^2 &= (5293.20 \text{ ft})x \\ x &= \frac{13{,}886{,}928.00 \text{ ft}^2}{5293.20 \text{ ft}} \\ &= 2623.54 \text{ ft} \end{aligned}$$

This is the proportionate distance from the $N\frac{1}{4}$ corner of section 9 to the temporary stake on the latitudinal line. To find the azimuth of the same line, inverse between the coordinates given for the $N\frac{1}{4}$ corner of section 9 and the $N\frac{1}{4}$ of section 10.

$N\frac{1}{4}$ corner of section 9		$N\frac{1}{4}$ corner of section 10
7618.8200	87°58′00″	7805.4567
2321.8000	5260.20 ft	7578.6879

Using this azimuth and the proportionate distance calculated, find the coordinates of the temporary stake on the latitudinal line.

$N\frac{1}{4}$ corner of section 9		temporary stake on the latitudinal line
7618.8200	87°58′00″	7711.9056
2321.8000	2623.54 ft	4943.6892

By intersecting cardinal lines passing through the two temporary stakes, as shown in the following illustration, the coordinates of the reestablished corner common to sections 3, 4, 9, and 10 can be found.

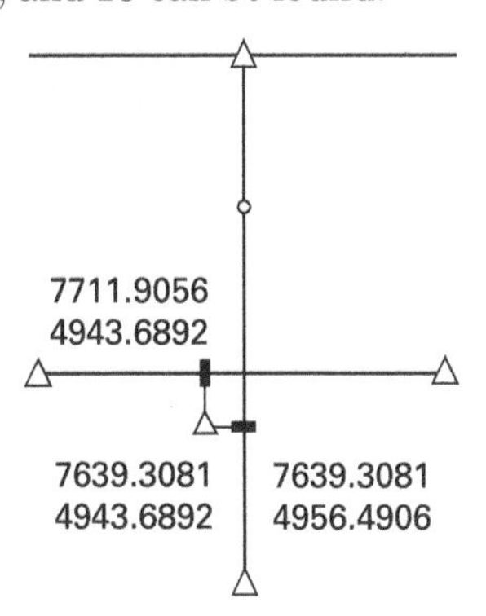

The answer is (A).

91.3. First, inverse between the coordinates given for the NW corner of section 3 and the reestablished corner at the SW corner of section 3.

NW corner of section 3		reestablished SW corner of section 3
12,917.9242	179°11′40″	7639.3081
4869.4718	5279.1378 ft	4943.6892

The record distance for the same line is 81.00 chains, or 5346.00 ft.

$$\begin{array}{r} 40.00 \text{ chains} \\ +41.00 \text{ chains} \\ \hline 81.00 \text{ chains} \end{array}$$

$$(81.00 \text{ chains})\left(66.00 \ \frac{\text{ft}}{\text{chain}}\right) = 5346.00 \text{ ft}$$

Use single proportionate measurement to find the required distance along the west line of section 3.

$$\begin{aligned} \text{record} \quad & \quad \text{measured} \\ \frac{2640.00 \text{ ft}}{5346.00 \text{ ft}} &= \frac{x}{5279.1378 \text{ ft}} \\ 13{,}936{,}923.79 \text{ ft}^2 &= (5346.00 \text{ ft})x \\ x &= \frac{13{,}936{,}923.79 \text{ ft}^2}{5346.00 \text{ ft}} \\ &= 2606.9816 \text{ ft} \end{aligned}$$

The proportional distance from the SW corner of section 3 to the $W\frac{1}{4}$ corner is 2606.9816. It is advisable to check the work by proportioning the remainder of the line as well.

PLSS

$$\begin{array}{c} \text{record} \quad \text{measured} \\ \frac{2706.00 \text{ ft}}{5346.00 \text{ ft}} = \frac{x}{5279.1378 \text{ ft}} \end{array}$$

$$14{,}285{,}346.89 \text{ ft}^2 = (5346.00 \text{ ft})x$$

$$x = \frac{14{,}285{,}346.89 \text{ ft}^2}{5346.00 \text{ ft}} = 2672.1562 \text{ ft}$$

The sum of the two proportional distances along the west line of section 3 should equal the total measured length.

$$\begin{array}{r} 2606.9816 \text{ ft} \\ +2672.1562 \text{ ft} \\ \hline 5279.1378 \text{ ft} \end{array}$$

Finally, the coordinates of the reestablished W$\frac{1}{4}$ corner of section 3 can be found by proceeding north along the west line of the section from its SW corner, using the azimuth and proportional distance found.

SW corner of section 3		W$\frac{1}{4}$ corner of section 3
7639.3081	359°11′40″	10,246.0321
4943.6892	2606.9816 ft	4907.0386

The answer is (A).

SOLUTION 92

The lost corner is on a township boundary and must be restored by single proportionate measurement from the nearest identified regular corners to the north and south. The record shows that the township line is straight; the retracement of the line indicates that its bearing is N0°07′W. The record distance from the missing point to the E$\frac{1}{4}$ corner of section 25 is 40.00 chains.

The distance to the E$\frac{1}{4}$ corner of section 24 is also 40.00 chains. Therefore, the missing point lies halfway between these two known points.

$$\frac{5309.12 \text{ ft}}{2} = 2654.56 \text{ ft}$$

The answer is (C).

SOLUTION 93

93.1. Under normal circumstances, a lost township corner such as that illustrated would be restored by double proportionate measurement. However, the latitudinal township line has not been established east of the lost corner and double proportionment is not possible. The correct procedure in such a case is to use three-point control. The meridional line is divided proportionally and a temporary stake set, as in double proportionment. However, the record bearing and distance is used to establish the second temporary stake along the latitudinal line. Cardinal lines through the temporary stakes intersect at the restored corner.

The answer is (D).

93.2. The first step in restoring the NE corner of section 1 is to set a temporary stake on the meridional township line at the proper proportional distance. The record indicates that the line between the E$\frac{1}{4}$ of section 1 and the E$\frac{1}{4}$ corner of section 36 is 80.00 chains, or 5280.00 ft. However, the retracement of the line shows it to be 5225.61 ft. The record establishes that the proper proportion between the line E$\frac{1}{4}$ to NE of section 1 and the line from NE section 1 to E$\frac{1}{4}$ of section 36 is half and half.

$$\begin{array}{c} \text{record} \quad \text{measured} \\ \frac{2640.00 \text{ ft}}{5280.00 \text{ ft}} = \frac{x}{5225.61 \text{ ft}} \end{array}$$

$$13{,}795{,}610.40 \text{ ft}^2 = (5280.00 \text{ ft})x$$

$$x = \frac{13{,}795{,}610.40 \text{ ft}^2}{5280.00 \text{ ft}} = 2612.80 \text{ ft}$$

The distance from the E$\frac{1}{4}$ corner of section 1 to the temporary stake on the meridional line is 2612.80 ft, or one-half of 5225.61 ft. The bearing from the E$\frac{1}{4}$ corner of section 1 and this temporary stake is the bearing measured in the retracement of the line N00°04′W.

E$\frac{1}{4}$ corner of section 1		temporary stake on the meridional line
5000.00	N00°04′W	7612.80
5000.00	2612.80 ft	4996.96

Next, the temporary stake on the latitudinal line is placed at the record bearing and distance from the N$\frac{1}{4}$ corner of section 1.

N$\frac{1}{4}$ corner of section 1		temporary stake on the meridional line
7612.80	S89°56′E	7609.73
2384.15	2640.00 ft	5024.15

Cardinal lines through these temporary stakes intersect at the correct location of the restored corner.

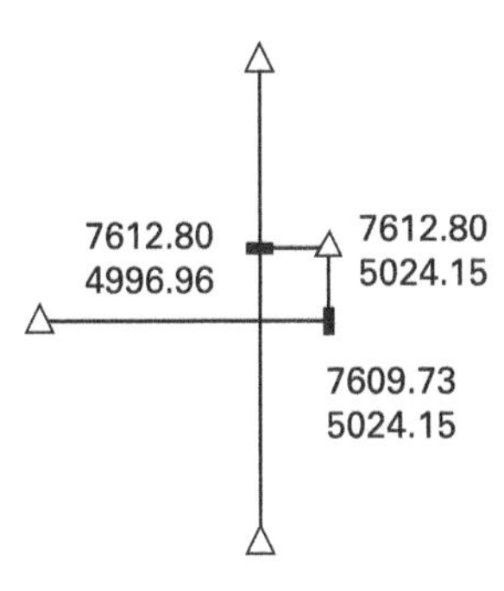

The answer is (C).

SOLUTION 94

94.1. The lost section corner cannot be restored by either single or double proportionment, since the lines north and east of the lost corner have not been established. The record bearing and distance to the $E\frac{1}{4}$ and $N\frac{1}{4}$ corners of the section must be used to restore the corner. Temporary stakes should be set north of the $E\frac{1}{4}$ corner and east of the $N\frac{1}{4}$ corner at the record bearings and distances. Cardinal lines established through these stakes will intersect at the restored section corner. The procedure is known as *two-point control*.

The answer is (B).

94.2. The illustration indicates that section 8 has been lotted on the north and east. Therefore, the distance from the NW corner to the $N\frac{1}{4}$ corner of section 8 is 40.00 chains, and the distance from the $N\frac{1}{4}$ corner to the $N\text{-}E\frac{1}{16}$ is 20.00 chains. The closing distance to the NE corner of the section is therefore 19.76 chains from the west, and 18.92 chains from the south.

The answer is (D).

SOLUTION 95

95.1. The closing corner at the NW corner of section 5 must be restored by single proportionate measurement. Standard parallels and township exteriors have precedence over subdivisional lines.

The answer is (C).

95.2. The measured distance from the $S\frac{1}{4}$ corner to the SW corner of section 32 is 2629.10 ft. The record distance for the same line is half of 80.03 chains, or 2640.99 ft. The proportionate distance should be calculated according to the falling recorded to the nearest regular corner 0.96 chains, or 63.36 ft.

$$\begin{aligned} \text{record} \quad & \quad \text{measured} \\ \frac{63.36\text{ ft}}{2640.99\text{ ft}} &= \frac{x}{2629.10\text{ ft}} \\ 166{,}579.78\text{ ft}^2 &= (2640.99\text{ ft})x \\ x &= \frac{166{,}579.78\text{ ft}^2}{2640.99\text{ ft}} \\ &= 63.1\text{ ft} \end{aligned}$$

The answer is (C).

SOLUTION 96

A dependent resurvey is done on the foundation created by the original survey; it does not supersede the original survey. A dependent resurvey does not include the segregation of private holdings into tracts that were previously aliquot parts. In law, the aliquot parts established by the original survey are reestablished unchanged by a dependent resurvey.

The answer is (A).

SOLUTION 97

Retracement refers to a survey of public lands that is done to determine the length and direction of the lines previously established; it does not include the restoration of lost corners. Remeasurement refers only to the measurement of the lengths of lines. Relocation generally refers to the movement of a mining claim within public lands.

The answer is (D).

SOLUTION 98

An independent resurvey is not done on the foundation created by the original survey. It supersedes the original survey. An independent resurvey includes the segregation of private holdings into tracts that were previously aliquot parts. The aliquot parts established by the original survey are not reestablished by the independent resurvey.

The answer is (D).

SOLUTION 99

The first segregated tract is numbered 37 to avoid confusion with the original sections, which are numbered 1 to 36.

The answer is (C).

SOLUTION 100

The corners of the tracts in an independently resurveyed township are known as angle points, and the abbreviation AP appears on the monuments of those corners. The angle points are numbered consecutively. Number 1 is usually assigned to the NE corner of the tract, with the numbers increasing counterclockwise around the tract. A corner of each tract is tied to a corner of the resurvey.

The answer is (D).

SOLUTION 101

Old monuments are destroyed during an independent resurvey, since such a survey is intended to supersede the original work. However, such found monuments are definitely not destroyed during the work of a dependent resurvey.

The answer is (C).

SOLUTION 102

Subdivisions of sections that are not aliquot parts and that bear assigned lot numbers on the official township plat are considered lots in the Public Land Surveying System.

The answer is (D).

SOLUTION 103

A retracement of a section does not lead to the the creation of lots in that section.

The answer is (D).

SOLUTION 104

An elongated section may contain regular quarter-quarter sections. For example, a section may be elongated north of its meridional quarter line and still contain a regular SE and SW quarter.

The answer is (D).

SOLUTION 105

The creation of lots in an area covered by an independent resurvey is somewhat different than in areas not covered by such a resurvey. The rules governing the lots' positions and numbering are designed to avoid any possible confusion in description with the parts of the original survey.

The answer is (D).

SOLUTION 106

106.1. The numbering of lots in a section is similar to the numbering of sections in a township. It begins at the most northeasterly lot and increases consecutively to the west in the first tier. The next tier of lots south is numbered from west to east, the next is numbered from east to west, and the most southerly tier of lots is numbered from east to west.

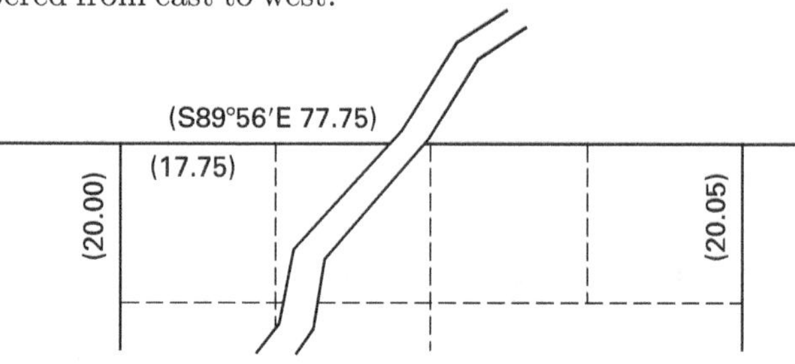

(S89°56′E 77.75) record (in chains)

The answer is (D).

106.2. The river that crosses the section is not navigable, but it is more than 3 chains wide and, therefore, meanderable. The parts of the section that have been invaded by the river are considered lots, along with the lots created by the excess and deficiency of measurement from the rest of the township. There are 10 lots in the section.

The answer is (D).

SOLUTION 107

107.1. The most probable explanation is that an area was created that was considered too small or irregular to constitute a lot. The protraction of the section was likely arranged to add the small area to another adjoining lot.

The answer is (C).

107.2. The required length is the average of the parenthetical distances given along the northern and southern section lines for the western range of lots. The western section line has a continuous bearing, and the meridional lengths of each lot are equal at 20.00 chains. Therefore, the average of the parentheticals at each end will be correct.

$$\begin{array}{r} 17.75 \text{ chains} \\ +17.87 \text{ chains} \\ \hline 35.62 \text{ chains} \end{array}$$

$$\frac{35.62 \text{ chains}}{2} = 17.81 \text{ chains}$$

The answer is (C).

107.3. One method of solving this problem is to find the change in the parenthetical distances per the unit of length of the adjacent section line. For example, looking first at the western range of lots, the difference between the given parenthetical distances is 0.12 chains. Dividing the difference between these parentheticals by the total length of the western section line yields the rate of change.

$$\begin{array}{rr} \text{southernmost parenthetical} = & 17.87 \text{ chains} \\ \text{northernmost parenthetical} = & \underline{-17.75 \text{ chains}} \\ & 0.12 \text{ chains} \end{array}$$

$$\frac{0.12 \text{ chains}}{80.00 \text{ chains}} = 0.0015$$

This ratio can be used to determine the difference between the consecutive parenthetical distances along the western range of lots. Multiply this ratio by the meridional length of lot 5 to find the record length.

$$(20.00 \text{ chains})(0.0015) = 0.03 \text{ chains}$$

The difference in the record lengths of the southern line and the northern line of lot 5 is 0.03 chains.

$$\begin{array}{rr} \text{length of the northern line of lot 5} = & 17.75 \text{ chains} \\ \text{length of the southern line of lot 5} = & \underline{+0.03 \text{ chains}} \\ & 17.78 \text{ chains} \end{array}$$

The same approach can be used to determine the record length of the eastern line of lot 5.

$$\begin{array}{rr} \text{easternmost parenthetical} = & 20.05 \text{ chains} \\ \text{westernmost parenthetical} = & \underline{-20.00 \text{ chains}} \\ & 0.05 \text{ chains} \end{array}$$

$$\frac{0.05 \text{ chains}}{77.75 \text{ chains}} = 0.000643$$

Multiply this ratio by the length of the northern line of lot 5.

$$(17.75 \text{ chains})(0.000643) = 0.01 \text{ chains}$$

$$\begin{array}{rr} \text{length of the western line of lot 5} = & 20.00 \text{ chains} \\ \text{length of the eastern line of lot 5} = & \underline{+0.01 \text{ chains}} \\ & 20.01 \text{ chains} \end{array}$$

The answer is (B).

107.4. The meander line was not established to define the limits of ownership of patent holders, but to indicate the sinuosities of the banks of the stream. Since the stream is not navigable, it is most likely that the thread of the stream is the boundary of the owner of lot 3.

The answer is (D).

SOLUTION 108

Extreme lengths and narrow widths are avoided in the creation of lots, and a lot should not extend beyond a section or quarter line. The lots along meander lines and other broken boundaries are generally not smaller than 5 ac or larger than 45 ac in modern practice. The lots with a full, normal width are generally not smaller than 10 ac or larger than 50 ac.

The answer is (D).

SOLUTION 109

109.1. One convenient aspect of the use of the chain as a unit of measurement is that 10 sq chains are equal to 1 ac. The area of lot 10 computed by coordinates is approximately 35.68 ac. However, it is unlikely that the record area would be so computed, since the boundaries of the lot are regular. One simple method that might be employed is to multiply the averages of the opposite lot boundaries.

$$\frac{17.84 \text{ chains} + 17.87 \text{ chains}}{2} = 17.855 \text{ chains}$$

$$\frac{20.00 \text{ chains} + 20.00 \text{ chains}}{2} = 20.00 \text{ chains}$$

$$(20.00 \text{ chains})(17.855 \text{ chains}) = 357.10 \text{ sq chains}$$

$$(357.10 \text{ sq chains})\left(\frac{1 \text{ ac}}{10 \text{ sq chains}}\right) = 35.71 \text{ ac}$$

An even simpler method of finding the acreage in cases where the lot is 20.00 chains wide is to add the lengths of the opposite lot boundaries.

$$\begin{array}{r} 17.84 \text{ chains} \\ \underline{+17.87 \text{ chains}} \\ 35.71 \text{ ac} \end{array}$$

The answer is (B).

109.2. To find the area, average the lengths of the opposite sides of the lot and multiply the results.

$$\frac{17.78 \text{ chains} + 17.75 \text{ chains}}{2} = 17.765 \text{ chains}$$

$$\frac{20.00 \text{ chains} + 20.01 \text{ chains}}{2} = 20.005 \text{ chains}$$

$$(20.005 \text{ chains})(17.765 \text{ chains}) = 355.39 \text{ sq chains}$$

$$(355.39 \text{ sq chains})\left(\frac{1 \text{ ac}}{10 \text{ sq chains}}\right) = 35.54 \text{ ac}$$

The answer is (A).

SOLUTION 110

The area of a lot with a width of 20.00 chains and regular boundaries is equal to the sum of the parenthetical distances. Since the area and one parenthetical distance are known, the other parenthetical distance can be derived.

$$\text{parenthetical distance} = \frac{\begin{array}{r} 39.72 \text{ ac} \\ -\,19.84 \text{ chains} \end{array}}{19.88 \text{ chains}}$$

The answer is (B).

SOLUTION 111

The Bureau of Land Management has the authority to conduct an official government resurvey when more than half of the area of a township remains in the public domain. Otherwise, the resurvey must be requested by at least three-quarters of the entrymen, and the estimated cost of the resurvey deposited.

The answer is (D).

SOLUTION 112

When land has been omitted from the public lands survey by a fraudulent meander line or one that is in gross error, the meander line may be made a fixed boundary, as between surveyed and unsurveyed land.

The answer is (D).

SOLUTION 113

The discrepancies between the design and the reality of the monumented corners may be larger than would be expected if the work were done today with modern equipment. Despite that caveat, the original work was remarkably good. However, regardless of an original work's quality, the original work will always take precedence over subsequent retracements or resurveys. The monuments set during the execution of the original survey, and the boundary lines they describe, are correct and inviolable by law.

The answer is (D).

SOLUTION 114

Reserved public lands may be acquired by adverse possession through satisfaction of strict conditions under the Color of Title Act, 43 C.F.R. § 2540. The patents, which fall under two classifications, cannot be for more than 160 ac. Where valuable improvements or cultivation have been maintained for more than 20 years or taxes paid since January 1, 1901, under color of title, in good faith, peacefully, and by a claimant, his ancestors, or grantors, a patent may be issued.

The answer is (D).

SOLUTION 115

The meander line along a permanent natural body of water should follow the mean high water line.

The answer is (D).

SOLUTION 116

The individual courses of the traverses that constitute a meander line are not monumented. Meander corners are established only where the meandering intersects standard, township, or section lines.

The answer is (D).

SOLUTION 117

Where aliquot lines have been surveyed through a section, their intersection with a meander line is monumented by a special meander corner. A meander line around a lake larger than 50 ac and enclosed entirely within a section will frequently include special meander corners. The position of special meander corners in such circumstances is calculated along the section centerline.

The answer is (D).

SOLUTION 118

An auxiliary meander corner is established to provide a method of connecting a meander line to a regular section corner. A small island that is meandered, but otherwise not tied to a section line, and a meandered lake not crossed by any quarter line both can be handled by the establishment of an auxiliary meander corner.

The answer is (D).

SOLUTION 119

According to the *Manual of Surveying Instructions*, lakes greater than or equal to 50 ac should be meandered. Swamp and overflow lands should not be meandered. Islands that have been in existence since before a state was admitted to the United States should be meandered, but not islands that have arisen after statehood. All navigable rivers, streams, and bayous should be meandered. Well-defined nonnavigable watercourses more than 3 chains in average right angle width should be meandered on both banks.

The answer is (A).

SOLUTION 120

The Act of May 10, 1872, is also known as the Mining Act and R.S. 2331 (30 U.S.C. 35). The provisions of this law still guide the survey of mineral claims on public lands.

The answer is (D).

SOLUTION 121

A survey for the patent of a mineral claim on public land must be conducted by a U.S. deputy mineral land surveyor. The Bureau of Land Management will provide a list of approved mineral surveyors to any claimant requiring such services.

The answer is (C).

SOLUTION 122

The Mining Act of 1872 provides that a mineral lode claim may not exceed 600 ft by 1500 ft. It need not be aliquot.

The answer is (C).

SOLUTION 123

The sidelines of a mineral claim are often parallel with the outcrop of the vein on the surface, and the end lines are parallel to one another. This arrangement allows the miner to follow the dip of the ore into the earth, even when the pursuit carries the work beyond the boundaries of the claim. This is known as an extralateral right. The Mining Act of 1872 specifically calls for the end lines to be parallel.

The answer is (C).

SOLUTION 124

A claim can be made upon public lands for the purpose of placer mining. A placer claim may include minerals such as particles in sands or gravel, but not veins in place, which are the subjects of lode claims.

The answer is (D).

SOLUTION 125

Placer claims, unlike lode claims, must be surveyed in conformance with the aliquot lines of the Public Land Surveying System. Latitude is given regarding this specification in areas where placer claims are filed in advance of the rectangular survey.

The answer is (D).

SOLUTION 126

Under the provisions of the Mining Act of 1872, each claimant may hold 20 ac as a placer claim.

The answer is (B).

SOLUTION 127

Five acres may be included within a mill site for the processing of minerals; however, the mill site must be located on lands that do not contain valuable minerals.

The answer is (B).

SOLUTION 128

A U.S. deputy mineral surveyor may survey a claim for patent that does not conform to the original mineral location certificate. The claimant may have moved the claim and obtained an amended location certificate any number of times.

The answer is (D).

Surveying Instruments and Procedures

DEFINITIONS

PROBLEM 1

Which of the following statements is true of plane surveying?

(A) Plane surveying disregards the curvature of the earth.

(B) As the size of the area increases, the accuracy and precision of plane surveying decreases.

(C) The term *plane surveying* is reserved for surveys whose results are expressed in state plane coordinates.

(D) Both A and B are true.

PROBLEM 2

The four points of the compass—north, south, east, and west—are known by what common name?

(A) occidental directions

(B) collimation directions

(C) oriental directions

(D) cardinal directions

PROBLEM 3

Which of the following scales is NOT included on a standard triangular engineer's scale?

(A) 1 in = 10 ft

(B) 1 in = 40 ft

(C) 1 in = 50 ft

(D) 1 in = 100 ft

PROBLEM 4

The standard gauge of railroad tracks in the United States is normally measured between points $\frac{5}{8}$ in below the top of the rails. What is the standard railroad gauge within the United States?

(A) 4 ft, $6\frac{1}{4}$ in

(B) 4 ft, $8\frac{1}{2}$ in

(C) 4.60 ft

(D) 5.0 ft

PROBLEM 5

Which of the following quantities may be properly measured in miner's inches?

(A) depth of a shaft

(B) weight of ore

(C) density of a mineral

(D) rate of flow of water

FIELD NOTES

PROBLEM 6

When a number has been recorded in a field book in error, the correction of the entry is best accomplished by which of the following procedures?

(A) erasing the error and rewriting

(B) writing the correct number over the error

(C) voiding the entire page

(D) drawing a line through the error and rewriting the proper value above

PROBLEM 7

A measurement of 3.90 ft is recorded in a surveyor's field notes. Which of the following assumptions is supported by this entry?

(A) The measurement was made to the nearest hundredth.

(B) The measurement was made to the nearest tenth.

(C) The measurement was estimated.

(D) The measurement was made to the nearest hundredth but is only significant to the nearest tenth.

THE COMPASS

PROBLEM 8

To which of the following charts should a surveyor refer to determine the approximate magnetic declination in a particular location?

(A) isogradient chart

(B) isohaline chart

(C) isomagnetic chart

(D) isogonic chart

PROBLEM 9

Which of the following variations does NOT affect the magnetic declination at a place over time?

(A) secular variation

(B) annual variation

(C) daily variation

(D) constant variation

PROBLEM 10

A line observed in 1911 had a magnetic bearing of N10°24′E, and the magnetic declination at that time was 04°E. What will the magnetic bearing of the line be today if the declination is 00°25′W?

(A) N05°59′E

(B) N09°50′E

(C) N14°31′E

(D) N14°49′E

OPTICS

PROBLEM 11

The following three problems refer to the information given.

In 1814, the magnetic bearing of line XY was S49°27′E, and the magnetic declination was 4°00′W. Today, the magnetic bearing of line XY is S50°15′E.

11.1. What is the geodetic bearing of line XY?

(A) S45°27′E

(B) S46°15′E

(C) S50°15′E

(D) S53°27′E

11.2. What is the secular variation from 1814 to today?

(A) N00°48′E

(B) N00°50′E

(C) N03°12′E

(D) N04°49′E

11.3. What is the magnetic declination at line XY today?

(A) 00°48′W

(B) 03°12′W

(C) 03°20′W

(D) 04°00′W

PROBLEM 12

Which of the following statements correctly defines the effect of atmospheric refraction?

(A) Generally, refraction makes objects appear lower than they actually are.

(B) As the density of the air increases, refraction decreases.

(C) Generally, refraction makes objects appear higher than they actually are.

(D) Both A and B are true.

PROBLEM 13

A surveyor is looking through an instrument at a target 2 mi away. The instrument and the target are both standing in warm air masses, but the line of sight passes over a lake between them that is $1/4$ mi wide. The air over the lake is considerably cooler than the air at the instrument or the target. Which statement best describes the effect of the cooler air on the line of sight?

(A) There is no effect—air pressure influences refraction, not temperature.

(B) The line of sight will be subject to less refraction in the cooler air than in the warmer air.

(C) The line of sight will be refracted equally throughout its length.

(D) The line of sight will be subject to more refraction in the cooler air than in the warmer air.

PROBLEM 14

When looking through the telescope of a particular instrument after focusing, an observer finds that if the eye is shifted slightly, the cross hairs appear to move with respect to the object being sighted. What is the cause of this effect and what should be done about it?

(A) The effect is due to normal refraction and nothing can be done to eliminate it.

(B) The effect is known as *parallax* and the instrument should be adjusted to eliminate it.

(C) The effect is known as *halation* and the instrument should be adjusted to eliminate it.

(D) The effect is caused by heat waves and nothing can be done to eliminate it.

PROBLEM 15

Two types of eyepieces are generally used in surveying instruments: erecting and inverting. Which of the following correctly represents the arrangement of lenses appropriate to each?

(A) two convex lenses: inverting
one convex lens: erecting

(B) four concave lenses: erecting
two convex lenses: inverting

(C) two concave lenses: inverting
four convex lenses: erecting

(D) two convex lenses: inverting
four convex lenses: erecting

PROBLEM 16

The image produced by a particular instrument is obscured by various hues of light. What is the name given to this condition?

(A) coma

(B) lacuna distortion

(C) chromatic aberration

(D) precession

PROBLEM 17

What objective lens design is used to minimize both spherical and chromatic aberrations in the optical systems of many surveying instruments?

(A) double-convex objective lens

(B) single-coated objective lens

(C) compound lens of crown glass and flint glass

(D) polarizing filter

PROBLEM 18

Which of the following ranges of magnification includes most surveying telescopes?

(A) 5 to 12 diameters

(B) 5 to 35 diameters

(C) 18 to 42 diameters

(D) 46 to 71 diameters

THE LEVEL AND THE ROD

PROBLEM 19

Which of the following statements is true of self-reading level rods?

(A) The graduations on a self-reading rod increase from the top to the bottom of the rod.

(B) A self-reading rod requires the rodman to set and read movable targets in accordance with signals from the instrument operator.

(C) Self-reading rods are graduated to allow the instrument operator to read them clearly.

(D) The graduated strip on a self-reading rod is movable and may be set at a particular rod reading to allow reading of elevations directly without calculation.

PROBLEM 20

One of the most common level rods used today is the Philadelphia rod. Which of the following properties applies to the Philadelphia rod?

(A) The Philadelphia rod usually has three sections.

(B) When not extended, both the front and back of a Philadelphia rod are graduated.

(C) The Philadelphia rod accommodates readings up to 7 ft without extension.

(D) Both B and C are true.

PROBLEM 21

After rough level is achieved with the three screws of a third-order automatic, or self-leveling, level, how is the instrument brought to more precise level?

(A) The instrument is tilted about a ball and socket fulcrum by means of one screw.

(B) The line of sight through the instrument is kept level by means of a compensator mechanism consisting of prisms on a pendulum.

(C) The instrument is brought into precise level by manipulating a micrometer until the two halves of a coincidence bubble are brought together.

(D) The instrument is kept level by slight adjustments of the three screws of the leveling head so as to keep the bull's-eye bubble within its etched circle.

GRADING AND LEVELING

PROBLEM 22

What is the purpose of the procedure known as *waving the rod*?

(A) The rod is waved to make sure that the joint clamps have been properly tightened.

(B) The rodman finds the vertical position of the rod by waving it before the instrument operator makes any attempt to read it. Once the vertical position is found, the rod is held steady for reading.

(C) The procedure causes the rod reading to rise and fall so that its lowest value may be read by the instrument operator.

(D) The rod is waved to check the focus of the instrument.

PROBLEM 23

When running differential levels, it is good practice to balance the sights. The purpose of balancing the lengths of the foresights and backsights in a level circuit is to

(A) minimize the effect of a line of sight that is not perfectly horizontal

(B) compensate for the effect of parallax

(C) minimize the effect of any inconsistency in plumbing the rod

(D) minimize the effect of refraction

PROBLEM 24

How can the elevation of an intermediate turning point be calculated from the usual note format of differential levels?

(A) subtract the backsight from the height of instrument

(B) add the foresight to the height of instrument

(C) subtract the foresight from the height of instrument

(D) add the backsight to the previous benchmark

PROBLEM 25

A level circuit must cross a river with a width of approximately 800 ft. It is not possible to set up the instrument anywhere but on the banks. What method would be useful in such a situation?

(A) The observations are best made on an overcast day when the temperature and atmospheric conditions are stable.

(B) Two rods, or even two instruments, used to make simultaneous observations would improve the results.

(C) Readings taken to turning points on both sides of the river from first one, and then the other, bank would allow two determinations of the difference in elevations of the two points.

(D) All of the above are useful.

PROBLEM 26

What main advantage might be gained by using two rods and two rodmen in running a level circuit?

(A) The systematic error is reduced.

(B) The walking distance is less for the rodmen.

(C) The need to mark turning points on the ground is eliminated.

(D) The need to balance the backsights and foresights is eliminated.

PROBLEM 27

How far must the top of a fully extended 13.00 ft rod deviate from plumb to cause an error of +0.02 ft in a 7.00 ft rod reading?

(A) 0.23 ft

(B) 0.51 ft

(C) 0.98 ft

(D) 1.06 ft

PROBLEM 28

A level is set midway between BM A and BM B. The rod readings from this position are A = 1.35 ft and B = 4.14 ft. The level is then moved very close to A and the rod readings from this position are A = 4.46 ft and B = 7.12 ft. What rod reading at BM B from the second position would indicate that the level is in proper adjustment?

(A) 6.95 ft

(B) 6.99 ft

(C) 7.12 ft

(D) 7.25 ft

PROBLEM 29

Which of the following series of numbers would correctly replace those indicated by the letters X, Y, and Z on the field notebook page shown?

STA	BS	HI	FS	ELEV	
BM A				210.01	
	U	211.99			
TP 1			3.61	V	
	7.53	215.91			
TP 2			W	211.61	
	5.77	X			
TP 3			6.15	211.23	
	Y	214.74			
BM B			Z	209.43	
	6.35	215.78			
TP 4			7.55	208.23	
	2.21	210.44			
TP 5			4.90	205.54	
	7.33	212.87			
TP 6			1.21	211.66	
	5.10	216.76			
BM A			6.69	210.07	

(A) X = 216.21
Y = 2.09
Z = 5.51

(B) X = 217.38
Y = 3.51
Z = 5.31

(C) X = 217.38
Y = 4.65
Z = 5.51

(D) X = 230.38
Y = 3.12
Z = 4.10

Instruments/ Procedures

PROBLEM 30

Given the following two observations, what are the rod readings in each case? Are the rods an equal distance from the observer?

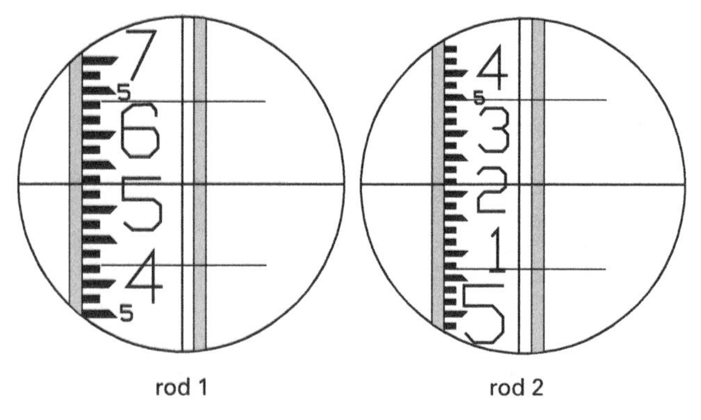

(A) rod 1 = 5.21

rod 2 = 5.53

Rod 1 is closer to the observer than rod 2.

(B) rod 1 = 5.21

rod 2 = 5.54

The rods are an unequal distance from the observer.

(C) rod 1 = 5.53

rod 2 = 5.21

Rod 1 is closer to the observer than rod 2.

(D) rod 1 = 5.35

rod 2 = 5.64

The rods are an equal distance from the observer.

PROBLEM 31

The following three problems refer to the illustration shown. The work was done with a rod calibrated in feet. The level's stadia interval factor was 100.

THREE-WIRE LEVELING

STA	BS	THREAD INT	FS	THREAD INT	ELEV
BM X					1609.350
	6.739		7.529		
	5.365	1.374	6.328	1.201	
	3.988	1.377	5.133	1.195	
3	16.092	2.751 3	18.990	2.396	
	5.364		6.330		
TP 1	3.898		8.486		
	2.375	1.523	6.990	1.496	
	0.850	1.525	5.491	1.499	
3	7.123	3.048 3	20.967	2.995	
	2.374		6.989		

31.1. What is the elevation of TP1?

(A) 1608.384 ft

(B) 1608.385 ft

(C) 1608.387 ft

(D) 1610.316 ft

31.2. The first backsight and foresight are not balanced. What is the difference between their lengths?

(A) 3.55 ft

(B) 17.75 ft

(C) 25.73 ft

(D) 35.50 ft

Instruments/ Procedures

31.3. Which of the readings shown in the field notes should have been repeated before the work proceeded?

(A) Since the backsight at the first turning point has a lowest wire reading of 0.850, it is too close to the bottom of the rod.

(B) In three of the sets of readings, the interval between the middle and the upper wire differs by more than 0.002 from the interval between the middle and lower wire. They should all have been repeated.

(C) The second backsight and foresight on the page should have been repeated since they are so far out of balance.

(D) Both A and C are true.

PROBLEM 32

Spot elevations are shown on a map of an existing street. What type of work on such a map would involve interpolation?

(A) spacing contours between the spot elevations

(B) calculating the grade of the street's centerline

(C) preparing the plan of a survey to further densify the available elevation information

(D) preparing a profile drawing of the street

PROBLEM 33

The following three problems refer to the information given.

Trigonometric leveling was used to find the difference in elevation between two monuments known as Thornton and 300. The slope distance measured between them was 6782.955 ft. The following data were also measured.

station	Thornton to 300	300 to Thornton
zenith		
forward	91°06′16″	
back		88°55′15″
height of		
EDM	4.70 ft	5.60 ft
prism	4.50 ft	3.80 ft
instrument	4.00 ft	5.00 ft

The heights are measured in feet and the elevation of Thornton is known to be 5440.6 ft.

33.1. From the given information, what is the elevation of the monument at 300 to the nearest foot?

(A) 5311 ft

(B) 5315 ft

(C) 5353 ft

(D) 5439 ft

33.2. Suppose that only the forward observation was taken from Thornton to 300. What compensation would be necessary in the calculation of the elevation of 300?

(A) The cosine of the zenith angle measured at Thornton multiplied by the slope distance will yield the difference in the elevation between Thornton and 300.

(B) The zenith angle measured at Thornton must be corrected for parallax.

(C) The zenith angle at Thornton must be measured to a target at the same HI.

(D) The zenith angle measured at Thornton must be corrected for the effects of curvature and refraction.

33.3. Suppose that only the forward observation was taken from Thornton to 300. When the appropriate corrections are applied to the forward zenith angle, what elevation is determined for point 300, to the nearest foot?

(A) 5308 ft

(B) 5309 ft

(C) 5310 ft

(D) 5311 ft

PROBLEM 34

Atmospheric pressure is sometimes used to determine the difference in elevation between points. Assume that the air pressure was simultaneously measured by barometers at two monuments whose elevations differed by 1000 ft. If the atmosphere was stable at the time of the readings, approximately what difference would be expected in the air pressure at the two points, and which would have the larger reading?

(A) The difference between the readings would be approximately 0.1 in of Hg and the higher point would have the larger reading.

(B) The difference between the readings would be approximately 1.0 in of Hg and the higher point would have the smaller reading.

(C) The difference between the readings would be approximately 0.01 in of Hg and the higher point would have the larger reading.

(D) The difference between the readings would be approximately 5.0 in of Hg and the higher point would have the smaller reading.

PROBLEM 35

The invert of a concrete sewer line, constructed of pipe with a 36 in inner diameter and a 40 in outer diameter was measured at sta 25+15.30 and found to be 1892.49 ft. If the grade of the line is −2½%, what should the elevation of the top of the pipe be at sta 28+82.75?

(A) 1879.97 ft

(B) 1881.63 ft

(C) 1883.30 ft

(D) 1886.47 ft

PROBLEM 36

The invert of an 8 in main sewer line at manhole 10 is found to be 389.00 ft. The grade of the line between manhole 10 and manhole 11 is +0.40%. A new 6 in house service is to be connected to the line between manholes 10 and 11. At the house, the elevation of the flow line of the service is 393.00 ft, and 115 ft of pipe will be used to run between the house and the sewer. If the grade of the service is −2.10%, what is the distance along the sewer from manhole 10 to the service connection? The flowline of the house service must connect at the top (or crown) of the 8 in main sewer line.

(A) 152.3 ft

(B) 193.7 ft

(C) 230.0 ft

(D) 306.9 ft

PROBLEM 37

A 2500 ft level loop is intended to satisfy the Federal Geodetic Control Committee (FGCC) specifications for third order. Which of the following elements would disqualify the work from meeting these specifications?

(A) a wooden or metal level rod

(B) only second order benchmarks available for control

(C) reading only the center wire

(D) sight lengths of 325 ft

PROBLEM 38

A differential level line was run ascending a grade. The sights were not balanced; the backsights were consistently 325 ft, and the foresights were 210 ft. The elevation difference between BM 1 at the bottom of the grade and BM 2 at the top of the grade was calculated to be 96.32 ft. The run required 24 setups.

Later, the level was discovered to be out of adjustment. During the level work, the line of sight of the instrument was inclined downward 0.02 ft per 300 ft. In light of this information, what is the adjusted difference in elevation between BM 1 and BM 2?

(A) 96.14 ft

(B) 96.23 ft

(C) 96.41 ft

(D) 96.50 ft

PROBLEM 39

A level is set up between a BM and a turning point (TP) with the foresight and backsight balanced. The elevation of the BM at the backsight is 4107.22 ft, and the reading on the rod is 1.47 ft. The reading on the rod at the foresight is 4.69 ft. If the rod at the foresight was inclined toward the instrument by 15° at the moment it was read, what is the elevation of the TP?

(A) 4103.84 ft

(B) 4104.00 ft

(C) 4104.16 ft

(D) 4104.32 ft

TRANSITS AND THEODOLITES

PROBLEM 40

When an angle has been determined by repetition on a repeating theodolite, which of the following statements is true?

(A) The plate of the instrument used to measure the angle was set to zero before each backsight.

(B) Each successive measurement of the angle was added to the sum of those already measured.

(C) The angle was measured each time with the telescope in the inverted position.

(D) The angle was turned an odd number of times.

PROBLEM 41

What is a difference between a repeating theodolite and a directional theodolite?

(A) Directional theodolites have automatic vertical circle compensators and repeating theodolites do not.

(B) Repeating theodolites are equipped with interior optics for reading the circles and directional theodolites are not.

(C) Directional theodolites have three leveling screws or cams and repeating theodolites have four.

(D) Repeating theodolites are equipped with upper and lower motions and directional theodolites are not.

PROBLEM 42

What is the name of the component of an instrument that includes the stadia hairs, and what is the most commonly used stadia interval factor?

(A) The component is the reticle, and the stadia interval most often used is 100.

(B) The component is the micrometer, and the stadia interval most often used is 300.

(C) The component is the dioptric scale, and the stadia interval most often used is 200.

(D) The component is the telescope objective, and the stadia interval most often used is 100.

PROBLEM 43

Verniers are often used in reading both the horizontal and vertical circles of transit. A particular vernier scale is attached to a circle that is graduated to 20′. The vernier has 40 increments that, taken together, equal 39 on the circle. What is this type of vernier, and what is its least count?

(A) It is a folding vernier with a least count of 01′.

(B) It is a retrograde vernier with a least count of 30″.

(C) It is a direct vernier with a least count of 01′.

(D) It is a direct vernier with a least count of 30″.

PROBLEM 44

Which of the following statements does NOT describe a portion of the procedure known as *double-centering*?

(A) A backsight is taken once with the telescope of the instrument in the direct position and once with the telescope in the inverted position.

(B) One point is set on the line described with the telescope in the direct position and another with the telescope in the inverted position.

(C) The angle is measured between the two points, subtracted from 180°, and the resulting angle is set off from the backsight to correctly prolong the line.

(D) Midway between the two temporary points, a third point is set on the correct prolongation of a straight line.

PROBLEM 45

Which of the following procedures is known as *closing the horizon*?

(A) turning a small deflection angle to clear an obstacle, moving forward to a point on that line, and doubling the same deflection in the opposite direction

(B) setting the instrument between two points and shifting it laterally until it stands on the line described by the two points

(C) turning a horizontal angle with the telescope of the instrument in both the direct and the inverted positions

(D) measuring horizontal angles from the backsight to the foresight and from the foresight to the backsight and checking to see that their sum is approximately 360°

PROBLEM 46

Which of the following methods would provide the most accurate check of the adjustment of an optical plummet on a modern tribrach?

(A) Set the tribrach on a tripod in a room and place a piece of paper beneath it. Draw a cross on the paper and, using a plumb bob, move the instrument directly over the cross. Remove the plumb bob and, with the tribrach level, check the cross hairs of the optical plummet against the cross on the piece of paper.

(B) Set the tribrach on a tripod in a room and place a piece of paper beneath it. With the tribrach level, draw a cross on the paper, and, using the optical plummet, move the instrument until the cross is exactly coincident with the cross hairs. Hang a plumb bob and check it against the cross on the piece of paper.

(C) Set the tribrach on a tripod in a room and place a piece of paper beneath it. Draw the outline of the base plate of the tribrach directly on the tripod head. With the tribrach level, mark the position of the plummet's cross hair on the paper. Rotate the tribrach, place the base plate carefully within the penciled outline, and repeat the procedure. If the marks on the paper from three such rotations coincide, the optical plummet is in adjustment.

(D) Set a tribrach known to be in good adjustment on a tripod in a room and place a piece of paper beneath it. Draw the outline of the base plate of the tribrach directly on the tripod head. With the tribrach level, mark the position of the plummet's cross hair on the paper. Remove the tribrach and replace it with another that must be checked. Carefully place the second tribrach within the outline of the first and check it against the cross on the piece of paper.

TAPING

PROBLEM 47

Any unsupported portion of a steel tape sags in an arc known by which of the following names?

(A) cycloid

(B) parabola

(C) helical arc

(D) catenary

PROBLEM 48

Which of the following specifications would typically achieve a standard length from a 100 ft steel tape?

(A) a tape supported throughout its length under 25 lb of tension at 72°F

(B) a tape supported only at its end under 25 lb of tension at 72°F

(C) a tape supported only at its end under 15 lb of tension at 45°F

(D) a tape supported throughout its length under 12 lb of tension at 68°F

PROBLEM 49

A steel tape that is used when the temperature is colder or warmer than that for which its length was standardized is subject to contraction or expansion. The coefficient of thermal expansion of steel is a critical value for quantifying such changes. What is this coefficient?

(A) 6.5×10^{-6} per °F

(B) 1.16×10^{-5} per °C

(C) 0.0000065 per °F

(D) all of the above

PROBLEM 50

A steel tape that is 100.00 ft long when supported only at its ends under a tension of 20 lb is used to measure a line. The temperature is 68°F, the pull on the tape is 40 lb, and the line is measured as 469.40 ft. The tape used is 0.33 in wide × 0.02 in thick. If one assumes the modulus of elasticity of steel to be 29,000,000 psi, what is the actual length of the measured line?

(A) 469.36 ft

(B) 469.39 ft

(C) 469.42 ft

(D) 469.45 ft

PROBLEM 51

Invar tapes are noted for their diminished thermal expansion and contraction. What is the percentage of decreased sensitivity to temperature of invar compared to steel?

(A) 3%

(B) 12%

(C) 20%

(D) 66%

PROBLEM 52

A line is measured with a steel tape using standard tension and supported throughout its length on an incline with a vertical angle of 27°. The temperature of the tape is 83°F. The slope distance is measured as 132.67 ft. What is the true length of the line to the nearest hundredth of a foot?

(A) 115.63 ft

(B) 118.22 ft

(C) 120.54 ft

(D) 125.09 ft

PROBLEM 53

The measurement of a line was recorded as 1315.72 ft. Later, the tape used to make the measurement was compared to a standard at 68°F and found to have a length of 100.02 ft rather than 100.00 ft. Also, the measurement was made when the temperature of the tape was 45°F rather than the standard 68°F. What is the true length of the line?

(A) 1315.79 ft

(B) 1315.98 ft

(C) 1316.09 ft

(D) 1316.18 ft

PROBLEM 54

An unsupported 3 lb, 100.00 ft tape at its standard temperature under 10 lb of tension was used to establish stakes intended to be 600.00 ft apart. In the end, the established stakes were not exactly 600.00 ft apart. The distance between the same stakes was measured a second time with the same tape, unsupported, in 25.00 ft increments. Each segment was under 10 lb of tension. What was the observed length of the last segment, expressed to the nearest tenth of a foot?

(A) 22.9 ft

(B) 24.3 ft

(C) 25.0 ft

(D) 27.1 ft

STADIA

PROBLEM 55

The following three problems refer to the observation described.

An instrument with an internal focus and a stadia interval factor of 100 occupies station A. The HI is 5.27 ft and the elevation of station A is 2487.91 ft. The stadia interval observed on a rod at station B is 7.51 ft. The rod reading at the center hair is 6.85 ft at a zenith angle of 87°37′.

55.1. What is the horizontal distance from station A to station B?

(A) 749.70 ft

(B) 750.70 ft

(C) 751.00 ft

(D) 752.00 ft

55.2. What is the vertical distance between the height of the instrument at station A and the reading at the middle cross hair on the rod at station B?

(A) +16.15 ft

(B) +31.20 ft

(C) +32.20 ft

(D) +62.40 ft

55.3. What is the elevation of station B to the nearest tenth of a foot?

(A) 2513.8 ft

(B) 2517.5 ft

(C) 2519.1 ft

(D) 2545.0 ft

EDM

PROBLEM 56

Concerning a modern EDM that uses coherent laser light, which of the following atmospheric factors will be the least critical to the accuracy of its measurement of distance?

(A) air temperature

(B) barometric pressure

(C) relative humidity

(D) each is equally significant

PROBLEM 57

While the temperature correction applied to EDM measurements is not of great consequence for short distances, it is more important for longer distances. Assume an error of +18°F has been made in the determination of the temperature at the time of an EDM measurement. What magnitude of error in parts per million might result?

(A) −20 ppm

(B) −18 ppm

(C) +5 ppm

(D) +10 ppm

PROBLEM 58

Assume that an error of +25 mm of Hg has been made in the determination of the air pressure at the time of an EDM measurement. What magnitude of error in parts per million might result?

(A) −15 ppm

(B) −10 ppm

(C) +5 ppm

(D) +10 ppm

PROBLEM 59

What is the maximum error one might expect in a distance measurement of 4932.48 ft made with an EDM whose specified reliability is ±(5 mm + 5 ppm)?

(A) ±0.03 ft

(B) ±0.06 ft

(C) ±0.09 ft

(D) ±0.15 ft

PROBLEM 60

A surveyor using the same instruments throughout occupied point 1 and measured 1186.42 ft to point 3 and 355.93 ft to point 2. The surveyor then occupied point 2 and measured 830.61 ft to point 3. Assuming the measurements are properly corrected for slope and atmospheric conditions, what is the reflector constant of the prisms used in the measurements?

1 2 3

(A) 0.03 ft

(B) 0.06 ft

(C) 0.07 ft

(D) 0.12 ft

PROBLEM 61

An EDM with a standard error of ±(7 mm + 7 ppm) was used in a survey that was to have a relative precision of 1 part in 10,000. The work included a measurement of 200.00 ft. Considering only the standard error, what was most nearly the relative precision of that particular measurement?

(A) 1 part in 9000

(B) 1 part in 10,000

(C) 1 part in 10,100

(D) 1 part in 14,000

ROUTE STAKING

PROBLEM 62

A route survey is conducted to

(A) gather information about natural features for making a topographic map

(B) locate the final position of new construction for documentation and evaluation of the work

(C) establish property corners and lines

(D) facilitate the planning, design, and construction of a pipeline, roadway, powerline, or any of a number of linear projects

PROBLEM 63

The process of prolonging a straight line by double centering is accomplished by backsighting first in the direct position, then

(A) plunging to reverse, and setting a point forward, then backsighting in reverse, plunging to direct, and setting another point forward. Halving the distance between the two forward points yields the correct prolongation of the line

(B) turning an angle of 180°, and setting a point forward, then backsighting in reverse, turning an angle of 180°, and setting another point forward. Halving the distance between the two forward points yields the correct prolongation of the line

(C) plunging to reverse, and setting a point forward, then backsighting the set forward point in direct and plunging to reverse on the backsight. Halving the angle between the resulting line of sight and the backsight point yields the correction to apply at the forward point for the true prolongation of the line

(D) plunging to reverse, and setting a point forward, then backsighting in reverse, turning an angle of 180°, and setting another point forward. Halving the distance between the two forward points yields the correct prolongation of the line

PROBLEM 64

Which of the following techniques is NOT a well-known method of establishing a straight route around an obstacle on the line?

(A) the measured offset method

(B) the equilateral triangle method

(C) the equal angle method

(D) the single directional method

PROBLEM 65

Periodic astronomic observations are taken in surveying routes of great length to

(A) correct for accumulated error in azimuth

(B) establish geographic coordinates

(C) facilitate ties to NGS control stations

(D) quantify changes in magnetic declination

PROBLEM 66

What role, if any, does geodesy play in route surveying?

(A) Route surveys are plane surveys and the shape of the earth need not be considered.

(B) Route surveys of great length should take the curvature of the earth into account to avoid otherwise unresolvable discrepancies.

(C) Only route surveys that have great length in a north-south orientation need consider the curvature of the earth.

(D) The only route surveys that call for geodetic considerations are those that are conducted for underground mine work.

PROBLEM 67

Centerline stakes set every 100 ft are standard for many types of route surveys; however, more closely spaced stakes are necessary under some conditions. Which of the following situations would likely require more closely spaced centerline stakes?

(A) When the intersection of one roadway with another occurs at a point other than a full station, a stake is often set at the point of intersection to indicate the stationing equation.

(B) When a route includes horizontal or vertical curves, stakes are set at the beginning and end of the curves, which seldom occur at full stations.

(C) When the requirements for a pipeline route call for a specified number of stakes per pipe length, the spacing is likely to be less than 100 ft between stakes.

(D) All of the above are true.

PROBLEM 68

Which of the following statements concerning the grading of stakes along a route survey is NOT true?

(A) The grade of a water line is generally more critical than that of a sewer line.

(B) Offset stakes or guard stakes are usually marked with cuts or fills, rather than centerline stakes.

(C) It is generally not necessary to turn through every stake that is to be graded.

(D) Grades for earthwork and pipelines are generally calculated to the nearest tenth of a foot.

PROBLEM 69

A blue top is a

(A) stake set at the catch point of the side slope of a template with the natural ground

(B) guard stake driven so its top is directly over the line stake

(C) control point monument located well away from the route being surveyed

(D) stake driven so its top is at the designed grade, or with the cut or fill to that grade written on the stake

PROBLEM 70

A shiner is a

(A) tack sometimes set in a line stake for more precise location than the stake alone provides

(B) hook from which a plumb bob may be hung in surveying tunnels and mines

(C) 1 in × 1 in stake, 12–18 in long

(D) nail set through a bottle cap or some other type of metal disk in a hard surface such as pavement

PROBLEM 71

A batter board is a

(A) horizontal board, usually 1 in × 6 in, nailed between two 2 in × 4 in posts that have been driven into the ground near a control point to prevent destruction of the point

(B) lath set to establish the clearing limits along a route

(C) pile that has been driven at an angle other than vertical; such piles are frequently used to brace others

(D) horizontal board, usually 1 in × 6 in, nailed between two 2 in × 4 in posts that have been driven into the ground across a trench or near a building corner

PROBLEM 72

Slope stakes are usually set to indicate the

(A) point where a roadway's pavement ends and its side slope begins

(B) beginning of superelevation along a highway curve

(C) point where a side slope intersects the natural ground

(D) maximum embankment allowed per the angle of repose

PROBLEM 73

The information usually written on a slope stake is the

(A) cut or fill at the centerline

(B) ratio of the slope, the cut or fill at the centerline, and the distance of the stake from the centerline

(C) station of the stake, the cut or fill at the centerline, the distance of the stake from the centerline, and the ratio of the slope

(D) ratio of the slope, the cut or fill at the centerline, the distance of the stake from the centerline, the station of the stake, and the width of the roadbed

PROBLEM 74

Slope stakes are to be set along a particular roadway. The grade at the centerline at sta 14+00 is designed to be 115.4 ft and the HI of a level arranged to set slope stakes is 120.6 ft. In this situation, what is the value known as the *grade rod*?

(A) −5.2 ft

(B) −2.6 ft

(C) 2.6 ft

(D) 5.2 ft

PROBLEM 75

The width of a particular roadbed is 66 ft, the side slope is designed to be 2:1, the grade rod is 5.2 ft, and the rod is held at a trial position for a slope stake 45 ft from the centerline. The cross section is in cut. The rod is read through the level and the ground rod is found to be 1.1 ft. Which of the following statements is correct under these circumstances?

(A) Another point should be tried farther away from the centerline.

(B) Another point should be tried closer to the centerline.

(C) The slope stake should be set at the trial point.

(D) The stake should be set at the trial point as an offset stake, with an offset of 3 ft written on it.

PROBLEM 76

The width of a particular roadbed is 50 ft, the side slope is designed to be 3:1, the grade rod is 2.8 ft, and the rod is held at a trial position for a slope stake 32 ft from the centerline. The cross section is in fill. The rod is read through the level and the ground rod is found to be 10.3 ft. The natural ground is relatively level. Which of the following statements is correct under these circumstances?

(A) Another point should be tried farther away from the centerline.

(B) Another point should be tried closer to the centerline.

(C) The slope stake should be set at the trial point.

(D) The stake should be set at the trial point as an offset stake, with an offset of 10 ft written on it.

PROBLEM 77

What is the usual position of a slope stake and the information written on it?

(A) A slope stake is usually driven so that the cut or fill information is facing the centerline of the route; the stationing is written on the side away from the centerline.

(B) A slope stake is usually driven so that the cut or fill information is facing away from the centerline of the route; the stationing is written on the side facing the centerline.

(C) A slope stake is usually driven so that the cut or fill information is parallel with the centerline of the route; the stationing is written on the opposite side.

(D) A slope stake is usually driven so that the cut or fill information and the stationing are written on the side facing the centerline.

PROBLEM 78

Superelevation is

(A) another term used to describe compound vertical curves

(B) the practice of raising the outer edge of pavement or the outer rail of a railway above the elevation of the inner edge or rail along a curve

(C) the embankment of earth above the angle of repose

(D) the gradual increase of the centerline radius of a curve along a roadway or railway

PROBLEM 79

What type of curve is frequently used in conjunction with superelevation in high-speed highway work?

(A) horizontal circular curve

(B) vertical curve

(C) compound vertical curve

(D) transition spiral curve

PROBLEM 80

Which of the following will NOT typically be provided by a proper route survey?

(A) a representation of existing parcels affected and their current ownership

(B) an established project coordinate system for subsequent right-of-way and construction layout

(C) a control alignment based on the original project plans

(D) the creation of parcels for acquisition

PROBLEM 81

Which of the following statements correctly describes an offset from the baseline of a route survey?

(A) An offset is the perpendicular horizontal distance from the baseline, either right (plus) or left (minus) from the point of view of a person facing up-station.

(B) An offset is the perpendicular horizontal distance from the baseline, either right (plus) or left (minus) from the point of view of a person facing down-station.

(C) An offset is the slope distance from the baseline, either right (plus) or left (minus) from the point of view of a person facing up-station.

(D) An offset is the slope distance from the baseline, either right (plus) or left (minus) from the point of view of a person facing down-station.

PROBLEM 82

A route survey has the following stages. Which would be first?

(A) location survey

(B) reconnaissance

(C) preliminary survey

(D) construction survey

ERRORS AND STATISTICS

PROBLEM 83

A particular line was measured with the same tape on two separate occasions. Crew 1 measured the distance three times and found it to be 537.96 ft, 538.02 ft, and 537.94 ft. Later, crew 2 measured the same line and found the distance to be 538.072 ft, 538.145 ft, and 538.223 ft. Which of the two crews had the better precision of measurement, and why?

(A) Crew 1 had better precision in its measurements since the mean and the median of its readings are closer together than are those of crew 2.

(B) Crew 2 had better precision in its measurements since its readings have a smaller range than those of crew 1.

(C) Crew 2 had better precision in its measurements since its readings are expressed to the thousandth of a foot, while those of crew 1 are expressed to the hundredth of a foot.

(D) Crew 1 had better precision in its measurements since its readings have a smaller scatter and range than those of crew 2.

PROBLEM 84

A particular line was measured with the same tape on two separate occasions. Crew 1 measured the distance three times and found it to be 1972.85 ft, 1972.79 ft, and 1973.01 ft. Later, crew 2 measured the same line and found the distance to be 1973.03 ft, 1972.83 ft, and 1973.16 ft. The line was subsequently measured with EDM equipment and found to be 1973.01 ft. Which of the taped measurements was the most accurate, and why?

(A) Crew 1 provided the most accurate measurements since one of its measurements matched that of the EDM.

(B) Crew 2 provided the most accurate measurements since the arithmetic mean of its three measurements is the same as that of the EDM.

(C) The measurements of crew 1 are the most precise, so its measurements are the most accurate.

(D) The measurements of crew 2 are the most accurate since the scatter and range of its observations are smaller than those of crew 1.

PROBLEM 85

Several errors and mistakes hampered a particular project. The EDM equipment was used throughout the survey with a reflector constant of 0.12 ft when it should have been 0.10 ft. The line of sight through the telescope of the level was inclined by 0.005 ft per 100 ft. The instrument operator was able to read all of the horizontal and vertical angles directly in degrees, minutes, and thirds of minutes (20 sec). However, the angles were also estimated to the nearest 5 sec by the instrument operator, and these estimates were sometimes too high or too low. Which of these errors and mistakes should be eliminated from the field recorded data, to the degree that is possible, before the final adjustment of the project is attempted?

(A) All of the described errors are systematic. Systematic errors should be minimized before any adjustment procedure is implemented.

(B) The error of the reflector constant is systematic and should be minimized before any adjustment procedure is implemented. However, the other two are random errors and may be properly distributed in adjustment.

(C) The discrepancies in the field-recorded data resulting from the incorrect reflector constant and the inclined level sight should both be minimized before any adjustment procedure is implemented. However, the errors resulting from the estimation of the seconds in the angle measurements are random errors and may be properly distributed in an adjustment.

(D) All of the described errors are random. Systematic errors should be eliminated before any adjustment procedure is implemented, but random errors may be properly distributed in an adjustment.

PROBLEM 86

The manipulation of random errors in surveying measurement rests on the application of assumptions from the theory of probability. Which of the following answers best describes these assumptions?

(A) Small errors occur more often than large ones, large errors occur infrequently, and there is an equal opportunity for the occurrence of positive and negative errors.

(B) There is an equal opportunity for the occurrence of large and small errors, errors cancel each other, and the error of closure of a traverse is the sum of its accumulated random errors.

(C) Large errors occur more often than small ones, and there is an equal opportunity for the occurrence of random and systematic errors.

(D) Errors remain the same under the same set of conditions, large mistakes and small mistakes occur with equal frequency, and the error of closure of a traverse is the sum of its accumulated random errors.

PROBLEM 87

The following five problems refer to the astronomic azimuths shown, which were determined from the station Gravel to station Radio Beacon North.

$59°18'37''$	$59°18'25''$	$59°18'39''$	$59°18'27''$
$59°18'47''$	$59°18'21''$	$59°18'20''$	$59°18'49''$
$59°18'22''$	$59°18'49''$	$59°18'17''$	$59°18'46''$
$59°18'22''$	$59°18'27''$	$59°18'26''$	$59°18'23''$
$59°18'24''$	$59°18'19''$	$59°18'40''$	$59°18'22''$
$59°18'27''$	$59°18'26''$	$59°18'10''$	$59°18'37''$
$59°18'34''$	$59°18'21''$	$59°18'30''$	$59°18'41''$

87.1. What is the mean, expressed to the nearest tenth of a second, and what are the modes, expressed to the nearest second?

(A) mean $= 59°18'27.4''$

mode $= 59°18'29''$

(B) mean $= 59°18'29.6''$

modes $= 59°18'27''$ and $59°18'22''$

(C) mean $= 59°18'30.0''$

modes $= 59°18'23''$ and $59°18'22''$

(D) mean $= 59°18'30.1''$

modes $= 59°18'49''$ and $59°18'10''$

87.2. What are the largest and smallest residuals of the group, expressed in absolute value to the nearest second?

(A) largest residual = 3″

smallest residual = 5″

(B) largest residual = 10″

smallest residual = 5″

(C) largest residual = 20″

smallest residual = 0″

(D) largest residual = 20″

smallest residual = 19″

87.3. What is the standard deviation of a single measurement calculated from the list of azimuths given, expressed to the nearest tenth of a second?

(A) ±0.8″

(B) ±2.0″

(C) ±5.9″

(D) ±10.5″

87.4. Which of the following statements best describes the standard deviation of a single measurement?

(A) Theoretically, the average of a set of observations has a 68.3% probability of approaching the true value, within the limit specified by the standard deviation of a single measurement.

(B) Theoretically, each of a set of observations has a 99.7% probability of approaching the mean value of the set, within the limit specified by the standard deviation of a single measurement.

(C) The standard deviation of a single measurement is a test of the fit of the actual frequencies of the distribution of data compared to the theoretical frequencies.

(D) Theoretically, each of a set of observations has a 68.3% probability of approaching the mean value of the set, within the limit specified by the standard deviation of a single measurement.

87.5. What is the standard error of the mean calculated from the list of azimuths given, expressed to the nearest tenth of a second?

(A) ±0.8″

(B) ±2.0″

(C) ±5.9″

(D) ±8.4″

PROBLEM 88

Which of the following statements best describes the standard error of the mean?

(A) Theoretically, the average of a set of observations has a 68.3% probability of approaching the true value, within the limit specified by the standard error of the mean.

(B) Theoretically, each of a set of observations has a 99.7% probability of approaching the mean value of the set, within the limit specified by the standard error of the mean.

(C) The standard error of the mean is a class interval test of the distribution of data.

(D) Theoretically, each of a set of observations has a 95.5% probability of approaching the true value of the set, within the limit specified by the standard error of the mean.

PROBLEM 89

A manufacturer specified the standard error of the distances measured with a particular EDM as ±(2 mm + 4 ppm). A distance of 7283.91 ft was measured with the equipment. Presuming that all atmospheric corrections have been made, what is the range of a 95.5% certainty that may be attached to this measurement?

(A) ±0.03 ft

(B) ±0.06 ft

(C) ±0.10 ft

(D) ±0.17 ft

PROBLEM 90

The directions for a particular traverse are being determined by a theodolite with a least count of 1.0″. The standard deviation of the mean of each angle at each point is not to exceed 1.2″. What is most nearly the standard deviation of the mean for the following four angles, and does it fulfill the specification?

32°14′57″
32°14′58″
32°15′00″
32°14′52″

(A) The standard deviation of the mean is 3.4″, and it does not fulfill the specification.

(B) The standard deviation of the mean is 1.2″, and it does fulfill the specification.

(C) The standard deviation of the mean is 1.7″, and it does not fulfill the specification.

(D) The standard deviation of the mean is 1.0″, and it does fulfill the specification.

PROBLEM 91

Standard deviation is also known as which of the following?

(A) root mean square error

(B) cumulant

(C) kurtosis

(D) sample size

PROBLEM 92

In statistics, the term *variance* means the

(A) square of the standard deviation

(B) difference between an independent measurement in a set and the mean of those measurements

(C) sum of the values of individual measurements of a quantity divided by the number of measurement repetitions

(D) standard deviation of the probability density function for a random variable

PROBLEM 93

When analyzing the weighting of a network, the optimal value of the standard deviation of unit weight is closest to

(A) 0.0

(B) 0.5

(C) 1.0

(D) 2.0

PROBLEM 94

Presuming that blunders have been removed, which of the following is the best general explanation for when the standard deviation of unit weight is less than, or more than, the optimal value?

(A) If the standard deviation of unit weight is *more than* the optimal value, the measurements are *inferior* to their weights. If the standard deviation of unit weight is *less than* the optimal value, the measurements are *superior* to their weights.

(B) If the standard deviation of unit weight is *more than* the optimal value, the measurements are *superior* to their weights. If the standard deviation of unit weight is *less than* the optimal value, the measurements are *inferior* to their weights.

(C) If the standard deviation of unit weight is *more than* the optimal value, many residuals are large. If the standard deviation of unit weight is less than the optimal value, the constrained adjustment will probably pass the chi-square test.

(D) If the standard deviation of unit weight is more than the optimal value, most residuals are small. If the standard deviation of unit weight is *less than* the optimal value, the constrained adjustment will probably not pass the chi-square test.

PROBLEM 95

Which of the following is NOT another name for the term *standard deviation of unit weight*?

(A) standard error of unit weight

(B) error total

(C) network reference factor

(D) redundancy

PROBLEM 96

An angle has been measured four times. In a statistical sense, how many degrees of freedom does this set of measurements have?

(A) 1

(B) 2

(C) 3

(D) 4

PROBLEM 97

Which of the following correctly characterizes the best relationship between the weights, variances, and size of corrections used in an adjustment of uncorrelated surveying observations?

(A) weights are inversely proportional to variances and the size of corrections

(B) weights are inversely proportional to variances and directly proportional to the size of corrections

(C) weights are directly proportional to variances and inversely proportional to the size of corrections

(D) weights are directly proportional to variances and the size of corrections

SOLUTION 1

Plane surveying differs from geodetic surveying in the treatment of the curvature of the earth. Plane surveying is conducted with a horizontal plane as its reference surface, whereas geodetic surveying takes curvature into account.

The answer is (D).

SOLUTION 2

The four principal points of the compass are known as the *cardinal directions.*

The answer is (D).

SOLUTION 3

The standard scales available on the triangular engineer's scale include 10, 20, 30, 40, 50, and 60, but not 100.

The answer is (D).

SOLUTION 4

While narrow- and broad-gauge railroad tracks exist, 4 ft, $8\frac{1}{2}$ in is standard and is sometimes marked on steel tapes for reference.

The answer is (B).

SOLUTION 5

A miner's inch is an approximate measurement of rate of flow. Generally speaking, a miner's inch equals a rate of flow of about 1.5 ft^3 per minute.

The answer is (D).

SOLUTION 6

Erasing entries in a field book is not good practice. Information recorded in a field book should remain legible even if it appears to be in error. The technique of lining through incorrect entries is widely accepted. The correct values are then written above those in error.

The answer is (D).

SOLUTION 7

Even though the last digit is zero, the entry indicates that the measurement is significant to the nearest hundredth.

The answer is (A).

SOLUTION 8

An isogonic chart is a map connecting points of equal magnetic declination. Reference to such a chart is one way to determine approximate magnetic declination.

The answer is (D).

SOLUTION 9

Secular variation is the very gradual change in magnetic declination that follows cycles of approximately 300 years. The annual, or yearly, variation is generally less than a minute of arc throughout the United States. The daily variation is an easterly movement early in the day, followed by a corresponding westerly movement, reaching a maximum change of about 8 min of arc. Irregular variations are caused by magnetic storms that in most instances result from sunspot activity. Constant variation is an invention. There is no category of magnetic variation called *constant.*

The answer is (D).

SOLUTION 10

The true bearing of the line in 1911 can be found by adding the declination to the magnetic bearing.

$$\text{N}10°24'\text{E} + 04°00' = \text{N}14°24'\text{E}$$

The magnetic bearing today must be larger than the true bearing since the declination is now west of north. The magnetic bearing can be found by adding the present declination to the true bearing.

$$\text{N}14°24'\text{E} + 00°25' = \text{N}14°49'\text{E}$$

The answer is (D).

SOLUTION 11

11.1. Since line XY is a bearing in the southeastern quadrant, the geodetic bearing of line XY is found by summing the magnetic bearing and the magnetic declination of line XY in 1814.

$$\begin{aligned} \text{magnetic bearing from 1814} &= \text{S}49°27'\text{E} \\ \text{western magnetic declination from 1814} &= \underline{+\text{S}4°00'\text{E}} \\ \text{geodetic bearing} &= \text{S}53°27'\text{E} \end{aligned}$$

The answer is (D).

11.2. The secular variation from 1814 to today is the difference between today's magnetic bearing and the magnetic bearing in 1814.

$$\begin{aligned} \text{today's magnetic bearing} &= \text{S}50°15'\text{E} \\ \text{magnetic bearing from 1814} &= \underline{-\text{S}49°27'\text{E}} \\ \text{secular variation from 1814 to today} &= \text{N}00°48'\text{E} \end{aligned}$$

The answer is (A).

11.3. The magnetic declination at line XY today is the difference between the geodetic bearing of line XY and today's magnetic bearing of line XY.

$$\begin{aligned} \text{geodetic bearing} &= 53°27'\text{E} \\ \text{today's magnetic bearing} &= \underline{-50°15'\text{E}} \\ \text{today's magnetic declination} &= 03°12'\text{W} \end{aligned}$$

The answer is (B).

SOLUTION 12

The arc of a ray of light passing from one point to another is generally not the same as the curve of the earth's surface between them. The radius of the arc of the ray of light is nearly seven times greater than the radius of the curve of the earth. The resulting effect, known as *refraction*, tends to make objects appear higher than they really are. Refraction increases as the density of the medium through which light passes increases.

The answer is (C).

SOLUTION 13

When light moves from a less dense to a more dense medium, the refraction increases. Cool air is more dense than warm air; therefore, refraction will increase over the lake.

The answer is (D).

SOLUTION 14

If the cross hairs appear to move with respect to the target when an observer looks through the telescope of an instrument, the condition is known as *parallax.* Parallax occurs when the image created by the lens lies in front of or behind the plane of the cross hairs. It can be remedied by adjustment.

The answer is (B).

SOLUTION 15

Usually, four convex lenses constitute an erecting eyepiece and two convex lenses constitute an inverting eyepiece. Because it has fewer lenses, an inverting eyepiece provides slightly superior optical acuity, but is not widely used in the United States.

The answer is (D).

SOLUTION 16

Chromatic aberration is caused by different elements of the spectrum being diverted differently as they pass through a lens.

The answer is (C).

SOLUTION 17

The combination of a double-convex lens of crown glass and an inner concavo-convex lens of flint glass tends to minimize the effects of both spherical and chromatic aberration.

The answer is (C).

SOLUTION 18

Surveying instruments tend to have rather low magnification in order to minimize increasing the effect of heat waves, turbulence, and vibration on observations. Greater magnification also reduces the field of view and is, therefore, undesirable in most surveying work. Most instruments have magnification between 18 and 42 diameters.

The answer is (C).

SOLUTION 19

Self-reading rods are sometimes called *speaking rods*. They are graduated to allow the instrument operator to read them. Target level rods are read by the rodman and are not widely used.

The answer is (C).

SOLUTION 20

A Philadelphia rod has two sections and is graduated on both sides up to 7 ft. Extended, the rod usually reaches 13 ft.

The answer is (D).

SOLUTION 21

The compensator built into most automatic levels consists of an arrangement of three prisms that is allowed to swing like a pendulum. This dampened mechanism maintains a horizontal line of sight even when the telescope barrel is not precisely level.

The answer is (B).

SOLUTION 22

When no rod bubble is available, waving the rod is one technique for verifying that the reading is taken with the rod in a vertical position. While facing the instrument operator, the rodman slowly rocks the rod forward and back. The lowest reading occurs when the rod is vertical.

The answer is (C).

SOLUTION 23

Errors due to the lack of a perfectly horizontal line of sight tend to cancel each other out if the foresight and backsight are roughly equal in length.

The answer is (A).

SOLUTION 24

The reading on the foresight (FS) is subtracted from the height of the instrument (HI) to find the elevation of an intermediate turning point (TP).

The answer is (C).

SOLUTION 25

The procedure known as *reciprocal leveling* is useful in such situations. The level is set up on one side of the river, near a turning point on which a rod is held. Simultaneously, another rod is held on a turning point on the far side of the river. Observing the rods in close succession allows the calculation of the difference in elevation between the two turning points while minimizing the effects of refraction. By moving the instrument to the far bank of the river and repeating the procedure, the difference in elevation is determined again. These two differences will likely be unequal. The mean of the two differences will closely approximate the actual change in elevation between the two turning points.

The answer is (D).

SOLUTION 26

The use of two rods and two rodmen, known as *double-rodding*, can reduce the effect of systematic error in a level circuit. The instrument operator can take the back-sight and the foresight in more rapid succession if a rod is held on both turning points simultaneously. The atmospheric effects on the line of sight are thereby reduced. If the rodmen alternate on the foresights and backsights, the error in each rod is also compensated for.

The answer is (A).

SOLUTION 27

To cause an error of 0.02 ft at 7.00 ft, the angle by which the rod deviates from vertical can be found by

$$\cos\alpha = \frac{7.00 \text{ ft}}{7.02 \text{ ft}}$$

Angle α is the angle of the rod.

$$\cos\alpha = 0.997151$$
$$\alpha = 04°19'34''$$

This angle can now be used to find the distance from plumb at the top of the 13.00 ft rod.

$$(\sin 04°19'34'')(13.00 \text{ ft}) = 0.98 \text{ ft}$$

The answer is (C).

SOLUTION 28

The rod readings taken with the level midway between BM A and BM B show that the true difference between their elevations is

$$4.14 \text{ ft} - 1.35 \text{ ft} = 2.79 \text{ ft}$$

This is the best indication of the true difference because the sights are balanced.

When the instrument is moved very close to BM A, the difference between the readings is

$$7.12 \text{ ft} - 4.46 \text{ ft} = 2.66 \text{ ft}$$

The error is presumed to occur in the longer sight, to BM B. The error is

$$\begin{array}{r} 2.79 \text{ ft} \\ -2.66 \text{ ft} \\ \hline 0.13 \text{ ft} \end{array}$$

Therefore, if the instrument were in perfect adjustment, the expected reading at BM B would be

$$\begin{array}{r} 7.12 \text{ ft} \\ +0.13 \text{ ft} \\ \hline 7.25 \text{ ft} \end{array}$$

The answer is (D).

SOLUTION 29

Backsights are added to the previous elevation to find the height of the instrument (HI). Foresights are subtracted from the HI to find the elevation. Reversing the process in each case will allow the determination of the BS from the HI and the FS from the elevation.

STA	BS	HI	FS	ELEV	
BM A				210.01	
	1.98	211.99			
TP 1			3.61	208.38	
	7.53	215.91			
TP 2			4.30	211.61	
	5.77	217.38			
TP 3			6.15	211.23	
	3.51	214.74			
BM B			5.31	209.43	
	6.35	215.78			
TP 4			7.55	208.23	
	2.21	210.44			
TP 5			4.90	205.54	
	7.33	212.87			
TP 6			1.21	211.66	
	5.10	216.76			
BM A			6.69	210.07	

The answer is (B).

SOLUTION 30

The rod reading at the center hair is 5.21 for rod 2 and 5.53 for rod 1. The approximate distance to the rod can be determined by the interval between the upper and the lower stadia hairs. The interval is smaller for rod 1. Therefore, rod 1 is closer to the observer.

The answer is (C).

SOLUTION 31

31.1. Find the HI for the first backsight using 5.364, the mean value, rather than 5.365, the reading taken at the middle cross hair.

$$1609.350 \text{ ft} + 5.364 \text{ ft} = 1614.714 \text{ ft}$$

Subtract the mean value of the readings at the foresight to find the elevation of TP1.

$$1614.714 \text{ ft} - 6.330 \text{ ft} = 1608.384 \text{ ft}$$

The answer is (A).

31.2. The stadia interval factor of the instrument is 100, and the interval between the upper and lower cross hairs is 2.751 for the backsight and 2.396 for the foresight. The distance to the backsight is approximately

$$(2.751)(100) = 275.1 \text{ ft}$$

The distance is approximately

$$(2.396)(100) = 239.6 \text{ ft}$$

The difference to the foresight is approximately

$$\begin{array}{r} 275.1 \text{ ft} \\ -\,239.6 \text{ ft} \\ \hline 35.5 \text{ ft} \quad (35.50 \text{ ft}) \end{array}$$

The answer is (D).

31.3. The usual limit placed on the difference between the three-wire readings is two of the smallest units being recorded. The only set of readings that satisfies this limit is the backsight taken at TP1.

The answer is (B).

SOLUTION 32

Interpolation is a method of finding intermediate values between fixed values. The plotting of contours on a map with spot elevations involves interpolation since the spots seldom fall precisely on the contour elevations.

The answer is (A).

SOLUTION 33

33.1. A small angular correction must be applied to the measured zenith angles at each end of the line to compensate for the difference in the height of the instrument and the EDM.

$$\frac{\sin(\text{measured zenith})}{\text{slope distance}} = \frac{\sin(\text{angular correction})}{\text{height of EDM} - \text{height of instrument}}$$

The zenith forward is

$$\frac{\sin 91°06'16''}{6782.955 \text{ ft}} = \frac{\sin \alpha}{4.70 \text{ ft} - 4.00 \text{ ft}}$$
$$\alpha = 00°00'21.3''$$

$$\begin{array}{c}\text{measured zenith forward} \\ +\text{ angular correction}\end{array} = \text{corrected angle}$$

$$91°06'16'' + 00°00'21.3'' = 91°06'37.3''$$

The zenith back is

$$\frac{\sin 88°55'15''}{6782.955 \text{ ft}} = \frac{\sin \alpha}{5.00 \text{ ft} - 5.60 \text{ ft}}$$
$$\alpha = 00°00'18.2''$$

$$\begin{array}{c}\text{measured zenith forward} \\ +\text{ angular correction}\end{array} = \text{corrected angle}$$

$$88°55'15'' + 00°00'18.2'' = 88°55'33.2''$$

Next, find the difference in elevation forward (from Thornton to 300) and back (from 300 to Thornton) using the following expression.

$$(\cos \text{corrected zenith})(\text{slope distance}) + \text{height EDM} - \text{height prism}$$

The difference in elevation forward is

$$\begin{aligned} \Delta \text{ elev} &= (\cos 91°06'37.3'')(6782.955 \text{ ft}) \\ &\quad + 4.70 \text{ ft} - 4.50 \text{ ft} \\ &= -131.44 \text{ ft} + 4.70 \text{ ft} - 4.50 \text{ ft} \\ &= -131.24 \text{ ft} \end{aligned}$$

The difference in elevation back is

$$\begin{aligned} \Delta \text{ elev} &= (\cos 88°55'33.2'')(6782.955 \text{ ft}) \\ &\quad + 5.60 \text{ ft} - 3.80 \text{ ft} \\ &= 127.15 \text{ ft} + 5.60 \text{ ft} - 3.80 \text{ ft} \\ &= +128.95 \text{ ft} \end{aligned}$$

The average of these two values is used to calculate the elevation at 300.

$$\begin{array}{r} 128.95 \text{ ft} \\ +\,131.24 \text{ ft} \\ \hline 260.19 \text{ ft} \end{array}$$

The mean difference in elevation is

$$\frac{260.19 \text{ ft}}{2} = 130.10 \text{ ft}$$

$$\begin{array}{rl} \text{elev of Thornton} = & 5440.6 \text{ ft} \\ \Delta \text{ elev} = & +(-130.1 \text{ ft}) \\ \hline \text{elev of 300} = & 5310.5 \text{ ft} \quad (5311 \text{ ft}) \end{array}$$

The answer is (A).

33.2. Differential leveling with a spirit level may neglect the effects of curvature and refraction since the sights involved are seldom long enough or high enough to require such corrections. However, trigonometric leveling must take note of these corrections. When reciprocal observations are available, averaging will correct for both effects, but an approximation must be included in the calculation when reciprocal observations are not available.

The answer is (D).

33.3. A small angular correction must be applied to the measured zenith angles at each end of the line to compensate for the difference in the height of the instrument and the EDM.

$$\frac{\sin(\text{measured zenith})}{\text{slope distance}} = \frac{\sin(\text{angular correction})}{\text{height of EDM} - \text{height of instrument}}$$

The zenith forward is

$$\frac{\sin 91°06'16''}{6782.955 \text{ ft}} = \frac{\sin\alpha}{4.70 \text{ ft} - 4.00 \text{ ft}}$$
$$\alpha = 00°00'21.3''$$

$$\begin{array}{c}\text{measured zenith forward} \\ + \text{ angular correction}\end{array} = \text{corrected angle}$$

$$91°06'16'' + 00°00'21.3'' = 91°06'37.3''$$

$$\begin{array}{c}(\cos \text{corrected zenith})(\text{slope distance}) \\ + \text{height EDM} - \text{height prism}\end{array}$$

The apparent difference in elevation forward is

$$\begin{aligned}\Delta \text{ elev} &= (\cos 91°06'37.3'')(6782.955 \text{ ft}) \\ &\quad +4.70 \text{ ft} - 4.50 \text{ ft} \\ &= -131.44 \text{ ft} + 4.70 \text{ ft} - 4.50 \text{ ft} \\ &= -131.24 \text{ ft}\end{aligned}$$

The approximate elevation of point 300 is

$$\begin{array}{rl}\text{elev of Thornton} = & 5440.60 \text{ ft} \\ \Delta \text{ elev} = & +(-131.24 \text{ ft}) \\ \hline \text{elev of 300} = & 5309.36 \text{ ft}\end{array}$$

However, this value is only approximate. The effect of curvature tends to make point 300 appear lower than it is, while the effect of refraction tends to make it appear higher. It is important to note that the effect of refraction is about one-seventh the effect of the curvature. The total effect of curvature, c, and refraction, r, is given by the formula

$$c \text{ and } r = 0.0206M^2$$

M is the distance between the point in thousands of feet, and c and r are expressed in feet. Therefore,

$$M = \frac{6782.955 \text{ ft}}{1000} = 6.782955$$

$$\begin{aligned}c \text{ and } r &= 0.0206M^2 \\ &= (0.0206)(6.782955)^2 \\ &= 0.95 \text{ ft}\end{aligned}$$

The correction is added to the elevation since the largest component is attributable to curvature and not to refraction.

$$\begin{array}{r}5309.36 \text{ ft} \\ +0.95 \text{ ft} \\ \hline 5310.31 \text{ ft}\end{array}$$

Rounded to the nearest foot, the elevation of 300 is 5310 ft.

The answer is (C).

SOLUTION 34

As the elevation increases, the barometric pressure decreases. As a rule of thumb, a difference in altitude of 1000 ft will cause a difference in atmospheric pressure of 1 in of mercury (Hg).

The answer is (B).

SOLUTION 35

First, find the distance between the stations.

$$\begin{array}{r}2882.75 \text{ ft} \\ -2515.30 \text{ ft} \\ \hline 367.45 \text{ ft}\end{array}$$

Next, find the fall of the pipe over that distance at a grade of −2½%.

$$(367.45 \text{ ft})(-0.025) = -9.19 \text{ ft}$$

The pipe is 9.19 ft lower at sta 28+82.75 than it was at sta 25+15.30. Therefore, the flowline of the pipe (the invert) is that much lower as well.

elevation of the invert at sta 25+15.30 =	1892.49 ft
fall across 367.45 ft at $-2\frac{1}{2}\%$ =	+ (−9.19 ft)
elevation of the invert at sta 28+82.75 =	1883.30 ft

Finally, to find the elevation of the top of the pipe at sta 28+82.75, add the inside diameter and the thickness of the pipe to the elevation of the invert.

1883.30 ft	invert
3.00 ft	inside diameter
0.17 ft	thickness
1886.47 ft	

The answer is (D).

SOLUTION 36

The grade of the service is −2.10% for 115 ft. The fall can be computed as

$$(115.0 \text{ ft})(-0.021) = -2.42 \text{ ft}$$

The elevation of the flowline at the house is 393.00 ft, so the elevation of the flowline of the service at its connection with the sewer is

$$\begin{array}{r} 393.00 \text{ ft} \\ -2.42 \text{ ft} \\ \hline 390.58 \text{ ft} \end{array}$$

Since the connection of the service must be at the top of the 8 in main sewer, the vertical distance calculated must increase by 8 in (0.66 ft) to accommodate the width of the main sewer line.

$$\begin{array}{r} 390.58 \text{ ft} \\ -0.66 \text{ ft} \\ \hline 389.92 \text{ ft} \end{array}$$

Next, find the difference in elevation between the invert at manhole 10 and the invert at the point of connection.

elevation of the flowline of the main at the connection =	389.92 ft
elevation of the invert of manhole 10 =	− 389.00 ft
rise from the invert to the connection =	0.92 ft

The distance between the two points can be found by dividing the difference in the elevation by the grade along the main sewer.

$$\frac{0.92 \text{ ft}}{0.004} = 230.0 \text{ ft}$$

The answer is (C).

SOLUTION 37

The acceptable misclosure for a third-order loop is

$$E = 12\sqrt{K}$$

K is the distance in kilometers.

For a loop of 2500 ft, or 0.762 km,

$$\begin{aligned} \text{allowable misclosure in mm} &= 12\sqrt{0.762} \\ &= 10.5 \text{ mm} \end{aligned}$$

$$(10.5 \text{ mm})\left(0.00328 \ \frac{\text{ft}}{\text{mm}}\right) = 0.034 \text{ ft}$$

The element that would disqualify the level run from the FGCC definition of third order is the sight length of 325 ft. The maximum allowable sight limit is 90 m, or 295 ft.

The answer is (D).

SOLUTION 38

If the sights had been balanced on the level run, the inclination of the line of sight would not have altered the measurement of the difference in elevation between the BMs. However, the sights were out of balance by 115 ft at each setup.

$$\begin{array}{r} 325 \text{ ft} \\ -210 \text{ ft} \\ \hline 115 \text{ ft} \end{array}$$

At each of the 24 setups, the inclination of the line of sight acted to make the HI appear lower than it actually was. First, to find the total distance over which the error applied, multiply the number of setups by the feet out of balance.

$$(115 \text{ ft})(24 \text{ setups}) = 2760 \text{ ft}$$

The magnitude of the error is 0.02 ft per 300 ft. The total error is

$$\left(\frac{0.02 \text{ ft}}{300 \text{ ft}}\right)(2760 \text{ ft}) = 0.184 \text{ ft}$$

The previously calculated difference in elevation between the two BMs is in error by 0.184 ft. Since the inclination was downward and the foresights were shorter than the backsights, the calculated difference in elevation was too small. The difference should be increased by 0.184 ft.

$$\begin{array}{r} 96.32\ \text{ft} \\ +0.184\ \text{ft} \\ \hline 96.50\ \text{ft} \end{array}$$

The answer is (D).

SOLUTION 39

Since the rod at the TP was inclined toward the instrument when the reading was taken, the reading was too high and the elevation determined from it was too low. The correct rod reading can be found by multiplying 4.69 ft by the cosine of 15°.

$$(4.69\ \text{ft})(\cos 15°) = (4.69\ \text{ft})(0.96592) = 4.53\ \text{ft}$$

The actual elevation of the TP can now be found.

$$\begin{array}{rr} \text{elevation of the BM} = & 4107.22\ \text{ft} \\ \text{backsight HI} = & +1.47\ \text{ft} \\ \cline{2-2} & 4108.69\ \text{ft} \\ \text{foresight} = & -4.53\ \text{ft} \\ \cline{2-2} & 4104.16\ \text{ft} \end{array}$$

The answer is (C).

SOLUTION 40

A repeating instrument allows the operator to accumulate the measurements of an angle on the plate, adding each successive angle to the sum of those already turned.

The answer is (B).

SOLUTION 41

Using the lower clamp and tangent screw for backsighting and the upper motion for foresighting allows the operator to accumulate several angular readings on the plate of a repeating theodolite. The technique can not be used with a directional theodolite, which has no lower motion.

The answer is (D).

SOLUTION 42

In the past, the reticle, sometimes called the *reticule*, was often constructed of threads from the web of a brown spider. Today it is usually composed of lines etched in glass. The stadia interval factor most frequently used in the reticle is 100.

The answer is (A).

SOLUTION 43

The vernier described is of the direct type, as are nearly all transit verniers. The increments on the direct vernier are slightly smaller than those on the circle; in this case, 40 spaces on the vernier correspond to 39 spaces on the circle. On the other hand, a retrograde vernier is characterized by increments on the vernier that are larger than those on the circle.

The least count of the vernier may be found by the expression

$$\text{least count} = \frac{s}{n}$$

s is the length of each graduation of the circle, and n is the number of vernier spaces. In this case, the least count is found by

$$\frac{20'}{40} = 0.5' = 30''$$

The answer is (D).

SOLUTION 44

There is no need to measure the angle between the two temporary points. The middle position between two points set with the telescope direct and inverted will lie on the prolongation of the desired line.

The answer is (C).

SOLUTION 45

Closing the horizon is a term that means the procedure of measuring angles around a point such that the last angle terminates on the sight where the first angle began. The sum of such a series of angles should be 360°.

Choice (A) is a partial description of one method of traversing around an obstacle. Choice (B) describes a procedure known as *balancing in*. Choice (C) is sometimes known as *doubling the angle*.

The answer is (D).

SOLUTION 46

Checking an optical plummet against a plumb bob is likely to be accurate within approximately ±1 mm. The procedure described in choice (C) will provide the best check, within an error circle of approximately ±½ mm.

The answer is (C).

SOLUTION 47

A steel tape not supported throughout its length will assume the form of a catenary.

The answer is (D).

SOLUTION 48

Many 100 ft steel tapes are manufactured to provide their standard length when they are horizontal, fully supported, and under 12 lb of tension at 68°F. The standard tension is somewhat more variable than the standard temperature, the latter being almost universally accepted as 68°F, or 20°C.

The answer is (D).

SOLUTION 49

Steel expands and contracts at a rate of approximately 0.0000065 per unit of length for each degree Fahrenheit and 0.0000116 per unit of length for each degree Celsius.

The answer is (D).

SOLUTION 50

The formula to calculate the correction for a tape used without the correct tension is

$$C_p = \frac{(P - P_s)L}{AE}$$

C_p is the elongation of the tape, P is the pull on the tape, P_s is the standard pull, A is the cross-sectional area of the tape, E is the modulus of elasticity of steel, and L is the length measured.

$$\begin{aligned} C_p &= \frac{(40 \text{ lbf} - 20 \text{ lbf})(469.40 \text{ ft})}{(0.0066 \text{ in}^2)\left(29{,}000{,}000 \frac{\text{lbf}}{\text{in}^2}\right)} \\ &= \frac{(20 \text{ lbf})(469.40 \text{ ft})}{191{,}400 \text{ lbf}} \\ &= \frac{9388 \text{ ft-lbf}}{191{,}400 \text{ lbf}} \\ &= 0.05 \text{ ft} \end{aligned}$$

The correction for the measured line should be added to the recorded measurement, since the tape was stretched by the excess pull and reads short.

$$\begin{array}{r} 469.40 \text{ ft} \\ +0.05 \text{ ft} \\ \hline 469.45 \text{ ft} \end{array}$$

A rule of thumb that may be used to estimate the effect of a greater than standard pull is that the length of a heavy, 100 ft steel tape is increased by approximately 0.001 ft for each 3 lb increase in the pull.

The answer is (D).

SOLUTION 51

The effect of temperature on the length of an invar tape is negligible. The coefficient of thermal expansion of the nickel and steel alloy is 0.0000002 per degree Fahrenheit, or approximately one-thirtieth that of steel.

The answer is (A).

SOLUTION 52

The measurement should be corrected for temperature and slope. It is best to consider the temperature correction first. The tape is probably standard at 68°F, so the following formula applies.

$$C_t = k(T_1 - T)L$$

C_t is the correction for the change in temperature, k is the coefficient of thermal expansion for steel, T_1 is the temperature of the tape, T is the standard temperature for the tape, and L is the recorded length of the measurement.

$$\begin{aligned} C_t &= \left(\frac{0.0000065}{°\text{F}}\right)(83°\text{F} - 68°\text{F})(132.67 \text{ ft}) \\ &= \left(\frac{0.0000065}{°\text{F}}\right)(15°\text{F})(132.67 \text{ ft}) \\ &= (0.0000975)(132.67 \text{ ft}) \\ &= 0.01 \text{ ft} \end{aligned}$$

$$\begin{array}{r} 132.67 \text{ ft} \\ +0.01 \text{ ft} \\ \hline 132.68 \text{ ft} \end{array}$$

Next, the correction for slope can be calculated using the cosine function.

$$\begin{aligned}\text{horizontal distance} &= (\text{slope distance}) \\ &\quad \times (\cos \text{vertical angle}) \\ &= (132.68 \text{ ft})(\cos 27°) \\ &= 118.22 \text{ ft}\end{aligned}$$

The answer is (B).

SOLUTION 53

The temperature and the length of the tape are not standard. The tape is found to be 100.020 ft at 68°F. Use the following formula to find one length of the tape at 45°F.

$$C_t = k(T_1 - T)L$$

C_t is the correction for the change in temperature, k is the coefficient of thermal expansion for steel, T_1 is the temperature of the tape, T is the standard temperature for the tape, and L is the recorded length of the measurement.

$$\begin{aligned}C_t &= \left(\frac{0.0000065}{°\text{F}}\right)(45°\text{F} - 68°\text{F})(100.020 \text{ ft}) \\ &= \left(\frac{0.0000065}{°\text{F}}\right)(-23°\text{F})(100.020 \text{ ft}) \\ &= (-0.00015)(100.020 \text{ ft}) \\ &= -0.015 \text{ ft}\end{aligned}$$

$$\begin{array}{r} 100.020 \text{ ft} \\ +(-0.015 \text{ ft}) \\ \hline 100.005 \text{ ft} \end{array}$$

At the colder temperature, the tape will have contracted to nearly its nominal length. The following formula is useful for correcting a measurement made with a tape that is not the correct length.

$$\begin{aligned}C_a &= \frac{\begin{array}{c}(\text{actual length} - \text{nominal length}) \\ \times (\text{recorded measurement})\end{array}}{\text{nominal length}} \\ &= \frac{(100.005 \text{ ft} - 100.000 \text{ ft})(1315.72 \text{ ft})}{100.000 \text{ ft}} \\ &= \frac{(0.005 \text{ ft})(1315.72 \text{ ft})}{100.000 \text{ ft}} \\ &= \frac{6.5786 \text{ ft}^2}{100.000 \text{ ft}} \\ &= 0.066 \text{ ft}\end{aligned}$$

The true length of the measured line will be longer than the recorded length by 0.07 ft.

$$\begin{array}{r} 1315.72 \text{ ft} \\ +0.07 \text{ ft} \\ \hline 1315.79 \text{ ft} \end{array}$$

The answer is (A).

SOLUTION 54

The measurements should be corrected for the sag of the tape before they are compared. The formula for the sag correction is

$$C_s = \frac{W^2 L}{24P^2}$$

C_s is the correction between the points of support in feet, W is the weight of the tape between supports in pounds, L is the distance between the supports in feet, and P is the applied pull in pounds of force.

$$\begin{aligned}C_s &= \frac{(3 \text{ lbf})^2(100.00 \text{ ft})}{(24)(10 \text{ lbf})^2} = \frac{900.00 \text{ lbf}^2\text{-ft}}{2400.00 \text{ lbf}^2} \\ &= 0.375 \text{ ft}\end{aligned}$$

The distance measured with the unsupported tape is greater than the actual distance by 0.375 ft for every full tape length. The correction for six full tape lengths is

$$(6)(0.375 \text{ ft}) = 2.25 \text{ ft}$$

The actual distance is

$$\begin{array}{r} 600.00 \text{ ft} \\ -2.25 \text{ ft} \\ \hline 597.75 \text{ ft} \end{array}$$

The same procedure can be used to discover the true length of each 25.00 ft segment.

$$\begin{aligned}C_s &= \frac{(0.75 \text{ lbf})^2(25.00 \text{ ft})}{(24)(10 \text{ lbf})^2} = \frac{14.0625 \text{ lbf}^2\text{-ft}}{2400.00 \text{ lbf}^2} \\ &= 0.006 \text{ ft}\end{aligned}$$

The distance measured with the unsupported tape is greater than the actual distance by 0.006 ft for every segment length. The correction for 24 segment lengths is

$$(24)(0.006 \text{ ft}) = 0.14 \text{ ft}$$

The actual distance is

$$\begin{array}{r} 600.00\text{ ft} \\ -0.14\text{ ft} \\ \hline 599.86\text{ ft} \end{array}$$

The difference between the two measurements can now be found.

$$\begin{array}{r} 599.86\text{ ft} \\ -597.75\text{ ft} \\ \hline 2.11\text{ ft} \end{array}$$

The last segment length will then be less than 25.00 ft.

$$\begin{array}{r} 25.00\text{ ft} \\ -2.11\text{ ft} \\ \hline 22.89\text{ ft} \end{array}$$

Expressed to the nearest tenth of a foot, the answer is 22.9 ft.

The answer is (A).

SOLUTION 55

55.1. The fact that the instrument has an internal focus is important to the solution of this problem. The factor most often symbolized by C in stadia calculations is equal to zero in such instruments. Observations made with external focusing instruments most often consider the C factor to be equal to 1. However, in this case the formula for reducing the stadia observation for horizontal distance can be expressed as

$$H = Ks\sin^2 Z$$

H is the horizontal distance, K is the stadia interval factor, s is the stadia interval, and Z is the observed zenith angle.

$$\begin{aligned} H &= (100)(7.51)(\sin^2 87^\circ 37') \\ &= (751)(0.99827) \\ &= 749.70\text{ ft} \end{aligned}$$

The answer is (A).

55.2. The formula for solving the vertical distance is

$$V = \tfrac{1}{2}Ks\sin 2Z$$

V is the vertical distance, K is the stadia interval factor, s is the stadia interval, and Z is the observed zenith angle.

$$\begin{aligned} V &= \left(\frac{1}{2}\right)(100)(7.51\text{ ft})\big((\sin(2)(87^\circ 37')\big) \\ &= \left(\frac{1}{2}\right)(751\text{ ft})(0.08309) \\ &= +31.20\text{ ft} \end{aligned}$$

The answer is (B).

55.3. The elevation of station A is 2487.91 ft and the HI is 5.27 ft; therefore the elevation of the instrument itself is

$$\begin{array}{r} 2487.91\text{ ft} \\ +5.27\text{ ft} \\ \hline 2493.18\text{ ft} \end{array}$$

The vertical distance to the rod at station B, as determined in the solution to Prob. 55.2, is +31.20 ft. Therefore, the elevation of the rod reading at station B is

$$\begin{array}{r} 2493.18\text{ ft} \\ +31.20\text{ ft} \\ \hline 2524.38\text{ ft} \end{array}$$

The rod reading at station B is 6.85 ft.

$$\begin{array}{r} 2524.38\text{ ft} \\ -6.85\text{ ft} \\ \hline 2517.53\text{ ft} \end{array}$$

The answer is (B).

SOLUTION 56

Relative humidity has a relatively small effect on the speed of light waves. However, in using a microwave EDM, an error of 3°F in the difference between the temperature of the wet and dry bulbs of a psychrometer can account for an error of nearly 10 ppm.

The answer is (C).

SOLUTION 57

An error of +18°F in the determination of the air temperature would cause a corresponding error in the calculation of the apparent velocity of light of approximately 9000 ft per sec. The resulting error in the measured distance would be approximately +10 ppm.

The answer is (D).

SOLUTION 58

While an increase in air temperature allows the apparent velocity of light to increase, an increase in air pressure has the opposite effect. An error of +25 mm of Hg (torr) would result in an error in the calculation of the measured distance of approximately −10 ppm.

The answer is (B).

SOLUTION 59

The formula used to calculate the maximum error for the specified EDM is

$$\pm 3\sqrt{\left(\frac{5\text{ mm}}{304.8\ \frac{\text{mm}}{\text{ft}}}\right)^2 + \left(\frac{5d}{1{,}000{,}000}\right)^2}$$

The data provided by the manufacturer includes a constant value (5 mm) and a value that is relative to the distance measured (5 times the distance).

The division of the 5 mm by 304.8, the number of millimeters in a foot, converts the first part of the expression to decimals of a foot. The distance measured is multiplied by 5 and divided by 10^6 to find the length of 5 ppm. The two elements of the standard error are then squared, and the square root of their sum is multiplied by three to find the maximum error of the particular EDM measurement.

$$\begin{aligned}
&\pm 3\sqrt{\left(\frac{5\text{ mm}}{304.8\ \frac{\text{mm}}{\text{ft}}}\right)^2 + \left(\frac{(5)(4932.48\text{ ft})}{1{,}000{,}000}\right)^2} \\
&= \pm 3\sqrt{(0.0164\text{ ft})^2 + (0.0246624\text{ ft})^2} \\
&= \pm 3\sqrt{0.000877234\text{ ft}^2} \\
&= \pm 0.09\text{ ft}
\end{aligned}$$

The answer is (C).

SOLUTION 60

The reflector constant is the distance the light beam travels within the reflector itself. The distance is included in the measured distance, so a correction must be made. Most modern EDMs allow such a correction to be preset. However, the determination of the length of the correction is necessary from time to time when new or different reflectors are used on a project.

The sum of the measurements of lines 1-2 and 2-3 includes the reflector constant twice. The measurement of line 1-3 includes it once. Therefore, the difference between the sum of two line segments and the total measured length is the reflector constant.

$$\begin{aligned}
\text{line 1-2} &= \quad 355.93\text{ ft} \\
\text{line 2-3} &= \underline{+\,830.61\text{ ft}} \\
&\quad\ 1186.54\text{ ft}
\end{aligned}$$

$$\begin{aligned}
&\quad\ 1186.54\text{ ft} \\
\text{line 1-3} &= \underline{-\,1186.42\text{ ft}} \\
&\quad\ 0.12\text{ ft}
\end{aligned}$$

The answer is (D).

SOLUTION 61

$$\begin{aligned}
&\pm\sqrt{\left(\frac{7\text{ mm}}{304.8\ \frac{\text{mm}}{\text{ft}}}\right)^2 + \left(\frac{7d}{1{,}000{,}000}\right)^2} \\
&= \pm\sqrt{\left(\frac{7\text{ mm}}{304.8\ \frac{\text{mm}}{\text{ft}}}\right)^2 + \left(\frac{1400.00\text{ ft}}{1{,}000{,}000}\right)^2} \\
&= \pm\sqrt{(0.0230\text{ ft})^2 + (0.0014\text{ ft})^2} \\
&= \pm\sqrt{0.000531\text{ ft}^2} \\
&= \pm 0.023\text{ ft}
\end{aligned}$$

$$\frac{200.00}{0.023} = 8696 \quad (9000)$$

The relative precision is 1 part in 8696 (i.e., 1 part in 9000). Generally, the greatest precision is achieved over longer distances with EDM equipment.

The answer is (A).

SOLUTION 62

The term *route survey* is generally used to describe surveying for a project that deals with a path for transportation of people or commodities from place to place. In practice, a route survey frequently includes aspects from most of the other categories of surveys described.

The answer is (D).

SOLUTION 63

Establishing two forward points, one plunged in the direct position and the other plunged in the reverse position, will provide an immediate indication of the maladjustment of the instrument. The true prolongation of the line will lie halfway between the two points.

The answer is (A).

SOLUTION 64

The right angle offset method and the measured offset method rely on the establishment of a line parallel to the route of the survey. The equilateral triangle method and the equal angle method rely on the establishment of a point or points at known angles from the route.

The answer is (D).

SOLUTION 65

Astronomic observations are often used in long route surveys to correct for error accumulated in azimuths, or to quantify that error.

The answer is (A).

SOLUTION 66

Some route surveys extend for hundreds of miles. Depending on the accuracy required, surveys of such great distances should take the shape of the earth into account.

The answer is (B).

SOLUTION 67

These are only a few of the circumstances that call for stakes to be spaced to be more closely than every full station. In fact, it is rare that a route survey does not require at least some stakes set at odd stations.

The answer is (D).

SOLUTION 68

Generally, the grade of a sewer line is more critical since it relies on gravity flow.

The answer is (A).

SOLUTION 69

The term *blue top* is derived from the practice of marking the top of a stake driven to grade with blue keel. The term is also sometimes used for stakes marked for grade.

The answer is (D).

SOLUTION 70

It is not unusual for a route survey to cross areas where stakes cannot be driven; a shiner is one of several options for establishing the location of a point. A nail driven through an aluminum or other type of metal disk is convenient and permanent enough in most circumstances.

The answer is (D).

SOLUTION 71

Batter boards are used most frequently in the construction of pipelines and at building corners to provide grade and location. For a pipeline, a horizontal board, usually 1 in × 6 in, is established across the trench by nailing it between two 2 in × 4 in posts that have been driven into the ground. A nail is driven into the horizontal board at the centerline of the pipe. The top of the horizontal board is usually established at a whole number of feet above the grade of the pipe's flowline. A wire or string stretched from nail to nail provides the reference for the location and grade of the pipe.

The answer is (D).

SOLUTION 72

Slope stakes are usually set where the cuts or fills necessary for the construction of a roadway or other route exceed approximately 3 ft. In such situations, the points at which the designed side slopes intersect the natural ground are indicated by slope stakes.

The answer is (C).

SOLUTION 73

The usual data written on a slope stake include the station of the stake, the cut or fill at the centerline, the distance of the stake from the centerline, and the ratio of the slope.

The answer is (C).

SOLUTION 74

The grade rod is the rod reading that would occur if the bottom of the rod was held precisely at the finished grade at the centerline.

$$\text{HI} - \text{elevation of the grade} = \text{grade rod}$$

$$\begin{array}{r} 120.6\ \text{ft} \\ -115.4\ \text{ft} \\ \hline 5.2\ \text{ft} \end{array}$$

The answer is (D).

SOLUTION 75

The first step in solving this problem is to find the difference between the grade rod and the ground rod.

$$\begin{aligned} h &= \text{grade rod} - \text{ground rod} \\ &= 5.2\ \text{ft} - 1.1\ \text{ft} \\ &= 4.1\ \text{ft} \end{aligned}$$

This value indicates the cut that would be necessary to bring the trial point to the grade elevation of the roadway.

The next step is to determine whether the distance from the centerline, 45 ft, is the correct distance to accomplish the 4.1 ft cut at a 2:1 slope. The applicable formula is

$$d = \frac{w}{2} + hs$$

d is the distance from the centerline, w is the width of the roadbed, h is the difference between the grade rod and the ground rod, and s is the run of the designed slope.

$$\begin{aligned} d &= \frac{66 \text{ ft}}{2} + (4.1 \text{ ft})(2) \\ &= 33 \text{ ft} + 8.2 \text{ ft} \\ &= 41.2 \text{ ft} \end{aligned}$$

The rod is currently 45 ft from the centerline, so it needs to be moved closer to the centerline for another shot.

The answer is (B).

SOLUTION 76

The first step in solving this problem is to find the difference between the grade rod and the ground rod.

$$\begin{aligned} h &= \text{grade rod} - \text{ground rod} \\ &= 2.8 \text{ ft} - 10.3 \text{ ft} \\ &= -7.5 \text{ ft} \end{aligned}$$

This value indicates the fill that would be necessary to bring the trial point to the grade elevation of the roadway. The next step is to determine if the distance from the centerline, 32 ft, is the correct distance to accomplish the 7.5 ft fill at a 3:1 slope. Use the same formula as in the previous problem to find the distance.

$$\begin{aligned} d &= \frac{w}{2} + hs \\ &= \frac{50 \text{ ft}}{2} + (7.5 \text{ ft})(3) \\ &= 25 \text{ ft} + 22.5 \text{ ft} \\ &= 47.5 \text{ ft} \end{aligned}$$

The rod is currently 32 ft from the centerline, so the rod needs to be moved farther from the centerline for another shot.

The answer is (A).

SOLUTION 77

The cut or fill information is usually written on the broad part of the stake and faces the centerline. The stationing is usually written on the opposite side and faces away from the centerline.

The answer is (A).

SOLUTION 78

The raising of the outer edge of pavement along a curve is called *superelevation*. It helps to control the centrifugal force exerted on a vehicle.

The answer is (B).

SOLUTION 79

The transition spiral curve is frequently used in conjunction with superelevation. The constant rate of change of its curvature makes it ideal for this purpose.

The answer is (D).

SOLUTION 80

A route survey does typically show control alignments based on the original project plans, the current ownership of the existing parcels affected, and the established coordinate system. However, route surveys are not considered original or retracement surveys, and therefore are not used to create parcels for acquisition.

The answer is (D).

SOLUTION 81

An offset in a route survey is a horizontal distance perpendicular to the baseline, reckoned right or left from the point of view of a person facing up-station.

The answer is (A).

SOLUTION 82

A route survey supplies the information necessary for the design and construction of various projects, such as roads, railroads, pipelines, and utilities. The typical order of the surveying is reconnaissance, followed by the location survey, then the preliminary survey, then the construction survey. The first stage, reconnaissance, is a rough survey through the expected pathway of the project, with the purpose of gathering data to inform the choice of the best route and perhaps the approximate cost of the project.

The answer is (B).

SOLUTION 83

The mean of a series of measurements of the same quantity is the sum of the measurements divided by the number of repetitions.

For crew 1,

$$\begin{array}{r} 537.96 \text{ ft} \\ 537.94 \text{ ft} \\ +538.02 \text{ ft} \\ \hline 1613.92 \text{ ft} \end{array}$$

$$\frac{1613.92 \text{ ft}}{3} = 537.973 \text{ ft (arithmetic mean)}$$

For crew 2,

$$\begin{array}{r} 538.072 \text{ ft} \\ 538.145 \text{ ft} \\ +538.223 \text{ ft} \\ \hline 1614.440 \text{ ft} \end{array}$$

$$\frac{1614.440 \text{ ft}}{3} = 538.147 \text{ ft (arithmetic mean)}$$

The median of a series of observations of the same quantity is the middle observation, after the measurements have been arranged in ascending or descending order.

For crew 1,

$$\begin{array}{rl} & 537.94 \text{ ft} \\ \text{median} = & 537.96 \text{ ft} \\ & 538.02 \text{ ft} \end{array}$$

For crew 2,

$$\begin{array}{rl} & 538.072 \text{ ft} \\ \text{median} = & 538.145 \text{ ft} \\ & 538.223 \text{ ft} \end{array}$$

The comparison of the arithmetic mean of a series of observations with its median is not a test for the precision of the observations themselves. However, the scatter and the range of such a series is revealing. Generally, the larger the scatter and range of repeated observations, the less precise are the measurements. The scatter may be defined as the dispersal around the arithmetic mean.

$$v = \text{reading} - \text{mean}$$

crew 1	v
537.96 ft	−0.013
537.94 ft	−0.033
538.02 ft	+0.047

$$537.973 \text{ ft} = \text{arithmetic mean}$$

crew 2	v
538.072 ft	−0.075
538.145 ft	−0.002
538.223 ft	+0.076

$$538.147 \text{ ft} = \text{arithmetic mean}$$

The scatter of the observations is somewhat larger in the measurements of crew 2, even though the difference between the mean and the median is less.

The range may be defined as the difference between the largest and the smallest values in the series of measurements.

For crew 1,

$$\begin{array}{r} 538.02 \text{ ft} \\ -537.96 \text{ ft} \\ \hline 0.06 \text{ ft} \end{array}$$

For crew 2,

$$\begin{array}{r} 538.223 \text{ ft} \\ -538.072 \text{ ft} \\ \hline 0.151 \text{ ft} \end{array}$$

The range of the observations is larger in the measurements of crew 2.

The measurements of crew 2 are less precise than the measurements of crew 1 because the range and scatter of the observations of crew 1 are smaller.

It is important to note that the precision and the uniformity of these measurements does not determine their accuracy.

The answer is (D).

SOLUTION 84

The precision of a series of measurements of the same quantity is not indicative of the accuracy of those measurements. The EDM measurement provides a more accurate standard than the taped distances. The comparison of the arithmetic mean of each series of taped observations with the EDM measurements shows which crew provided the most accurate distance.

For crew 1,

$$\begin{array}{r} 1972.85 \text{ ft} \\ 1972.97 \text{ ft} \\ +1973.01 \text{ ft} \\ \hline 5918.65 \text{ ft} \end{array}$$

$$\frac{5918.65}{3} = 1972.88 \text{ ft (arithmetic mean)}$$

For crew 2,

$$\begin{array}{r} 1973.03 \text{ ft} \\ 1972.83 \text{ ft} \\ +1973.16 \text{ ft} \\ \hline 5919.02 \text{ ft} \end{array}$$

$$\frac{5919.02 \text{ ft}}{3} = 1973.01 \text{ ft (arithmetic mean)}$$

$$\text{EDM measurement} = 1973.01 \text{ ft}$$

The answer is (B).

SOLUTION 85

Mistakes that are the result of carelessness, inattention, bad habits, or faulty technique should not be confused with errors in surveying measurement. Errors are often divided into two categories.

Systematic errors remain the same under the same set of conditions and vary with those conditions. For example, the error introduced by the inclined sight of the level would grow as the distance of the level rod from the level increased. Systematic errors may be quantified and should be eliminated from survey data before it is adjusted.

Random errors do not follow any physical or mathematical law. The estimation of the seconds in the readings of the angles described in the question are random errors. The properties of random errors are imagined to be statistical in nature and are, therefore, subject to adjustment.

The answer is (C).

SOLUTION 86

Random errors, unlike systematic errors and mistakes, cannot be avoided and are generally beyond the realm of detection and removal. Therefore, surveyors rely on some basic assumptions from the laws of probability to predict their occurrence and behavior.

The answer is (A).

SOLUTION 87

87.1. The mean is the sum of the individual measurements divided by the number of observations. The sum of the seconds in the azimuths given is 828. There are a total of 28 observations.

$$\frac{828 \text{ sec}}{28} = 29.6 \text{ sec}$$

The arithmetic mean to the nearest tenth of a second is 29.6″.

The modes are the values that occur most frequently in a series of observations of the same value.

59°18′49″	59°18′37″	59°18′26″	59°18′22″
59°18′49″	59°18′37″	59°18′26″	59°18′21″
59°18′47″	59°18′34″	59°18′25″	59°18′21″
59°18′46″	59°18′30″	59°18′24″	59°18′20″
59°18′41″	59°18′27″	59°18′23″	59°18′19″
59°18′40″	59°18′27″	59°18′22″	59°18′17″
59°18′39″	59°18′27″	59°18′22″	59°18′10″

In this series of observations, 49 sec occurs twice, 37 sec occurs twice, 27 sec occurs three times, 26 sec occurs twice, 22 sec occurs three times, and 21 sec occurs twice.

Therefore, the modes are 59°18′27″ and 59°18′22″.

The answer is (B).

87.2. A residual is the difference between an individual reading in a series of observations and the arithmetic mean of that series. Residuals are frequently symbolized by v. The largest value for the azimuth shown is 59°18′49″, and the smallest is 59°18′10″. The arithmetic mean of the series is 59°18′30″.

$$\begin{aligned} \text{residual} &= \text{observation} - \text{arithmetic mean} \\ +19'' &= 59^\circ 18' 49'' - 59^\circ 18' 30'' \\ -20'' &= 59^\circ 18' 10'' - 59^\circ 18' 30'' \end{aligned}$$

The residuals of these two observations are very close in terms of absolute value. The 20″ is the largest. There is one observation in the series that is equal to the arithmetic mean.

$$0'' = 59^\circ 18' 30'' - 59^\circ 18' 30''$$

The smallest residual of the series is 0″.

The answer is (C).

87.3. The calculation of the standard deviation of a single measurement begins with the calculation of the arithmetic mean. After the mean is determined, the residuals are found. These residuals are squared and the sum of those squares, symbolized by v^2, is calculated.

Instruments/ Procedures

The following formula may be used to find the standard deviation of a single measurement.

$$\sigma = \sqrt{\frac{\sum v^2}{n-1}}$$

σ is the standard deviation of a single measurement, v^2 is the sum of the squares of the residuals, and n is the number of individual observations.

azimuths	v	v^2	azimuths	v	v^2
59°18′49″	+19.4″	376.36	59°18′26″	−03.6″	12.96
59°18′49″	+19.4″	376.36	59°18′26″	−03.6″	12.96
59°18′47″	+17.4″	302.76	59°18′25″	−04.6″	21.16
59°18′46″	+16.4″	268.96	59°18′24″	−05.6″	31.36
59°18′41″	+11.4″	129.96	59°18′23″	−06.6″	43.56
59°18′40″	+10.4″	108.16	59°18′22″	−07.6″	57.76
59°18′39″	+09.4″	88.36	59°18′22″	−07.6″	57.76
59°18′37″	+07.4″	54.76	59°18′22″	−07.6″	57.76
59°18′37″	+07.4″	54.76	59°18′21″	−08.6″	73.96
59°18′34″	+04.4″	19.36	59°18′21″	−08.6″	73.96
59°18′30″	+00.4″	0.16	59°18′20″	−09.6″	92.16
59°18′27″	−02.6″	6.76	59°18′19″	−10.6″	112.36
59°18′27″	−02.6″	6.76	59°18′17″	−12.6″	158.76
59°18′27″	−02.6″	6.76	59°18′10″	−19.6″	384.16
					$\Sigma v^2 = 2990.88$

$$\begin{aligned}\sigma &= \sqrt{\frac{\sum v^2}{n-1}} \\ &= \sqrt{\frac{2990.88}{28-1}} \\ &= \sqrt{110.773} = \pm 10.5''\end{aligned}$$

The answer is (D).

87.4. The standard deviation of a single measurement is given by the formula

$$\sigma = \sqrt{\frac{\sum v^2}{n-1}}$$

σ is the standard deviation of a single measurement, $\sum v^2$ is the sum of the squares of the residuals, and n is the number of individual observations.

The standard deviation is expressed as a plus or minus range, such as ±10.6″. When attached to the arithmetic mean of a series of observations, for example 59°18′29.6″ ±10.6″, it means that each set of observations has a 68.3% probability of falling within the specified range of the mean value of the set.

In other words, 68.3% of the azimuths in the given list are likely to fall between an upper limit of 59°18′19.0″ and a lower limit of 59°18′40.2″. In fact, 71% of the observations fall between those values, indicating that the list is close to a truly random sampling.

The answer is (D).

87.5. The standard error of the mean of a series of measurements of a single quantity is given by the formula

$$\sigma_m = \frac{\sigma}{\sqrt{n}}$$

σ_m is the standard error of the mean, σ is the standard deviation of a single measurement, and n is the number of individual observations.

$$\begin{aligned}\sigma_m &= \frac{\sigma}{\sqrt{n}} \\ &= \frac{\pm 10.6''}{\sqrt{28}} \\ &= \pm 2.0''\end{aligned}$$

The answer is (B).

SOLUTION 88

One way to understand the standard error of the mean is to realize that as the denominator of the expression $\sigma_m = \sigma/\sqrt{n}$ approaches infinity (∞), the standard error of the mean (σ_m) approaches zero.

The standard error of the mean, unlike the standard error of a single observation, is a statement of the uncertainty of the mean, not of a single observation. The uncertainty of the range is expressed with respect to the true value of the quantity being measured.

For example, the standard error of the mean for the list of azimuths given is ±2.0″. Therefore, it may be said with 68.3% certainty that the unknowable true value of the azimuth, which has been subject to 28 measurements, lies somewhere within the range of 59°18′29.6″ ± 2.0″.

The answer is (A).

SOLUTION 89

The standard error specified by the manufacturer is given by the formula

$$\begin{aligned}\sigma &= \pm\sqrt{\left(\frac{2\ \text{mm}}{304.8\ \frac{\text{mm}}{\text{ft}}}\right)^2 + \left(\frac{4d}{1{,}000{,}000}\right)^2} \\ &= \pm\sqrt{\left(\frac{2\ \text{mm}}{304.8\ \frac{\text{mm}}{\text{ft}}}\right)^2 + \left(\frac{(4)(7283.91\ \text{ft})}{1{,}000{,}000}\right)^2} \\ &= \pm\sqrt{\left(\frac{2\ \text{mm}}{304.8\ \frac{\text{mm}}{\text{ft}}}\right)^2 + \left(\frac{29{,}135.64\ \text{ft}}{1{,}000{,}000}\right)^2} \\ &= \pm 0.030\ \text{ft}\quad \text{(standard error)}\end{aligned}$$

This is the range of certainty within which 68.3% of a set of measurements may be expected to fall. However, the value sought is not the range for 68.3% certainty, but for 95.5% certainty. The 95.5% certainty range is twice the standard error.

$$(\pm 0.030 \text{ ft})(2) = \pm 0.06 \text{ ft}$$

The answer is (B).

SOLUTION 90

First, find the arithmetic mean of the four angles. Note that this calculation can be simplified by taking the arithmetic mean of the seconds and adding that value to 32°14′.

The sum of the seconds is

$$\begin{array}{r} 57'' \\ 58'' \\ 60'' \\ +\,52'' \\ \hline 227'' \end{array}$$

The arithmetic mean is

$$\begin{aligned} m &= 32°14' + \frac{\sum \text{seconds}}{\text{number of angles}} \\ &= 32°14' + \frac{227''}{4} \\ &= 32°14'56.75'' \end{aligned}$$

Next, find the residuals, the difference between the measured angle and the arithmetic mean, then determine the sum of their squares.

	v	v^2
32°14′57″	0.25″	0.062
32°14′58″	1.25″	1.562
32°14′60″	3.25″	10.562
32°14′52″	−4.75″	22.562
		$\sum = 34.748$

Now find the standard deviation of a single measurement.

$$\begin{aligned} \sigma &= \sqrt{\frac{\sum v^2}{n-1}} \\ &= \sqrt{\frac{34.748}{4-1}} \\ &= \sqrt{\frac{34.748}{3}} \\ &= \sqrt{11.583} \\ &= \pm 3.4'' \end{aligned}$$

Finally, calculate the standard deviation of the mean.

$$\begin{aligned} \sigma_m &= \frac{\sigma}{\sqrt{n}} \\ &= \frac{\pm 3.4''}{\sqrt{4}} \\ &= \pm 1.7'' \end{aligned}$$

The answer is (C).

SOLUTION 91

The standard deviation is also known as the root mean square error (RMSE). It is the square root of the average of the squares of deviations about the mean of a set of data. It may be said that the standard deviation is a statistical measure of spread or variability.

The answer is (A).

SOLUTION 92

Variance is the square of the standard deviation and is a value that gives the precision of a data set. The alternate answer choices are definitions for the terms *residual* (the difference between an independent measurement in a set and the mean of those measurements), *mean* (the sum of values of individual measurements of a quantity divided by the number of measurement repetitions), and *standard error of the mean* (the standard deviation of the probability density function for a random variable).

The answer is (A).

SOLUTION 93

The closer the standard deviation of unit weight is to 1.0, the better the weighting of the network.

The answer is (C).

SOLUTION 94

The value of the standard deviation of unit weight is a unitless parameter. It is an indication of the fit of the model to the data. In other words, the closer this value is to 1.0, the better a network is weighted. Also, if the standard deviation of unit weight is *more than* the optimal value, the measurements are *inferior* to their weights. If the standard deviation of unit weight is *less than* the optimal value, the measurements are *superior* to their weights.

The answer is (A).

SOLUTION 95

The term *standard deviation of unit weight* is also known as the standard error of unit weight, the error total, or the network reference factor. Redundancy generally refers to measurements that are beyond the minimum required to uniquely determine a quantity.

The answer is (D).

SOLUTION 96

If an angle is measured four times, one measurement is enough to establish the angle; the remaining three are redundant. The redundant observations can be used to expose the variation in the other measurements and adjustment becomes possible. The estimation of the mean value for the measurement removes one degree of freedom from the total number of measurements. Said another way, the degrees of freedom equal the number of observations less the number of parameters estimated. Therefore, an angle measured n times will produce $n-1$ degrees of freedom. Applying $n-1$ to this problem, an angle measured four times will produce $4-1$, or 3, degrees of freedom. This establishes the acceptable value range for the angle.

The answer is (C).

SOLUTION 97

The relative worth of an observation as compared to other measurements is reflected in the weight it is given. In this way, weights are used to control the size of corrections applied to measurements in an adjustment. Higher weights are given to more precise observations. In other words, the smaller the variance, the higher the weight. Therefore, weights should be inversely proportional to both the variances and correction sizes.

The answer is (A).

Legal Descriptions

DEFINITIONS

PROBLEM 1

A description of land by the consecutive reference to courses and distances around its perimeter is called a

(A) bounds description

(B) plat description

(C) aliquot description

(D) metes and bounds description

PROBLEM 2

The first point described in a metes and bounds description is often called the *point of beginning.* Which of the following would be characteristic of a good point of beginning?

(A) Its identity is clear and certain in the field.

(B) It has greater standing than any other corner recited in the description.

(C) The destruction of its monumentation will void the description.

(D) It is a standard section corner.

PROBLEM 3

When the phrase "more or less" follows a measurement in a description, which of the following best describes its meaning?

(A) The measurement has never been made.

(B) The measurement is intended to extend to the next recited monument whether or not, in fact, it is sufficient to do so.

(C) The words are a warning signal that the measurement is suspected to be in conflict with other information.

(D) The words are of no real significance.

PROBLEM 4

A latent ambiguity in a legal description can frequently be cured by evidence not contained in the instrument itself. Which of the following terms can be used to describe such evidence?

(A) inculpatory evidence

(B) extrinsic evidence

(C) autotopic evidence

(D) adminicular evidence

PROBLEM 5

The northern and southern boundaries of lot 3 of the Valhalla Acres subdivision are found to be parallel, while the eastern and western boundaries are not.

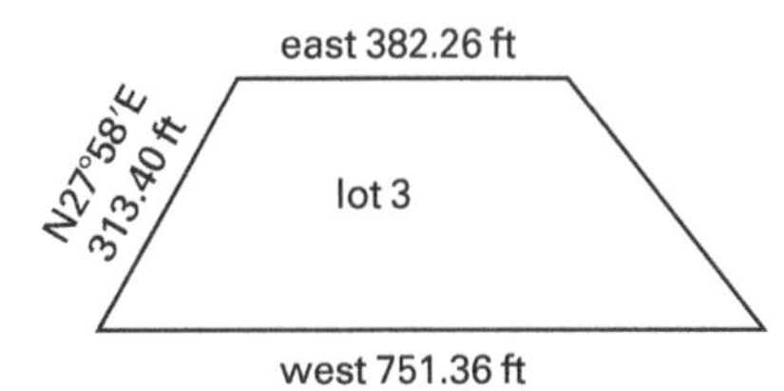

The owner of this trapezoidal lot wishes to convey the northern half. Which of the following statements best reflects the meaning of the northern half in this context?

(A) The southern boundary of the northern half would connect the midpoints of the eastern and western boundaries of the lot.

(B) The southern boundary of the northern half would be the cardinal line, parallel with the northern and southern boundaries of the lot, which divides the area into halves.

(C) The area of the lot is not a consideration in this division.

(D) Any line may divide the lot into halves, so the phrase "the northern half" is insufficient for a satisfactory determination of the division.

PROBLEM 6

The word *adjoining* is correctly used in a description when the intended meaning is which of the following?

(A) adjacent to

(B) parallel with

(C) coincident with

(D) prolonging

PROBLEM 7

A legal description reads, in part, "the N$\frac{1}{2}$ of the NW$\frac{1}{4}$ of section 9, T12N, R68W, of the 6th PM." Which answer best describes the meaning of one-half in this circumstance?

(A) It should be construed without consideration of area.

(B) The southern boundary of the north one-half should be established by quarter-quarter corners.

(C) The dividing line depends on the establishment of the midpoint of the east and west boundaries of the northwest quarter of section 9.

(D) All of the above are true.

PROBLEM 8

The phrase "bounded on the south by the Zupta Tract, as recorded in book 51, page 18, of maps" appears in a legal description. Which statement would describe this call?

(A) It is a call for a senior boundary on the south.

(B) It is a call for a record monument that includes the Zupta Tract in its entirety.

(C) It is a call for a record monument that includes only the line common to both the Zupta Tract and the property being described.

(D) It is a call that depends on the location of monuments in the field marking the common line.

PROBLEM 9

Consider a four-sided parcel, where none of the sides are parallel or perpendicular with one another. Which of the following phrases would describe a tract with a width of 50 ft whose easterly boundary would be parallel with the parcel's westerly line?

(A) the westerly 50 ft

(B) the westerly 50 ft measured along the north boundary

(C) the westerly 50 ft measured along the north and south boundaries

(D) both A and B

PROBLEM 10

A legal description that contains the phrase "to a point of cusp" refers to which of the following?

(A) a point of intersection between two curves of opposite direction

(B) a point of intersection of a quarter section line with a meander line

(C) a point where a corner monument has been lost

(D) the point of beginning of an easement

PRIORITY OF CALLS

PROBLEM 11

Conflicting elements in a description sometimes cause difficulty in determining which calls are locative and which are merely informative. What is the intent of a description that contains the call "... thence N89°45′ W664.57 ft along the south line of said section 15 to the SW corner thereof"?

(A) The described line must be N89°45′W, but the distance may be shortened or lengthened to terminate at the SW corner of section 15.

(B) The distance must be 664.57 ft, but the bearing of the south line of the section may vary from N89°45′W, and the section line must be followed.

(C) The line must follow the section line and end at the section corner, regardless of the bearing and distance.

(D) The line must proceed at the described bearing, N89°45′W, for the described distance, 664.57 ft, regardless of the location of the section line or the section corner.

PROBLEM 12

On its face, a particular metes and bounds description is clear, concise, and unambiguous. However, a subsequent survey of the parcel uncovers several conflicts between the elements of the description and the measurements. In general, the most subordinate element would be a call for which of the following?

(A) natural monument

(B) direction

(C) distance

(D) area

PROBLEM 13

A grantor of land in adjoining sections 9 and 10 wished to convey his nonaliquot holdings in section 9 alone. This was stated in the general language of the deed description; however, the specific metes and bounds description that followed included some of his land in section 10 as well. The deed has been properly executed and the grantee claims property in both sections. Which statement most correctly describes this situation?

(A) The grantor will find that the specific description will be given greater effect than the general one. The grantee was probably deeded land in section 10.

(B) The deed description will be construed in favor of the grantee, since it is presumed that the grantor chose the language used. The grantee was probably deeded land in section 10.

(C) If the grantor did not intend to convey the land in dispute, he cannot be forced to do so. The grantee probably was not deeded land in section 10.

(D) Both A and B are true.

PROBLEM 14

A legal description, even one containing conflicting elements, may be said by a court to be sufficient. Which answer best defines this meaning of the word *sufficient*?

(A) considered adequate by the parties to the deed

(B) acceptable for recordation by the county clerk

(C) capable of being located on the ground by a competent surveyor

(D) appropriate for adjudication

PROBLEM 15

In the event of a conflict within a legal description containing the following elements, which would most probably be presumed superior to the others?

(A) a called-for monument found undisturbed

(B) a specified state plane coordinate

(C) a direction expressed to the nearest second

(D) a distance recited to the thousandth of a foot

PROBLEM 16

A surveyor writing the description of a newly created parcel adjoining a long established right-of-way noted the following.

> ... thence N45°15′W 152.67 ft to a 4 in by 4 in concrete monument on the southeasterly line of the Southern Pacific Railroad right-of-way; thence S45°33′W 459.20 ft along said railroad right-of-way line to a 4 in by 4 in concrete monument; thence northeasterly to the point of beginning. Tract contains 0.82 ac.

Should conflicts arise between the aspects of this portion of the description, which will most likely outweigh the others?

(A) the 0.82 ac

(B) the Southern Pacific Railroad right-of-way

(C) the bearings N45°15′W and S45°33′W

(D) the distances 152.67 ft and 459.20 ft

PROBLEM 17

While retracing the following description, a surveyor finds and verifies the called-for aluminum caps, as well as the truck axle.

> ... thence from a found no. 5 rebar with a 2 in aluminum cap stamped "Tract Corner 1," N52°10′E 495.20 ft to an old truck axle; thence N35°58′W 597.31 ft to a found no. 5 rebar with a 2 in aluminum cap stamped "Tract Corner 3."

The truck axle is found to be located N52°10′E 505.20 ft from tract corner 1 and S35°58′E 597.29 ft from tract corner 3. What should the surveyor do?

(A) Intersect the distances.

(B) Use the bearings and distances and ignore the truck axle.

(C) Use the 495.20 ft distance, which best retraces the footsteps of the original surveyor.

(D) Use the location of the truck axle where it has been found.

PROBLEM 18

Although exceptions may arise, which of the following reflects the priority of calls most often used in interpreting legal land descriptions?

(A) artificial monuments, natural monuments, direction, distance, and area

(B) natural monuments, artificial monuments, direction, distance, and area

(C) date of survey, area, artificial monuments, natural monuments, and coordinates

(D) direction, distance, area, coordinates, and artificial monuments

PROBLEM 19

During the retracement of a description in the field, a surveyor finds a fence that appears to be of long standing nearly 12 ft east of the western boundary described in the client's deed. This apparent encroachment extends along the entire length of the property.

An investigation uncovers clear evidence of an unwritten right in property. If the encroachment is subsequently found to be a valid right of possession, what would be its relative importance with respect to the calls of the written description?

(A) It would be inferior to all the elements of the written description.

(B) It would be superior to all the elements of the written description, except a senior right.

(C) It would be superior to all the elements of the written description, as well as a senior right.

(D) The fence would be considered superior to any distances and directions in the description, but inferior to any called-for found monuments.

PROBLEM 20

When two adjoining descriptions are unambiguous and correct on their faces but are found to be overlapping in the field, which solution is most likely correct? One parcel is junior to the other.

(A) The overlap belongs to the senior parcel.

(B) The overlap is equitably proportioned between the parcels, regardless of seniority.

(C) The solution is completely dependent on the extent of the overlap.

(D) The overlap belongs to the senior parcel, unless monuments called for in the junior description would be violated.

BASIS OF BEARINGS AND DIRECTIONS

PROBLEM 21

A description written in 1873 using magnetic bear-ings calls for a direction of N65°15′W on its first course. The magnetic declination was 5°10′W at the time the description was written. The magnetic declination today at the same place is 1°05′E. What would be the magnetic bearing of the line today?

(A) N55°00′W

(B) N61°10′W

(C) N68°15′W

(D) N71°30′W

PROBLEM 22

The body of a description opens, "Beginning at the SE corner of said section 12, thence north along the east section line to the NE corner of the SE$\frac{1}{4}$ thereof, thence ..." Which answer best describes the meaning of *north* in the description?

(A) magnetic north at the time of the original survey of section 12

(B) astronomic north

(C) geodetic north

(D) the line described by the section corner and the quarter corner

PROBLEM 23

Directions in a description may be expressed in azimuths. Which statement correctly applies to this use of azimuths?

(A) The direction of a line is measured from 0° through 360° from a reference meridian.

(B) A description will express directions with azimuths or bearings, but never both.

(C) An azimuth may be measured from north or south.

(D) Both A and C are true.

PROBLEM 24

While writing a land description, a surveyor wishes to express the direction of a line that is known only generally. The line has an azimuth somewhere between 280° and 285°. Which of the following words expresses the direction most clearly and concisely?

(A) westerly

(B) an azimuth of 282°30′

(C) southwesterly

(D) northwesterly

PROBLEM 25

A description includes the phrase "... thence deflecting to the right 43°31′." Which answer best describes the meaning of the phrase?

(A) an angle of 43°31′ measured counterclockwise from the preceding course to the next one

(B) an angle of 223°31′ measured from north, clockwise to the next course

(C) an angle of 43°31′ measured clockwise from the preceding course to the next one

(D) an angle of 43°31′ measured to the right from the prolongation of the preceding line

"OF" DESCRIPTIONS

PROBLEM 26

The following two problems refer to the illustration shown.

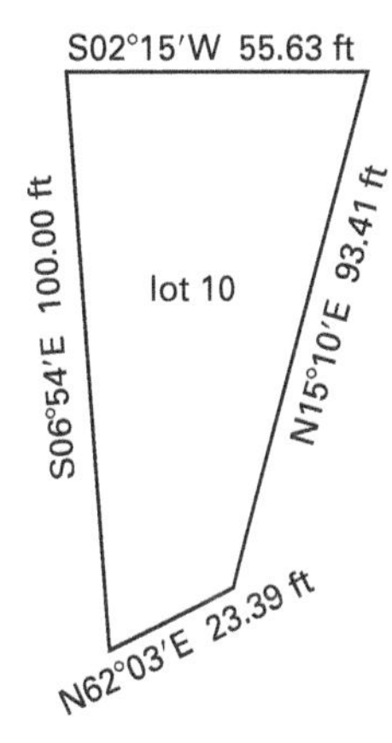

26.1. The owner of the lot shown intends to convey the property described as "the southerly 50 ft of lot 10." To best satisfy the intention of the description, the dividing line should

(A) begin at the midpoint of the westerly lot line

(B) split the lot into two parts of equal area

(C) be parallel with the southerly line of the lot and 50 ft from it, measured along the westerly line

(D) be parallel with the southerly line of the lot and 50 ft from it, measured perpendicularly

26.2. After conveying the southerly 50 ft of lot 10, the owner wishes to sell the remaining portion of the lot. Which language would best express this intention?

(A) the northerly 50.36 ft of lot 10, measured along the westerly line

(B) the north 50 ft of lot 10

(C) the north half of lot 10

(D) all of lot 10, except the southerly 50 ft

PROBLEM 27

The following lot was divided using the description "the easterly 360 ft of the westerly 610 ft measured along the south line of the lot, except the northerly 300 ft."

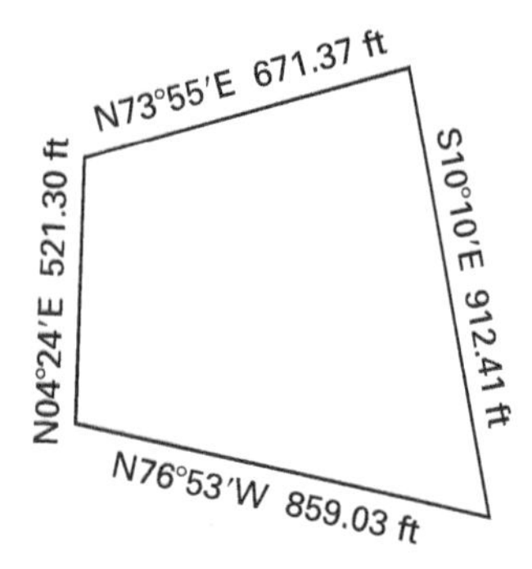

Which of the following illustrations indicates the correct partitioning?

(A)

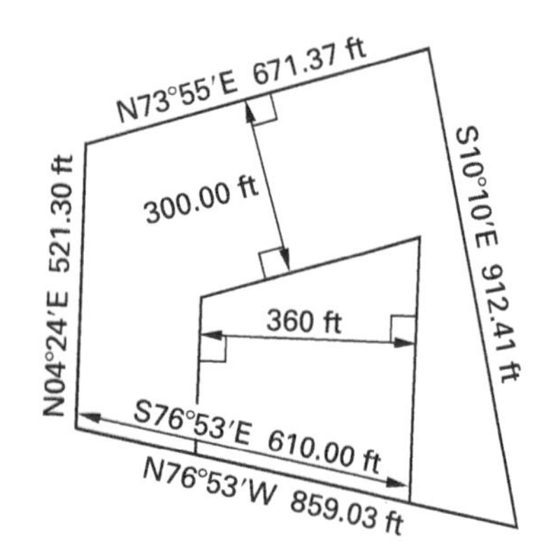

(B)

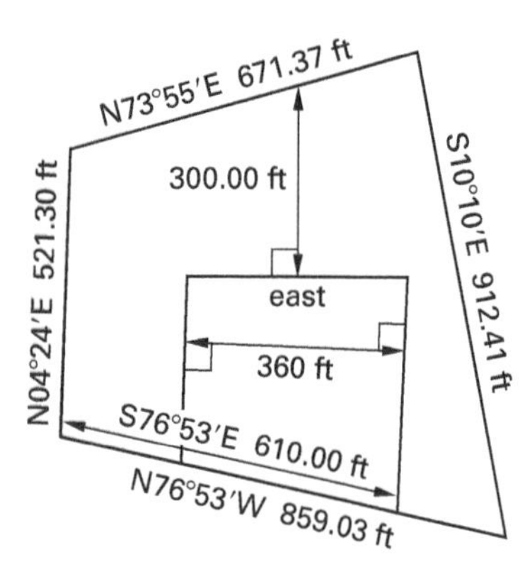

(C)

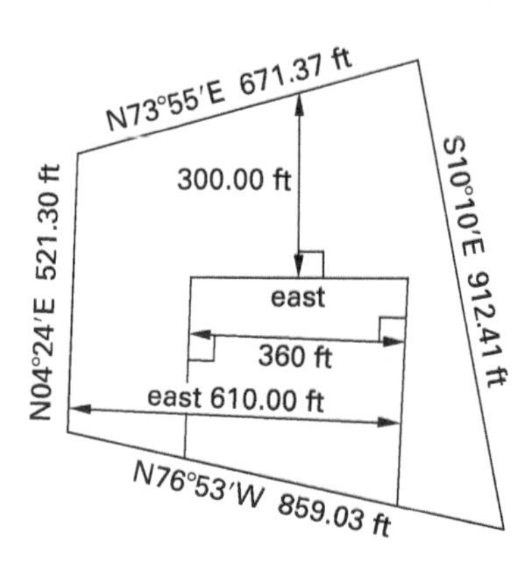

(D)

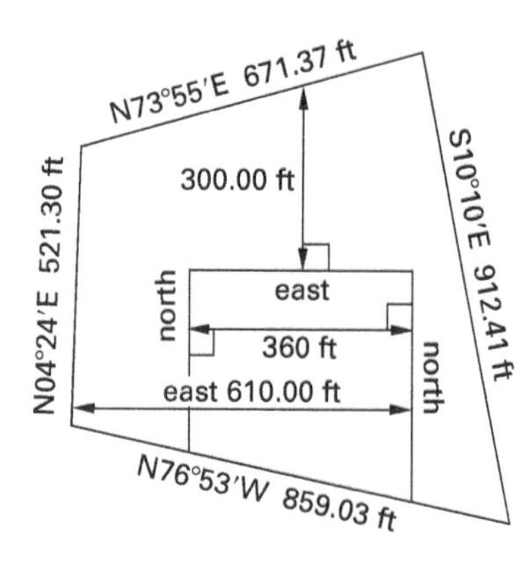

PROBLEM 28

Which of the following statements best defines the meanings of the phrase "the east half of" as it would be used to describe a partitioning in private lands and its use to describe such partitioning in public, as opposed to private, lands?

(A) In private lands, it would mean one-half of the area of the parcel involved.

(B) In public lands, it would indicate the connection of the midpoint of the northern boundary with the midpoint of the southern boundary, in most cases.

(C) Strictly speaking, any of a number of lines may divide private land into halves, whereas the dividing line is strictly defined in public lands.

(D) All of the above are true.

PROBLEM 29

In a four-sided lot that contains no right angles and no curved sides, which language would convey the maximum area to a grantee?

(A) the west 50 ft of ...

(B) the west 50 ft measured along the north line of ...

(C) the west 50 ft measured along the south line of ...

(D) the west 50 ft measured along the north and south line of ...

PUBLIC LAND SURVEYING SYSTEM DESCRIPTIONS

PROBLEM 30

The following five problems refer to the illustration shown.

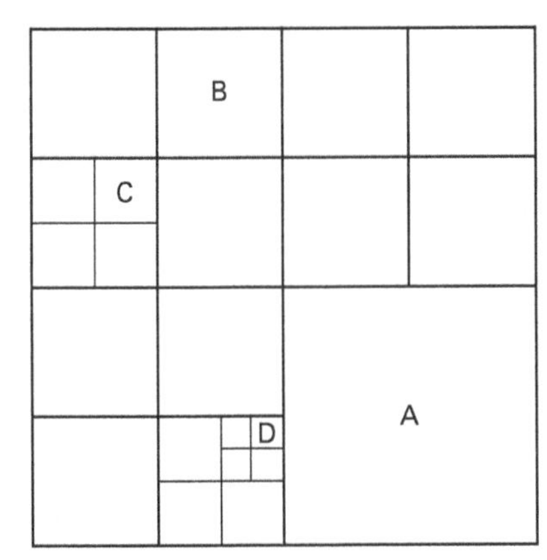

Sec. 15, T3N, R10W, 6th PM

30.1. Which of the following best describes the part of section 15 labeled A?

(A) SE$\frac{1}{4}$ Sec. 15, T3N, R10W, 6th PM

(B) Beginning at the E$\frac{1}{4}$ corner of Sec. 15, T3N, R10W, of the 6th PM; thence south along the east line of said section to the SE corner thereof; thence west along the south line of said section to the S$\frac{1}{4}$ corner thereof; thence north along the centerline of said section to the C$\frac{1}{4}$ thereof; thence east to the point of beginning.

(C) E$\frac{1}{2}$ of the E$\frac{1}{2}$ Sec. 15, T3N, R10W, of the 6th PM

(D) SW$\frac{1}{4}$ of Sec. 15, T3N, R10W, of the 6th PM

30.2. Which of the following best describes the area labeled B?

(A) NE$\frac{1}{4}$ of the NE$\frac{1}{4}$ of Sec. 15, T3N, R10W, of the 6th PM

(B) NW$\frac{1}{4}$ of the NW$\frac{1}{4}$ of Sec. 15, T3N, R10W, of the 6th PM

(C) NE$\frac{1}{4}$ of the NW$\frac{1}{4}$ of Sec. 15, T3N, R10W, of the 6th PM

(D) NE$\frac{1}{4}$ NW$\frac{1}{4}$ Sec. 15, T3N, R10W, 6th PM

30.3. Which of the following fractions is the correct unit of subdivision for the area labeled C?

(A) $\frac{1}{4}$

(B) $\frac{1}{8}$

(C) $\frac{1}{16}$

(D) $\frac{1}{64}$

30.4. Which of the following best describes the area labeled D?

(A) NE$\frac{1}{4}$ NE$\frac{1}{4}$ SE$\frac{1}{4}$ SW$\frac{1}{4}$ Sec. 15, T3N, R10W, 6th PM

(B) NE$\frac{1}{4}$ NW$\frac{1}{4}$ SW$\frac{1}{4}$ SE$\frac{1}{4}$ Sec. 15, T3N, R10W, 6th PM

(C) NE$\frac{1}{4}$ NE$\frac{1}{4}$ SE$\frac{1}{4}$ SE$\frac{1}{4}$ Sec. 15, T3N, R10W, 6th PM

(D) NE$\frac{1}{4}$ NW$\frac{1}{4}$ SE$\frac{1}{4}$ SW$\frac{1}{4}$ Sec. 15, T3N, R10W, 6th PM

30.5. Assuming a standard section, what is the approximate total area of the aliquot parts labeled in the illustration?

(A) 106 ac

(B) 212 ac

(C) 425 ac

(D) 530 ac

PROBLEM 31

The description "the $E\frac{1}{2}$ of lot 2, Sec. 6, T6S, R96W, 6th PM" is best characterized by which of the following statements?

(A) It is a description of an aliquot part of section 6.

(B) It is a description that calls for the division of lot 2 by area.

(C) It is not a description of an aliquot part of section 6.

(D) Both B and C are true.

PROBLEM 32

Which of the following best categorizes the description "the $SW\frac{1}{4}$ Sec. 4, T8N, R33W, SBM"?

(A) metes and bounds description

(B) strip description

(C) acreage description

(D) subdivisional description

PROBLEM 33

A description of a parcel within Sec. 4, T2S, R3W Ute Meridian is said to be "bounded on the south by Original Creek, a nonnavigable stream." The 1974 township plat indicates that Original Creek has been meandered. Which statement best describes the location of the south boundary of the parcel?

(A) the thread, or middle, of the stream

(B) the meander line

(C) the high water line; nonnavigable streams are never meandered

(D) a straight line connecting the meander corners

PROBLEM 34

In researching a chain of title, a surveyor finds that land patented in a particular township in 1919 as an aliquot part is now designated by the tract number 37. An official resurvey of the township was conducted in 1980. The type of survey was

(A) a retracement

(B) a dependent resurvey

(C) a contract survey

(D) an independent resurvey

PROBLEM 35

A surveyor has recently measured the perimeter of the NW quarter of a particular section in order to prepare a legal description. The description will cover only a portion of a quarter-quarter section. The record courses and distances returned decades ago by the original surveyor differ significantly. How would this discrepancy be reconciled in the description?

(A) The recent measurements are more apt to be correct; there is no need to use the original measurements at all.

(B) The establishment of the quarter-quarter section will require using a proportional relationship between the recent measurements and those of the original survey.

(C) The original measurements are officially accepted as correct and cannot be changed. The original survey information must be used.

(D) The discrepancy cannot be resolved.

DESCRIPTIONS WITHIN A SUBDIVISION

PROBLEM 36

The following two problems refer to the description given.

> lot 3, block 12 in the Sedgwick Subdivision in the city of Huron, county of Beadle, state of South Dakota, as per map recorded in book 17, page 32, of Maps in the office of the County Recorder of said County.

36.1. The language given satisfies which of the following ideas?

(A) a description that is unlikely to be subject to senior rights of an adjoiner

(B) an adequate caption for a metes and bounds survey that should be followed by perimeter description of the lot

(C) a complete description of a lot that includes, by reference, all of the information available on the subdivision plat

(D) both A and C

36.2. Lot 3 is bounded on the south by Utah Street. The street was originally established by the Sedgwick Subdivision plat. There is some talk that Utah Street may have been formally abandoned. What rights, if any, will the owner of the lot have in the roadbed of the old public way if the street has been abandoned?

(A) None; there is no mention of Utah Street in the description. If the grantor had wanted to convey title to the street, it should have been clearly expressed.

(B) Unless the contrary is clearly expressed, lot 3 has always held, and still holds, title to the middle of the street, whether Utah Street has been abandoned or not.

(C) The owner of lot 3 could acquire title to the roadbed only after it has been abandoned by the public.

(D) A metes and bounds description may have extended ownership to the center of the street, but the current description cannot possibly do so.

METES AND BOUNDS DESCRIPTIONS

PROBLEM 37

The fundamental parts of a metes and bounds description, the caption and the body, are often followed by qualifications, exceptions, and/or reservations. Which statement correctly describes the functions of these parts?

(A) The caption of the description is general, and the more detailed portion of the description is the body.

(B) An exception is used to exclude a portion of what has been described.

(C) A reservation is used when a grantor wishes to retain a right that would otherwise pass. A new encumbrance is created on the property by a reservation.

(D) All of the above are true.

PROBLEM 38

Consider the following portion of the description of lot 1 of the Newmark Tract.

> ...thence S10°10′E along the easterly line of lot 1 912.41 ft to a point; thence N76°53′W 859.03 ft to a point; thence N04°24′E 521.30 ft to a point; thence S73°55′W 671.37 ft to the point of beginning.

Which of the following statements points to errors in this portion of the description?

(A) The direction of travel is wrong on the last course; it should be north and east in order to maintain the clockwise direction of the other calls.

(B) The distance 912.41 ft should follow the direction S10°10′E, without the phrase "along the easterly line of lot 1" standing between them. The distance could be misread as 1912.41 ft.

(C) The phrase "to a point" is redundant and unnecessary, and should be eliminated.

(D) All of the above are true.

PROBLEM 39

A metes and bounds description was prepared, inadvertently including a small area of land the grantor did not own. The grantor did have clear title to most of the land described, but not all. The deed containing this description had already been executed when the error was discovered. Which of the following statements applies to this situation?

(A) The deed will be considered void.

(B) The deed will most likely not be considered void.

(C) The instrument will probably pass title to the portion of the land that the grantor owned, but not to the land he did not own.

(D) Both B and C are true.

DESCRIBING CURVES

PROBLEM 40

The elements of a plane circular curve include its length, radius, tangent, and central (delta) angle. How many of these elements are needed to define a unique curve?

(A) one

(B) any two

(C) any three

(D) all four

PROBLEM 41

A description includes the phrase " ... thence easterly 83.91 ft along said curve to the beginning of a compound curve." What is the relationship between the compound curves in the description?

(A) They share a common tangent line.

(B) They share a common radial line.

(C) Their shared radial and tangent lines are perpendicular to each other.

(D) All of the above are true.

PROBLEM 42

The following three problems refer to the illustration shown.

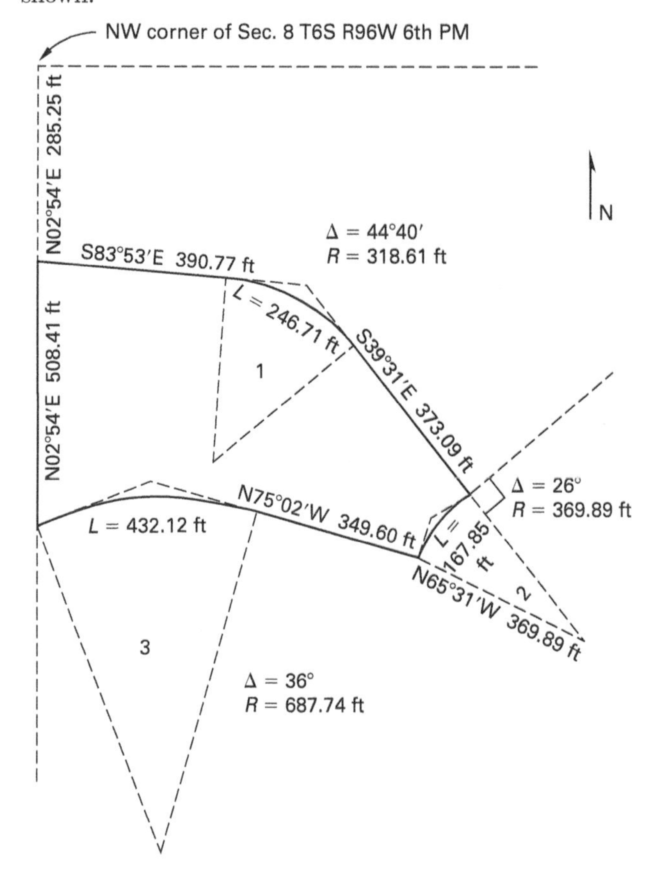

42.1. Which of the following descriptions best represents curve 1, the course preceding it, and the course following it?

(A) S83°53′E 390.77 ft to the beginning of a curve; thence easterly along said curve 246.71 ft; thence S39°31′E 373.09 ft

(B) south 83°53 minutes east 390.77 ft to the beginning of a tangential curve to the right; thence along said curve through an angle of 44 degrees 22 minutes; thence south 39°31 minutes 373.09 ft

(C) S83°53′E 390.77 ft to the beginning of a tangent curve concave to the SW having a radius of 318.61 ft; thence easterly and southeasterly along said curve; thence S39°31′E 373.09 ft

(D) S83°53′E 390.77 ft to the beginning of a tangent curve concave to the SW having a radius of 318.61 ft; thence easterly and southeasterly along said curve 246.71 ft; thence S39°31′E 373.09 ft

42.2. Which of the following statements is true of this description of curve 2?

> S39°31′E 373.09 ft to the beginning of a nontangent curve concave to the SE having a radius of 369.89 ft; thence southwesterly and southerly along said curve through an angle of 26°00′ to the beginning of a nontangent line; thence along said line N75°02′W 349.60 ft.

(A) It is sufficient to adequately describe the curve and its relation to the two lines mentioned.

(B) The curve is described as nontangent to the first line, when it is in fact tangent.

(C) The curve is described as concave to the SE when it is concave to the east.

(D) Insufficient data are given by which to relate the curve to the lines.

42.3. Which of the following statements is true of this description of curve 3?

> ... thence N75°02′W 349.60 ft to the beginning of a tangent curve concave to the south having a radius of 687.74 ft to which point a radial line bears N14° 58′E; thence westerly 432.12 ft along said curve to its intersection with the west line of said section 8.

(A) There is no mention of the central angle of the curve, so the description is insufficient.

(B) The length of the curve, 432.12 ft, is in contradiction with the central angle of 36° and the radius of 687.74 ft shown in the illustration.

(C) It is unnecessary to write "to which a radial line bears N14°58′E″when the curve is known to be tangent.

(D) The call "to its intersection with the west line of section 8" may contradict the length of the curve given, 432.12 ft, and create an ambiguity that would void the deed.

DESCRIBING EASEMENTS

PROBLEM 43

The following five problems refer to the description and illustration given.

> An easement over a strip of land 80 ft wide over a portion of lot 20 of the Oldham Tract in the county of Gunnison, state of Colorado, as per the plat filed in book 34, page 28, in the office of the county recorder of said county, the centerline of which is described as follows.
>
> Beginning at a point on the south line of said lot N75° 02′W 142.84 ft from the most southerly corner of said lot; thence N19°12′W 94.44 ft; thence N49°07′W 370.26 ft to the beginning of a curve concave to the south, which has a radius of 252.30 ft; thence along said curve 320.97 ft.

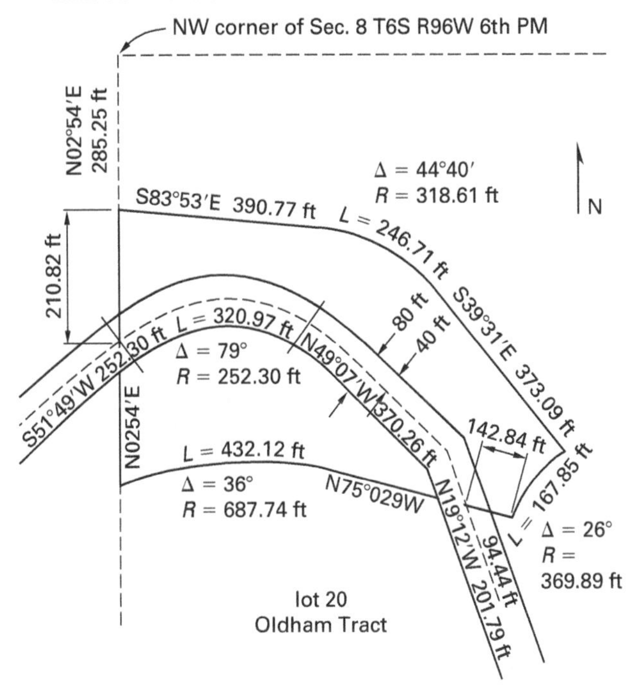

43.1. Which statement is correct concerning the caption of the description given?

(A) The caption should express the purposes and uses to which the easement is limited.

(B) The caption should mention the NW$\frac{1}{4}$ of section 8.

(C) The phrase "the centerline of which" is not as good as "40 ft on either side of a line."

(D) The caption should end with the phrase "being more particularly described in detail as follows, to wit:"

Legal Descriptions

43.2. Which statement is correct concerning the beginning of the easement through the course, which is written "...thence N19°12′W 94.44 ft"?

(A) The phrase "the sidelines of said strip to be shortened or prolonged to terminate at the southerly line of said lot" should be included in the description to prevent the overlap and gap that will otherwise result at the beginning.

(B) The body of the description must commence at the SE corner of the lot, and begin at the centerline of the easement.

(C) The call "N19°12′W 94.44 ft" must end with "to a point" to be correct.

(D) Both A and B are true.

43.3. Which statement is correct concerning the description through the course, which is written "to the beginning of a curve concave to the south, which has a radius of 252.30 ft"?

(A) The curve must be described as a curve to the left, not as "concave to the south."

(B) The width of the easement should be mentioned in the body of the description.

(C) The phrase "the sidelines of said easement to be shortened or prolonged to meet at angle point intersections" should appear somewhere in the description to prevent the overlap and gap that will otherwise result.

(D) Both A and B are true.

43.4. Concerning the description as a whole, which statement is correct?

(A) The phrase "the sidelines of said easement to be shortened or prolonged to terminate at the westerly line of said lot" should be included in the description.

(B) The phrase "to the west line of section 8, T6S, R96W, of the 6th PM" should be added to assure the termination of the centerline curve at the west line of the section.

(C) The last direction used in the description should be "...thence S51°49′W" with a distance carrying the easement beyond the west line of section 8.

(D) B and C are true.

43.5. The previous description was rewritten to begin, "An easement for pipeline purposes over a strip of land 80 ft wide." However, the grantor of the easement wants to ensure continued personal use of the strip as a roadway to transport timber on trucks across the property. The grantor has asked a surveyor to rewrite the easement again to guarantee personal ingress and egress. Which statement is correct concerning the grantor's request?

(A) It cannot be done; it would constitute an overburdening of the easement.

(B) An easement can be written for only one purpose, not several.

(C) The use would conflict with the pipeline maintenance and cannot be written into the description.

(D) A property owner cannot hold an easement across personal property.

SOLUTION 1

While the term *metes and bounds* might be considered archaic, it is still widely used and understood. Metes means measurement or the assignment of measurement, and bounds refers to boundary monuments, both physical and legal.

The answer is (D).

SOLUTION 2

The choice of the point of beginning is based on the ease and certainty of its identification, not a desire to give it more standing than any other corner in the description. Both its legal relationship to other monuments and its own physical attributes should be part of its description to aid in its identification.

The answer is (A).

SOLUTION 3

More or less is a cautionary signal in a description. The scrivener has reason to believe that the measurement may not be accurate. When a tie is made in description to a monument or is a matter of record, the tie holds without the phrase "more or less."

The answer is (C).

SOLUTION 4

Evidence not contained in the description is external or extrinsic. Another appropriate term would be *evidence aliunde.*

The answer is (B).

SOLUTION 5

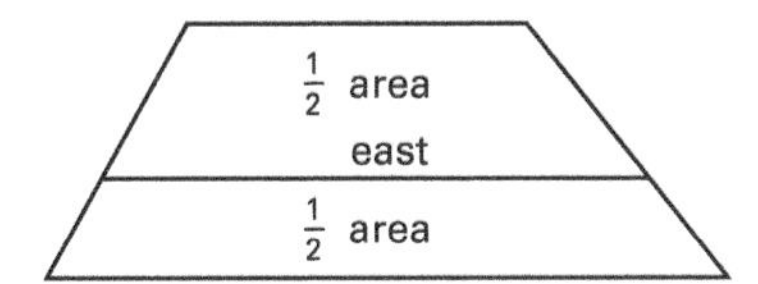

In such proportional descriptions, the area of the parcel is the controlling consideration. While it is true that many lines may divide the lot in question into halves, the phrase "the northern half" is sufficient in such a circumstance. It is generally understood to mean that the line will be parallel with the parallel sides and will divide the lot into equal areas.

The answer is (B).

SOLUTION 6

Adjoining is correctly used to indicate actual contact, unlike adjacent, which implies proximity but not contact. This distinction can be significant in construing the intended meaning of the words used in a description, especially before a court of law.

The answer is (C).

SOLUTION 7

The sectionalized lands in the Public Land Surveying System are subdivided by the rules set forth in the *Manual of Surveying Instructions.* These rules provide for the establishment of subdivisions of aliquot parts, halves, and quarters, without regard to area.

The answer is (D).

SOLUTION 8

The reference to an adjoiner of record is often characterized as a call for a record monument. A record monument frequently has senior standing. The call for a record monument includes the entire adjoining tract, not only the shared line.

The answer is (B).

SOLUTION 9

The westerly 50 ft is, by definition, measured at right angles from the westerly line of the parcel. Furthermore, the eastern boundary of the westerly 50 ft would be assumed to be parallel with the western boundary of the original parcel. Since the boundaries of the parcel are not perpendicular, measurement along the north and south boundaries will not yield parallel lines.

The answer is (A).

SOLUTION 10

A point of cusp is the intersection of two curves or a curve and a line of opposite direction.

The answer is (A).

SOLUTION 11

In most jurisdictions, both distance and direction are subordinate to verifiable corner monuments. The underlying principle is the belief that monuments are more likely than measurements to reflect the intention of the description unambiguously. The section line is a record monument and is also superior to a measured bearing.

The answer is (C).

SOLUTION 12

While it is possible to cause a specified area to become a controlling element of a description, such as "the north 10 ac of," generally the area is subordinate. Courts have held that where both a metes and bounds description and a quantity of land are given, the former will control. The principle is based upon the idea that area by itself does not indicate the location of specific boundaries, as a detailed perimeter description does.

The answer is (D).

SOLUTION 13

Since the property in question was not aliquot, the most specific description—the metes and bounds description—will have greater weight than the general one. While the intentions of the parties to a deed is the paramount consideration in resolving conflicting title elements, those intentions must be clearly discernable from the written words of the instrument. The clearest words are those that are most specific. It is also a well-settled principle that private grants are interpreted in favor of the grantee.

The answer is (D).

SOLUTION 14

Written descriptions are considered sufficient when a competent surveyor can locate the property certainly and unambiguously on the ground, with or without extrinsic evidence.

The answer is (C).

SOLUTION 15

The superiority of called-for, identifiable, and undisturbed monuments over the other elements is based upon the assumption that they are the most certain in fixing the location of a line or corner. There are limitations to the control of monuments when they can be shown to be disturbed or are not called for in the description. Generally, however, they are considered superior to distance, direction, area, and coordinates.

The answer is (A).

SOLUTION 16

Bearings and distances are generally held to be superior to a recited area, and identified, undisturbed artificial monuments are superior to bearings and distances. However, all of these are considered subordinate to a senior adjoiner. In the description quoted, the southeasterly line of the Southern Pacific Railroad would outweigh the other elements in case of a conflict.

The answer is (B).

SOLUTION 17

A found, verified original monument is generally considered a more reliable indication of the intentions expressed in the description than any measurements associated with it.

The answer is (D).

SOLUTION 18

Any ranking of the priority of calls is an oversimplification; exceptions can always be found. However, the guiding principle of the superiority of elements that are most likely to reflect the original intentions supports the order given in choice (B).

The answer is (B).

SOLUTION 19

A valid right of possession that has ripened by unwritten means is considered superior to all other private rights.

The answer is (C).

SOLUTION 20

A senior right is superior to all other written rights when interpreting a legal description.

The answer is (A).

SOLUTION 21

A sketch is useful for solving this problem.

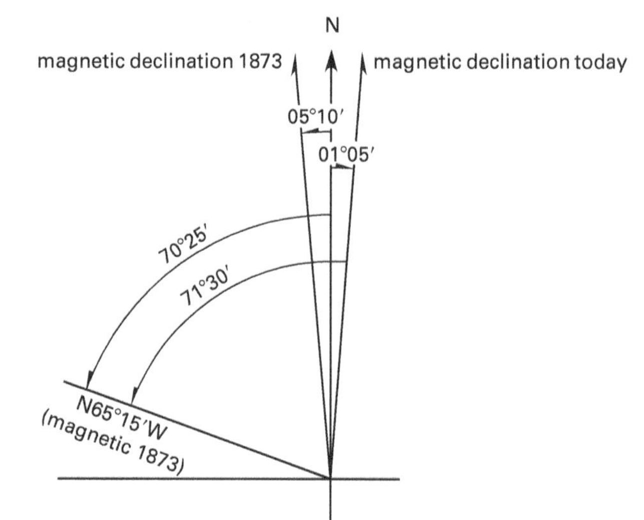

The illustration indicates that the bearing of the line is found by adding the 1873 magnetic bearing to its declination.

$$\begin{array}{l} \text{N}65^\circ15'\text{W} \\ \underline{+05^\circ10'} \\ \text{N}70^\circ25'\text{W} \end{array}$$

Today's easterly declination is then added to the true bearing to find the magnetic bearing today.

$$\begin{array}{l} \text{N}70^\circ25'\text{W} \\ \underline{+01^\circ05'} \\ \text{N}71^\circ30'\text{W} \end{array}$$

The answer is (D).

SOLUTION 22

The use of the word *north* is open to many possible interpretations, and without other qualifications, can easily lead to ambiguity in a description. In this problem, there are other qualifications that make the word definite. The SE corner of the section and the NE corner of the SE quarter of the section define the line clearly.

The answer is (D).

SOLUTION 23

Azimuths are measured through 360° from a reference meridian, either north or south. It is quite possible to find a combination of several methods of defining direction in one description.

The answer is (D).

SOLUTION 24

Westerly, northwesterly, southwesterly, etc., are used to define sectors of the compass embracing approximately 45° of arc—22½° on each side of the named direction. Since the direction to be expressed is within 15° of west, the best word to use is westerly.

The answer is (A).

SOLUTION 25

The word *deflecting* is meant to indicate a deflection angle. A deflection angle is measured from the prolongation of the previous course, to the right or left, by the specified amount.

The answer is (D).

SOLUTION 26

26.1. When a description calls for a partition with a linear dimension, it is assumed to indicate that the measurement is at right angles to the boundary from which the measurement is made. The division line is assumed to be parallel with the line from which it is described. These assumptions are based on the principle of construction, which gives greatest favor to the grantee.

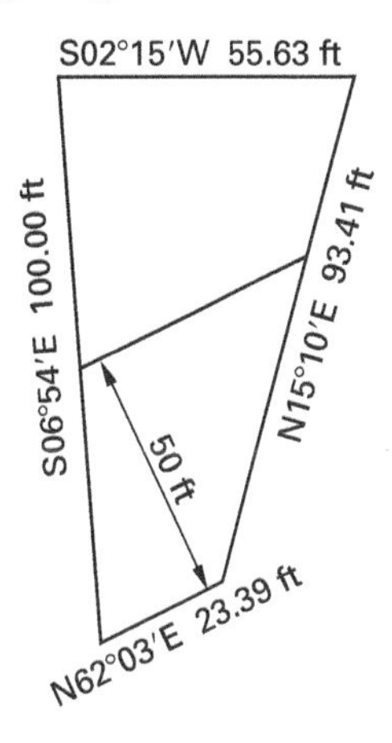

The answer is (D)

26.2. Describing the remainder of lot 10 in relation to the northern line of the lot would create a line parallel with the northern lot line, which would be in conflict with the partition already established. A safer approach is to describe the remainder by exception. Describing the conveyance as lot 10, except that which has already been separated, minimizes the chances of conflict.

The answer is (D).

SOLUTION 27

The eastern and western boundaries of the cutout must be parallel with the western boundary of the lot. The northern boundary of the cutout parcel must be parallel with the northern boundary of the lot. The description requires that the measurements be perpendicular to the sides, except the 610.00 ft, which is specifically described as being measured along the southerly line.

The answer is (A).

SOLUTION 28

Public lands are divided into aliquot parts, or halves and quarters, in accordance with strictly defined rules. Private lands are divided by area into proportional parts. The direction of the dividing line in private lands is less certain than it is in public lands.

The answer is (D).

SOLUTION 29

"The west 50 ft of" indicates a measurement of 50 ft perpendicular to the western boundary. Since the lot contains no right angles, measuring 50 ft along the north or south line would convey a smaller area.

The answer is (A).

SOLUTION 30

30.1. Brevity is a virtue in writing legal descriptions, as long as clarity is not sacrificed. Aliquot parts of the Public Land Surveying System are better described by abbreviation than by metes and bounds. The rules of partition of public lands are so well defined that reference to aliquot parts is both concise and unambiguous.

The answer is (A).

30.2. Both (C) and (D) are correct; however, (D) is more concise.

The answer is (D).

30.3. The fractions commonly used to define the aliquot parts of a section are based on proportional area. The area labeled C is approximately 10 ac of area. A full standard section contains 640 ac. Therefore, C would be called a 1/64th.

The answer is (D).

30.4. The aliquot part labeled D is a 1/256th. It is correctly described by choice (A).

The answer is (A).

30.5. The aliquot parts labeled in the illustration have the following approximate areas.

$$A = 160 \text{ ac}$$
$$B = 40 \text{ ac}$$
$$C = 10 \text{ ac}$$
$$D = 2.5 \text{ ac}$$

Their sum is 212.5 ac, which is approximately 212 ac.

The answer is (B).

SOLUTION 31

An aliquot part of a section in the Public Land Surveying System is one-half or one-quarter of the larger division. By definition, lots created from fractional sections are not aliquot parts, nor are they proportional parts. The description given calls for the division of lot 2 by area.

The answer is (D).

SOLUTION 32

Rectangular surveys were begun in 1785 to subdivide the public domain into manageable parts. The process continues today with well-defined procedures by which that subdivision is accomplished. One of the benefits of the system is the ease and certainty with which those subdivisions can be described.

The answer is (D).

SOLUTION 33

Meander lines are not legal boundaries, unless expressly made so. They are intended to determine the area of bodies of water, not to define title. Nonnavigable streams over three chains wide are meandered under current instructions, but the boundary, in most cases such as this, is the thread of the stream.

The answer is (A).

SOLUTION 34

A dependent resurvey would not have resulted in changing the aliquot designation of patented lands. However, an independent resurvey would, most often, protect bona fide rights within the township by creating tracts. Their numbers would begin at 37 to avoid confusion with the section numbers.

The answer is (D).

SOLUTION 35

Use proportional measurement. The original survey is presumed to be a correct representation of the distances and directions found using the instrumentation available at the time. The standard of measurement may be different, but the measurements are assumed to have been made between the same corners as those of the original survey.

The answer is (B).

SOLUTION 36

36.1. Lot 3 is part of a simultaneous conveyance in which all of the lots of the Sedgwick Subdivision were created at the same moment. The recording of the subdivision plat provided a record monument to which the description refers. That reference incorporates all the information available on the plat into the description given; it is not necessary to recount it to make the description sufficient. The description is sufficient as it stands.

The answer is (D).

36.2. A conveyance bounded by a public street carries title to the center of that street. If the contrary is intended and the street is to be excluded, it must be clearly stated. Were Utah Street abandoned as a public street, the underlying fee interest of the adjoining owners would remain and the roadbed would revert to them.

The answer is (B).

SOLUTION 37

The caption of a description limits the title, and the body provides more detailed information. Exceptions and reservations are not synonymous: the first cuts off, while the second holds back. Reservations usually create new encumbrances to be retained by the grantor; an easement right would be an example of this.

The answer is (D).

SOLUTION 38

The direction of travel should be consistent in a metes and bounds description. It is good practice to bring the direction and distance together to avoid unnecessary ambiguities. The phrase "to a point" is unnecessary. If a monument stands at the intersection of the lines, it should be described.

The answer is (D).

SOLUTION 39

When a description includes land to which the grantor has title along with land to which he has none, the former will be conveyed and the latter will not. Generally, a deed will be construed to pass title when any reasonable construction will do so.

The answer is (D).

SOLUTION 40

Any two of the elements given will describe a unique curve. Frequently, three elements are given, but two are adequate.

The answer is (B).

SOLUTION 41

Compound and reverse curves both share common radial and tangent lines, which are perpendicular to each other.

The answer is (D).

SOLUTION 42

42.1. It is important to read carefully when considering alternative language for descriptions. While it is possible to describe the curve under consideration in several ways, only one of those given is complete. Missing data disqualify the remaining answers.

The answer is (D).

42.2. The description is not adequate. Some phrase such as "to which point a radial line bears N39°31′W" is required at the end of the description of the first line in order to provide orientation for the curve. Without more information about its relative position, all that can be learned from the description is that the curve is not tangent to either line.

The answer is (D).

42.3. The call to the intersection of the west line of the section obviates the need to recite the central angle of the curve. If the intersection does in fact contradict the length of the curve given, the record monument, the west line of the section, will prevail since it is superior in the normal priority of calls. The bearing of a radial line is redundant, since the curve has already been described as being tangent to the previous course.

The answer is (C).

SOLUTION 43

43.1. The caption could most certainly be improved by defining the uses that the grantor of the easement intends to allow. The existing language could be construed to convey a fee interest in the strip of land, even though the use of the word *easement* contradicts that idea.

The phrase "40 ft on either side of a line" would confuse the arrangement of the easement. It would indicate that the 40 ft would fall on one or the other side of the described line, not both.

The two remaining answers would add redundant and unnecessary words to the description.

The answer is (A).

43.2. It is not vital that the description have a point of commencing and a point of beginning; the method used is adequate. There is no need to add the phrase "to a point" at the intersection of each pair of lines; it is self-evident that such an intersection is a point. However, the sidelines must be shortened or prolonged to close on the south line of the lot. The termination of an easement described by its centerline will otherwise be considered normal to the centerline, creating undesirable gaps and overlaps.

The answer is (A).

43.3. While curves are sometimes described by reciting their direction—that is, to the left or to the right—this is not the only sufficient method. The width of the easement is mentioned in the caption and need not also be mentioned in the body. However, language extending the sidelines to meet at the angle points is important and should be included in the description.

The answer is (C).

43.4. Carrying the easement beyond the west line of the section and then terminating it with an exception is an effective way to achieve the desired ending of the lines.

The answer is (D).

43.5. Easements can be written to include many purposes in one instrument. The considerations of overburdening and conflicting uses need not even be discussed since the grantor's request is not possible. The grantor cannot own an easement over personal property; an easement right must be in the property of another.

The answer is (D).

8 Photogrammetry

GENERAL DEFINITIONS

PROBLEM 1

The term *fiducial marker*, as it is used in photogrammetry, is best defined as

(A) one of the (usually) two images on a photograph that define the axis of the tilt

(B) a premarked target set on the ground to balance the stereo model

(C) one of the (usually) four objects connected to the camera's interior that form images on the negative as each photograph is taken

(D) the plane that defines the camera's focal length

PROBLEM 2

Which of the following statements best describes the Porro-Koppe principle?

(A) Distortion caused by imperfections in the lens system of a plotter can be minimized if the imperfections have the same distortion as the lenses of the camera used to create the photographs.

(B) The image plane, object plane, and the plane of the lens of a direct-focusing projector intersect in the same line to ensure sharp focus.

(C) Irregular changes in brightness are proportional to the refractive index between the object and the image.

(D) The focal length of the camera, the flight height of the aircraft, and the elevation of the point determine the scale.

PROBLEM 3

The term *principal point*, as it is used in photogrammetry, is best described as

(A) a point on the ground used as the basis for the determination of scale and orientation

(B) synonymous with focal point

(C) a point that appears on both photographs of a stereo pair

(D) the point at which the lines through opposing sets of fiducial marks intersect

PROBLEM 4

An orthophotograph is best described as

(A) a stereo pair of aerial photographs viewed together to represent the three-dimensional nature of the area photographed

(B) a mosaic of aerial photographs at the scale of a standard quad, 1:24,000

(C) a perspective photograph from which the displacements caused by tilt and relief have been removed

(D) an aerial photograph in which the horizon is visible

PROBLEM 5

A vertical photograph is best described as

(A) an aerial photograph taken with a camera whose optical axis is virtually vertical

(B) an aerial photograph taken with a camera whose optical axis is virtually horizontal

(C) an aerial photograph in which the horizon is visible

(D) an aerial photograph taken with a camera whose optical axis has been tilted but in which the horizon is not visible

PROBLEM 6

What is the meaning of the term *air base* as it is used in photogrammetry?

(A) the mounting of the aerial camera in the aircraft

(B) the line between two air stations where overlapping photographs were taken

(C) the absolute stereoscopic parallax of a point appearing on a stereo pair of photographs as determined by the algebraic difference of the distance of the points from their respective principal points

(D) the rate of change of parallax with respect to the changes in flight height

PROBLEM 7

Which of the following best describes the *K*-factor?

(A) the smallest contour interval that can be plotted at the required mapping accuracy

(B) the degree to which an image is displaced radially toward or away from the nadir point of the photograph

(C) the apparent inward or outward displacement of an image from the principal point due to lens design

(D) the ratio of the air base length to the altitude at which a pair of aerial photographs were taken

PROBLEM 8

What is meant when an aerial photograph is said to have been rectified?

(A) the effects of tilt have been removed

(B) the photograph has been evaluated for vertical map accuracy in a stereoplotter

(C) the photograph has been transferred to a transparent positive on a glass plate

(D) the photograph has been reduced to an orthophotograph

PROBLEM 9

In a true vertical photograph made by a perfectly adjusted camera, which two points would be identical?

(A) the principal point and the nadir point

(B) the detail point and the pass point

(C) the tie point and the base point

(D) the picture point and the photograph center

PROBLEM 10

What is a stereoscope?

(A) a portable optical device used to view two photographs simultaneously, creating the impression of a three-dimensional image

(B) a device used to measure the separation between two index marks on a stereo pair of aerial photographs

(C) an optical device used for measuring coordinates, both rectangular and polar, on a photographic plate

(D) a composite from a pair of slotted templates representing a stereoscopic model

PROBLEM 11

Tilt in photogrammetry is best defined as

(A) the angle between the nadir point and the principal point of an aerial photograph

(B) the apparent deflection of the image of a straight line on the ground due to relief displacement

(C) the dihedral angle between a truly horizontal plane and the plane of the aerial photograph

(D) referring to the angle used in the radial line method of relating the isocenter to other points on the photograph

PROBLEM 12

What is the difference between x-tilt and y-tilt?

(A) x-tilt is the angle of rotation around the axis perpendicular to the flight line, while y-tilt is the angle of rotation around the axis of the flight line.

(B) x-tilt is the angle of rotation around the axis of the flight line, while y-tilt is the angle of rotation around the axis perpendicular to the flight line.

(C) x-tilt is the angle of rotation around the axis described by the principal point and the perspective center, while y-tilt is the angle of rotation around the axis perpendicular to the flight line.

(D) x-tilt is the angle of rotation around the axis perpendicular to the flight line, while y-tilt is the angle of rotation around the axis described by the principal point and the perspective center.

PROBLEM 13

Which of the following assumptions is the basis for the radial line method of plotting on aerial photographs?

(A) The photographs have such insignificant angles of tilt that they are assumed to be truly vertical photographs.

(B) The principal point, the plumb point, and the pass point are identical.

(C) The relief displacement in the photograph is equal at each edge.

(D) The x-parallax in the photograph is zero.

PROBLEM 14

What is the isocenter of an aerial photograph?

(A) the point on the ground directly beneath the optical center of the lens at the moment of the exposure

(B) the point at the intersection of three planes: the truly horizontal plane, the plane of the tilted photograph, and the vertical plane that contains the optical axis of the camera

(C) a point whose horizontal and vertical positions are known from photogrammetric methods that is used for the absolute orientation of other photographs

(D) the foot of the perpendicular from the perspective center to the plane of the photograph

PROBLEM 15

What sort of composite of aerial photographs involves trimming the edges of the individual photographs?

(A) controlled mosaic

(B) planimetric map

(C) uncontrolled mosaic

(D) topographic map

PROBLEM 16

Given the following parameters, which feature would be the best photo-identifiable ground control point (GCP) to be collected by GPS for the rectification of digital imagery from either satellites or aircraft? Consider that the imagery has already been collected and has a 60 cm pixel at the ground.

- unambiguously visible in the image
- sufficient size and clarity in relation to the size of the imagery's pixels
- relatively high contrast when compared with its surroundings
- durable and unchanging over time
- located in a public place and accessible in the future
- provides a clear view of the sky
- not subject to multipath during GPS collection
- as close as possible to the level of the surrounding ground

(A) center of a small tree in an arid area

(B) intersection of two unpaved roads on a military base

(C) corner of a concrete slab on grade surrounded by grass

(D) corner of a building at the ground level

PROBLEM 17

A digital image is comprised of pixels. The position of each pixel in an image can be defined by row and column, also known as row and sample. Which of the following is the standard position of the origin (0, 0) in this system as it is used in photogrammetry and remote sensing?

(A) upper-right pixel

(B) upper-left pixel

(C) lower-right pixel

(D) lower-left pixel

PROBLEM 18

What is the difference between an 8-bit digital image and a 16-bit digital image?

(A) There are more possible digital number values available for each pixel in a 16-bit digital image than there are in an 8-bit image.

(B) There is better radiometric resolution in an 8-bit digital image than there is in a 16-bit image.

(C) There are more possible digital number values available for each pixel in an 8-bit digital image than there are in a 16-bit image.

(D) There is no difference between the two images.

PROBLEM 19

Which of the following statements concerning LiDAR is NOT correct?

(A) It is an acronym that can stand for Light Detection and Ranging.

(B) LiDAR systems are often used on aircraft with instruments such as GPS receivers and Inertial Measurement Systems onboard for orientation.

(C) It usually uses portions of the electromagnetic spectrum in the neighborhood of near infrared, ultraviolet or visible light.

(D) LiDAR data collection is not affected by cloud cover, smoke, or other aerosol particles in the atmosphere.

FORMATS AND CAMERAS

PROBLEM 20

What are the typical dimensions of the portion of the negative within the focal plane of an aerial camera used for mapping?

(A) 4 in × 5 in

(B) 8 in × 10 in

(C) 9 in × 9 in

(D) 12 in × 12 in

PROBLEM 21

What is the term used to describe a positive aerial photograph printed on a transparent plate?

(A) dioptric

(B) trimetrogon

(C) strip photograph

(D) diapositive

PROBLEM 22

Which of the following is NOT a typical component of an aerial camera?

(A) magazine

(B) cone

(C) diaphragm

(D) collimator

PROBLEM 23

In a photogrammetric camera lens, which focal length specification would most emphasize the effects of parallax?

(A) 6

(B) $8\frac{1}{4}$

(C) 10

(D) 12

PROBLEM 24

What sort of shutter is used on most aerial cameras?

(A) capping

(B) focal-plane

(C) louvre

(D) between-the-lens

PROBLEM 25

The function of an intervalometer is to control the

(A) loading of the film magazine

(B) relationship between the object plane, lens, and image plane of the camera

(C) splitting of the light passing through the lens of the camera into two beams

(D) shutter of the camera

PROBLEM 26

A component unique to a reseau camera that is used to minimize the distortion in its aerial photographs is the

(A) super-aviogon lens

(B) vacuum pump film flattening system

(C) goniometer

(D) rectangular grid etched on glass

PROBLEM 27

For a given format size, as the focal length of a camera decreases, which of the following occurs?

(A) Radial distortion increases.

(B) Tangential distortion decreases.

(C) Spherical aberration increases.

(D) The angular field of view increases.

PROBLEM 28

Why will an aerial camera that has not been corrected for the crab angle cause a reduction in the ground coverage of the resulting photographs?

(A) The sides of the photographs will not be parallel to the flight line.

(B) The endlap of the photographs will exceed 60%.

(C) The photographs will have an excessive x-tilt.

(D) The photographs will have an excessive y-tilt.

PROBLEM 29

Which of the following types of aerial cameras is most often used for topographic mapping purposes?

(A) continuous-strip cameras

(B) multilens convergent cameras

(C) multicamera systems

(D) single-lens cameras

PROBLEM 30

What is the angular field of view for a single lens aerial camera with a 6 in fixed focal length and a 9 in × 9 in format?

(A) 56°

(B) 75°

(C) 93°

(D) 122°

SCALE

PROBLEM 31

The following two problems refer to the illustration shown.

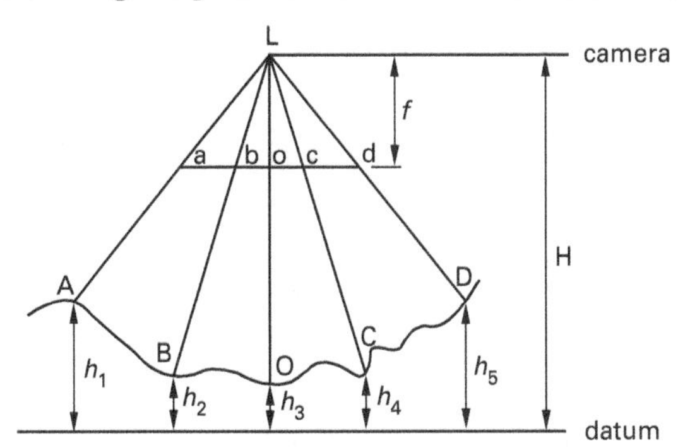

31.1. A camera with a focal length of 6 in was used to make a vertical photograph at an altitude of 10,000 ft above sea level. The height of point B and point C is approximately 4500 ft above sea level in both cases. What is the approximate scale along line BC in the photograph?

(A) 1:750

(B) 1:1000

(C) 1:11,000

(D) 1:20,000

31.2. A camera with a focal length of 6 in was used to make a vertical photograph at an altitude of 3000 ft above sea level. Point B is 150 ft above sea level, and point A is 800 ft above sea level. What is the scale at points A and B in the photograph?

(A) A: 1 in = 367 ft
B: 1 in = 475 ft

(B) A: 1 in = 442 ft
B: 1 in = 363 ft

(C) A: 1 in = 475 ft
B: 1 in = 367 ft

(D) A: 1 in = 546 ft
B: 1 in = 824 ft

PROBLEM 32

What flight height would be required to produce vertical photographs with a scale of 1:12,000 over level terrain if the average elevation is 3300 ft and the focal length of the camera is 8 in?

(A) 8200 ft

(B) 10,000 ft

(C) 11,300 ft

(D) 15,300 ft

PROBLEM 33

The length of a one-story building is measured as 3.52 in on a vertical photograph. A map, which has a scale of 1:24,000, shows the same building as having a length of 2.67 in. What is the engineer's scale of the photograph?

(A) 1 in = 1136 ft

(B) 1 in = 1517 ft

(C) 1 in = 2400 ft

(D) 1 in = 4321 ft

PROBLEM 34

A distance between two road intersections on a vertical photograph is 5.12 in. The same distance on a map with a scale of 1:24,000 measures 3.00 in. The elevation of the essentially level terrain in the area is 1200 ft above sea level. If the focal length of the lens in the camera used to make the photograph was 6 in, what was the altitude of the aircraft above sea level when the photograph was taken?

(A) 7197 ft

(B) 8230 ft

(C) 12,000 ft

(D) 14,060 ft

PROBLEM 35

The scale of two diapositives used to compile a manuscript in a Kelsh plotter is 1 in = 500 ft. If the focal length of the camera is 6 in and the optimum projection distance of the plotter is 760 mm, what is the optimum model scale?

(A) 1 in = 100 ft

(B) 1 in = 200 ft

(C) 1 in = 500 ft

(D) 1 in = 1000 ft

PROBLEM 36

A multiplex plotting instrument with a fixed plotting scale has an optimum projection distance of 360 mm. If plotting is anticipated on a multiplex instrument at a scale of 1:10,000 and the average elevation of the terrain is 3000 ft, what flight height above sea level should be used?

(A) 10,000 ft

(B) 12,870 ft

(C) 14,810 ft

(D) 16,310 ft

RELIEF DISPLACEMENT

PROBLEM 37

Which of the following best defines the term *relief displacement* as it is used in photogrammetry?

(A) the radial movement of the images on a tilted photograph with respect to the isocenter when they are on the high or low side of the isometric parallel

(B) the range of points that can be focused sharply in an aerial photograph

(C) the angle at a point on the ground between the direction of gravity and the line through the point normal to a reference surface

(D) the radial displacement of a point on a photograph toward or away from the nadir caused by the elevation of the object above or below a given datum

PROBLEM 38

Which variables determine the extent of relief displacement on an aerial photograph?

(A) the size of the format, the flight height above the datum, and the radial distance of the image point from the principal point on the photograph

(B) the overlap and the flight height above the datum

(C) the elevation of the point from the datum, the flight height over the datum, and the radial distance of the image point from the principal point on the photograph

(D) the elevation of the point from the datum, the size of the format, the flight height above the datum, and the radial distance from the principal point on the photograph

PROBLEM 39

A vertical photograph made at a flight height of 2000 ft above sea level shows a radio tower with a base elevation 540 ft above the same datum. The image of the tower has a relief displacement of 1.33 in. The distance from the photograph's principal point to the top of the tower is 5.97 in. What is the height of the tower to the nearest foot?

(A) 240 ft

(B) 325 ft

(C) 433 ft

(D) 512 ft

PROBLEM 40

At the bottom of a valley, the scale of a particular vertical photograph is 1:8000. The focal length of the lens used to make the photograph is 6 in. A road intersection on the same photograph is 495 ft above the valley floor and 3.99 in from the principal point. What is the relief displacement of the road intersection with respect to the bottom of the valley?

(A) 0.50 in

(B) 0.75 in

(C) 1.52 in

(D) 3.21 in

OVERLAP AND THE BASE-HEIGHT RATIO

PROBLEM 41

What is the typical end lap and side lap between stereo pairs?

(A) 10% end lap
30% side lap

(B) 40% end lap
60% side lap

(C) 50% end lap
50% side lap

(D) 60% end lap
30% side lap

PROBLEM 42

Which of the following statements is NOT a reason for overlapping aerial photographs?

(A) The appearance of the center of each photograph on one or more other photogaphs facilitates the construction of mosaics and minimizes the effects of relief displacement.

(B) Overlapping is necessary for stereoscopic viewing.

(C) Poor photographs may sometimes be eliminated since each area is photographed at least twice.

(D) Overlapping is necessary in order to eliminate the Scheimpflug condition.

PROBLEM 43

The base-height ratio is the ratio of the air base, or ground distance between exposure stations, to the flight height above the average terrain elevation. As this ratio increases, which of the following also increases?

(A) the vertical exaggeration in the stereoscopic image from a stereo pair

(B) the sharpness of the focus of the plotting instrument projection

(C) the distance between the isocenter and the principal point

(D) the x-tilt of the photograph

PROBLEM 44

What is the base-height ratio of an aerial photograph taken with a 6 in focal length lens at 4500 ft above the average elevation of the terrain on a 9 in × 9 in format with a 60% end lap?

(A) 0.50

(B) 0.60

(C) 0.70

(D) 0.80

GROUND COVERAGE

PROBLEM 45

What is the acreage covered by an aerial photograph taken at a height of 12,000 ft above the average ground with a camera having a focal length of 6 in and a 9 in × 9 in format?

(A) 1861 ac

(B) 4649 ac

(C) 6974 ac

(D) 7438 ac

PROBLEM 46

If aerial photographs have a 9 in format, maintain a 60% end lap and a 30% side lap, and are taken at a height of 10,000 ft above the average ground with a camera having a focal length of 8.25 in, the acreage covered by the neat model is most nearly

(A) 440 ac

(B) 770 ac

(C) 1300 ac

(D) 2700 ac

PLANNING AERIAL PHOTOGRAPHY

PROBLEM 47

The following nine problems refer to the illustration shown.

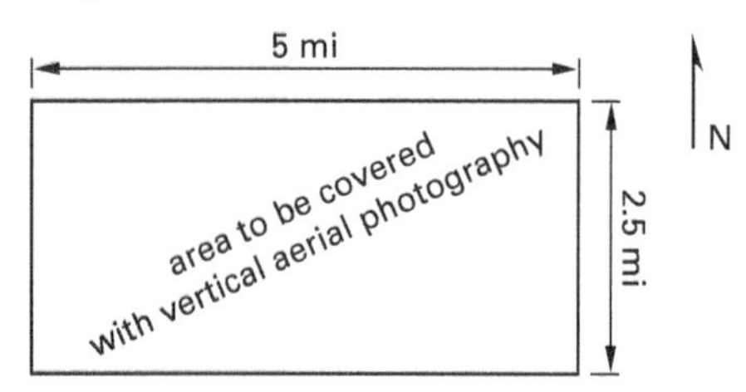

60% end lap and 30% side lap
6 in focal length camera
9 in × 9 in format
scale 1:10,000

Consider the area shown as being virtually level. The coverage of the project with vertical aerial photography should be sufficient to produce a photo scale of 1:10,000. The end lap is to be 60% and the side lap is to be 30% at a minimum. The camera is to have a 6 in focal length and the format is 9 in × 9 in.

47.1. Given the previous parameters, what is the ground dimension along the side of a single aerial photograph?

(A) 830 ft

(B) 3800 ft

(C) 5300 ft

(D) 7500 ft

47.2. What would be the maximum spacing between adjacent flight lines for the project illustrated?

(A) 2250 ft

(B) 3750 ft

(C) 5250 ft

(D) 7500 ft

47.3. What arrangement of the flight lines would provide the most efficient coverage of the area illustrated?

(A) The flight lines can be arranged in cardinal directions where convenient, particularly in public lands states where the improvements along section lines provide a guide for the pilot.

(B) It is frequently economical to arrange flight lines parallel with the project boundaries; in this case, the flight lines should be east and west.

(C) A north-south orientation of the flight lines would be most efficient in this instance to accommodate the 60% end lap.

(D) Both A and B are true.

47.4. To best accommodate the necessary photo coverage required of the northernmost flight line of the project, the first, or northernmost, strip should be photographed flying either east or west along

(A) the northern boundary of the project

(B) a line about 1500 ft south of the northern boundary of the project

(C) a line about 3750 ft north of the northern boundary of the project

(D) a line about 3750 ft south of the northern boundary of the project

47.5. If the maximum spacing between east-west flight lines is 5250 ft, and the northern and southern flight lines are each to be 1500 ft inside the project perimeter, what number of flight lines will cover the project, and what will be the adjusted distance between the adjacent flight lines?

(A) two flight lines spaced 5250 ft apart

(B) three flight lines spaced 5100 ft apart

(C) four flight lines spaced 3300 ft apart

(D) four flight lines spaced 4275 ft apart

47.6. Allowing for a 60% end lap, what is most nearly the linear advance necessary for each successive photo in the project?

(A) 3000 ft

(B) 3800 ft

(C) 5300 ft

(D) 7500 ft

47.7. Allowing for four photos outside of the project boundary along each flight line, two on the east and two on the west, how many aerial photographs will be needed to cover the project area?

(A) 21

(B) 39

(C) 47

(D) 52

47.8. A guide for a pilot is prepared by plotting the flight lines on a quad sheet with a scale of 1:24,000. What would be the spacing between the flight lines?

(A) 1.67 in

(B) 2.55 in

(C) 4.00 in

(D) 5.10 in

47.9. Assuming that the aircraft used in photographing a project travels at a constant speed of 100 mph, how long should the intervalometer setting allow between exposures?

(A) 12 sec

(B) 18 sec

(C) 20 sec

(D) 24 sec

PROBLEM 48

Aerial photography for topographic mapping is in the planning stage. The work will be compiled on an optical plotter that can accommodate relief variations in the stereoscopic model of 20% of the projection distance. What is most nearly the minimum flight height available in planning the photography of an area that has a maximum difference in elevation of 750 ft between its highest and lowest point?

(A) 3000 ft

(B) 3800 ft

(C) 5300 ft

(D) 5500 ft

PROJECTION PROCEDURES

PROBLEM 49

The optical plotting instrument that will be used to compile the aerial photography for a planned project has a *C*-factor of 800. The project requires a contour interval of 5 ft on the final map. What flight height above the average terrain should be used?

(A) 4000 ft

(B) 5000 ft

(C) 5500 ft

(D) 6000 ft

PROBLEM 50

A wide-angle projection plotter achieves its optimum focus at a projection distance of 30 in when the principal distance is set at 153 mm. What is the approximate focal length of the projector's objective lens?

(A) 5 in

(B) 8 in

(C) 200 mm

(D) 220 mm

PROBLEM 51

An aerial photograph having a tilt of 20° was exposed with a 6 in focal length camera. What is the distance between the principal point and the nadir point on the photograph?

(A) 0.25 in

(B) 1.53 in

(C) 2.18 in

(D) 3.62 in

PROBLEM 52

In the projection of direct optical stereoplotters, an effort is made to exactly reproduce the relative angular relationship between the two photographs at the moment of exposure. Generally, of the six motions possible with each projector, three are rotations about the mutually perpendicular axes x, y, and z. What is the origin of this system of axes?

(A) the principal point

(B) the focal point

(C) the nadir point

(D) the upper nodal point

PROBLEM 53

Which of the following statements best defines the three linear translations generally available in direct optical stereoplotters?

(A) y-translation changes the spacing between the projectors, z-translation moves the projectors perpendicular to the line between them, and x-translation moves the projectors in relation to the reference table.

Photogrammetry

(B) x-translation changes the spacing between the projectors, y-translation moves the projectors perpendicular to the line between them, and z-translation moves the projectors in relation to the reference table.

(C) z-translation changes the spacing between the projectors, y-translation moves the projectors perpendicular to the line between them, and x-translation moves the projectors in relation to the reference table.

(D) z-translation changes the spacing between the projectors, x-translation moves the projectors perpendicular to the line between them, and y-translation moves the projectors in relation to the reference table.

PROBLEM 54

The following six problems refer to the illustrations shown.

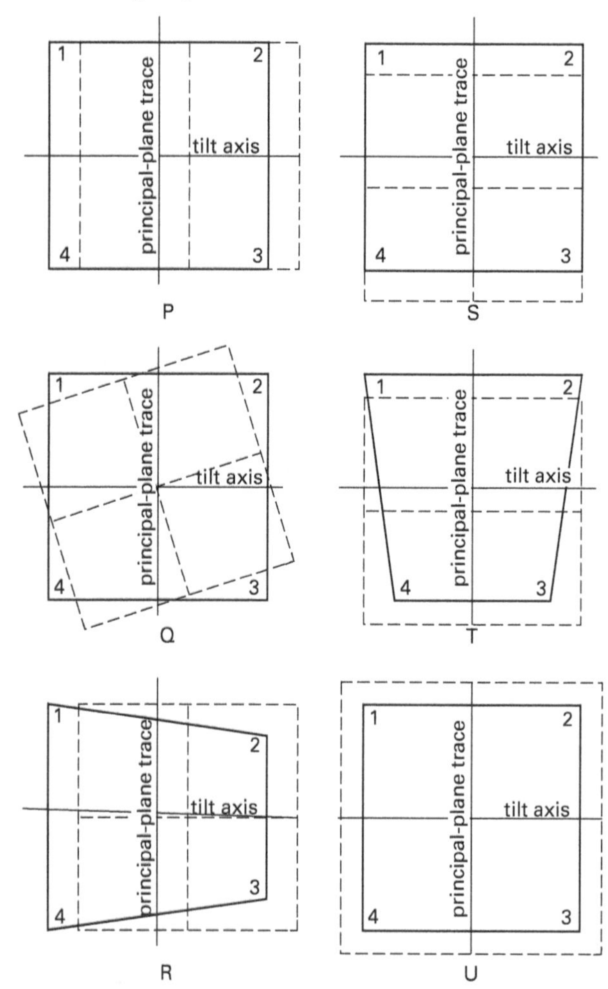

A square figure is shown to illustrate the effects of six possible rectifier motions on the projection of a negative or stereomodel, using the one-projector method.

54.1. Which of the six lettered figures illustrates the effect known as *x-displacement*?

(A) P

(B) S

(C) Q

(D) T

54.2. Which of the six lettered figures illustrates the effect known as *swing*?

(A) P

(B) Q

(C) T

(D) U

54.3. Which of the six lettered figures illustrates rotation about the y-axis?

(A) T

(B) Q

(C) R

(D) U

54.4. Which of the six lettered figures illustrates the effect of magnification?

(A) S

(B) Q

(C) P

(D) U

54.5. Which of the six lettered figures illustrates rotation about the x-axis?

(A) Q

(B) U

(C) P

(D) T

54.6. Which of the six lettered figures illustrates the effect known as *y-displacement*?

(A) P

(B) S

(C) Q

(D) T

ARTIFICIAL TARGETS

PROBLEM 55

Which of the following considerations would be involved in planning the sort of panels, or pre-marks, to use in an aerial photography project?

(A) Does the project include tree-covered areas?

(B) Are there portions of the area to be mapped that are covered with snow or sand?

(C) What is the intended photo-scale?

(D) All of the above are appropriate.

PROBLEM 56

The following two problems refer to the illustration shown.

56.1. If L is the width of a leg in a typical panel, how long should each leg be?

(A) $2L$

(B) $5L$

(C) $10L$

(D) $20L$

56.2. The dimensions of the square in the center of the pre-mark panel as shown on the photo negative are required to be 0.02 mm × 0.02 mm. If the focal length of the camera is 6 in and the flight height is 5000 ft above the average terrain elevation, what would be the dimensions of the square on the ground?

(A) 1.3 ft × 1.3 ft

(B) $5\frac{3}{4}$ in × $5\frac{3}{4}$ in

(C) $7\frac{7}{8}$ in × $7\frac{7}{8}$ in

(D) 240 mm × 240 mm

PROBLEM 57

It is common practice for a field crew using a set of aerial photos to make positive identification of ground control points. Which of the following procedures is appropriate for such work?

(A) The location of the control point on the photograph should be pricked with a pin.

(B) The photograph should not have holes poked in it; instead, a written description of the position of the control point should be used.

(C) The best identification of a control point is its assigned number and coordinates.

(D) The photographs need not be marked in any way since they are for the reference of the field crew to verify that the control point has not been moved.

PROBLEM 58

The overlapping aerial photographs shown have the typical symbols used to identify control points in photogrammetry. Which of the following correctly identifies the number and kind of control points illustrated?

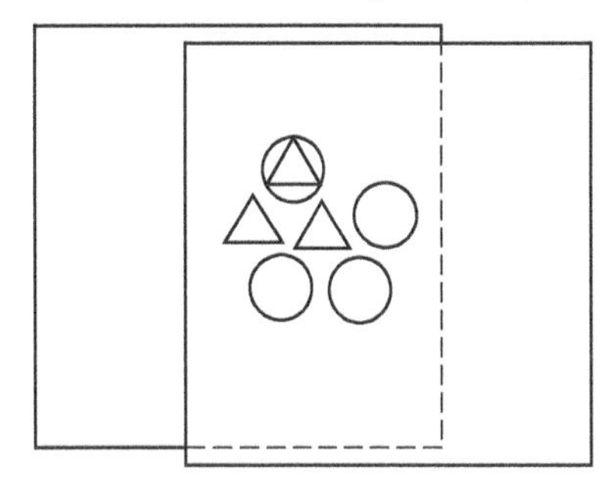

(A) 3 horizontal control points
2 vertical control points
1 NGS control point

(B) 3 horizontal control points
2 vertical control points
1 combined horizontal and vertical control point

(C) 3 vertical control points
2 horizontal control points
1 combined horizontal and vertical control point

(D) 3 combined horizontal and vertical control points
2 vertical control points
1 horizontal control point

PROBLEM 59

The following figures illustrate a pair of aerial photographs intended for use in topographic mapping. The photographs show an end lap of 60%. Which of the figures indicates the best number and configuration of ground control points?

(A)

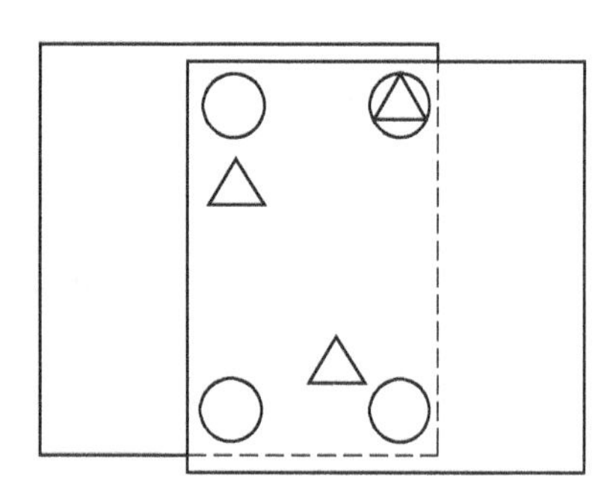

(B)

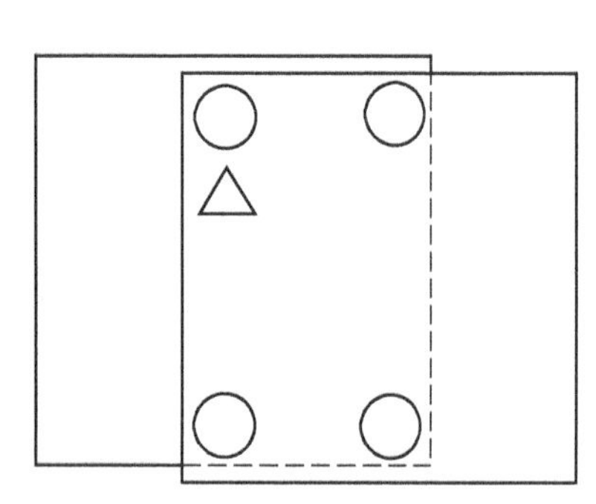

(C)

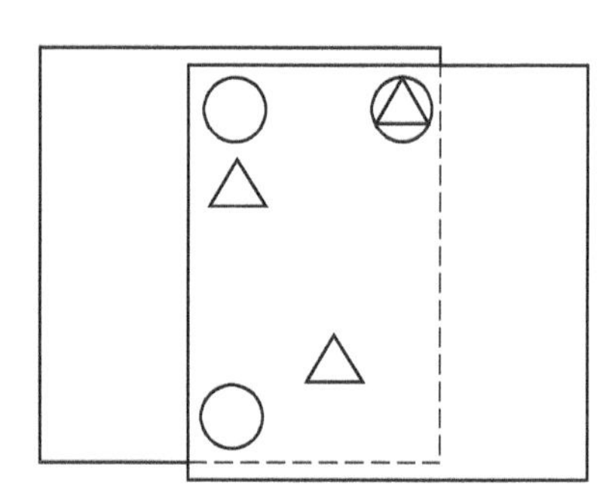

(D)

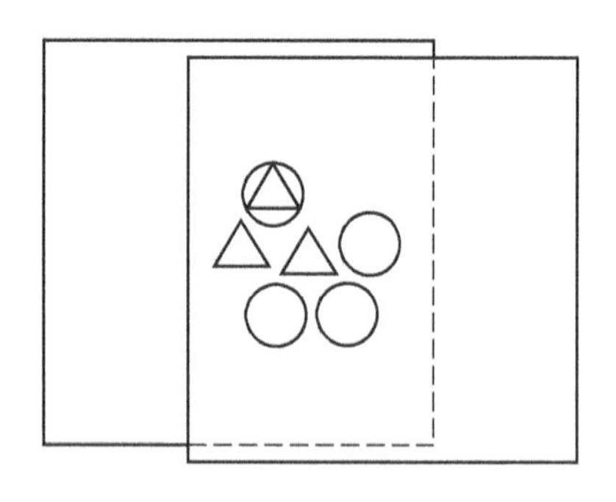

PROBLEM 60

National map standards require that the principal planimetric features on a map of a scale less than 1:20,000 be shown within one-thirtieth of an inch of their true position. Consider a map plotted at a scale of 1 in to 100 ft. What is most nearly the maximum allowable error in a ground measurement?

(A) 0.11 ft

(B) 1.6 ft

(C) 2.5 ft

(D) 3.3 ft

PROBLEM 61

A widely used rule of mapping is that the photo control should include no error greater than 0.005 of the map scale when that scale is greater than 1:20,000. Considering this rule, what magnitude of error in ground control may most nearly be tolerated for a map of the scale 1:24,000?

(A) 1 ft

(B) 7 ft

(C) 10 ft

(D) 15 ft

PROBLEM 62

Which of the following specifications is good practice in preparing a pre-mark panel to function as a vertical control point for photogrammetry?

(A) The panel should be set in a flat area, and the elevation of the monument at its center determined.

(B) A description and sketch of the panel should be prepared.

(C) The vertical measurement from the monument to the panel itself should be made if the latter is above or below the monument.

(D) All of the above are true.

PROBLEM 63

National map standards require that the elevations of 90% of the points tested be correct within what part of its contour interval?

(A) one-sixth

(B) one-fifth

(C) one-fourth

(D) one-half

PROBLEM 64

Which of the following best describes the meaning of the term *picture point* as it is used in photogrammetry?

(A) a particularly clear pre-marked artificial panel point on an aerial photograph

(B) a pin prick on the emulsion of an aerial photograph made during control verification

(C) a sharp and unambiguous natural feature used to supplement ground control

(D) any one of a number of points imagined to exist on aerial photographs, including the nadir point, the principal point, and the isocenter, among others

SOLUTION 1

The images of four fiducial markers fixed inside the camera appear on each photograph. These markers allow consistent orientation of the photographs.

The answer is (C).

SOLUTION 2

According to the law of reversibility of light paths, the effect of lens distortion in the camera is reversed in the plotting instruments if identical imperfections exist in both lens systems.

The answer is (A).

SOLUTION 3

The principal point is the intersection of the optical axis of the camera and the principal surface and is indicated by the intersection of the lines through the fiducial markers.

The answer is (D).

SOLUTION 4

A stereoscopic plotting instrument theoretically presents a model of the terrain photographed when a stereo pair is properly oriented. Optical and electronic manipulation can eliminate the displacement of the points in the photographs due to tilt and relief. The result is known as an orthophotograph.

The answer is (C).

SOLUTION 5

A high oblique aerial photograph is one in which the horizon is visible. A low oblique photograph is taken with a tilted camera, but the horizon is not visible. A vertical photograph is made with a camera whose optical axis is nearly vertical.

The answer is (A).

SOLUTION 6

The air base is also known as the *baseline* or the *model base.* It is the distance the plane flew between overlapping exposures. As reconstructed in the plotting instrument, it is the distance, at scale, between adjacent perspective centers.

The answer is (B).

SOLUTION 7

The base-height ratio is also known as the *K-factor*.

The answer is (D).

SOLUTION 8

An aerial photograph can be translated, rotated, and scaled so as to reverse the effects of the original picture-taking process, eliminating the effects of tilt, but not of relief displacement.

The answer is (A).

SOLUTION 9

In a photograph taken with the optical axis of the camera aligned with gravity, the point at which the line through the perspective center of the lens system intersects the photograph would coincide with the principal point.

The answer is (A).

SOLUTION 10

A stereoscope is a small optical device that allows a viewer, with some practice, to view two-dimensional photographs simultaneously and form a mental image of the three-dimensional surface they represent.

The answer is (A).

SOLUTION 11

The tilt of an aerial photograph is caused by the impossibility of maintaining a truly vertical camera axis when making the exposures. The inclination of the optical axis of the camera from the vertical is usually expressed as the angle between the plane of the photograph and the horizontal plane. The effects of tilt can be compensated for in stereoscopic plotting instruments by the rotation and movement of projectors to reflect the position of the aerial camera when the photographs were made.

The answer is (C).

SOLUTION 12

The x-axis is assumed to be the line described by fiducial marks that most closely matches the flight line. The y-axis is nearly perpendicular to the flight line. In other words, when the nose of the aircraft moves down, the result would be described as a positive y-tilt.

The answer is (B).

SOLUTION 13

In the radial line method of plotting, the angles drawn from the radial center to other points on an aerial photograph are assumed to equal the corresponding angles on the ground. The method is based on the assumption that the camera was perpendicular to the horizontal plane at the moment of exposure. This condition is almost never met, but the magnitude of the tilt is often not significant enough to negate the plotting method.

The answer is (A).

SOLUTION 14

The isocenter is approximately midway between the nadir point and the principal point on a tilted aerial photograph. Its distance from the principal point is equal to the focal length multiplied by the tangent of half the angle of the tilt.

The answer is (B).

SOLUTION 15

The construction of a controlled mosaic or photo map requires cutting the edges of the prints along features that would reproduce as lines anyway.

The answer is (A).

SOLUTION 16

The larger a ground control point (GCP) feature, the more visible it will be in the digital imagery. Of the options provided, the intersection of unpaved roads is the largest feature; however, the intersection would be hard to resolve to an unambiguous point on a 60 cm image pixel and accessibility may be difficult on a military base. The contrast at the ground near the corner of a building may be good, but as the sun moves, the shadow of the building may restrict future usefulness. There is also a high likelihood of multipath disrupting the GPS observation.

A small tree may stand out against an arid background, but it presents several problems as a GCP feature. For instance, a tree grows and changes and may be trimmed or cut down. It may be difficult to collect the tree's position at ground level, and as the sun moves it can become indistinct by its own shadow. Therefore, a tree's applicability as a GCP is questionable. The feature most likely to satisfy the most criteria for a photo-identifiable GCP is the corner of the concrete slab on grade surrounded by grass.

The answer is (C).

SOLUTION 17

The standard position of the origin in a digital image is the upper left pixel.

The answer is (B).

SOLUTION 18

The pixels in an 8-bit image can have an integer digital number (DN) value from 0 to 255, meaning that there are 256 (28) possible shades, or intensities, available for each pixel. The DN can be thought of as intensity or shade, where 0 is black and 255 is white. The pixels in a 16-bit image can have an integer digital number value from −32768 to 32767, meaning there are 65,536 (216) possible shades, or intensities, available for each pixel.

The answer is (A).

SOLUTION 19

LiDAR, which can stand for Light Detection and Ranging, is often used on aircraft that have GPS receivers and inertial measurement systems. LiDAR usually uses portions of the electromagnetic spectrum, such as near infrared, ultraviolet, or visible light and is affected by cloud cover, smoke, or other aerosol particles in the atmosphere.

The answer is (D).

SOLUTION 20

The typical dimensions for the portion of the negative within the focal plane of an aerial camera used in topographic mapping is 9 in × 9 in (230 mm × 230 mm).

The answer is (C).

SOLUTION 21

Aerial photographs are printed on glass plates in preparation for their projection in a stereoscopic plotting instrument or comparator. These diapositives are sometimes produced at a reduced scale, using lenses that compensate for the distortion of the camera lens used to make the photographs.

The answer is (D).

SOLUTION 22

A collimator is used to calibrate the camera, but is not part of the camera itself.

The answer is (D).

SOLUTION 23

The magnification of parallax is a desirable characteristic in aerial photographs used for topographic purposes. The measurement of parallax is a critical factor in plotting topographic maps. A relatively short focal length and a wide-angle field is preferred for such work. The 6 in focal length wide-angle lens is typical.

The answer is (A).

SOLUTION 24

While the focal-plane shutter is capable of faster speeds than the between-the-lens shutter, this efficiency is offset by the distortion introduced when such a shutter is used in a moving aircraft. The between-the-lens shutter exposes the whole negative at once and is preferred over the focal-plane shutter, which exposes the negative from one end to the other by moving across it.

The answer is (D).

SOLUTION 25

Once it is set, the intervalometer operates the shutter of the camera automatically, making exposures that take the speed of the airplane into account to assure photographs with a predetermined overlap.

The answer is (D).

SOLUTION 26

In a reseau camera, a rectangular grid is superimposed on the image produced by the exposure. The distortion of the photograph can then be minimized by restoring the grid to its proper shape and size.

The answer is (D).

SOLUTION 27

The diagonal or angular field of view for a given format increases as the focal length of the camera lens decreases. This is significant in photogrammetry since the shorter the focal length of the lens used, the wider the ground coverage that can be achieved on a single photograph.

The answer is (D).

SOLUTION 28

Side winds can cause the heading of an aircraft to deviate from its actual direction of travel. The angle between the longitudinal axis of the aircraft and its flight line is known as the *yaw* or *crab angle*. When an aerial camera is not corrected for this effect, the sides of the resulting photographs are not parallel with the direction of flight, and the ground coverage can be significantly reduced.

The answer is (A).

SOLUTION 29

A single-lens precision camera is preferred for topographic mapping. It provides a combination of short focal length and wide angle for high geometric picture quality.

The answer is (D).

SOLUTION 30

Given the format size and focal length of a camera, the formula for calculating the diagonal field of view is

$$\alpha = 2\tan^{-1}\left(\frac{d}{2f}\right)$$

d is the length of the diagonal across the format, and f is the focal length.

$$\begin{aligned}\alpha &= 2\tan^{-1}\left(\frac{\sqrt{(9\text{ in})^2 + (9\text{ in})^2}}{(2)(6\text{ in})}\right)\\ &= 2\tan^{-1}(1.0607)\\ &= 93°\end{aligned}$$

The answer is (C).

SOLUTION 31

31.1. The formula for the scale at a particular elevation is

$$S_e = \frac{f}{H - h}$$

S_e is the scale at a given elevation, f is the focal length, H is the altitude of the aircraft above a datum, and h is the height of the ground station above the same datum.

$$\begin{aligned}S_e &= \frac{6\text{ in}}{10{,}000\text{ ft} - 4500\text{ ft}}\\ &= \frac{0.5\text{ ft}}{5500\text{ ft}}\\ &= \frac{1}{11{,}000}\end{aligned}$$

The solution relies on solving the similar triangles Lbc and LBC and on presuming that line BC lies across essentially level terrain. The answer means that 1 unit on the photograph is equal to approximately 11,000 of the same units on the ground along line BC.

The answer is (C).

31.2. The formula for the scale at a particular elevation is

$$S_e = \frac{f}{H - h}$$

S_e is the scale at a given elevation, f is the focal length, H is the altitude of the aircraft above a datum, and h is the height of the ground station above the same datum.

For point B,

$$\begin{aligned}S_e &= \frac{6\text{ in}}{3000\text{ ft} - 150\text{ ft}}\\ &= \frac{6\text{ in}}{2850\text{ ft}}\\ &= \frac{1\text{ in}}{475\text{ ft}}\end{aligned}$$

Repeat the same calculation for point A.

$$\begin{aligned}S_e &= \frac{6\text{ in}}{3000\text{ ft} - 800\text{ ft}}\\ &= \frac{6\text{ in}}{2200\text{ ft}}\\ &= \frac{1\text{ in}}{367\text{ ft}}\end{aligned}$$

The answer is (A).

SOLUTION 32

To find the altitude of the aircraft, solve the equation

$$S_{\text{ave}} = \frac{f}{H - h_{\text{ave}}}$$

S_{ave} is the average scale stipulated, f is the focal length, H is the altitude of the aircraft above a datum, and h_{ave} is the average elevation of the terrain above the same datum.

$$\begin{aligned}\frac{1}{12{,}000} &= \frac{0.67\text{ ft}}{H - 3300\text{ ft}}\\ H - 3300\text{ ft} &= (0.67\text{ ft})(12{,}000)\\ H &= 8040\text{ ft} + 3300\text{ ft}\\ &= 11{,}340\text{ ft}\quad (11{,}300\text{ ft})\end{aligned}$$

The answer is (C).

SOLUTION 33

The scale of the photograph can be defined as

$$\begin{aligned}\text{scale of the photograph} &= \left(\frac{\text{distance on the photo}}{\text{distance on the map}}\right) \\ &\quad \times(\text{map scale}) \\ &= \left(\frac{3.52 \text{ in}}{2.67 \text{ in}}\right)\left(\frac{1}{24{,}000}\right) \\ &= (1.3183)\left(\frac{1}{24{,}000}\right) \\ &= \frac{1}{18{,}205}\end{aligned}$$

This scale may be read as

$$\begin{aligned}1 \text{ in} &= 18{,}205 \text{ in} \\ &= (18{,}205 \text{ in})\left(\frac{1 \text{ ft}}{12 \text{ in}}\right) \\ &= 1517 \text{ ft}\end{aligned}$$

The answer is (B).

SOLUTION 34

The scale of the photograph can be expressed as

$$\begin{aligned}\text{scale of the photograph} &= \left(\frac{\text{distance on the photo}}{\text{distance on the map}}\right) \\ &\quad \times(\text{map scale}) \\ &= \left(\frac{5.12 \text{ in}}{3.00 \text{ in}}\right)\left(\frac{1}{24{,}000}\right) \\ &= (1.707)\left(\frac{1}{24{,}000}\right) \\ &= \frac{1}{14{,}063}\end{aligned}$$

To find the altitude of the aircraft, solve the equation

$$S_{\text{ave}} = \frac{f}{H - h_{\text{ave}}}$$

S_{ave} is the average scale, f is the focal length, H is the altitude of the aircraft above a datum, and h_{ave} is the average elevation of the terrain above the same datum.

$$\begin{aligned}\frac{1}{14{,}060} &= \frac{0.5 \text{ ft}}{H - 1200 \text{ ft}} \\ H - 1200 \text{ ft} &= (0.5 \text{ ft})(14{,}063) \\ H &= 7030 \text{ ft} + 1200 \text{ ft} \\ &= 8232 \text{ ft} \quad (8230 \text{ ft})\end{aligned}$$

The answer is (B).

SOLUTION 35

The optimum projection distance, which is the plane of best focus, is used to fix the model scale for a particular stereo plotter. The formula used is

$$\begin{aligned}\text{model scale} &= \left(\frac{\text{projection distance}}{\text{camera focal length}}\right)(\text{photo scale}) \\ &= \left(\frac{760 \text{ mm}}{6 \text{ in}}\right)\left(\frac{1 \text{ in}}{500 \text{ ft}}\right)\left(\frac{1 \text{ in}}{25.4 \text{ mm}}\right) \\ &= 1 \text{ in}/100 \text{ ft}\end{aligned}$$

The answer is (A).

SOLUTION 36

The plotting scale is fixed in the multiplex instrument and therefore is the same as the model scale.

$$\begin{array}{c}\text{optimum} \\ \text{model} \\ \text{scale}\end{array} = \frac{\text{optimum projection distance}}{\text{flight height} - \text{average terrain elevation}}$$

The model scale is known in this case; it is 1:10,000. The optimum projection distance is more conveniently expressed in feet.

$$(360 \text{ mm})\left(\frac{1 \text{ in}}{25.4 \text{ mm}}\right)\left(\frac{1 \text{ ft}}{12 \text{ in}}\right) = 1.181 \text{ ft}$$

$$\begin{aligned}\frac{1}{10{,}000} &= \frac{1.181 \text{ ft}}{\text{flight height} - 3000 \text{ ft}} \\ \text{flight height} - 3000 \text{ ft} &= (10{,}000)(1.181 \text{ ft}) \\ &= 11{,}810 \text{ ft} \\ \text{flight height} &= 11{,}810 \text{ ft} + 3000 \text{ ft} \\ &= 14{,}810 \text{ ft}\end{aligned}$$

The answer is (C).

SOLUTION 37

Points above or below each other on the ground are displaced from each other on an aerial photograph. This displacement is along a line from the image points to the principal point or nadir. On the photograph, points below the datum are displaced toward the nadir.

The answer is (D).

SOLUTION 38

The extent of relief displacement can be defined as

$$d = \frac{rh}{H}$$

d is the relief displacement, r is the radial distance of the ground point image from the principal point, h is the elevation of the point from the datum, and H is the flight height above the datum.

The answer is (C).

SOLUTION 39

The most convenient datum for the calculation is the bottom of the tower at 540 ft. The flight height can then be expressed relative to that base.

$$\begin{aligned} \text{flight height} &= 2000 \text{ ft} - 540 \text{ ft} \\ &= 1460 \text{ ft} \quad \left(\begin{array}{c}\text{relative to the base}\\ \text{of the tower}\end{array}\right) \end{aligned}$$

d is the relief displacement, r is the radial distance of the ground point image from the principal point, h is the elevation of the point from the datum in feet, and H is the flight height above the datum in feet. The formula for relief displacement ($d = rh/H$) may be rearranged.

$$\begin{aligned} h &= \frac{Hd}{r} = \frac{(1.33 \text{ in})(1460 \text{ ft})}{5.97 \text{ in}} \\ &= \frac{1941.80 \text{ in-ft}}{5.97 \text{ in}} \\ &= 325.25 \text{ ft} \quad (325 \text{ ft}) \end{aligned}$$

The answer is (B).

SOLUTION 40

Using the valley floor as the datum ($h = 0$), the formula may be written as

$$S = \frac{f}{H}$$

S is the scale at the valley floor, f is the focal length of the lens, and H is the altitude of the aircraft.

$$\begin{aligned} \frac{1}{8000} &= \frac{0.5 \text{ ft}}{H} \\ H &= (0.5 \text{ ft})(8000) \\ &= 4000 \text{ ft} \end{aligned}$$

The altitude of the aircraft when the photograph was taken was 4000 ft above the valley floor.

Next, the relief displacement may be found using the formula

$$d = \frac{rh}{H}$$

d is the relief displacement, r is the radial distance of the ground point image from the principal point, h is the elevation of the point from the datum, and H is the flight height above the datum.

$$\begin{aligned} d &= \frac{(3.99 \text{ in})(495 \text{ ft})}{4000 \text{ ft}} \\ &= \frac{1975 \text{ in-ft}}{4000 \text{ ft}} \\ &= 0.50 \text{ in} \end{aligned}$$

The answer is (A).

SOLUTION 41

The overlapping of adjacent photographs along a flight strip is known as *end lap* and is usually 60%. The typical 30% overlapping of flight strips is known as *side lap.*

The answer is (D).

SOLUTION 42

There are several reasons for overlapping aerial photographs, but the Scheimpflug condition is not one of them. The Scheimpflug condition is achieved when the object plane, the image plane, and the plane of the lens intersect along the same line, thus ensuring the sharpness of a projection.

The answer is (D).

SOLUTION 43

As the base-height ratio increases, the disparity between the apparent horizontal and vertical scales also increases. The result is that the stereoscopic view of a stereo pair affords an exaggerated depth of the image.

The answer is (A).

SOLUTION 44

First, calculate the scale of the photograph.

$$S = \frac{f}{H}$$

S is the scale of the photograph, f is the focal length of the lens, and H is the altitude of the aircraft above the average terrain elevation.

$$S = \frac{6 \text{ in}}{4500 \text{ ft}} = \frac{1 \text{ in}}{750 \text{ ft}}$$

Next, find the ground length covered by a single photograph.

$$\text{ground length} = (9 \text{ in})\left(750 \ \frac{\text{ft}}{\text{in}}\right) = 6750 \text{ ft}$$

However, with a 60% end lap, there is a 40% advance between exposures, so the air base is

$$\text{air base} = (0.40)(6750 \text{ ft}) = 2700 \text{ ft}$$

The base-height ratio is

$$\frac{2700 \text{ ft}}{4500 \text{ ft}} = 0.60$$

The answer is (B).

SOLUTION 45

The first step in solving this problem is to find the scale of the photograph. The formula is

$$S = \frac{f}{H}$$

S is the scale at the valley floor, f is the focal length of the lens, and H is the altitude of the aircraft.

$$S = \frac{6 \text{ in}}{12{,}000 \text{ ft}} = \frac{1 \text{ in}}{2000 \text{ ft}}$$

Next, find the ground dimension shown by a 9 in × 9 in photograph.

$$\text{ground dimension} = \left(2000 \ \frac{\text{ft}}{\text{in}}\right)(9 \text{ in}) = 18{,}000 \text{ ft}$$

Therefore, the required area in acres is

$$A = (18{,}000 \text{ ft})^2\left(\frac{1 \text{ ac}}{43{,}560 \text{ ft}^2}\right) = 7438 \text{ ac}$$

The answer is (D).

SOLUTION 46

The neat model is the portion of a pair of aerial photographs actually used in a mapping project. It is the stereoscopic area between successive principal points, extended perpendicular to the flight direction and ending in the middle of the side laps.

To find the area of the neat model, begin by finding the scale of the photograph. The formula for the scale is

$$S = \frac{f}{H}$$

S is the scale of the photograph, f is the focal length of the lens, and H is the altitude of the aircraft above the average ground elevation.

$$S = \frac{8.25 \text{ in}}{10{,}000 \text{ ft}} = \frac{1 \text{ in}}{1212 \text{ ft}}$$

Next, find the ground dimension of a side of the 9 in × 9 in photograph.

$$\text{ground dimension} = \left(1212 \ \frac{\text{ft}}{\text{in}}\right)(9 \text{ in}) = 10{,}908 \text{ ft}$$

The region of the overlap used as the neat model is a rectangle 40% of the length and 70% of the width of the photograph. These dimensions can now be translated into ground dimensions.

$$\text{ground dimension } (b) = (0.4)(10{,}908 \text{ ft}) = 4363 \text{ ft}$$
$$\text{ground dimension } (w) = (0.7)(10{,}908 \text{ ft}) = 7636 \text{ ft}$$

The area of the neat model can now be calculated.

$$\begin{aligned} A = bw &= (4363 \text{ ft})(7636 \text{ ft}) \\ &= 33{,}315{,}868 \text{ ft}^2 \end{aligned}$$

The area can be converted to acreage.

$$\frac{33{,}315{,}868 \text{ ft}^2}{43{,}560 \dfrac{\text{ft}^2}{\text{ac}}} = 765 \text{ ac} \quad (770 \text{ ac})$$

The answer is (B).

SOLUTION 47

47.1. The specified photo scale is 1:10,000, or

$$\begin{aligned} 1 \text{ in} &= 10{,}000 \text{ in} \\ &= (10{,}000 \text{ in})\left(\frac{1 \text{ ft}}{12 \text{ in}}\right) \\ &= 833.33 \text{ ft} \end{aligned}$$

The format of the photographs is 9 in × 9 in.

$$(9)(833.33 \text{ ft}) = 7500 \text{ ft}$$

Therefore, the ground dimension along the side of a single photograph will be 7500 ft.

The answer is (D).

47.2. The ground dimension along the side of a single photograph, symbolized as G, is 7500 ft, as shown in Prob. 47.1. Allowing for 30% sidelap, the spacing, symbolized as W, is the remaining 70% of G.

$$\begin{aligned} W &= 0.7G \\ &= (0.7)(7500 \text{ ft}) \\ &= 5250 \text{ ft} \quad (5300 \text{ ft}) \end{aligned}$$

The answer is (C).

47.3. In this case, the flight lines would probably be most economically arranged parallel with the longer project boundaries at the north and south.

The answer is (D).

47.4. The aerial photography should include coverage outside the perimeter of the project, using the minimum number of flight lines. In this case, the first strip can be made by allowing 30% of the photograph to be outside the project boundary. In this instance, G, the ground dimension along one side of the 9 in × 9 in photograph, is 7500 ft.

$$0.3G = (0.3)(7500 \text{ ft}) = 2250 \text{ ft}$$

If 2250 ft of the photograph is outside of the project, then 7500 ft − 2250 ft, or 5250 ft, of the photograph remains inside the project. However, the flight line itself is in the center of the 7500 ft wide strip, so

$$\begin{aligned} 0.5G - 0.3G &= (0.5)(7500 \text{ ft}) - (0.3)(7500 \text{ ft}) \\ &= 3750 \text{ ft} - 2250 \text{ ft} \\ &= 1500 \text{ ft} \end{aligned}$$

The first, or northernmost, flight line should be 1500 ft south of the northern project boundary.

The answer is (B).

47.5. The dimension of the project north to south is 2.5 mi, or 13,200 ft. However, the northern and southern flight lines are each 1500 ft inside the perimeter.

The distance between the outside flight lines will be

$$\begin{aligned} 13{,}200 \text{ ft} - (2)(1500 \text{ ft}) &= 13{,}200 \text{ ft} - 3000 \text{ ft} \\ &= 10{,}200 \text{ ft} \end{aligned}$$

Since the maximum spacing between flight lines is 5250 ft, one additional flight line is required.

For greatest efficiency, the centers of the flight lines can be adjusted to

$$\frac{10{,}200 \text{ ft}}{2} = 5100 \text{ ft}$$

The total number of flight lines is three, and the two spaces between the centerlines of the adjacent strips are each 5100 ft.

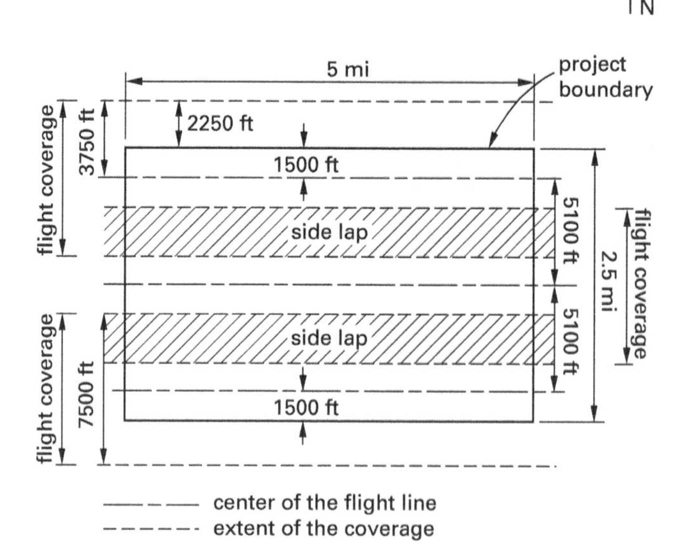

The answer is (B).

47.6. The value of G for this project is 7500 ft, which is the ground dimension of a side of the 9 in × 9 in format at the photo scale of 1:10,000. However, with a 60% end lap, just 40% of the photo is left. This constitutes the length of the air base.

$$\begin{aligned} B &= \text{linear advance} = 0.4G \\ &= (0.4)(7500 \text{ ft}) \\ &= 3000 \text{ ft} \end{aligned}$$

The answer is (A).

47.7. The east-west extent of the project is 5 mi, or 26,400 ft. The linear advance for each successive photo along each flight line is 3000 ft, 40% of G, and two photos should be added at the beginning and the end of each flight line to ensure complete coverage. The number of photos per flight line is

$$\frac{26{,}400 \text{ ft}}{3000 \text{ ft}} + 2 + 2 = 8.8 + 2 + 2 = 12.8$$

The closest integral to 12.8 is 13 photographs per strip. There are three flight lines in the project. The total number of photos required is

$$(13)(3) = 39$$

The answer is (B).

47.8. The distance between flight lines is 5100 ft. At a scale of 1:24,000, this distance would be

$$\left(\frac{1}{24{,}000}\right)(5100 \text{ ft})\left(12 \ \frac{\text{in}}{\text{ft}}\right) = 2.55 \text{ in}$$

The answer is (B).

47.9. The rate of the aircraft is more conveniently expressed in feet per second.

$$\left(100 \ \frac{\text{mi}}{\text{hr}}\right)\left(5280 \ \frac{\text{ft}}{\text{mi}}\right)\left(\frac{1 \text{ hr}}{3600 \text{ sec}}\right) = 147 \text{ ft/sec}$$

The linear advance (air base) necessary for each successive photo is 3000 ft. The rate of the aircraft is approximately 147 ft/sec, so

$$\frac{3000 \text{ ft}}{147 \ \frac{\text{ft}}{\text{sec}}} = 20.41 \text{ sec}$$

The intervalometer should allow 20 sec between successive photographs.

The answer is (C).

SOLUTION 48

The height of the aircraft above the ground can be treated similarly to the projection distance on the plotter. Where the limitation of the system in handling variations of relief is 20% of the projection distance, the flight height should be at least five times the maximum difference in the elevation of the terrain.

$$(750 \text{ ft})(5) = 3750 \text{ ft} \quad (3800 \text{ ft})$$

The answer is (B).

SOLUTION 49

The C-factor is an indication of the possible precision available in particular photogrammetric equipment. The larger the C-factor, the higher the available flying heights. The relationship can be expressed as

$$\text{flight height} = (\text{contour interval})(C\text{-factor})$$

In this case, the contour interval is 5 ft and the C-factor is 800.

$$\begin{aligned} \text{flight height} &= (5 \text{ ft})(800) \\ &= 4000 \text{ ft} \end{aligned}$$

The answer is (A).

SOLUTION 50

The principal distance is equal to the calibrated focal length of the camera with which the original photograph was taken. In this case, it is 153 mm, or 6 in.

$$(153 \text{ mm})\left(0.03937 \ \frac{\text{in}}{\text{mm}}\right) = 6.02 \text{ in} \quad [\text{round to 6.0 in}]$$

The projection distance is the distance from the lower nodal point of the projector's objective lens to the plane of the optimum focus and, in this case, is 30 in. The projection distances vary with plotters of different design; however, 30 in is typical for wide-angle projectors, and approximately 33 in is typical for normal-angle projectors.

The relationship between the principal and projection distances and the focal length of the projector lens can be expressed as

$$\frac{1}{f} = \frac{1}{p} + \frac{1}{h}$$

f is the focal length of the objective lens of the projector, p is the principal distance, and h is the projection distance.

$$\begin{aligned}\frac{1}{f} &= \frac{1}{6\text{ in}} + \frac{1}{30\text{ in}} = \frac{5}{30\text{ in}} + \frac{1}{30\text{ in}} \\ &= \frac{6}{30\text{ in}} \\ &= \frac{1}{5\text{ in}} \\ f &= 5\text{ in}\end{aligned}$$

The answer is (A).

SOLUTION 51

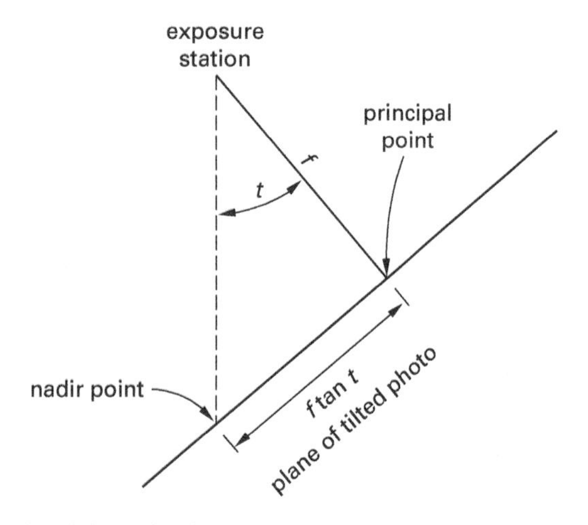

The focal length, f, is the distance from the exposure station to the plane of the photograph. If angle t is the tilt, the distance between the nadir point and the principal point is $f\tan t$.

$$(6\text{ in})(\tan 20°) = 2.18\text{ in}$$

The answer is (C).

SOLUTION 52

The origin of the axis system in the stereoplotter is the upper nodal point of the projector lens.

The answer is (D).

SOLUTION 53

Most direct optical stereoplotters have three angular rotations but not necessarily three linear translations. However, x-translation is essential; it provides the ability to change the spacing between the projectors. z-translation is the adjustment of the distance between the projectors and the reference table.

The answer is (B).

SOLUTION 54

54.1. x-displacement is illustrated in figure P, where the displacement is along the x-axis parallel to the tilt axis.

The answer is (A).

54.2. The rotation around the principal point in figure Q is called *swing*. Another way of describing this motion is rotation about the z-axis.

The answer is (B).

54.3. The apparent convergence of the sides parallel to the tilt axis in figure R is an effect of rotation about the y-axis.

The answer is (C).

54.4. The radial displacement of all of the corners of the figure relative to the principal point shown in figure U is the effect of magnification.

The answer is (D).

54.5. The apparent convergence of the sides parallel to the principal-plane axis in figure T is an effect of rotation about the x-axis.

The answer is (D).

54.6. y-displacement is illustrated in figure S, where the displacement is along the y-axis parallel to the principal plane.

The answer is (B).

SOLUTION 55

Contrast is one of the primary considerations in choosing the material used in panel points. Light-colored backgrounds require dark pre-marks. Areas of timber or brush raise the possibility that the panels may be obscured by shadows or displacement. The size of the panels is dependent on the intended photo-scale.

The answer is (D).

SOLUTION 56

56.1. A length of five times the width of the leg of a panel is generally held to be sufficient.

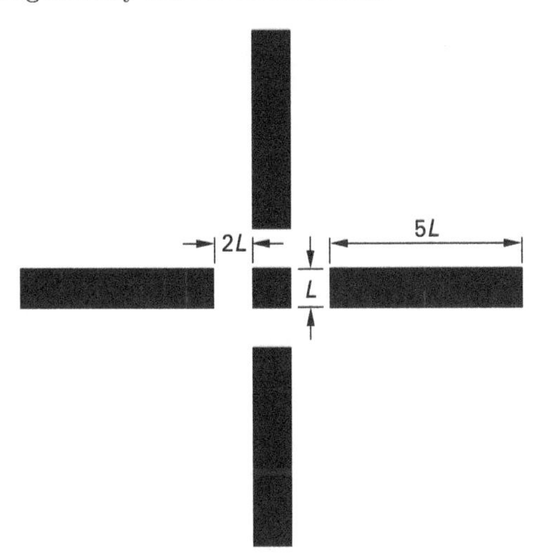

The answer is (B).

56.2. The scale of the photograph can be expressed as

$$S = \frac{f}{H}$$

S is the scale of the photograph, f is the focal length of the camera, and H is the flight height above the average terrain elevation.

$$S = \frac{0.5\ \text{ft}}{5000\ \text{ft}} = \frac{1}{10{,}000}$$

The scale of the photo will be 1:10,000. If the square must be 0.02 mm × 0.02 mm, then

$$(10{,}000)(0.02\ \text{mm}) = 200\ \text{mm}$$

The ground dimension of the square must be 200 mm × 200 mm.

$$(200\ \text{mm})\left(\frac{1\ \text{in}}{25.4\ \text{mm}}\right) = 7\tfrac{7}{8}\ \text{in} \quad (7\tfrac{7}{8}\ \text{in} \times 7\tfrac{7}{8}\ \text{in})$$

The answer is (C).

SOLUTION 57

Experience has shown that the most certain method of marking photo control for clear identification is a pin prick penetrating the emulsion at the precise position of the control point. A magnifying glass is useful in this process. Descriptions and sketches of the control are also desirable, but a pin prick is the method least susceptible to misidentification.

The answer is (A).

SOLUTION 58

The usual symbols used to identify control for photogrammetry are the circle, the triangle, and the circle and triangle combined. The circle is used to indicate vertical control. The triangle is used to indicate horizontal control. The combination of the triangle and the circle is used when horizontal and vertical controls are available from the same point.

The answer is (C).

SOLUTION 59

Topographic mapping requires at least two horizontal and four vertical control points in the overlapping area of adjacent photographs. However, it is best to allow for three horizontal control points rather than the minimum of two. The arrangement of the control is also important. Control in the area that will be included in the side lap of adjoining flight lines allows for greater efficiency.

The answer is (A).

SOLUTION 60

The standard requires the position of features be within one-thirtieth of an inch on the map of their true position on the ground, so

$$\begin{aligned}\left(\frac{1\ \text{in}}{30}\right)\left(12\ \frac{\text{in}}{\text{ft}}\right)\left(100\ \frac{\text{ft}}{\text{in}}\right) &= (40\ \text{in})\left(\frac{1\ \text{ft}}{12\ \text{in}}\right)\\ &= 3.33\ \text{ft} \quad (3.3\ \text{ft})\end{aligned}$$

The answer is (D).

SOLUTION 61

The scale 1:24,000 may also be stated as 1 in = 2000 ft.

$$(2000\ \text{ft})(0.005) = 10\ \text{ft}$$

The answer is (C).

SOLUTION 62

Finding the difference in elevation between the monument set at the center of a pre-mark panel point and the panel material itself is sometimes neglected. The panel is the basis of the photogrammetrist's determination of elevation. If its elevation differs from that of the monument, the difference should be known.

The answer is (D).

SOLUTION 63

The national map standards call for vertical accuracy to be within one-half of the contour interval shown on the map. A more typical standard for vertical control in photogrammetry is one-fifth of the contour interval.

The answer is (D).

SOLUTION 64

Picture points are generally natural features that are well defined on aerial photographs under magnification. They must be accessible and capable of positive identification in the field as well. When such natural features fall in an appropriate position on the aerial photographs, picture points are frequently used to augment artificially set pre-marked photogrammetric control.

The answer is (C).

Geodetic and Control Surveys

NORTH AMERICAN DATUMS

PROBLEM 1

Which of the following statements correctly describes the origin of the North American Datum of 1927 (NAD27)?

(A) The origin of NAD27 is at Meades Ranch in Kansas.

(B) The origin of NAD27 is geocentric.

(C) The geoidal height of the origin of NAD27 is assumed to be zero.

(D) Both A and C are true.

PROBLEM 2

Which of the following statements correctly describes the North American Datum of 1983 (NAD83)?

(A) The origin of NAD83 is intended to be geocentric.

(B) NAD83 is the horizontal control datum used throughout the western hemisphere.

(C) NAD83 is no longer used in the United States.

(D) Both A and B are true.

PROBLEM 3

NAD83 uses a figure to represent the earth that closely approximates its shape, but may be described mathematically by a few definite parameters. The shape is a

(A) biaxial ellipsoid rotated about its minor axis

(B) mean earth ellipsoid

(C) biaxial ellipsoid rotated about its major axis

(D) triaxial ellipsoid

PROBLEM 4

Which of the following statements correctly characterizes the transformation of coordinates from NAD27 to NAD83?

(A) There is not one simple, consistent method to reliably transform coordinates from NAD27 to NAD83.

(B) The transformation of ellipsoidal coordinates in terms of $\Delta\phi$, $\Delta\lambda$, and Δh from NAD27 to NAD83 can be accurately accomplished with a simple, one-step mathematical formula, but the transformation of Cartesian coordinates cannot.

(C) The transformation of Cartesian coordinates in terms of Δx, Δy, and Δz from NAD27 to NAD83 can be accurately accomplished with a simple, one-step mathematical formula, but transformation of ellipsoidal coordinates cannot.

(D) The transformation of state plane coordinates expressed in NAD27 to NAD83 involves only a change in the origin of the coordinates.

PROBLEM 5

Flattening and the length of the semimajor axis are two parameters often used to define the reference ellipsoids associated with geodetic datums. What are the dimensions used to define the reference ellipsoids associated with NAD27 and NAD83?

	datum	reference ellipsoid	semimajor	flattening
(A)	NAD83	GRS 80	6,378,137.0 m	$\frac{1}{298.257222101}$
	NAD27	Clarke 1866	6,378,206.4 m	$\frac{1}{294.9786982}$
(B)	NAD83	Fischer 1960	6,378,166.0 m	$\frac{1}{298.36714202}$
	NAD27	Clarke 1858	6,378,361.4 m	$\frac{1}{294.26982312}$
(C)	NAD83	WGS 84	6,378,137.0 m	$\frac{1}{298.257223563}$
	NAD27	Clarke 1866	6,378,206.4 m	$\frac{1}{294.9786982}$
(D)	NAD83	WGS 72	6,378,135.0 m	$\frac{1}{298.2623209}$
	NAD27	Clarke 1858	6,378,206.4 m	$\frac{1}{294.9786982}$

PROBLEM 6

What is the geoid?

(A) a surface defined by gravity

(B) a surface that is flattened without tearing or stretching

(C) a surface that would closely resemble the oceans' surfaces if they were completely still

(D) Both A and C are true.

PROBLEM 7

How is the flattening, f, of a reference ellipsoid defined in terms of the semimajor axis, a, and the semiminor axis, b?

(A) $f = \frac{b-a}{b}$

(B) $f = \frac{a-b}{a}$

(C) $f = b - a$

(D) $f = \frac{a+b}{b}$

PROBLEM 8

What is the cause of geoidal undulation?

(A) The presence of magnetically attractive ores within the earth cause the geoid to undulate.

(B) The center of mass of the earth does not coincide with the center of the geoid.

(C) The ellipsoid is not coincident with the geoid.

(D) The mass of the earth is irregularly distributed and, therefore, the geoid is also irregular.

PROBLEM 9

Which of the following statements about NGVD29 is NOT true?

(A) It was first called the Sea Level Datum of 1929.

(B) It involved more than 100,000 km of leveling.

(C) It represents mean sea level.

(D) none of the above

PROBLEM 10

In considering surveys of a very large extent, can a level surface at mean sea level be considered parallel with a level surface 100.0 m above mean sea level?

(A) No, a level surface is perpendicular everywhere to the direction of gravity.

(B) Yes, every level surface is parallel to every other level surface.

(C) No, level surfaces that are 100.0 m apart at the equator would be approximately 99.5 m apart at the poles, so they cannot be considered parallel.

(D) Both A and C are true.

PROBLEM 11

Which of the following statements about NAVD88 is NOT true?

(A) It included redoing over 50,000 mi of leveling.

(B) It addressed benchmarks across the United States.

(C) Its elevations are the same as NGVD29.

(D) none of the above

PROBLEM 12

Proceeding northward from the equator, which of the following will NOT occur?

(A) The acceleration of gravity increases.

(B) Meridians converge.

(C) Latitudinal lines are parallel.

(D) The distance along the earth's surface represented by a degree of latitude remains constant.

PROBLEM 13

If the numbers to the right of the decimal are significant, which of the following is the nearest to the actual precision of the following coordinate?

$$\phi = 39°30'15.3278''\text{ N}$$
$$\lambda = 109°54'11.1457''\text{ W}$$

(A) ± 0.01 ft

(B) ± 0.10 ft

(C) ± 1.0 ft

(D) ± 10.0 ft

PROBLEM 14

Which statement about biaxial and triaxial ellipsoids is correct?

(A) The reference ellipsoid for NAD83 is triaxial and the reference ellipsoid for NAD27 is biaxial.

(B) The length of the semimajor axis is constant in both a biaxial ellipsoid and a triaxial ellipsoid.

(C) The equator of a triaxial ellipsoid is elliptical; the equator of a biaxial ellipsoid is a circle.

(D) Both triaxial and biaxial ellipsoids have flattening at the equator.

PROBLEM 15

What constitutes the *realization* of a geodetic datum?

(A) the assignment of a semimajor axis and another parameter, such as the first eccentricity, flattening, and/or semiminor axis

(B) the monumentation of the physical network of reference points on the earth's surface that is provided with known coordinates in the subject datum

(C) an initial point at a known latitude and longitude, and a geodetic azimuth from there to another known position, along with two parameters of the associated ellipsoid, such as its semimajor and semiminor axes

(D) the transformation of its coordinates into another datum using the seven parameters of the Bursa-Wolfe process

PROBLEM 16

Which of the following statements about reference ellipsoids is NOT correct?

(A) The semimajor axis and flattening can be used to completely define an ellipsoid.

(B) The Clarke 1866 spheroid is the reference ellipsoid of the North American Datum of 1927 (NAD27), but it is not the datum itself.

(C) NAVD is the reference ellipsoid of the North American Datum of 1983 (NAD83).

(D) Six elements are required if a particular reference ellipsoid is to be used in a geodetic datum.

PROBLEM 17

Which of these was NOT intended to be a geocentric datum?

(A) NAD27

(B) Australian Geodetic Datum 1966

(C) South American Datum 1969

(D) all of the above

PROBLEM 18

Which of the following are aspects of a good High Accuracy Reference Networks (HARNs) station?

(A) inclusion in an NGS sanctioned readjustment

(B) vehicle accessibility

(C) good overhead visibility

(D) all of the above

PROBLEM 19

Which of the following statements about the geoid is NOT true?

(A) The mean sea level (MSL) and the geoid are not equivalent.

(B) It does not precisely correspond to the topography of the earth's dry land.

(C) It is bumpy.

(D) It never departs from true ellipsoidal form.

MAP PROJECTIONS

PROBLEM 20

Which of the following best defines map projection?

(A) the systematic transformation of state plane coordinates into Cartesian coordinates

(B) the rigorous mathematical transference of points on a spheroid to the points on a plane surface

(C) the function of relating an ellipsoid of revolution to the geoid

(D) the mathematical development of the graticule representing the geographical lines on a globe

PROBLEM 21

Which of the following map projections is/are used as the basis of state plane coordinate systems in the United States?

(A) the Lambert projection

(B) the oblique Mercator projection

(C) the transverse Mercator projection

(D) all of the above

PROBLEM 22

A developable surface is best described as a

(A) portion of the earth's surface that corresponds to mean sea level

(B) polar gnomonic map projection

(C) plane figure upon which the information from a spheroid may be projected and then flattened without gross distortion or tearing

(D) surface that a map of a portion of the earth's surface may be drawn upon without any distortion

PROBLEM 23

What characteristic of a map projection preserves the shape of small figures from the ellipsoid to the map?

(A) propagation

(B) standardization

(C) spatiality

(D) conformality

PROBLEM 24

A rhumb line, not a cardinal line, has been drawn between two points on a map produced from a Mercator projection. Does the line represent the shortest distance between those two points on the earth?

(A) No, while the rhumb line would appear straight on a Mercator projection, it would not follow the arc of a great circle.

(B) Yes, the described loxodrome would follow the shortest path between two points on a conformal projection.

(C) No, the rhumb line would follow the arc of a great circle on the Mercator projection and therefore would have a constant geodetic direction.

(D) Yes, the rhumb line would follow the arc of a great circle on the Mercator projection.

PROBLEM 25

Which of the following curves appears as a straight line on a gnomonic map projection?

(A) the parallels of latitude

(B) rhumb lines

(C) the arcs of great circles

(D) the arcs of small circles

PROBLEM 26

In the State Plane Coordinate System 1983, also known as SPCS83, the convergence is symbolized by which of the following?

(A) phi, ϕ

(B) omega, ω

(C) mu, μ

(D) gamma, γ

PROBLEM 27

Which of the following statements does NOT properly describe a difference between SPCS27, the State Plane Coordinate System based on NAD27, and SPCS83, the State Plane Coordinate System based on NAD83?

(A) There have been some changes in the distances between the SPCS zones and their origins.

(B) SPCS83 coordinates are expressed in more units than were SPCS27 coordinates.

(C) There are fewer SPCS83 zones than there were SPCS27 zones in some states.

(D) For the first time, the scale factor limit of 1 part in 10,000 changed in SPCS83.

THE LAMBERT PROJECTION

PROBLEM 28

Which of the following terms may be correctly applied to the Lambert map projection that is used as the foundation for state plane coordinates in the United States?

(A) conformal and conic

(B) azimuthal and equal-area

(C) cylindrical and conformal

(D) conic and sinusoidal

PROBLEM 29

The SPCS83 Lambert projection employs the Greek letter gamma, γ, as the symbol for what important quantity?

(A) scale factor

(B) standard parallel

(C) central meridian

(D) mapping angle

PROBLEM 30

The mapping plane used in the Lambert projection is often called a *secant cone*. Which of the following characteristics describes such a developable surface?

(A) The central axis of the cone is imagined to be a projection of the polar axis of the earth.

(B) The bottom of the cone is imagined to terminate at the earth's equator.

(C) The relationship between the altitude of the cone and its radius is always 1.55.

(D) The cone is imagined to intersect the ellipsoid along two standard parallels.

PROBLEM 31

In a Lambert projection, the scale factor varies with which of the following quantities?

(A) longitude

(B) elevation

(C) latitude

(D) mapping angle

PROBLEM 32

The Lambert projections used as the basis of state plane coordinates in the United States employ two lines of strength, or standard parallels. The scale along a standard parallel in such a projection is

(A) the smallest in the zone of a Lambert projection

(B) variable in the Lambert projection

(C) the largest in the zone of a Lambert projection

(D) 1 in a Lambert projection

PROBLEM 33

Meridians of longitude and parallels of latitude are represented differently in the Lambert projection. Which of the following statements best describes the difference between these two categories of lines?

(A) Meridians of longitude are represented by straight lines converging at the pole; parallels of latitude are represented by the arcs of concentric circles.

(B) Meridians of longitude are represented by the arcs of concentric circles; parallels of latitude are represented by straight lines.

(C) Parallels of latitude are represented by complex curves; meridians of longitude are represented by unequally spaced straight lines.

(D) Parallels of latitude are represented by curves concave toward the pole; meridians of longitude are represented by complex curves concave toward the central meridian.

PROBLEM 34

How many states use the Lambert conformal projection exclusively as the basis of their state plane coordinate systems?

(A) 18

(B) 20

(C) 29

(D) 33

PROBLEM 35

State plane coordinate systems based on Lambert conformal projections have been designed to minimize the distortion of grid lengths with respect to geodetic lengths. Where coverage by a single grid with a latitudinal extent of 158 statute miles is possible, which ratio most nearly represents the maximum difference between the grid and geodetic lengths of a line?

(A) 1:5000

(B) 1:10,000

(C) 1:40,000

(D) 1:80,000

PROBLEM 36

The following two problems refer to the illustration shown.

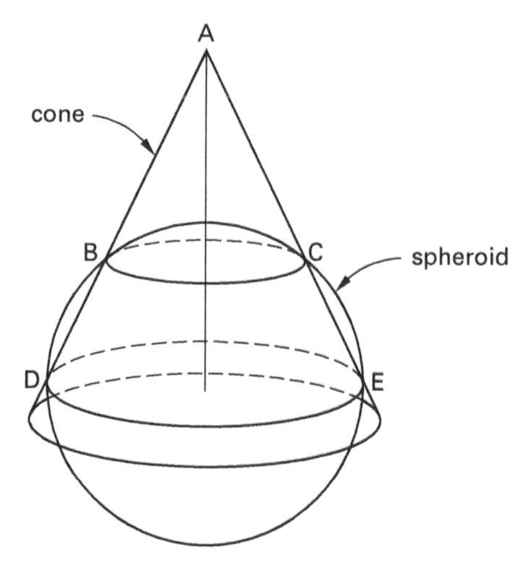

36.1. Only part of the cone represented in the illustration is actually used in the developed Lambert projection. What is the typical portion included in a zone?

(A) Three-fifths of the zone is made up of the area of the cone within the spheroid—that is, between the standard parallels. One-fifth of the zone is the area north of the northerly standard parallel, and one-fifth is south of the southerly standard parallel.

(B) Only the portion of the cone within the spheroid is included in the zone.

(C) All of the area of the cone outside of the standard parallels BC and DE is included in the zone.

(D) Four-sixths of the zone is made up of the area of the cone within the spheroid—that is, between the standard parallels. One-sixth of the zone is the area north of the northerly standard parallel (BC) and one-sixth is south of the southerly standard parallel (DE).

36.2. What is the usual alignment of the axis of the cone in the illustration?

(A) perpendicular to the plane of the ecliptic

(B) coincident with an extension of the earth's axis of rotation

(C) arbitrary and has no typical arrangement

(D) coincident with an extension of the earth's magnetic pole

THE TRANSVERSE MERCATOR PROJECTION

PROBLEM 37

Two mapping projections have been used in the development of nearly all of the state plane coordinate systems in the United States; one is the Lambert projection and the other is the transverse Mercator projection. What is the significance of the word *transverse* in the name of the transverse Mercator projection?

(A) It refers to the characteristic representation of a line of constant azimuth on the spheroid as a straight line.

(B) It is used to indicate that the axis of the cylinder employed in the projection is perpendicular to the axis of the spheroid.

(C) It is used because the only straight line on the projection is the central meridian.

(D) The cylindrical surface used in the transverse Mercator projection is tangent to the spheroid at the equator.

PROBLEM 38

Which of the following statements describes the treatment of scale in the Lambert and the transverse Mercator projections?

(A) The Lambert projection scale is variable with respect to changes in longitude; in the transverse Mercator projection, it varies with respect to latitude.

(B) In the transverse Mercator projection, the lines of exact scale are meridians of longitude; in the Lambert projection, they are called *standard parallels of latitude.*

(C) The transverse Mercator projections used as the foundation of state plane coordinate systems are designed to limit scale distortion to 1 part in 20,000; in Lambert-based systems, distortion is limited to 1 part in 10,000.

(D) The scale is constant up and down the central meridian of a transverse Mercator projection; it is variable along the central meridian of a Lambert projection.

PROBLEM 39

The grid imposed on a transverse Mercator projection indicating a particular NAD27 state plane coordinate system is composed of straight lines. However, the geographical lines—that is, the parallels of latitude and meridians of longitude—are not straight. What symbol is used in SPCS83 to indicate the angle between grid north through a particular point and the meridian of longitude through the same point?

(A) ϕ

(B) θ

(C) γ

(D) λ

PROBLEM 40

The cylinder is the developable surface employed in the transverse Mercator projection. Which of the following statements correctly describes the configuration of the cylinder used in the state plane coordinate systems based on the transverse Mercator projection?

(A) The axis of the cylinder is imagined to be perpendicular to the rotational axis of the spheroid; it intersects the spheroid along two meridians of longitude.

(B) The axis of the cylinder is imagined to be perpendicular to the rotational axis of the spheroid; it intersects the spheroid along two ellipses that are equally distant and parallel with the plane of the central meridian.

(C) The axis of the cylinder is imagined to be perpendicular to the rotational axis of the spheroid; it is tangent to the spheroid along the central meridian.

(D) The axis of the cylinder is imagined to be perpendicular to the rotational axis of the spheroid; it is tangent to the spheroid along the equator.

PROBLEM 41

Which of the following statements correctly describes some of the features of the projection known as the Universal Transverse Mercator (UTM)?

(A) UTM employs 60 zones; each includes 6° of longitude.

(B) The foundation of UTM is the Jacobi reference ellipsoid; the coordinates in each zone are expressed in meters.

(C) The foundation of UTM is WGS84.

(D) Both B and C are true.

PROBLEM 42

Which of the following does NOT correctly state a difference between the Transverse Mercator coordinate system used in SPCS and the Universal Transverse Mercator (UTM) coordinate system?

(A) The UTM system of coordinates was designed and established by the United States Army in cooperation with NATO member nations. The Transverse Mercator SPCS was established by the United States Coast and Geodetic Survey.

(B) The official unit of measurement in the UTM system of coordinates is the meter. The official unit in Transverse Mercator SPCS83 is always the international foot.

(C) The UTM system of coordinates is augmented by the Universal Polar Stereographic (UPS) to complete its worldwide coverage. The Transverse Mercator SPCS does not have such a broad scope.

(D) The scale factor in the UTM system of coordinates can reach 1 part in 2500. In the Transverse Mercator SPCS, the scale factor is usually no more than 1 part in 10,000.

PROBLEM 43

Which of the following heights is NOT measured along a line perpendicular to the ellipsoid?

(A) ellipsoidal height

(B) geoid height

(C) geodetic height

(D) orthometric height

PROBLEM 44

Which of the following parameters is NOT usually paired with the semimajor axis to define a reference ellipsoid?

(A) semiminor axes

(B) reciprocal of the flattening

(C) eccentricity

(D) reciprocal of the eccentricity

STATE PLANE COORDINATE SYSTEMS

PROBLEM 45

The following five problems refer to the following illustration, which shows the geometrical relationship between some of the significant lines comprising a state plane coordinate grid based on a transverse Mercator projection.

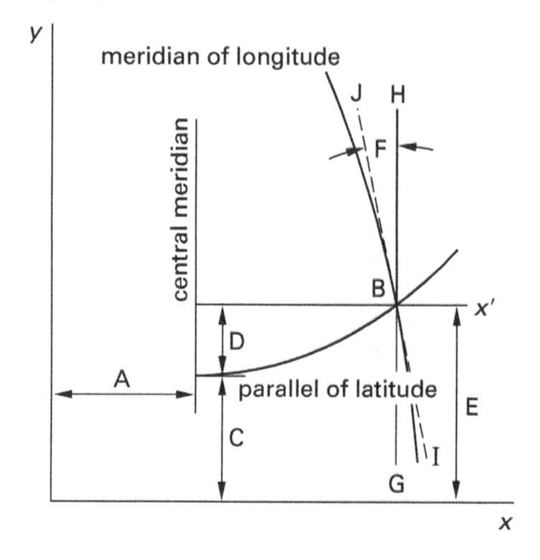

45.1. Considering SPCS27 state plane coordinate positions derived from a transverse Mercator projection, the arbitrary constant represented by the letter A is most nearly

(A) 150,000 m

(B) 300,000 ft

(C) 500,000 ft

(D) 700,000 ft

45.2. Considering SPCS83 state plane coordinate positions derived from a transverse Mercator projection, what arbitrary constant, represented by the letter A, will be used?

(A) 100,000 m, 300,000 m, and 900,000 m

(B) 200,000 m, 400,000 m, 600,000 m, and 800,000 m

(C) 213,360 m, 850,000 m, and 165,000 m

(D) all of the above and more

45.3. Line GH is parallel with the central meridian; the dashed line IJ is not parallel with the central meridian. Which of the following statements would be correct for an observer standing at station B and facing geodetic north?

(A) The observer would find line IJ on the left of geodetic north and line GH on the right.

(B) The observer would find line GH on the left of geodetic north and line IJ on the right.

(C) The observer would find both line GH and line IJ on the left of geodetic north.

(D) The observer would be looking along line IJ and line GH would be on the right.

45.4. Considering SPCS83, which of the following labels are commonly used for the distances labeled A and E?

(A) A = central meridian false easting
E = northing coordinate

(B) $A = \pm ab$
$E = \Delta\lambda$

(C) $A = H$
$E = y_o$

(D) A = northing coordinate
$E = \Delta\phi$

45.5. Considering SPCS83, which of the following expressions is correct for the angular value symbolized by the letter F?

(A) $\pm\gamma$

(B) γ'

(C) $+\gamma$

(D) $-\gamma$

PROBLEM 46

What approximation of the radius of the reference ellipsoid has traditionally been used as the basis for calculating the sea-level factor?

(A) 6,378,000 m

(B) 20,906,000 ft

(C) 20,925,000 ft

(D) 24,000,000 ft

PROBLEM 47

The following six problems refer to the illustration shown.

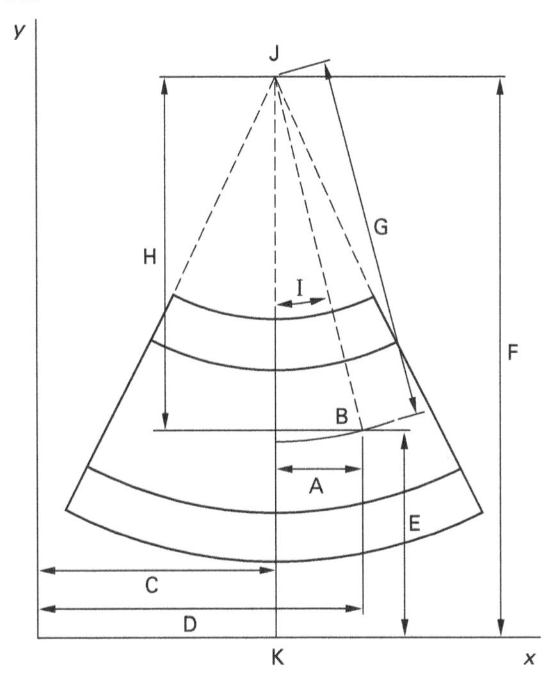

The illustration shows the geometrical relationship between some of the significant lines comprising a state plane coordinate grid based upon a Lambert projection.

47.1. Considering SPCS27 state plane coordinate positions derived from a Lambert projection, what is the arbitrary constant represented by the letter C?

(A) 100,000 ft

(B) 500,000 ft

(C) 1,000,000 m

(D) 2,000,000 ft

47.2. Considering SPCS83, which of the following symbols most completely describes the angular value labeled I?

(A) $-\gamma$

(B) $+\gamma$

(C) γ

(D) γ'

47.3. Which of the following expressions is commonly used to represent the distance labeled A in SPCS83?

(A) $R \sin \gamma$

(B) $R \cos \gamma$

(C) R_b

(D) both A and B

47.4. Using the most typical symbols for SPCS83, which of the formulas given is equivalent to F – E = H?

(A) $R - \text{E} = N$

(B) $R \cos \gamma - N = R_b$

(C) $P - R \sin \gamma = \text{C}$

(D) $R_b - N = R \cos \gamma$

47.5. What is the name usually given to the line labeled JK in the illustration?

(A) the y-axis

(B) ϕ_o

(C) the central meridian

(D) λ_1

47.6. The distance labeled D represents the east coordinate of station B, and the distance labeled E is its north coordinate. Which of the following formulas correctly represents these coordinates in SPCS83 terms?

(A) east $= 2{,}000{,}000 - x$

north $= R_b + N_b - R \cos \gamma$

(B) east $= \text{E}_o + R \sin \theta$

north $= R_b + N_b - R \cos \gamma$

(C) east $= 2{,}000{,}000 + x'$

north $= R \cos \gamma - R_b + N_b$

(D) east $= \text{E}_o + R \sin \gamma$

north $= R_b + N_b - R \cos \gamma$

PROBLEM 48

Some elements are constant for each zone of NAD27 state plane coordinates based on a Lambert projection, and some vary with each station. Which elements are constant?

(A) Each zone has a unique origin, a constant R_b, and a distance of 2,000,000 ft from the y-axis to the central meridian.

(B) Each zone has a unique central meridian, a constant x', and a distance of 2,000,000 ft from the y-axis to the central meridian.

(C) Each zone has a unique central meridian, a constant R, and a distance of 2,000,000 ft from the x-axis to the central meridian.

(D) Each zone has a unique central meridian, a constant R_b, and a distance of 2,000,000 ft from the x-axis to the central meridian.

PROBLEM 49

The following lists are made up of variables from NAD27 state plane coordinate systems (SPCS27) based on the Lambert projection and the transverse Mercator projection. Which lists show only those variables that vary with latitude?

(A) Lambert: $\Delta\alpha$, H, V and a

transverse Mercator: θ, R, scale factor, and b

(B) Lambert: ℓ, x', and θ

transverse Mercator: b, c, and y_o

(C) Lambert: R, scale factor, and θ

transverse Mercator: y_o, H, V, and a

(D) Lambert: y_o, H, V, and a

transverse Mercator: R, scale factor, and θ

PROBLEM 50

Most state plane coordinate systems are designed to limit scale distortion to 1 part in 10,000. If this goal has been achieved, within what limits should the scale factor fall?

(A) not less than 0.9994563;

not more than 1.0051020

(B) not less than 0.9999000;

not more than 1.0001000

(C) not less than 0.9999569;

not more than 1.0000839

(D) not less than 0.9999823;

not more than 1.0001787

PROBLEM 51

Between the standard parallels of a state plane coordinate zone based on the Lambert projection, the scale factor remains within what specific limits?

(A) The scale factor is always greater than 1 between the standard parallels.

(B) The scale factor is always greater than 1 and less than 2 between the standard parallels.

(C) The scale factor is always greater than zero and less than 1 between the standard parallels.

(D) The scale factor is always less than zero between the standard parallels.

PROBLEM 52

In SPCS27 state plane coordinate systems derived from a transverse Mercator projection, what value is the scale factor based on?

(A) $\Delta\lambda''$

(B) $\Delta\phi$

(C) y-coordinate

(D) x'

PROBLEM 53

Considering a state plane zone based on the transverse Mercator projection, the scale factor outside the lines of exact scale is always

(A) greater than 1 outside the lines of exact scale

(B) 1 outside the lines of exact scale

(C) less than zero outside the lines of exact scale

(D) greater than zero and less than 1 outside the lines of exact scale

PROBLEM 54

Which factor is used to directly reduce distances measured on the topographic surface of the earth to the state plane?

(A) the scale factor

(B) the K-factor

(C) the sea-level factor

(D) the combination factor

PROBLEM 55

It is sometimes appropriate to apply a weighted mean scale factor to a line of 5 mi or more. Which of the following circumstances would warrant such a refinement?

(A) a long, predominantly east-west line on the Lambert projection or a long, predominantly north-south line on the transverse Mercator projection

(B) a long, predominantly east-west line on the transverse Mercator projection or a long, predominantly north-south line on the Lambert projection

(C) a long line near the center of the zone on the transverse Mercator projection and a similar line on the Lambert projection

(D) a long line near the edge of the zone on the transverse Mercator projection and a similar line on the Lambert projection

PROBLEM 56

There is a parallel of latitude known as ϕ_o and B_o in each zone based on the Lambert projection. Which of the following statements correctly identifies some of the special characteristics of this parallel of latitude?

(A) All of the geodetic lines in the zone are slightly concave toward this parallel of latitude.

(B) The scale factor ratio is lowest along this parallel of latitude.

(C) The latitude of this line is equidistant from each of the standard parallels.

(D) Both A and B are true.

PROBLEM 57

Which of the following symbols is normally used to indicate the second-term correction applied to the direction of long lines in the SPCS83 Lambert state plane coordinate system?

(A) ω

(B) δ

(C) $\Delta\alpha'$

(D) σ

PROBLEM 58

What is the formula used to convert a geodetic azimuth to a grid azimuth in the Lambert conformal projection when a relatively short line is involved?

(A) grid azimuth = geodetic azimuth + θ

(B) grid azimuth = geodetic azimuth − θ

(C) grid azimuth = θ − geodetic azimuth

(D) grid azimuth = geodetic azimuth + $\theta - \theta'$

PROBLEM 59

Which of the following expressions is equal to the mapping angle in the transverse Mercator projection?

(A) $x' + 500{,}000$

(B) $y_o + V\left(\frac{\Delta\lambda''}{100}\right)^2 \pm c$

(C) $H\Delta\lambda'' \pm ab$

(D) $\Delta\lambda'' \sin\phi + g$

PROBLEM 60

What is the relationship between the mapping angle, γ, and the distance from the central meridian for a station in a zone of a state plane coordinate system based upon the Lambert projection?

(A) The absolute value of γ is always larger than the difference in longitude from the central meridian.

(B) γ is not always the same for a specific meridian of longitude.

(C) The absolute value of γ is always the same as the difference in longitude from the central meridian.

(D) The absolute value of γ is always smaller than the difference in longitude from the central meridian.

PROBLEM 61

Considering SPCS27, which of the following expressions is equal to the mapping angle, θ, in a state plane coordinate system based on the Lambert projection?

(A) $\Delta\lambda\ell$

(B) $\tan^{-1}\left(\frac{x'}{R_b - y}\right)$

(C) $\sin^{-1}\left(\frac{R_b - y}{R}\right)$

(D) both A and B

SHORT GEODETIC CALCULATIONS

PROBLEM 62

The geodetic azimuth (from south) from station Telo to station Klamath is 186°15′22.21″, and the mapping angle at station Telo is +00°40′21.85″. What is the grid azimuth from station Telo to station Klamath if the second term is not considered?

(A) 05°35′00.36″

(B) 06°55′44.06″

(C) 185°35′00.36″

(D) 186°55′44.06″

PROBLEM 63

The grid azimuth from station Exeter to station Abott is 212°32′14.1″, and the mapping angle at station Exeter is +00°27′45.2″. What is the geodetic azimuth from station Abott to station Exeter if the second term is not considered?

(A) 32°59′59.3″

(B) 212°04′28.9″

(C) 212°59′59.3″

(D) insufficient information given

PROBLEM 64

The following seven problems refer to the illustration shown. These problems are based on State Plane Coordinates 1927 (NAD27).

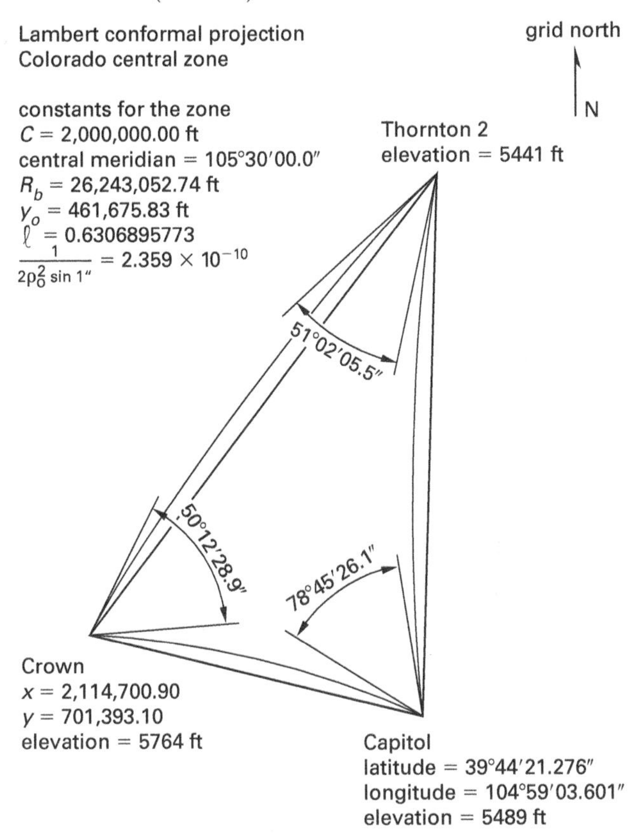

(not to scale)

Note that the concave lines connecting the three stations indicate the curvature of long lines, such as those in the illustration, on the state plane. The angles shown are observed values.

64.1. The following values are taken from the *Plane Coordinate Projection Tables—Colorado*, published by the U.S. Government Printing Office.

latitude	R (ft)	tabular difference for 1″ of latitude (ft)
39°41′	25,569,283.24	101.18217
42′	25,563,212.31	101.18283
43′	25,557,141.34	101.18350
44′	25,551,070.33	101.18400
45′	25,544,999.29	101.18467

What is the appropriate R value for the station labeled Capitol?

(A) 25,541,081.04 ft

(B) 25,547,152.08 ft

(C) 25,548,917.54 ft

(D) 25,554,988.58 ft

64.2. What is the x-coordinate of station Capitol?

(A) 145,021.38 ft

(B) 2,145,021.38 ft

(C) 25,548,507.95 ft

(D) 27,548,505.95 ft

64.3. What is the value of θ for the station labeled Capitol?

(A) −00°30′56.39″

(B) −00°20′10.92″

(C) +00°19′30.81″

(D) +00°19′35.35″

64.4. What is the y-coordinate of station Capitol?

(A) 609,803.31 ft

(B) 687,312.94 ft

(C) 694,546.79 ft

(D) 706,091.46 ft

64.5. What is the angular spherical excess in the figure created by Crown, Capitol, and Thornton 2?

(A) 0.2 sec

(B) 0.8 sec

(C) 1.2 sec

(D) 3.1 sec

64.6. What is the second-term, or θ', correction that would be applied at station Crown for the line Crown-Capitol?

(A) −01.2″

(B) −00.4″

(C) +00.9″

(D) +01.7″

64.7. What is the second-term, or θ', correction that would be applied at station Crown for the line Capitol-Crown?

(A) −01.7″

(B) −0.08″

(C) +01.7″

(D) +01.9″

PROBLEM 65

In the following illustration, the concave lines connecting the three stations indicate the curvature of long lines on the state plane. (This angular work is done in the symbology typical of State Plane Coordinates 1983 (NAD83).) The angles shown are observed values. The angular values circled are the second-term, or δ, values.

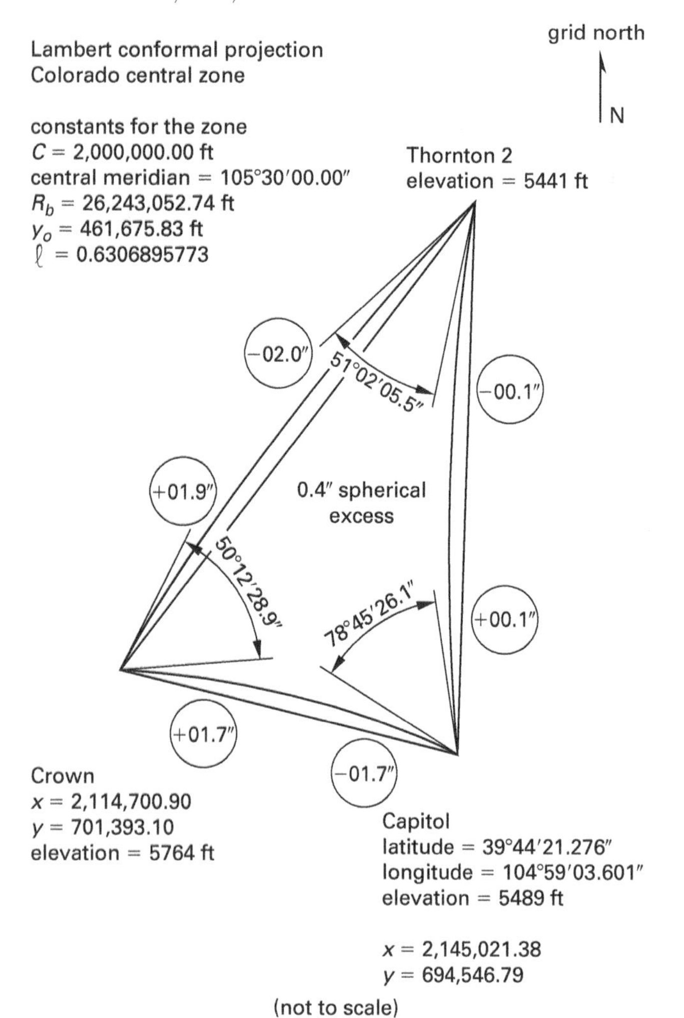

Given this information, what are the correct values for the three interior angles of the figure after both second terms and spherical excess have been taken into consideration?

(A) 51°02′02.3″
78°45′26.7″
50°12′27.5″

(B) 51°02′03.5″
78°45′27.8″
50°12′28.6″

(C) 51°02′03.6″
78°45′28.0″
50°12′28.8″

(D) 51°02′05.5″
78°45′26.1″
50°12′29.0″

PROBLEM 66

Note that this problem is based on State Plane Coordinates 1927 (NAD27). Given the coordinates of Capitol and Crown, what is the geodetic azimuth from station Capitol to station Crown as shown in the following illustration?

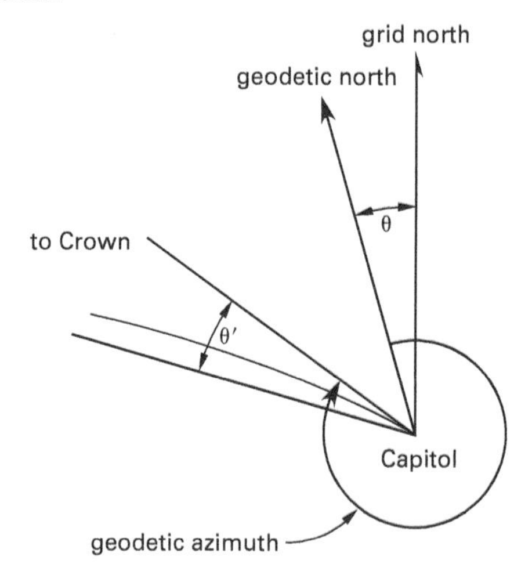

(A) 257°16′35.7″

(B) 257°36′06.5″

(C) 282°43′26.1″

(D) 283°02′58.6″

PROBLEM 67

Given the value of ℓ for the Colorado central zone of 0.6306895773, what is the parallel of latitude for that zone along which the scale factor ratio is smallest?

(A) 38°27′00.00″

(B) 39°06′03.66″

(C) 39°33′21.53″

(D) 39°45′00.00″

PROBLEM 68

Please note that this problem is based on State Plane Coordinates 1927 (NAD27). The state plane coordinates of Thornton 2 in the Colorado central zone are

$$x = 2{,}146{,}262.93$$
$$y = 742{,}549.22$$

The constants for the zone are

$$C = 2{,}000{,}000 \text{ ft}$$
$$\text{central meridian} = 105°30'00''$$
$$R_b = 26{,}243{,}052.74 \text{ ft}$$
$$y_o = 461{,}675.83$$
$$\ell = 0.6306895773$$

The applicable tabular values are

latitude	R (ft)	tabular difference for 1″ of latitude (ft)
39°51′	25,508,572.22	101.18867
52′	25,502,500.90	101.18933
53′	25,496,429.54	101.19000
54′	25,490,358.14	101.19083
55′	25,484,286.69	101.19133

What is the latitude of station Thornton 2?

(A) 39°48′20.99″

(B) 39°51′37.41″

(C) 39°52′15.59″

(D) 39°54′29.78″

PROBLEM 69

This problem is based on State Plane Coordinates 1927 (NAD27). The state plane coordinates of Thornton 2 in the Colorado central zone are

$$x = 2{,}146{,}262.93$$
$$y = 742{,}549.22$$

The constants for the zone are

$$C = 2{,}000{,}000 \text{ ft}$$
$$\text{central meridian} = 105°30'00''$$
$$R_b = 26{,}243{,}052.74 \text{ ft}$$
$$y_o = 461{,}675.83$$
$$\ell = 0.6306895773$$

The value of θ for the station is +00°19′43.06″.

What is the longitude of Thornton 2?

(A) 104°32′12.09″

(B) 104°44′21.87″

(C) 104°56′55.99″

(D) 104°58′44.18″

PROBLEM 70

This problem is based on State Plane Coordinates 1927 (NAD27). The elevations and latitudes of Thornton 2 and 400 for the Colorado central zone are as follows.

For Thornton 2,

$$\phi = 39°52'15.594''$$
$$\text{elev} = 5441 \text{ ft}$$

For 400,

$$\phi = 39°50'56.090''$$
$$\text{elev} = 5246 \text{ ft}$$

The applicable tabular information is

ϕ	R (ft)	tabular difference 1″ ϕ (ft)	scale as a ratio
39°46′	25,538,928.21	101.18533	1.0000033
47′	25,532,857.09	101.18600	1.0000068
48′	25,526,785.93	101.18667	1.0000103
49′	25,520,714.73	101.18717	1.0000139
50′	25,514,643.50	101.18800	1.0000175
39°51′	25,508,572.22	101.18867	1.0000213
52′	25,502,500.90	101.18933	1.0000252
53′	25,496,429.54	101.19000	1.0000291
54′	25,490,358.14	101.19083	1.0000331
55′	25,484,286.69	101.19133	1.0000372

The distance measured between the two stations is 10,206.50 ft.

What is the grid distance between Thornton 2 and 400?

(A) 10,203.89 ft

(B) 10,204.13 ft

(C) 10,204.92 ft

(D) 10,206.74 ft

GOVERNMENT ADMINISTRATION OF STATE PLANE COORDINATE SYSTEMS

PROBLEM 71

Which agency established the design of the state plane coordinate systems used in the United States?

(A) U.S. Geological Survey

(B) U.S. Coast and Geodetic Survey

(C) Bureau of Land Management

(D) Department of the Interior

PROBLEM 72

What organization is ultimately responsible for legalization of the components of the various state plane coordinate systems?

(A) National Geodetic Survey

(B) U.S. Geological Survey

(C) Federal Board of Surveys and Maps

(D) legislature of each state

PROBLEM 73

Where are state plane coordinate systems available?

(A) Every state and possession of the United States has a state plane coordinate system.

(B) Thirty-seven states have state plane coordinate systems.

(C) Every state except Hawaii has a state plane coordinate system.

(D) Every public lands state has a state plane coordinate system.

PROBLEM 74

Which of the following organizations publishes geographical and state plane coordinates for its national network of geodetic control stations?

(A) Bureau of Land Management

(B) Bureau of Reclamation

(C) Federal Geodetic Control Committee

(D) National Geodetic Survey

STANDARDS

PROBLEM 75

In the geometric relative positioning accuracy standards for three-dimensional surveys, using space system techniques in the *Geometric Geodetic Accuracy Standards and Specifications for the Use of GPS Relative Positioning Techniques* published by the FGCC in 1988, the previously established first-, second-, and third-order classifications are found grouped under one of four primary orders. Which of the new primary orders in this publication includes these older classifications?

(A) order AA

(B) order A

(C) order B

(D) order C

PROBLEM 76

Which of the following organizations publishes standards and specifications for the surveying of geodetic control networks?

(A) Federal Board of Surveys and Maps

(B) Bureau of Reclamation

(C) Federal Geographic Data Committee

(D) U.S. Forest Service

PROBLEM 77

Which of the following are parts of the U.S. Federal Geographic Data Committee (FGDC) Geospatial Positioning Accuracy Standards?

I. Reporting Methodology and Standards for Geodetic Networks

II. National Standard for Spatial Data Accuracy and Standards for Nautical Charting Hydrographic Surveys

III. Architecture, Engineering, Construction, and Facilities Management

IV. First-order; Second-order, class I; Second-order, class II; Third-order, class I; Third-order, class II

(A) I and II only

(B) II and IV only

(C) IV only

(D) I, II, and III only

SOLUTION 1

The North American Datum of 1927 (NAD27) is based on ellipsoidal parameters defined as the Clarke Spheroid of 1866.

A readjustment of the national network done to introduce Laplace azimuths yielded the geodetic positions now available on NAD27. The origin of that readjustment is designated as

$$\phi = 39^\circ 13' 26.686''\text{N}$$
$$\lambda = -98^\circ 32' 30.506''\text{E}$$

The origin is located at Meades Ranch in Kansas, and the azimuth from Meades Ranch to Waldo is defined as $75^\circ 28' 09.64''$. The geoidal height is assumed to be zero, and NAD27 is not geocentric.

The answer is (D).

SOLUTION 2

The origin of the North American Datum of 1983 (NAD83) is intended to be geocentric. 250,000 points, including 600 stations whose positions had been determined by satellite geodesy, were used to constrain the datum to its geocentric origin.

NAD83 has replaced NAD27 in the United States and is used in the United States, Canada, Mexico, and Central America.

The answer is (A).

SOLUTION 3

The shape of the earth is represented in NAD83 by an ellipsoid of revolution that is biaxial and rotated about its minor axis. The actual shape of the earth is oblate—that is, flattened at the poles. Rotation approximates the shape.

The answer is (A).

SOLUTION 4

Local distortions caused by low-quality original observations in NAD27 coordinates are one source of the difficulty involved in transforming coordinates from that datum to NAD83. Approximate methods of transformation can be designed by computing local translation components based on stations with three-dimensional coordinates common to both systems, but there is no universal one-step mathematical approach available.

The most reliable transformations are based on a return to the original observations and the performance of a readjustment of the entire region.

The answer is (A).

SOLUTION 5

Clarke 1866 is the reference ellipsoid for NAD27. GRS80 is the reference ellipsoid for NAD83. The WGS84 ellipsoid is virtually identical to the GRS80, but there is a slight difference in the flattening, probably due to the lack of sufficient significant figures in its original computation.

The answer is (A).

SOLUTION 6

One way to understand the geoid is to imagine that trenches have been cut through the land masses of the earth, and water has been allowed to flow into them. The surface assumed by the water would then approximate the surface known as the *geoid*, which would closely resemble the ocean's surface if they were completely still. The geoid is often called an *equipotential surface*. In other words, the direction of gravity is always perpendicular to the geoid.

The answer is (D).

SOLUTION 7

The flattening associated with a reference ellipsoid is conventionally expressed as

$$f = \frac{a - b}{a}$$

The answer is (B).

SOLUTION 8

The geoid is defined by the earth's gravity. The unequal distribution of the mass of the earth causes irregularity in the force and direction of gravity; consequently, the geoid is also irregular.

The answer is (D).

SOLUTION 9

The Sea Level Datum of 1929 was renamed in 1976 to the National Geodetic Vertical Datum of 1929 (NGVD29). The datum was renamed because it does not accurately represent mean sea level or any equipotential surface. The datum was based on observed heights of mean sea level at 26 tidal stations, 21 in the United States and 5 in Canada, along the Atlantic, Gulf, and Pacific coasts. The datum involved more than 100,000 km of leveling established benchmarks throughout the United States.

The answer is (C).

SOLUTION 10

While level surfaces may be considered parallel for surveying a limited area, in fact, they are not. The geoid, a close representation of a level surface, is not only discontinuous over variations of density in the earth, but is subject to a general convergence as well. The relative distance between two level surfaces will decrease from the equator to the poles as a consequence of the increase in gravity.

The answer is (D).

SOLUTION 11

The vertical readjustment, known as the North America Vertical Datum of 1988 (NAVD88), began in the 1970s. It addressed the elevations of benchmarks across the United States. The effort included field work that replaced destroyed and disturbed benchmarks, and over 50,000 mi of leveling was redone. The benchmark elevations determined in NGVD29 compared with the elevations of the same benchmarks in NAVD88 vary from approximately −1.3 ft in the east, to approximately +4.9 ft in the west for the 48 coterminous states of the United States.

The answer is (C).

SOLUTION 12

Concerning the acceleration of gravity, the average acceleration of a falling object is approximately 978 gals (978 cm/sec^2 or 32.09 ft/sec^2) at the equator. At the poles, the acceleration of a falling object increases to approximately 984 gals (984 cm/sec^2 or 32.28 ft/sec^2). The acceleration of a falling object at 45° latitude is between these two values at 980.6199 gals. This value is sometimes called *normal gravity*.

The length of a degree of longitude and the length of a degree of latitude is approximately the same in the vicinity of the equator—about 60 nautical mi (111 km or 69 mi). However, a degree of longitude gets shorter as it nears either pole. At two-thirds of the distance from the equator to the pole (i.e., 60° north or south latitude), a degree of longitude is about 55.5 km (34.5 mi) long—half the length it had been at the equator. As one proceeds northward or southward, a degree of longitude continues to shrink until it fades to nothing.

On the other hand, lines of latitude do not converge; they are always parallel with the equator. In fact, as one approaches the poles where a degree of longitude becomes small, a degree of latitude grows slightly. This small increase occurs because the ellipsoid gets flatter near the poles. If the earth was a sphere, this flattening would not occur and a degree of latitude would always be as long as it is at the equator—110.6 km (68.7 mi).

However, since the earth is an oblate spheroid as Newton predicted, a degree of latitude gets longer at the poles. It grows to about 111.7 km (69.4 mi).

The answer is (D).

SOLUTION 13

At the given coordinates, one second of latitude is approximately 100 ft, and one ten thousandth of a second of latitude is approximately 0.01 ft. One second of longitude is approximately 80 ft, and one ten thousandth of a second of longitude is approximately 0.008 ft.

The answer is (A).

SOLUTION 14

The reference ellipsoids for both NAD83 and NAD27 are biaxial; that is, they have two axes. A triaxial ellipsoid has three axes, an idea that has been around a long time. In 1860, Captain A. R. Clarke wrote to the Royal Astronomical Society stating, "The earth is not exactly an ellipsoid of revolution. The equator itself is slightly elliptic." Therefore, the length of the semimajor axis is not constant in triaxial ellipsoids. A triaxial ellipsoid has flattening at both the poles and the equator so that the length of the semimajor axis varies along the equator. The Krassovski ellipsoid, also known as Krasovsky ellipsoid, is triaxial.

The answer is (C).

SOLUTION 15

The concrete manifestation of a datum is known as its *realization.* The realization of a datum is the actual marking and collection of coordinates on stations throughout the region covered by the datum. In other words, it is the creation of the physical network of reference points on the earth's surface.

The answer is (B).

SOLUTION 16

The National Geodetic Survey (NGS) and the Geodetic Survey of Canada set about the task of attaching and orienting the GRS80 ellipsoid to the actual surface of the earth, as defined by the best positions available at the time. It took more than 10 years to readjust and redefine the horizontal coordinate system of North America into what is now the North American Datum of 1983 (NAD83). With the surveying capability of GPS and the new NAD83 reference system in place, NGS began the long process of a nationwide upgrade of their control networks. This upgrade, known as the National Geodetic Reference System (NGRS), includes three networks. A horizontal network provides geodetic latitudes and longitudes in the North American Datums. A vertical network furnishes heights, also known as elevations, in the National Geodetic Vertical Datums (NGVD). A gravity network supplies gravity values in the U.S.'s absolute gravity reference system. Any particular station may have its position defined in one, two, or all three networks.

The answer is (C).

SOLUTION 17

Once the initial point and directions were fixed, the entire orientation of NAD27 was established—including the center of the reference ellipsoid. Its center was imagined to reside somewhere around the earth's center of mass. However, the two points were not coincident, nor were they intended to be; therefore, NAD27 does not use a geocentric ellipsoid. In the period before space-based geodesy was tenable, such a regional datum was not unusual. The Australian Geodetic Datum 1966, the Datum Européen 1950, and the South American Datum 1969 among others, were also designed as nongeocentric systems.

The answer is (D).

SOLUTION 18

The creations of High Accuracy Reference Networks (HARNs) are cooperative ventures between NGS and individual states, and often include other organizations as well. With heavy reliance on GPS observations, these networks are intended to provide extremely accurate, vehicle accessible, regularly spaced control points with good overhead visibility. To ensure coherence, when GPS measurements are complete, they are submitted to NGS for inclusion in a statewide readjustment of the existing NGRS covered by the state. Coordinate shifts of 0.3 m to 1.0 m from NAD83 values have been typical in these readjustments.

The answer is (D).

SOLUTION 19

Unavoidable forces cause the mean sea level (MSL) to deviate up to one, even two meters from the geoid. This fact is frequently mentioned to emphasize the inconsistency of the geoid's original definition as offered by J.B. Listing in 1872. Listing thought the geoidal surface was equivalent to MSL; however, MSL and the geoid are not the same. While Listing's definition does not stand up to today's scrutiny, it can still be instructive.

Just as the geoid does not precisely follow MSL, neither does it exactly correspond with the topography of the earth's dry land. Instead, the geoid is similar to the earth's terrestrial surface, as it is bumpy. If the solid earth did not have internal density anomalies, the geoid would be smooth and almost exactly ellipsoidal. If this were the case, the reference ellipsoid would fit the geoid almost perfectly. However, like the earth itself, the geoid defies such mathematical consistency and departs from true ellipsoidal form by as much as 100 m in some places.

The answer is (D).

SOLUTION 20

The earth is curved, and maps are flat. If the graticule, or grid, of a map is to correspond with the parallels of latitude and meridians of longitude on the terrestrial surface, an orderly method of transference must be adopted. There are many map projections that accomplish this; each has unique characteristics.

The answer is (B).

SOLUTION 21

The Lambert and the transverse Mercator projections form the foundation for nearly all of the state plane coordinate systems. However, zone 1 of the Alaska system is based on an oblique Mercator projection.

The answer is (D).

SOLUTION 22

The difficulty of representing the earth's surface on a flat map can be solved by projecting the information from an ellipsoidal representation of the earth onto some surface that may be unrolled or developed, such as a cone or cylinder. These conic sections (if a cylinder may be so considered) can each be cut along an element and neatly flattened, minimizing distortion. Distortion can never be completely eliminated, however.

The answer is (C).

SOLUTION 23

Some map projections allow for the angles between two curves on the spheroidal surface to be preserved on the map. This property, known as *conformality*, provides for the maintenance of elementary shapes on the map.

The answer is (D).

SOLUTION 24

A rhumb line, or loxodrome, appears to be a straight line on the Mercator projection. It crosses each meridian at a constant angle, but does not follow a great circle—the shortest path between two points.

The answer is (A).

SOLUTION 25

The arc of a great circle will be a straight line on a gnomonic map projection, which is generated on a plane tangent to the ellipsoid at a single point.

The answer is (C).

SOLUTION 26

In the State Plane Coordinate System (SPCS) the direction known as grid north is always parallel to the central meridian for the zone. In SPCS, north is grid north and the lines of the grid are parallel to each other. They must also be parallel to one another and the central meridian of the zone, so clearly geodetic north and grid north are not the same. In fact, grid north and geodetic north only coincide on the central meridian—everywhere else in the zone they diverge from one another, creating an angle between them. In SPCS27, the angle for the Lambert Conic project was symbolized with the Greek letter theta, θ; in the Transverse Mercator projection, it was given the symbol delta alpha, $\Delta\alpha$. However, in SPCS83, convergence is symbolized with gamma, γ, in both the Lambert Conic and the Transverse Mercator projections.

The answer is (D).

SOLUTION 27

There have been changes in the distances between the SPCS zones and their origins. In the old SPCS27 arrangement, the y-axis was 2,000,000 ft west from the central meridian in the Lambert Conic projection, and 500,000 ft in the Transverse Mercator projection. In the SPCS83 design, those constants changed so that the most common values became 600,000 m for the Lambert Conic projection, and 200,000 m for the Transverse Mercator projection. However, there is a good deal of variation in these numbers from state to state and from zone to zone.

It is important to note that the fundamental unit for SPCS27 is the U.S. survey foot and for SPCS83 it is the meter. The original goal with SPCS27 was to keep each zone small enough to ensure that the scale distortion was 1 part in 10,000 or less. However, when the SPCS83 was designed, some states did not maintain that scale. In five states, some SPCS27 zones consolidated into one zone, or were added to adjoining zones. In three of those states, the result was one single large zone. Therefore, because the area covered by these single zones became so large, they were not limited by the 1 part in 10,000 standard.

This is not the first time the goal of 1 part in 10,000 was changed. In Texas, the original scale factor was allowed to be slightly above the ratio of 1 part in 10,000 so that the state was completely covered with five zones. In

1933, among the guiding principles was covering the states with as few zones as possible and having zone boundaries follow county lines.

The answer is (D).

SOLUTION 28

Conformality is the characteristic of the Lambert projection that preserves the shape of small figures as they are transformed from the ellipsoid to the mapping plane. The Lambert projection is a conic projection; that is, the developable surface employed is imagined to be a cone.

The answer is (A).

SOLUTION 29

The Greek letter γ is used in the Lambert projection to indicate the mapping angle in the State Plane Coordinate System 1983 (SPCS83).

The answer is (D).

SOLUTION 30

The secant cone intersects the ellipsoid at two lines of exact scale. This design allows the scale distortion to be minimized. Most of the mapping plane is actually between the standard parallels and, therefore, beneath the surface of the ellipsoid.

The answer is (D).

SOLUTION 31

The scale factor in a Lambert projection is dependent on the latitude of a position.

The answer is (C).

SOLUTION 32

The standard parallels of a secant Lambert projection may be visualized as the intersection of the plane of the cone with the surface of the spheroid. Any distance along a standard parallel would be exactly the same on the spheroid as on the map.

The answer is (D).

SOLUTION 33

The conic Lambert projection represents parallels of latitude as arcs of nearly equally spaced concentric circles. Meridians of longitude are shown as straight lines converging at the pole.

The answer is (A).

SOLUTION 34

Twenty-nine states use the Lambert projection exclusively as the basis of their state plane coordinate systems—Arkansas, California, Colorado, Connecticut, Iowa, Kansas, Kentucky, Louisiana, Maryland, Massachusetts, Michigan, Minnesota, Montana, Nebraska, North Carolina, North Dakota, Ohio, Oklahoma, Oregon, Pennsylvania, South Carolina, South Dakota, Tennessee, Texas, Utah, Virginia, Washington, West Virginia, and Wisconsin.

The answer is (C).

SOLUTION 35

The target for maximum scale distortion is 1 part in 10,000. States that extend less than 158 mi north-south meet this goal. For example, the maximum scale in the Connecticut Coordinate System is about 1:40,000 along its southern and northern boundaries. On the other hand, the coverage of Texas by five zones was accomplished only by allowing distortion to go slightly more than 1:10,000.

The answer is (B).

SOLUTION 36

36.1. Most frequently, the area between the standard parallels is approximately 105 mi, or four-sixths of the total zone. In other words, two-thirds of the area mapped is within the spheroid. One-sixth, or about 26 mi of the zone, is north of the northerly standard parallel and one-sixth is south of the southerly standard parallel. These latter areas are outside or above the spheroid.

The answer is (D).

36.2. The usual alignment of the axis of the imaginary cone used in the Lambert projection is along the extension of the earth's polar axis.

The answer is (B).

SOLUTION 37

The transverse Mercator projection is distinct from the Mercator projection in that the axis of the cylindrical surface is imagined to be perpendicular to the axis of the spheroid.

The answer is (B).

SOLUTION 38

The variation of the scale in a transverse Mercator projection is based on changes in longitude, rather than latitude. The scale is constant along the central meridian of a transverse Mercator projection.

The answer is (D).

SOLUTION 39

The symbol used in the transverse Mercator projection to indicate the mapping angle, or convergence, is γ, in the State Plane Coordinate System 1983 (SPCS83).

The answer is (C).

SOLUTION 40

The cylinder used to generate a transverse Mercator-based state plane coordinate system intersects the spheroid along two lines that describe ellipses on the spheroid. The lines are the lines of strength of the developed map. The scale factor is 1 along these lines; they are equidistant from, and parallel with, the central meridian of the zone.

The answer is (B).

SOLUTION 41

The UTM projection provides a worldwide system of coordinates in meters. The foundation of UTM is WGS84. Five different reference ellipsoids are used. They are the Clarke spheroids of 1866 and 1880, the international ellipsoid, the Bessel ellipsoid, and the Everest ellipsoid. The Clarke spheroid of 1866 is presently used in North America.

The answer is (D).

SOLUTION 42

The Universal Transverse Mercator (UTM) coordinate system differs significantly from the Transverse Mercator system. The Transverse Mercator coordinate system is used in the State Plane Coordinate System, which was first designed by Dr. Oscar Adams, of the Division of Geodesy at the United States Coast and Geodetic Survey, assisted by Charles Claire. UTM was originally a military system that covered the globe.

The UTM secant projection gives approximately 180 km between the lines of exact scale where the cylinder intersects the ellipsoid. The scale factor grows from 0.9996 along the central meridian of a UTM zone to 1.00000 at 180 km to the East and West. SPCS zones are usually limited to about 158 mi and, therefore, have a smaller range of scale factors than do the UTM zones. In state plane coordinates, the scale factor is usually about 1 part in 10,000. In UTM coordinates it can be as large as 1 part in 2500. The 60 UTM zones nearly cover the earth, except for the polar regions, which are covered by two azimuthal polar zones called the Universal Polar Stereographic (UPS) projection.

When NAD83 was established, the National Geographic Society (NGS), which replaced the U.S. Coast and Geodetic Survey (USC&GS), mandated that the meter become the official unit of all the published coordinate values. However, in 1986, NGS announced that it would augment the state plane coordinate publications that were in meters with coordinates for the same stations in feet. Deciding whether to use the international foot or the U.S. survey foot would be determined by the state legislation in which the station was found.

Currently, some states use the U.S. survey foot and some use the international foot.

The answer is (B).

SOLUTION 43

An ellipsoidal height, also known as geodetic height, and a geoid height are measured along a line perpendicular to the ellipsoid of reference to a point on the earth's surface. An orthometric height is measured along a plumb line from the geoid to a point on the surface of the earth.

The answer is (D).

SOLUTION 44

The reciprocal of the eccentricity is usually not paired with the semimajor axis to define a reference ellipsoid.

The definition of a reference ellipsoid is accomplished with two numbers. It usually includes the semimajor axis and one of the other answer choices given. The following are some common pairs of constants: the semimajor and semiminor axes in meters, the semimajor axis in meters with the flattening or its reciprocal, and the semimajor axis and the eccentricity.

The answer is (D).

SOLUTION 45

45.1. A constant of 500,000 ft is used in states employing the SPCS27 transverse Mercator projection. This arbitrary constant between the y-axis and the central meridian of each zone prevents negative x values in the coordinates. This constant differs in SPCS83.

The answer is (C).

45.2. The arbitrary distance between the y-axis and the central meridian of transverse Mercator zones in SPCS83 varies. The objective is to avoid any confusion of SPCS27 with SPCS83 coordinate positions. The conversion of feet to meters is a bit problematic since some states use the U.S. survey foot, which is 0.3048006 of a meter, while some prefer to use the international foot, which is 0.3048000 of a meter. No single value is used universally, as was the 500,000 ft value.

The answer is (D).

45.3. The line labeled IJ represents geodetic north at station B. Line GH represents grid north and is parallel to the central meridian.

The answer is (D).

45.4. The symbols most consistently used in the state plane coordinate systems based on the transverse Mercator projection are A = central meridian's false easting, and E = northing coordinate. Note that the parallel of latitude through station B intersects the central meridian somewhat to the south of the line through station B perpendicular to the central meridian.

The answer is (A).

45.5. East of the central meridian of a zone, γ, the convergence (or mapping angle) is positive (+); west of the central meridian, the value is negative (−).

The answer is (C).

SOLUTION 46

The mean radius of the spheroid over the coterminous United States has been considered as 20,906,000 ft for many years. This approximation is adequate for many applications.

The answer is (B).

SOLUTION 47

47.1. A constant of 2,000,000 ft was used in all of the states employing the SPCS27 Lambert projection. This arbitrary constant between the y-axis and the central meridian of each zone prevented negative x values in the coordinates of NAD27. However, as states convert coordinate positions to NAD83, this constant is changing.

The answer is (D).

47.2. The usual symbol for the mapping angle, or convergence, in the Lambert projection is the Greek letter gamma, γ. It is considered to be positive to the east of the central meridian and negative to the west.

The answer is (B).

47.3. The distance labeled A is known as $R \sin \gamma$.

The answer is (A).

47.4. The distance labeled E is the northing coordinate of station B. The distance from the apex of the cone to the grid line through station B is equal to $R \cos \gamma$. The distance labeled F is usually symbolized R_b.

The answer is (D).

47.5. The central meridian is the name usually given to the longitudinal line chosen near the center of the zone.

The answer is (C).

47.6. The value of the mapping angle, γ, is sometimes positive and sometimes negative. Therefore, the correct formulas for the east and north coordinates in the Lambert projection are

$$\begin{aligned} \text{east} &= \mathrm{E}_o + R \sin \gamma \\ \text{north} &= R_b + N_b - R \cos \gamma \end{aligned}$$

The answer is (D).

SOLUTION 48

The constant C is 2,000,000 ft for each zone in SPCS27, and each zone has a unique origin for its coordinates. A third constant value, known as R_b, is the distance from the apex of the cone to its x-axis.

The answer is (A).

SOLUTION 49

In SPCS27, the variables that are dependent on latitude in the Lambert projection are: R, the radius from the apex of the cone to a given station; θ, the convergence; and the scale factor. In a transverse Mercator projection, the geodetic multipliers H and V, y_o, and the small correction a vary with the changes in latitude.

The answer is (C).

SOLUTION 50

Distortion of 1 part in 10,000 implies a scale factor of 0.9999 at the lower limit and 1.0001 at the upper limit. While some state plane coordinate zones may exceed these limits, the objective is to stay within them whenever possible.

The answer is (B).

SOLUTION 51

The developable surface of the mapping plane is below the spheroid between the standard parallels of a Lambert secant projection. Therefore, the distances on the spheroid are always reduced when they are projected onto the mapping plane. In other words, the scale factor is always less than 1 between the standard parallels.

The answer is (C).

SOLUTION 52

In SPCS27, the distance from a given point to the central meridian, also known as the *spheroidal perpendicular*, is the basis for the derivation of the scale factor in the transverse Mercator projection. The spheroidal perpendicular is symbolized by x'.

The answer is (D).

SOLUTION 53

The cylinder is above the spheroid outside the lines of its intersection, which are the lines of exact scale. The distances on the spheroid are increased when they are projected onto the developable surface. Therefore, the scale factor is greater than 1.

The answer is (A).

SOLUTION 54

The combination, or grid, factor is used to move directly from the earth's surface to the state plane in one step. This factor is the product of the scale, or K, factor and the sea-level factor.

The answer is (D).

SOLUTION 55

The scale factor varies with the latitude in the Lambert projection and with the longitude in the transverse Mercator projection. Large changes in these elements of the scale factor, especially in long lines, may require the use of the weighted mean in calculating the scale factor.

The answer is (B).

SOLUTION 56

The ϕ_o, also known as B_o, latitude is near, but not precisely at, the center of the zone. The latitude of the line can be found as $\sin^{-1}\ell$. The scale factor ratio reaches its minimum along this line, and the geodetic lines projected onto the state plane in the zone are concave toward it.

The answer is (D).

SOLUTION 57

In the Lambert projection, the most common symbol for the second-term correction is δ. The second term is a compensation applied to the direction of long lines on the state plane that would otherwise retain a slight curvature toward the center of the zone—that is, toward the parallel of latitude known as ϕ_o, or B_o.

The answer is (B).

SOLUTION 58

The formula for conversion of a geodetic azimuth to a grid azimuth is

$$\text{grid azimuth} = \text{geodetic azimuth} - \theta$$

This conversion ignores the effect of the second term since a relatively short line is involved.

The answer is (B).

SOLUTION 59

The expression $\Delta\lambda''\sin\phi + g$ is equivalent to $\Delta\alpha$, the mapping angle, in the transverse Mercator projection. The change in longitude in seconds, $\Delta\lambda''$, multiplied by the sine of the latitude is adjusted by the small quantity g, normally found in published tables.

The answer is (D).

SOLUTION 60

γ is positive to the east of the central meridian and negative to the west. Its absolute value is always less than the change in longitude from the central meridian.

The answer is (D).

SOLUTION 61

The following formulas are usually given in state plane projection tables.

$$\Delta\lambda = \frac{\theta}{\ell}$$

$$\theta = \Delta\lambda\ell$$

$$\tan\theta = \frac{x'}{R_b - y}$$

$$\theta = \tan^{-1}\left(\frac{x'}{R_b - y}\right)$$

The answer is (D).

SOLUTION 62

Since the geodetic azimuth is measured clockwise from south, subtract 180° to find the geodetic azimuth from north.

$$\begin{array}{r} 186°15'22.21'' \\ -\ 180°00'00.00'' \\ \hline 06°15'22.21'' \end{array}$$

The formula for finding the grid azimuth of short lines is

$$\begin{aligned} \text{grid azimuth} &= \text{geodetic azimuth} - \theta \\ &= 06°15'22.21'' - (+00°40'21.85'') \\ &= 05°35'00.36'' \end{aligned}$$

The answer is (A).

SOLUTION 63

There is not enough information given to answer this question. While the grid azimuth will be precisely 180° different at station Exeter (212°32′14.1″) from its value at station Abott (32°32′14.1″), the same cannot be said of the geodetic azimuth. Without the value of the mapping angle at station Abott or some information from which it may be derived, the question cannot be answered.

The answer is (D).

SOLUTION 64

64.1. The R value for a particular station is the distance from the apex of the cone to that station in feet. One way to derive its value is to use the tabular difference for 1″ of latitude given.

First, find the difference in seconds of latitude from the next smaller minute given.

$$\begin{array}{rr} \text{latitude of station Capitol} = & 39°44'21.276'' \\ & -39°44'00.000'' \\ \hline \text{difference in seconds} = & 21.276'' \end{array}$$

Next, multiply this difference by the tabular difference for 1″ of latitude to find the difference between the R value at latitude 39°44′00.000″ and at latitude 39°44′21.276″.

$$(21.276'')(101.18400\ \text{ft}) = 2152.79\ \text{ft}$$

Finally, subtract this difference from the R value given for the next smaller latitude.

39°44′	25,551,070.33 ft	101.18400 ft
	−2152.79 ft	
39°44′21.276″	25,548,917.54 ft	
39°45′	25,544,999.29 ft	101.18467 ft

The answer is (C).

64.2. +00°19′30.8115″ is the value of θ for station Capitol, and the value of R is 25,548,917.54 ft. These two quantities are sufficient to find the value of the x-coordinate, using the following formula.

$$x = R\sin\theta + \text{C}$$

C is the constant 2,000,000 ft between the central meridian and the y-axis. The following figure illustrates these relationships.

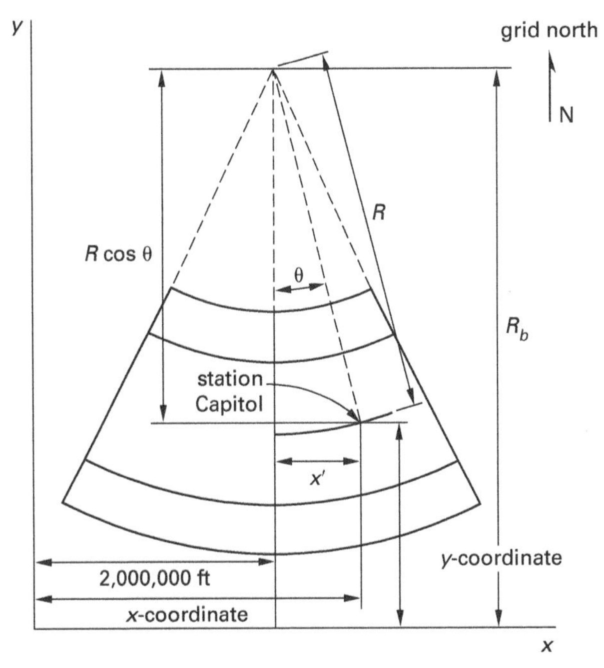

$$\begin{aligned} x &= R\sin\theta + C \\ &= (25{,}548{,}917.54\ \text{ft})(\sin +00°19'30.8115'') \\ &\quad +2{,}000{,}000\ \text{ft} \\ &= 145{,}021.38\ \text{ft} + 2{,}000{,}000\ \text{ft} \\ &= 2{,}145{,}021.38\ \text{ft} \end{aligned}$$

The answer is (B).

64.3. There is more than one way to calculate the θ angle from the longitude of a station. One way is to use the following relationship.

$$\theta = \Delta\lambda\ell$$

The symbol $\Delta\lambda$ refers to the change in longitude from the central meridian of the zone, which in this case is from 105°3′00″.

$$\begin{aligned} \text{central meridian} &= 105°30'00.000'' \\ \lambda \text{ of Capitol} &= -104°59'03.601'' \\ \hline \Delta\lambda &= +00°30'56.399'' \end{aligned}$$

The symbol ℓ indicates the sine of the apex angle for the zone. It is shown in the illustration as 0.6306895773.

$$\begin{aligned} \theta &= \Delta\lambda\ell \\ &= (+00°30'56.399'')(0.6306895773) \\ &= +00°19'30.81'' \end{aligned}$$

The answer is (C).

64.4. The value of θ for station Capitol is +00°19′30.8115″, and the value of R is 25,548,917.54 ft. These two quantities are sufficient to find the value of the y-coordinate, using the following formula.

$$y = R_b - R\cos\theta$$

R_b is the constant for the zone. The value of R_b for the central zone of Colorado is 26,243,052.74 ft, as shown in the illustration of stations Crown, Capitol, and Thornton 2.

$$\begin{aligned} y &= R_b - R\cos\theta \\ &= 26{,}243{,}052.74 \text{ ft} \\ &\quad -(25{,}548{,}917.54 \text{ ft})(\cos +00°19'30.8115'') \\ &= 26{,}243{,}052.74 \text{ ft} - 25{,}548{,}505.95 \text{ ft} \\ &= 694{,}546.79 \text{ ft} \end{aligned}$$

The answer is (C).

64.5. The area of the figure can be used to find the spherical excess. Each 75.6 mi^2 included in the figure corresponds to approximately 1 sec of spherical excess. The area of the figure can be found as follows.

For Crown,

$$\begin{aligned} x &= 2{,}145{,}021.38 \\ y &= 694{,}546.79 \end{aligned}$$

For Capitol,

$$\begin{aligned} x &= 2{,}114{,}700.90 \\ y &= 713{,}393.10 \end{aligned}$$

$$\begin{aligned} \begin{array}{c}\text{distance from} \\ \text{Crown to Capitol}\end{array} &= \text{distance } a \\ &= 31{,}083.81 \text{ ft} \quad \text{[grid]} \\ \text{angle at Crown} &= \text{angle B} = 50°12'28.9'' \\ \text{angle at Capitol} &= \text{angle C} = 78°45'26.1'' \\ \text{angle at Thornton 2} &= \text{angle A} = 51°02'05.5'' \end{aligned}$$

Therefore, the area of the figure is

$$\begin{aligned} \text{A} &= \frac{a^2 \sin \text{B} \sin \text{C}}{2 \sin \text{A}} \\ &= \frac{(31{,}083.81 \text{ ft})^2(\sin 50°12'28.9'')(\sin 78°45'26.1'')}{(2)(\sin 51°02'05.5'')} \\ &= \frac{(966{,}203{,}244.1 \text{ ft}^2)(0.753628089)}{1.555057444} \\ &= \frac{728{,}157{,}904.4 \text{ ft}^2}{1.555057444} \\ &= 468{,}251{,}450.9 \text{ ft}^2 \end{aligned}$$

Convert to square miles.

$$\frac{468{,}251{,}450.9 \text{ ft}^2}{\left(5280 \, \frac{\text{ft}}{\text{mi}}\right)^2} = 16.8 \text{ mi}^2$$

$$\text{spherical excess} = \frac{16.8 \text{ mi}^2}{75.6 \, \frac{\text{mi}^2}{\text{sec}}} = 0.2 \text{ sec}$$

The answer is (A).

64.6. The second-term correction is an angular value that is applied to long lines, generally lines of more than 5 mi in length. The line Crown-Capitol is nearly 6 mi long. The second-term correction is an accommodation of the residual curvature present in such lines even after the development of the mapping surface. The formula for finding the second-term correction in a Lambert projection is

$$\theta' = \left(\frac{x_2 - x_1}{2\rho_o^2 \sin 1''}\right)\left(y_1 - y_o + \frac{y_2 - y_1}{3}\right)$$

x_1, y_1 and x_2, y_2 are the coordinates of the beginning and the end of the line in question, respectively. y_o is the y-coordinate of the parallel of latitude near the center of the zone, which is a constant for each zone. The y_o value for the central zone of Colorado is 461,675.83 ft.

The quantity $1/(2\rho_o^2 \sin 1'')$ is also a constant for each zone and is 2.359×10^{-10} in the central zone of Colorado. Substitute the appropriate values into the formula.

For Crown,

$$\begin{aligned} x_2 &= 2{,}145{,}021.38 \\ y_2 &= 694{,}546.79 \end{aligned}$$

For Capitol,

$$\begin{aligned} x_1 &= 2{,}114{,}700.90 \\ y_1 &= 701{,}393.10 \end{aligned}$$

$$\theta' = \left(\frac{x_2 - x_1}{2\rho_o^2 \sin 1''}\right)\left(y_1 - y_o + \frac{y_2 - y_1}{3}\right)$$
$$= \left(\frac{2,145,021.4 - 2,114,700.9}{2\rho_o^2 \sin 1''}\right) \times\left(701,393.1 - 461,675.8 + \frac{694,546.8 - 701,393.10}{3}\right)$$
$$= (30,320.5)(0.0000000002359)\left(239,717.3 + \frac{-6846.3}{3}\right)$$
$$= (0.000007153)\big(239,717.3 + (-2282.1)\big)$$
$$= (0.000007153)(237,435.2)$$
$$= 01.7''$$

The answer is (D).

64.7. The calculation for finding the second-term correction for a long line can be simplified somewhat through expression of the numbers involved with exponents of 10. For example, the value of $(1/2\rho_o^2 \sin 1'')$ for the central zone of Colorado can be expressed as 2.359×10^{-10} instead of 0.0000000002359. The coordinates of the points Crown and Capitol can be simplified in the same manner.

For Capitol,

$$x_1 = 21.450 \times 10^5$$
$$y_1 = 6.945 \times 10^5$$

For Crown,

$$x_2 = 21.15 \times 10^5$$
$$y_2 = 7.014 \times 10^5$$

Similarly, the constant y_o can be expressed as 4.617×10^5.

$$\theta' = \left(\frac{x_2 - x_1}{2\rho_o^2 \sin 1''}\right)\left(y_1 - y_o + \frac{y_2 - y_1}{3}\right)$$
$$= \left(\frac{21.15 \times 10^5 - 21.450 \times 10^5}{2\rho_o^2 \sin 1''}\right) \times\left(6.945 \times 10^5 - 4.617 \times 10^5 + \frac{7.014 \times 10^5 - 6.945 \times 10^5}{3}\right)$$
$$= (-0.300 \times 10^5)(2.359 \times 10^{-10}) \times\left(2.328 \times 10^5 + \frac{0.069 \times 10^5}{3}\right)$$
$$= (-0.300 \times 10^5)(2.359 \times 10^{-10})(2.351 \times 10^5)$$

When multiplying, add exponents.

$$\theta' = (-0.300 \times 10^5)(5.546 \times 10^{-5})$$
$$= -01.66'' \quad (-01.7'')$$

The answer is (A).

SOLUTION 65

The second terms and the spherical excess are given in the illustration. Each angle may be corrected individually.

The angle at Thornton 2 is

$$\begin{aligned} \text{observed} &= 51°02'05.5'' \\ \delta &= +(-02.0'') \\ & -(-00.1'') \\ \text{one-third of spherical excess} &= -(+00.07'') \\ \hline & 51°02'03.53'' \end{aligned}$$

Rounded to the nearest tenth of a second, this value is 51°02′03.5″.

The angle at Capitol is

$$\begin{aligned} \text{observed} &= 78°45'26.1'' \\ \delta &= +(+00.1'') \\ & -(-01.7'') \\ \text{one-third of spherical excess} &= -(+00.07'') \\ \hline & 78°45'27.83'' \end{aligned}$$

Rounded to the nearest tenth of a second, this value is 78°45′27.8″.

The angle at Crown is

$$\begin{aligned} \text{observed} &= 50°12'28.9'' \\ \delta &= +(+01.7'') \\ & -(+01.9'') \\ \text{one-third of spherical excess} &= -(+00.07'') \\ \hline & 50°12'28.63'' \end{aligned}$$

Rounded to the nearest tenth of a second, this value is 50°12′28.6″.

The answer is (B).

SOLUTION 66

The grid azimuth from Capitol to Crown, 282°43′26.1″, may be found by inversing the grid coordinates.

For Crown,

$$x = 2{,}145{,}021.38$$
$$y = 694{,}546.79$$

For Capitol,

$$x = 2{,}114{,}700.90$$
$$y = 713{,}393.10$$

The mapping angle, θ, can be found from the expression

$$\theta = \Delta\lambda\ell$$

$\Delta\lambda$ is the change in longitude from the central meridian, and ℓ is a constant for the zone. 105°30′00″ is the central meridian for Colorado, and the constant ℓ is 0.6306895773.

$$\begin{aligned}\theta &= \Delta\lambda\ell \\ &= (105°30'00.0'' - 104°59'03.6'')(0.6306895773) \\ &= (00°30'56.4'')(0.6306895773) \\ &= 00°19'30.8''\end{aligned}$$

The second-term, or θ', value is −01.7″, as derived in Prob. 64.7.

$$\begin{aligned}\theta' &= \left(\frac{x_2 - x_1}{2\rho_o^2 \sin 1''}\right)\left(y_1 - y_o + \frac{y_2 - y_1}{3}\right) \\ &= \left(\frac{21.15 \times 10^5 - 21.450 \times 10^5}{2\rho_o^2 \sin 1''}\right) \\ &\quad \times \left(\begin{array}{l}6.945 \times 10^5 - 4.617 \times 10^5 \\ + \dfrac{7.014 \times 10^5 - 6.945 \times 10^5}{3}\end{array}\right) \\ &= (-0.300 \times 10^5)(2.359 \times 10^{-10}) \\ &\quad \times \left(2.328 \times 10^5 + \frac{0.069 \times 10^5}{3}\right) \\ &= (-0.300 \times 10^5)(2.359 \times 10^{-10})(2.351 \times 10^5) \\ &= (-0.300 \times 10^5)(5.546 \times 10^{-5}) \\ &= -01.7''\end{aligned}$$

The formula for the geodetic azimuth of a long line when the grid azimuth is available is

$$\begin{aligned}\text{geodetic azimuth} &= \text{grid azimuth} + \theta - \theta' \\ &= 282°43'26.1'' + (+00°19'30.8'') \\ &\quad -(-00°00'01.7'') \\ &= 283°02'58.6''\end{aligned}$$

The answer is (D).

SOLUTION 67

The latitude crossing the region where the spheroid is farthest outside the cone is the line where the scale factor ratio is smallest. The ℓ value, a constant for each zone, is the sine of that latitude.

$$\sin^{-1} 0.6306895773 = 39°06'03.66''$$

The answer is (B).

SOLUTION 68

There is more than one method of finding the latitude of a station, given the state plane coordinates. One method is to calculate θ, and then find the value of R, the distance from the apex of the cone to the station. With this data, the latitude of the station can be derived from the given tabular values.

First, find x' from

$$\begin{aligned}x' &= x - C \\ &= 2{,}146{,}262.93 - 2{,}000{,}000 \\ &= 146{,}262.93\end{aligned}$$

Next, find θ from

$$\begin{aligned}\tan\theta &= \frac{x'}{R_b - y} \\ &= \frac{146{,}262.93}{26{,}243{,}052.74 - 742{,}549.22} \\ &= \frac{146{,}262.93}{25{,}500{,}503.52} \\ &= 0.005735688 \\ \theta &= +00°19'43.06''\end{aligned}$$

Next, find the value of R from

$$\begin{aligned} R &= \frac{R_b - y}{\cos\theta} \\ &= \frac{26{,}243{,}052.74 \text{ ft} - 742{,}549.22 \text{ ft}}{\cos +00°19'43.06''} \\ &= \frac{25{,}500{,}503.52 \text{ ft}}{0.999983551} \\ &= 25{,}500{,}922.98 \text{ ft} \end{aligned}$$

Find the corresponding value of ϕ by interpolating from the tabular information.

latitude	R (ft)	tabular difference for 1″ of latitude (ft)
39°51′	25,508,572.22	101.18867
52′	25,502,500.90	101.18933
?	25,500,922.98	calculated value of R
53′	25,496,429.54	101.19000

Find the difference in R from that of the next smaller latitude.

$$\begin{array}{r} 25{,}502{,}500.90 \text{ ft} \\ -25{,}500{,}922.98 \text{ ft} \\ \hline 1577.92 \text{ ft} \end{array}$$

Divide the difference by the corresponding tabular difference for 1 sec of latitude.

$$\frac{1577.92 \text{ ft}}{101.18933} = 15.594''$$

Finally, add the seconds to the next smaller latitude to find the latitude of the station.

$$\begin{array}{rr} & 39°52'00.00'' \\ & 00°00'15.59'' \\ \hline \phi = & 39°52'15.59'' \end{array}$$

The answer is (C).

SOLUTION 69

When the mapping angle is given, the calculation of $\Delta\lambda$, or change in longitude, is fairly straightforward. The change in longitude can then be applied to the central meridian to find the longitude of the station.

Calculate $\Delta\lambda$ from

$$\begin{aligned} \Delta\lambda &= \frac{\theta}{\ell} \\ &= \frac{+00°19'43.06''}{0.6306895773} \\ &= +00°31'15.820'' \end{aligned}$$

Then apply the change in longitude to the central meridian.

$$\begin{aligned} \lambda &= \text{central meridian} - \Delta\lambda \\ &= 105°30'00.00'' - (+00°31'15.820'') \\ &= 104°58'44.18'' \end{aligned}$$

The answer is (D).

SOLUTION 70

The sea-level factor will contribute the largest change to the measured distance. The traditional calculation of the factor uses the approximate radius of 20,906,000 ft for the reference spheroid.

$$\begin{aligned} \text{sea-level factor} &= \frac{20{,}906{,}000 \text{ ft}}{\begin{array}{c}\text{average elevation} \\ \text{of stations}\end{array} + 20{,}906{,}000 \text{ ft}} \\ &= \frac{20{,}906{,}000 \text{ ft}}{5343.5 \text{ ft} + 20{,}906{,}000 \text{ ft}} \\ &= \frac{20{,}906{,}000.00 \text{ ft}}{20{,}911{,}343.50 \text{ ft}} \\ &= 0.99974447 \end{aligned}$$

The scale factor is calculated for the stations at each end of the line and an average of the two taken. The scale factor is based on the latitude in the Lambert projection and found by interpolation.

The scale factor for Thornton 2 is

39°51′	25,508,572.22	101.18867	1.0000213
52′	25,502,500.90	101.18933	1.0000252
39°52′15.594″		... ...	?
53′	25,496,429.54	101.19000	1.0000291

First, find the difference in scale factor for a full minute of latitude from 39°52′ to 39°53′.

$$\begin{array}{lr} 39°53' & 1.0000291 \\ 39°52' & -1.0000252 \\ \hline & 0.0000039 \end{array}$$

Next, find the proportional change in the scale factor for 15.594″ over the same interval.

$$\frac{x}{15.594''} = \frac{0.0000039}{60''}$$
$$60''x = (0.0000039)(15.594'')$$
$$x = \frac{0.00006082}{60''}$$
$$= 0.0000010$$

This is the increase appropriate to the scale factor over 15.594″.

39°51′	25,508,572.22	101.18867	1.0000213
52′	25,502,500.90	101.18933	1.0000252
39°52′15.594″		... ...	1.0000262
53′	25,496,429.54	101.19000	1.0000291

Repeating the same process for point 400 yields a scale factor for that point of 1.0000210.

The average of the two scale factors at each end of the line is the scale factor for the line.

$$\frac{1.0000262 + 1.0000210}{2} = 1.0000236$$

The scale factor for the line is 1.0000236, and the sea-level factor is 0.99974447. The product of these two factors is the grid, or combination, factor for the line.

$$\text{grid factor} = (\text{sea-level factor})(\text{scale factor})$$
$$= (0.99974447)(1.0000236) = 0.99976806$$

The product of the measured distance, 10,206.50 ft, and the grid factor, 0.99976806, is the grid distance.

$$(10{,}206.50 \text{ ft})(0.99976806) = 10{,}204.13 \text{ ft}$$

The answer is (B).

SOLUTION 71

The original request for a state plane coordinate system was submitted to the U.S. Coast and Geodetic Survey by an engineer from a state highway department in 1933. Within a year of the establishment of the North Carolina Coordinate System, a similar system had been designed for every state by Dr. O. S. Adams, a mathematician in the division of geodesy of the U.S. Coast and Geodetic Survey.

The answer is (B).

SOLUTION 72

The individual state legislatures have jurisdiction over the legal status of the use of state plane coordinate systems.

The answer is (D).

SOLUTION 73

Every state and possession of the United States has had a state plane coordinate system designed for it.

The answer is (A).

SOLUTION 74

Formerly known as the U.S. Coast and Geodetic Survey, the National Geodetic Survey maintains and publishes the coordinates of geodetic control stations across the United States.

The answer is (D).

SOLUTION 75

In *Geometric Geodetic Accuracy Standards and Specifications for the Use of GPS Relative Positioning Techniques*, the previously established first-order, second-order classes one and two, and third-order classifications are all found under order C.

The answer is (D).

SOLUTION 76

The Federal Geographic Data Committee (FGDC) is an interagency committee that promotes the coordinated development, use, sharing, and dissemination of geospatial data on a national basis. The FGDC also develops geospatial data standards.

The answer is (C).

SOLUTION 77

The U.S. Office of Management and Budget (OMB) established the Federal Geographic Data Committee in 1990, replacing the Federal Geodetic Control Committee (FGCC). In August 2002, the OMB revised Circular A-16, a government circular that provides guidelines for federal agencies involved in the creation, maintenance, or use of spatial data. The revision rechartered the FGDC, establishing five parts of Geospatial Positioning Accuracy Standards.

part 1 Reporting Methodology, FGDC-STD-007.1-1998

part 2 Standards for Geodetic Networks, FGDC-STD-007.2-1998

part 3 National Standard for Spatial Data Accuracy, FGDC-STD-007.3-1998

part 4 Architecture, Engineering, Construction, and Facilities Management, FGDC-STD-007.4-2002

part 5 Standards for Nautical Charting Hydrographic Surveys, FGDC-STD-007.5-2005

The answer is (D).

10 Plats and Mapping

MAP TYPES AND SYMBOLS

PROBLEM 1

Which of the following statements does NOT correctly describe a difference between a map and a plat?

(A) A plat may function as a legal description of the land it represents in lieu of a written description. A map rarely fulfills such a function.

(B) A plat is more likely to show property boundaries than is a map. A map's emphasis is more often on physical features.

(C) A map may contain omissions or generalizations of particular physical features. A plat generally shows all data necessary to accurately represent the land it delineates.

(D) A map is more likely to show dimensional data than is a plat. A plat usually is intended for uses where distances and quantities are determined by scaling.

PROBLEM 2

A planimetric map is best defined as a map

(A) showing the relief of an area by hachures or shading

(B) showing public areas and rights-of-way

(C) delineating different soils and geological features

(D) that shows only the relative horizontal positions of the features represented, without relief

PROBLEM 3

On maps, a full-headed north arrow is sometimes accompanied by an arrow with half a head. What is the usual significance of such a half-headed arrow?

(A) It is used to indicate the direction of astronomic north.

(B) It is used to indicate the direction of magnetic north.

(C) It is used to indicate the direction of north when the orientation of the map makes it necessary for north to be shown other than toward the top of the sheet.

(D) It is used to indicate the direction of the basis of bearings.

PROBLEM 4

An isogonic map is correctly described as a map

(A) that indicates the areas within which particular zoning regulations apply

(B) that shows the relative positions of the boundaries of several other maps

(C) with lines connecting points of equal magnetic declination

(D) with a scale of 1:600,000 or smaller

PROBLEM 5

What type of map shows contours that connect points of equal rainfall?

(A) a hydrological map

(B) bathymetric map

(C) isohyetal map

(D) Thiessen network

MAP SCALES

PROBLEM 6

The legend of a map indicates that its scale is 1:24,000. What name is usually given to this type of scale expression?

(A) engineer's scale

(B) graphical scale

(C) manuscript fraction

(D) representative fraction or scale ratio

PROBLEM 7

Which of the following scales is closest to 1 in = 1 mi?

(A) 1/24,000

(B) 1/62,500

(C) 1/63,400

(D) 1/100,300

PROBLEM 8

A map's legend shows that it has a scale of 1:62,500. A measurement between two features on the map is read as 27.20 on a scale that is incremented at 20 parts/in. Approximately what distance does the measured length represent on the ground?

(A) 7100 ft

(B) 14,000 ft

(C) 17,000 ft

(D) 170,000 ft

PROBLEM 9

What is the representative fraction or scale ratio for 1 cm = 500 m?

(A) 1:254

(B) 1:5000

(C) 1:24,000

(D) 1:50,000

PROBLEM 10

A map prepared for an engineering project has a scale of 1 in = 100 ft. A map prepared for a land subdivision has a scale of 1 in = 1 mi. Which of the following statements is true?

(A) Both maps have a large scale.

(B) The engineering map has the smaller scale of the two.

(C) Both maps have a small scale.

(D) The engineering map has the larger scale of the two.

PROBLEM 11

The United States National Map Accuracy Standards require that no more than 10% of the well-defined points tested will NOT have horizontal error more than which of the following?

(A) For maps on publication scales larger than 1:10,000, not more than 10% of the points tested shall be in error by more than 1/20 inch, measured on the publication scale; for maps on publication scales of 1:10,000 or smaller, 1/30 inch.

(B) For maps on publication scales larger than 1:24,000, not more than 10% of the points tested shall be in error by more than 1/10 inch, measured on the publication scale; for maps on publication scales of 1:24,000 or smaller, 1/20 inch.

(C) For maps on publication scales larger than 1:20,000, not more than 10% of the points tested shall be in error by more than 1/30 inch, measured on the publication scale; for maps on publication scales of 1:20,000 or smaller, 1/50 inch.

(D) one-half of the contour interval

TOPOGRAPHIC MAPPING

PROBLEM 12

A topographic map is best defined as a map

(A) that shows only the relative horizontal positions of the features represented, without relief

(B) delineating different soils and geological features

(C) that shows the relative positions of the boundaries of several other maps

(D) showing both the horizontal and vertical locations of the features it represents

PROBLEM 13

What is a contour interval?

(A) a line of constant elevation above the reference ellipsoid

(B) the horizontal distance between adjacent contour lines

(C) one-half the vertical spacing of the contour lines specified in the legend of the map

(D) the difference in elevation between adjacent contours

PROBLEM 14

What is the primary use of interpolation in constructing a topographic map?

(A) the connection of spot elevations by contour lines

(B) the determination of intermediate contour lines between points or lines of known elevation

(C) the designation of index contours

(D) drawing hachures

PROBLEM 15

In an area of a topographic map where the contour lines are equally spaced, a straight line crosses several contour lines. A consistent distance of 0.22 in is scaled on the map between each intersection of the straight line and the successive contour lines. If the scale of the map is 1:24,000 and the contour interval is 20 ft, what is the slope of the line?

(A) 0.38%

(B) 2.47%

(C) 3:1

(D) 4.55%

PROBLEM 16

What sort of topographic feature is indicated by the illustration shown?

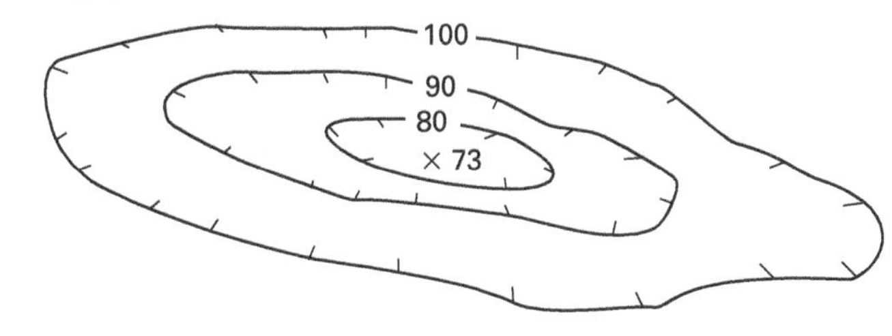

(A) lake or pond

(B) tailings pile

(C) glacier

(D) depression

PROBLEM 17

Which of the following statements is true of contour lines on a topographic map?

(A) Under some circumstances, a single contour line may split into two.

(B) Under some circumstances, two contour lines may join into one.

(C) A contour line cannot end within a map; it must either close on itself or terminate at the neat line of the map.

(D) One contour line never crosses another.

PROBLEM 18

The preparation of a topographic map requires that 5 ft contours be drawn. Two points, A and B, are 3.67 in apart on the map. Point A has an elevation of 773.34 ft, and point B has an elevation of 801.67 ft. On the map, what is the distance along the line between A and B from point A to the 790.00 ft contour?

(A) 1.66 in

(B) 1.91 in

(C) 2.16 in

(D) 2.72 in

PLATS

PROBLEM 19

Some general features of the laws governing platting are the same from state to state, but the details vary considerably. Which of the following specifications is most likely to differ from jurisdiction to jurisdiction?

(A) The boundary of the tract must be clearly shown with permanent monuments marking its corners.

(B) The surveyor's certification must include an official seal and signature.

(C) Enough information regarding the lengths and directions of the lines must be provided so that another surveyor can retrace them correctly.

(D) A surveyor must prepare a plat for every survey performed.

PROBLEM 20

Which of the following describes objectives that should be satisfied by a properly prepared plat?

(A) a representation of the correct shape and size of the property surveyed at scale

(B) the description and relative location of pertinent monuments, whether physical or record

(C) the identification of the appropriate evidence of title

(D) all of the above

PROBLEM 21

Which type of scale on a plat is most useful, and why?

(A) A representative fraction is most useful because it is dimensionless.

(B) A ratio scale, such as 1 in = 500 ft, is most useful since such a scale is universally understood.

(C) A bar scale is most useful since it remains valid despite enlargement or reduction of the plat.

(D) An architect's scale, such as $\frac{1}{8}$ in = 1 ft, is most useful since it allows the representation of the greatest detail.

PROBLEM 22

When do the monuments shown on a plat have equal standing with the monuments mentioned in a deed?

(A) when the plat is a matter of public record

(B) when the plat was drawn prior to the writing of the deed and the deed is unrecorded

(C) when the deed refers to the plat and the parties to the deed acted with reference to the plat

(D) when the monuments shown on the plat were set after the deed was written

PROBLEM 23

When a plat is filed by the county recorder, which of the following automatically ensues?

(A) Constructive notice is implied.

(B) The title insurer's liability ends.

(C) The bearings and distances shown supersede the previous measurements.

(D) The title is transferred.

PROBLEM 24

Which of the following will NOT be included in a post-construction as-built/record survey?

(A) features built as planned

(B) features not built as planned

(C) additions, deletions, and revisions in construction after the as-built/record survey's completion

(D) additions, deletions, and revisions in construction before the as-built/record survey's completion

PROBLEM 25

Which of the following does NOT describe a typical purpose of an as-built/record survey?

(A) to verify accordance with construction plans

(B) to assist with evaluation and payment

(C) to ensure that construction satisfies official requirements

(D) to establish slope stakes along a proposed road

PROBLEM 26

Which of the following usually requires an as-built/record survey?

(A) issuance of an occupancy permit for new construction

(B) creating a new parcel in a subdivision

(C) preparation of a deed description

(D) staking the corners of a building foundation for construction

ALTA/NSPS LAND TITLE SURVEY PLATS

PROBLEM 27

Which of the following is NOT a minimum standard detail requirement for the plat of an ALTA/NSPS Land Title Survey?

(A) North will be toward the top of the plat.

(B) A graphic scale will be included.

(C) The minimum size will be 18 in × 26 in with a 1 in border.

(D) The directional, distance, and curve data necessary to compute a mathematical closure will be shown.

PROBLEM 28

Which statement correctly describes the minimum requirement for delivering the plat of an ALTA/NSPS Land Title Survey to the client and the title insurance company?

(A) The surveyor must furnish copies of the plat and the associated boundary description, which must be attached to the face of the plat if feasible.

(B) The surveyor must furnish copies of the plat drawn in ink on mylar and copies of the associated boundary description on separate sheets.

(C) The surveyor must furnish copies of the plat or map of the survey on durable, dimensionally stable material of a quality standard acceptable to the insurer. Digital copies of the plat or map may be provided in addition to, or in lieu of, hard copies in accordance with the terms of the contract.

(D) The surveyor must furnish copies of the plat drawn in ink on mylar. The plat must be titled *ALTA/NSPS Land Survey Title Plat*, and the section, township, and range must be indicated on the plat.

PROBLEM 29

The location of all buildings on the parcel must be shown on the plat of an ALTA/NSPS Land Title Survey. The measured ties locating the buildings will be shown

(A) perpendicular to the boundaries

(B) directly to the nearest verified property corners

(C) along the lines established by the building's walls

(D) perpendicular to the building's wall lines

PROBLEM 30

The location and type of walls, fences, and buildings must be shown on the plat of an ALTA/NSPS Land Title Survey within what distance on either side of the boundary lines?

(A) 2 ft

(B) 3 ft

(C) 4 ft

(D) 5 ft

PROBLEM 31

A surveyor may note the date of the measurements taken and that a boundary shown on the plat of an ALTA/NSPS Land Title Survey is subject to change under which of the following circumstances?

(A) The boundary is apparently encumbered by an easement.

(B) The boundary abuts a railroad right-of-way.

(C) A natural water boundary is involved.

(D) The monuments found do not agree with the record.

PROBLEM 32

Which of the following statements regarding relative positional precision, as the term is used in the ALTA/NSPS Land Title Survey specifications, is correct?

(A) It is the length (in feet or meters) of the semimajor axis of the error ellipse.

(B) It represents the uncertainty due to random errors in measurements in the location of any survey point to any other point on the same survey at the 95% confidence level.

(C) It is estimated by the results of a correctly weighted least squares adjustment of the survey.

(D) All of the above.

PROBLEM 33

In an ALTA/NSPS Land Title Survey, when the record bearings, angles, or distances differ from the measured bearings, angles, or distances, which of the following must be shown?

(A) details at each corner showing their position according to the record bearings, angles, and distances, as well as their position according to the measured bearings, angles, and distances and the bearings and distances between them

(B) only the record bearings, angles, and distances

(C) only the measured bearings, angles, and distances

(D) both the record and measured bearings, angles, and distances

PROBLEM 34

In an ALTA/NSPS Land Title Survey, if the record description fails to form a mathematically closed figure, the surveyor should

(A) not mention it

(B) clearly indicate the misclosure

(C) refuse to perform the survey

(D) only mention the closed and adjusted boundary

PROBLEM 35

Which of the following would NOT be a typical, standard exception in the report prepared prior to issuance of a title insurance policy?

(A) easements, or claims of easements, not shown in the public records

(B) taxes or special assessments left off the public record

(C) unrecorded mechanic's liens

(D) matters arising before the effective date of the policy

PROBLEM 36

Which of the following statements regarding "Table A, Optional Survey Responsibilities and Specifications," from the *Minimum Standard Detail Requirements for ALTA/NSPS Land Title Surveys* is NOT correct?

(A) The optional survey responsibilities and specifications must be negotiated between the surveyor and the client.

(B) The surveyor may qualify or expand on the description of the items.

(C) The surveyor may provide the width and recording information of all plottable rights of way, easements, and servitudes burdening and benefitting the property surveyed as evidenced by Record Documents that have been provided to the surveyor.

(D) A space is provided for the addition of an item not listed on the printed form.

PROBLEM 37

If statutes, administrative rules, and/or ordinances established by states or local jurisdictions conflict with the Minimum Standard Detail Requirements for ALTA/NSPS Land Title Surveys, to the extent that the survey cannot be conducted in accordance with both, which will apply?

(A) Those of the state or local jurisdiction will apply.

(B) Those of ALTA/NSPS will apply.

(C) The more stringent will apply.

(D) Those designated by the insurer will apply.

PROBLEM 38

What is the maximum allowable relative positional precision for an ALTA/NSPS Land Title Survey, based on the direct distance between the two corners being tested?

(A) 1 cm (0.03 ft) plus 40 ppm

(B) 2 cm (0.07 ft) plus 50 ppm

(C) 3 cm (0.10 ft) plus 60 ppm

(D) 4 cm (0.13 ft) plus 70 ppm

PROBLEM 39

What does the surveyor do if the maximum allowable relative positional precision for an ALTA/NSPS Land Title Survey is exceeded?

(A) The surveyor adds a note to the plat or map that explains the site conditions that resulted in the exceedance.

(B) The surveyor repeats the survey until the maximum allowable relative positional precision is not exceeded.

(C) The surveyor readjusts the survey to bring the relative positional precision within allowable limits.

(D) The surveyor tells the client the land description provided is wrong.

PROBLEM 40

Which of the following will NOT be part of a request for an ALTA/NSPS Land Title Survey, according to the Minimum Standard Detail Requirements?

(A) permission to enter the property to be surveyed, adjoining properties, or offsite easements

(B) establishment of the optional items listed in Table A that are to be incorporated

(C) specification that an as-built is required

(D) a written authorization to proceed from the person or entity responsible for paying for the survey

PROBLEM 41

Must trees, bushes, and shrubs be included in an ALTA/NSPS Land Title Survey?

(A) not unless they are evidence of possession

(B) not unless specified by the contract

(C) yes, they must be included

(D) both A and B

PROBLEM 42

When a water feature forms a boundary of a property subject to an ALTA/NSPS Land Title Survey, which of the following does NOT need to be noted?

(A) the date it was measured

(B) that it is subject to change due to natural causes

(C) the attribute located (i.e., top of bank, edge of water, high water mark)

(D) whether the water feature is navigable or not navigable

PROBLEM 43

Is a surveyor performing an ALTA/NSPS Land Title Survey responsible for exact location of underground features, such as utilities existing on or serving the surveyed property?

(A) No, exact location of underground features requires excavation.

(B) Yes, but only if Table A, Item 11 is included in the contract for the survey.

(C) No, ALTA/NSPS Land Title Surveys can only include surface features.

(D) Yes, even if Table A, Item 11 is not included in the contract for the survey.

PROBLEM 44

Which of the following should NOT be addressed on the face of an ALTA/NSPS Land Title Survey plat or map?

(A) the source, date, and precision of photogrammetry, remote sensing, or laser scanning used in the location of features

(B) certification of professional liability insurance

(C) wetland delineation markers set by a qualified specialist

(D) the extent of natural or artificial realignments in water boundaries of which the surveyor is aware

FEMA

PROBLEM 45

According to the Federal Emergency Management Administration (FEMA) standards, the term *100-year flood* means that the chance of the flood being equaled or exceeded in any given year is

(A) 0.2%

(B) 1%

(C) 2%

(D) 10%

PROBLEM 46

Which of the following are NOT shown on Flood Insurance Rate Maps (FIRMs) administered by the Federal Emergency Management Administration (FEMA)?

(A) areas subject to flooding during a 1%-annual-chance flood event in a community

(B) flood insurance risk zones

(C) land area that would be inundated by the 1%-annual-chance flood based on hydrology of the future conditions

(D) in areas of detailed analyses, base flood elevations (BFEs)

PROBLEM 47

The owner of a parcel of land has recently completed the placement of fill on the property to elevate it above its original natural grade. The owner believes the property is no longer in the 1%-annual-chance flood floodplain and should not be shown inside the special flood hazard area (SFHA) on the Federal Emergency Management Administration's (FEMA's) mapping. Assuming the owner submits all appropriate property and elevation information to FEMA, FEMA may issue which of the following if its formal determination of the property's location relative to the SFHA agrees with the owner's contention?

(A) LOMR-F

(B) LOMA

(C) CLOMR

(D) LOMR

PROBLEM 48

A National Flood Insurance Program (NFIP) Elevation Certificate may be used for which of the following?

(A) to certify that a building is watertight (floodproofed) below the base flood elevation (BFE)

(B) to provide a waiver of the flood insurance purchase requirement

(C) to support a letter of map amendment (LOMA) or letter of map revision based on fill (LOMR-F) request

(D) to rate pre-flood insurance rate map (FIRM) buildings not being rated under the post-FIRM flood insurance rules

PROBLEM 49

A letter of map amendment (LOMA), an official amendment to an effective flood insurance rate map (FIRM), can be submitted via a Federal Emergency Management Agency (FEMA) web-based tool as an electronic letter of map amendment (eLOMA). Which of the following statements concerning eLOMA is correct?

(A) eLOMA accepts Letters of Map Revision (LOMR).

(B) eLOMA is only for FEMA-certified professionals, licensed land surveyors, and professional engineers.

(C) eLOMA is for processing requests for proposed structures.

(D) eLOMA is particularly appropriate when the subject of the request is on an alluvial fan (Zone V).

PROBLEM 50

Can a National Flood Insurance Program (NFIP) Elevation Certificate be completed while a building is under construction?

(A) Yes, if all of the floors are in place.

(B) No, the elevations for the Elevation Certificate must be based on a fully constructed building.

(C) Yes, and the flood insurance can be written, but a new Elevation Certificate will need to be done when construction is completed.

(D) No, it is not possible for flood insurance to be written while a building is under construction.

SOLUTION 1

A map is much more likely to depend on scaling for the derivation of distances, quantities, etc., than is a plat. A plat should show, among other things, sufficient dimensional information to make scaling unnecessary.

The answer is (D).

SOLUTION 2

A planimetric map is distinguished from a topographic map by the absence of any representation of the vertical orientation of the features shown.

The answer is (D).

SOLUTION 3

The most frequent use of the half-headed arrow is to indicate magnetic north. Often, two arrows are shown; the angle between them is the declination.

The answer is (B).

SOLUTION 4

The term *isogonic* refers to equal angles. An isogonic map includes lines of equal magnetic declination.

The answer is (C).

SOLUTION 5

Contours are used in several types of maps besides topographic maps. Contours are used to indicate the mapping of gravity on maps that delineate the geoid and are used to connect points of equal rainfall on isohyetal maps. Maps that use lines joining points with equal numerical values are known collectively as *isoplethic maps.*

The answer is (C).

SOLUTION 6

The name *representative fraction* or *scale ratio* is usually given to this sort of dimensionless scale.

The answer is (D).

SOLUTION 7

Scale expressed as a representative fraction is independent of units of measurement. One unit on the map is intended to represent the stated number of the same units on the ground.

$$\begin{aligned} 1\ \text{mi} &= (5280\ \text{ft})\left(12\ \frac{\text{in}}{\text{ft}}\right) \\ &= 63{,}360\ \text{in} \quad (63{,}400\ \text{in}) \end{aligned}$$

The scale 1/63,400 is closest to 1 in = 1 mi and is used as such on many maps.

The answer is (C).

SOLUTION 8

The scale of 1:62,500 may be interpreted as 1 ft = 62,500 ft, or as 1 in = 62,500 in.

$$\begin{aligned} 1\ \text{in} &= (62{,}500\ \text{in})\left(\frac{1\ \text{ft}}{12\ \text{in}}\right) \\ &= 5208.33\ \text{ft} \end{aligned}$$

The reading from a 20 parts/in scale can be divided by 20 to find the corresponding reading in inches.

$$\frac{27.20\ \text{parts}}{20\ \frac{\text{parts}}{\text{in}}} = 1.36\ \text{in}$$

To find the measured length in units of feet, use the proportion

$$\begin{aligned} \frac{1\ \text{in}}{5208.33\ \text{ft}} &= \frac{1.36\ \text{in}}{X\ \text{ft}} \\ (5208.33\ \text{ft})(1.36\ \text{in}) &= (X)(1\ \text{in}) \\ \frac{(5208.33\ \text{ft})(1.36\ \text{in})}{1\ \text{in}} &= 7083.33\ \text{ft} \quad (7100\ \text{ft}) \end{aligned}$$

The answer is (A).

SOLUTION 9

One method of converting to representative fractions or scale ratios is to express both sides of the equation in the same units. There are 100 cm in a meter. Therefore,

$$500\ \text{m} = 50\,000\ \text{cm}$$

The equivalent representative fraction is 1:50,000.

The answer is (D).

SOLUTION 10

The more area a particular map represents, the smaller its scale. In other words, a large-scale map covers less area than a small-scale map.

The answer is (D).

SOLUTION 11

United States National Map Accuracy Standards were established in 1941 by the U.S. Bureau of the Budget to set accuracy standards for all federally produced maps. The current standards were revised in 1947. The required horizontal accuracy for maps on publication scales larger than 1:20,000 is not more than 10 percent of the well-defined points tested shall be in error by more than 1/30 inch, measured on the publication scale; for maps on publication scales of 1:20,000 or smaller, 1/50 inch.

The answer is (C).

SOLUTION 12

A topographic map shows the vertical as well as the horizontal locations of the features it represents. Generally, contours are used to indicate the relative elevations of the valleys, mountains, and plains shown.

The answer is (D).

SOLUTION 13

A contour interval is the difference in elevation between adjacent contour lines. It is usually specified in the map's legend.

The answer is (D).

SOLUTION 14

Interpolation is the determination of intermediate values between two given values. Where a constant rate of change of the slope can be assumed between points of known elevation, the process is used to locate intermediate contour lines.

The answer is (B).

SOLUTION 15

The horizontal distance on the ground represented by 0.22 in on the map can be found as follows. According to the map's scale,

$$\begin{aligned} 1\ \text{in} &= 24{,}000\ \text{in} \\ &= 2000\ \text{ft} \\ (2000\ \text{ft})(0.22) &= 440\ \text{ft} \end{aligned}$$

The horizontal distance on the ground between successive intersections of the straight line and the contours it crosses is 440 ft.

The contour interval on the map is 20 ft. The division of the contour interval by the distance of the line between the contours, 440 ft, yields the slope of the line.

$$\frac{20\ \text{ft}}{440\ \text{ft}} = 0.0455 = 4.55\%$$

The answer is (D).

SOLUTION 16

Closed contours with tick marks or hachures inside the figure are used to indicate a depression. It is best to show the elevation of the lowest point of such a figure as well.

The answer is (D).

SOLUTION 17

Contour lines cannot join, nor can one contour line split into two. However, it is sometimes possible for them to cross, as in the case of overhanging ledges, where the lower contours may be shown as dashed lines. Contour lines either close on themselves or end at the edge of the map; they cannot stop within the boundaries of the map.

The answer is (C).

SOLUTION 18

The vertical distance between points A and B is 28.33 ft, and the horizontal distance between the same two points on the map is 3.67 in. Assuming that the grade between the two points is relatively constant, the rate of change in elevation between the two points is

$$\frac{3.67\ \text{in}}{28.33\ \text{ft}}$$

The vertical distance from point A to the 790.00 ft contour is 16.66 ft. The horizontal distance, x, on the map can then be found by proportion.

$$\begin{aligned}\frac{3.67\ \text{in}}{28.33\ \text{ft}} &= \frac{x}{16.66\ \text{ft}}\\ x &= \frac{(16.66\ \text{ft})(3.67\ \text{in})}{28.33\ \text{ft}}\\ &= 2.16\ \text{in}\end{aligned}$$

The answer is (C).

SOLUTION 19

While some states specify that a plat be recorded for every survey performed, many do not.

The answer is (D).

SOLUTION 20

While a plat may not settle questions of title, it should provide evidence of easements and encroachments, as well as other record and physical information affecting the title of the property.

The answer is (D).

SOLUTION 21

The bar scale, required by many platting laws, has the unique characteristic of retaining its validity despite reduction or enlargement of the plat.

The answer is (C).

SOLUTION 22

When parties to the deed act in reference to the plat, it gives status to the monuments shown on the plat. When the plat is mentioned in the writings of the deed, the plat and the monuments shown on it become an integral part of the deed itself.

The answer is (C).

SOLUTION 23

Recording a plat in the public registry automatically implies constructive notice. The plat becomes available to all who wish to know it.

The answer is (A).

SOLUTION 24

Construction is seldom completed entirely as the original plans anticipated. Along the way, variations from the plans are often necessary. An as-built/record survey is often done to show compliance of the construction of pavement, buildings, curb and gutter, utilities, etc., with the design, but its purpose is also to show the deviations. An as-built/record survey should include the results of the additions, deletions, and revisions. It will include the features that were built as originally intended, as well as those that were not. In other words, it can be used to illustrate both the fulfillment of plans and the departure from them. An as-built/record survey should show additions, deletions, and revisions that occur before the as-built survey's completion, but it cannot show those that occur after the as-built survey's completion.

The answer is (C).

Plats and Mapping

SOLUTION 25

As-built/record surveys are generally done post-construction when the improvements to a property have been completed. Therefore, establishing staking for a proposed road would not be the subject of an as-built/record survey.

The answer is (D).

SOLUTION 26

The creation of a new parcel subdivision, the preparation of a deed description, and the staking of the corners of a building foundation for construction can all be done without an as-built/record survey. However, municipalities do often require an as-built/record survey to show that a construction project is completed before issuing an occupancy permit to the owner.

The answer is (A).

SOLUTION 27

The size described in the minimum standard detail requirements for ALTA/NSPS Land Title Survey plats is $8\frac{1}{2}$ in × 11 in.

The answer is (C).

SOLUTION 28

The ALTA/NSPS Land Title Survey requirements do not specify the medium on which the plat must be drawn, except to say that it must be durable, dimensionally stable material of a quality acceptable to the insurer. The requirements also state that digital copies of the plat or map may be provided in addition to, or in lieu of, hard copies in accordance with the terms of the contract.

The answer is (C).

SOLUTION 29

Ties to buildings on the parcel are to be shown perpendicular to the property boundaries.

The answer is (A).

SOLUTION 30

All possible encroachments from adjoining properties must be shown. In addition, all fences, walls, and buildings within 5 ft on either side of the property boundary are to be shown.

The answer is (D).

SOLUTION 31

The location of property dependent on a water boundary is subject to change by natural causes and should be so noted on the plat of an ALTA/NSPS Land Title Survey.

The answer is (C).

SOLUTION 32

Relative positional precision is the length (in feet or meters) of the semimajor axis of the error ellipse. It represents the uncertainty due to random errors in measurements in the location of any survey point to any other point on the same survey at the 95% confidence level. Relative positional precision is estimated by the results of a correctly weighted least squares adjustment of the survey.

The answer is (D).

SOLUTION 33

In an ALTA/NSPS Land Title Survey when record bearings, angles, or distances are different than the measured bearings, angles, or distances, both the record and measured numbers are shown.

The answer is (D).

SOLUTION 34

In an ALTA/NSPS Land Title Survey, when the record description does not close mathematically, the surveyor must indicate the misclosure.

The answer is (B).

SOLUTION 35

The report prepared prior to issuance of a title insurance policy will show a numbered list of exceptions specific to the instant property. These may include liens, restrictions, and interests of others which are being excluded from coverage. There may also be easements and recorded restrictions which have been placed in a prior deed, or contained in covenants, conditions, and restrictions. The report will also show standard exceptions that generally apply. Among this latter category of exceptions, there are "gap" exceptions which exclude coverage for matters that show up in the public records after, not before, the effective date of the commitment.

The answer is (D).

SOLUTION 36

The surveyor must provide the width and recording information of all plottable rights of way, easements, and servitudes burdening and benefitting the property surveyed for an ALTA/NSPS Land Title Survey, as evidenced by Record Documents that have been provided to the surveyor. Since this is required, it is therefore not found in Table A.

The answer is (C).

SOLUTION 37

Where there is a conflict between statutes, administrative rules, and/or ordinances established by states or local jurisdictions and the Minimum Standard Detail Requirements for ALTA/NSPS Land Title Surveys, to the extent that the survey cannot be conducted in accordance with both, the more stringent will apply.

The answer is (C).

SOLUTION 38

Based on the direct distance between the two corners being tested, the maximum allowable relative positional precision for an ALTA/NSPS Land Title Survey is 2 cm (0.07 feet) plus 50 ppm (parts per million).

The answer is (B).

SOLUTION 39

Circumstances in the field can result in the maximum allowable relative positional precision being exceeded. In that case, a note is included on the plat or map explaining the site conditions that caused the exceedance.

The answer is (A).

SOLUTION 40

A proper request for an ALTA/NSPS Land Title Survey, according to the Minimum Standard Detail Requirements, does include permission to enter the property to be surveyed, adjoining properties, or offsite easements; the establishment of the optional items listed in Table A that are to be incorporated; and a written authorization to proceed from the person or entity responsible for paying for the survey. However, a request will specify that an ALTA/NSPS Land Title Survey is required, rather than simply an as-built.

The answer is (C).

SOLUTION 41

Trees, bushes, shrubs, and other natural vegetation need not be included in an ALTA/NSPS Land Title Survey other than as specified in the contract, unless they are deemed by the surveyor to be evidence of possession.

The answer is (D).

SOLUTION 42

When a water feature forms a boundary of a property subject to an ALTA/NSPS Land Title Survey, it is necessary to note the date, the attribute that was located, and that it is subject to change due to natural causes. It is not necessary to note whether the feature is navigable or not.

The answer is (D).

SOLUTION 43

When Table A, Item 11 is included in the contract for an ALTA/NSPS Land Title Survey, the surveyor is responsible for including the location of underground features, such as utilities existing on or serving the property. Their location, derived from plans and markings combined with observed evidence, will be inexact, but it will nevertheless be considered adequate, as exact location of underground features requires excavation.

The answer is (A).

SOLUTION 44

When Table A, Item 20 is included in the contract for performing an ALTA/NSPS Land Title Survey, the surveyor may furnish a certificate of professional liability insurance upon request, but it should not be addressed on the face of the plat or map.

The answer is (B).

SOLUTION 45

The chance of a 100-year-flood occurring is 1 in 100, or 1%. The 0.2% chance refers to the 500-year-flood (1 in 500). The 2% chance refers to the 50-year-flood (1 in 50). The 10% chance refers to the 10-year-flood (1 in 10). These definitions are taken from the April 2003 edition of *Guidelines and Specifications for Flood Hazard Mapping Partners: Glossary of Terms* published by the Federal Emergency Management Administration (FEMA).

The answer is (B).

SOLUTION 46

The land area that would be inundated by the 1%-annual-chance flood based on future-conditions hydrology is known as the future-conditions floodplain or flood hazard area. The future-conditions hydrology reveals the flood discharges associated with projected land-use conditions based on a community's zoning maps and/or comprehensive land-use plans and without consideration of projected future construction of flood detention structures or projected future hydraulic modifications within a stream or other waterway, such as bridge and culvert construction, fill, and excavation. Possible future conditions are not shown on Flood Insurance Rate Maps (FIRMs).

The answer is (C).

SOLUTION 47

When a property is incorrectly shown on the Flood Insurance Rate Map (FIRM) within a special flood hazard area (SFHA), the property owner or lessee may apply for a change. One of the instruments of such a change is a Letter of Map Amendment (LOMA), or a Letter of Map Revision. A LOMA is an official amendment, by letter, to an effective Federal Emergency Management Administration (FEMA) map. A LOMA establishes a property's location in relation to the SFHA. There is no appeal period. The letter becomes effective on the date sent.

Another instrument is a Letter of Map Revision (LOMR). This is an official revision, by letter, to an effective FEMA map. A LOMR may change flood insurance risk zones, floodplain and/or floodway boundary delineations, planimetric features, and/or base flood elevations (BFEs). Where new fill is involved, as in this problem, a Letter of Map Revision Base on Fill (LOMR-F) is issued. Any one of these documents may be issued by FEMA to officially remove a property and/or structure from the SFHA.

The procedure for receiving a LOMA or LOMR-F includes submitting to FEMA mapping and survey data for the property. Most often the applicant hires a land surveyor to prepare an Elevation Certificate for the property. A Conditional Letter of Map Revision (CLOMR) is FEMA's comment on a proposed project that would, upon construction, affect the hydrologic or hydraulic characteristics of a flooding source.

The answer is (A).

SOLUTION 48

A National Flood Insurance Program (NFIP) Elevation Certificate is used to certify building elevations and rate buildings constructed after the flood insurance rate maps (FIRMs) are published in Zones A1–A30, AE, AH, A (with base flood elevation (BFE)), VE, V1–V30, V (with BFE), AR, AR/A, AR/AE, AR/A1–A30, and AR/AH. The NFIP Elevation Certificate may be used to support a LOMA or LOMR-F request. Floodproofing certification requires a separate certificate. The NFIP Elevation Certificate cannot waive the necessity of flood insurance. Only a letter of map amendment (LOMA) or letter of map revision based on fill (LOMR-F) can amend the FIRM and remove the requirement that a lending institution compel the purchase of flood insurance. An NFIP Elevation Certificate is not required for pre-FIRM buildings unless the building is being rated under the optional post-FIRM flood insurance rules.

The answer is (C).

SOLUTION 49

eLOMA is a web tool specifically for licensed land surveyors, professional engineers, and FEMA-approved National Flood Determination Association (NFDA) Certified Professionals (CPs). The eLOMA tool does not accept letters of map revision (LOMR). It is not suitable for processing requests for proposed structures, or when the subject of the request is on an alluvial fan.

The answer is (B).

SOLUTION 50

By utilizing construction drawings, a National Flood Insurance Program (NFIP) Elevation Certificate can be completed while a building is under construction, whether all of the floors are in place or not, and the flood insurance can be written. However, in such cases a new Elevation Certificate will need to be done when construction is completed.

The answer is (C).

11 Global Positioning System (GPS)

SPACE SEGMENT

PROBLEM 1

The TRANSIT Doppler satellite positioning system was first released for commercial use in 1967. GPS technology represents an improvement over the earlier TRANSIT system in many respects. Which of the following statements correctly describes some of those improvements?

(A) The orbits of TRANSIT system satellites are more susceptible to variations in local gravity fields and atmospheric drag than GPS satellites.

(B) The TRANSIT system requires that line-of-sight be maintained between the stations of a network, while GPS does not.

(C) The accuracy and efficiency of GPS is significantly improved from the technology that was available when the TRANSIT system was established.

(D) Both A and C are true.

PROBLEM 2

What is the name given to the GPS satellites?

(A) OSCAR

(B) NOVA

(C) NAVSTAR

(D) ATLAS

PROBLEM 3

Which of the following orbital parameters best describes the final constellation of GPS satellites?

(A) six orbital planes inclined 55° with respect to the celestial equator

(B) nearly circular orbits at approximately 26,000 km

(C) nearly circular orbits at approximately 1100 km

(D) both A and B

PROBLEM 4

What is the mask angle in GPS, and what is its significance?

(A) Reflected signals received by a GPS antenna are eliminated by a ground plane that defines the mask angle.

(B) The mask angle is used in special circumstances to eliminate the reception of signals from satellites with unstable clocks.

(C) The mask angle is the zenith at which a particular antenna type ceases to receive signals efficiently and is different for each type.

(D) The mask angle is a vertical angle of 10–20° from the observer's horizon, below which signals from the GPS satellites are considered less reliable.

PROBLEM 5

What are the positioning signal frequencies used in GPS?

(A) 150 MHz and 400 MHz

(B) 420.9 MHz and 585.53 MHz

(C) 1575.42 MHz and 1227.60 MHz

(D) 1832.12 MHz and 3236.94 MHz

PROBLEM 6

The carriers L1 and L2 are modulated by PRNs, or pseudo-random noise codes. What are the names given to these codes?

(A) f_0, f_1, f_3, and f_4

(B) Ψ, Δ, π, and Υ

(C) P, C/A, M, and L2C

(D) IDOT, IODE, OODE, and LDOT

PROBLEM 7

Some PRN codes repeat themselves over specific intervals of time. What are the periods of those cycles, and what is their significance?

(A) The P code transmits for 38 weeks before it repeats itself, but each week of that cycle is assigned to a different satellite. The C/A code repeats itself every millisecond and each satellite transmits a different C/A code, allowing a receiver to distinguish signals received simultaneously from different satellites.

(B) The P code repeats itself every second, during which time it has over 10 million opportunities to switch from a +1 to a −1. The C/A code repeats itself every second and is, therefore, quite easily acquired.

(C) The P code repeats itself every nanosecond. The C/A code repeats itself every week and can be used as a moderately accurate navigation signal as well as a method of rapidly acquiring the P code.

(D) All of the above are true at selected intervals determined by the GPS control segment.

PROBLEM 8

The original GPS design broadcast a civil navigation message modulated onto both of the L1 and L2 carrier frequencies. It is now known as the legacy navigation message, or NAV. What is the frequency of the legacy navigation message, and what is its significance?

(A) The frequency of the legacy navigation message is 100 MHz, and the message contains information on the position of the GPS receiver.

(B) The frequency of the legacy navigation message is 50 Hz and contains information on the status of both the particular satellite and all satellites in the constellation.

(C) The frequency of the legacy navigation message is 1000 MHz and contains information on the clock corrections needed for the satellite timing systems and the ephemerides for all of the GPS satellites.

(D) The frequency of the legacy navigation message is 450 Hz, and the message contains information on the status of all GPS space vehicles, as well as messages, flags, and ionospheric delay parameters.

PROBLEM 9

The Block IIA GPS satellites are large and are equipped to perform several general functions. Which of the following statements correctly characterizes some of those functions?

(A) They receive transmissions from the receivers of GPS users and transmit timing and positioning information to the receivers of GPS users.

(B) They can function continuously for six months without intervention from the control ground station.

(C) They are expected to have an average lifespan duration of over 10 yrs.

(D) Both B and C are true.

PROBLEM 10

Timekeeping is critical to the operation of the GPS system. Which of the following statements best describes the devices used to maintain accurate time for positioning with GPS?

(A) Each GPS Block IIA satellite is equipped with two high-precision oscillators, and GPS receivers do not have on-board clocks.

(B) Most GPS receivers have quartz crystal oscillators as their internal clocks, and each Block IIA GPS satellite has four atomic oscillators.

(C) Most GPS receivers employ hydrogen masers as their timekeeping devices, and each Block IIA GPS satellite uses quartz crystal oscillators as its frequency standard.

(D) Each GPS Block IIA satellite and most GPS receivers employ hydrogen masers as their timekeeping devices.

CONTROL SEGMENT

PROBLEM 11

The GPS control segment consists of several government facilities that are responsible for operating the global positioning system. Which of the following procedures would NOT be executed by the control segment?

(A) continuous tracking of all GPS satellites

(B) creating updated navigation messages

(C) providing precise ephemeris data to the user community

(D) uploading corrected ephemeris data to the satellites

PROBLEM 12

There are several Block IIR(M) GPS satellites that have the capability to broadcast codes that other, older satellites can't broadcast. Which codes are they?

I. L1C

II. L2C

III. M

(A) I and II only

(B) I and III only

(C) II and III only

(D) I, II, and III

PROBLEM 13

What is the name of the carrier signal broadcast by Block IIF GPS satellites?

(A) L2

(B) L3

(C) L4

(D) L5

PROBLEM 14

The remote monitoring stations transfer their data to the Master Control Station, where computed satellite updates are generated and sent to the upload stations. Where is the Master Control Station located?

(A) Hawaii

(B) Ascension Island

(C) Diego Garcia

(D) Colorado Springs, Colorado

GPS RECEIVERS

PROBLEM 15

What is the purpose of a ground plane on a GPS antenna?

(A) The ground plane of a GPS antenna provides a convenient reference for its correct polarization.

(B) The ground plane of a GPS antenna improves its gain and coverage.

(C) The ground plane of a quadrifilar or volute antenna assists the adjustment of its element phasing.

(D) The ground plane of an antenna reduces the interference from reflected signals.

PROBLEM 16

GPS receivers are sometimes distinguished by the number of channels they include. What sorts of channels are involved, and what is their function?

(A) The channels of a GPS receiver are instructions in the software that define the phase center of the antenna.

(B) The GPS receiver must have the capability of isolating signals received simultaneously from several satellites. This isolation is accomplished through the use of dedicated channels.

(C) The channels of a GPS receiver are the number of its intermediate frequency (IF) stages.

(D) None of the above are true.

PROBLEM 17

What is an advantage of a dual-frequency receiver over a single-frequency receiver?

(A) Much of the error introduced by ionospheric delay can be eliminated with a dual-frequency receiver.

(B) Dual-frequency receivers are cheaper than single-frequency receivers.

(C) Dual-frequency receivers use the C/A code and single-frequency receivers must use the P code.

(D) Over long baselines, there is no significant advantage of a dual-frequency receiver over a single-frequency receiver.

PROBLEM 18

Which of the following are some of the major sources of error in GPS measurement that may be attributed to the GPS receiver?

(A) broadcast ephemeris error

(B) multipath

(C) quartz crystal clock biases

(D) both B and C

PROBLEM 19

A cycle slip

(A) is a reduction in the signal frequency caused by selective availability

(B) occurs when a satellite's signal is obstructed and cannot be tracked briefly

(C) is a discontinuity in the P code from a satellite that can be attributed to ionospheric delay

(D) is a discontinuity in the navigation message from a satellite that can be attributed to tropospheric disruption

PROBLEM 20

What is the meaning of the word *epoch* as it applies to GPS?

(A) a tabulation of the position of the GPS satellites

(B) the plane of any great circle including a GPS satellite

(C) an instant of time

(D) the cable that connects the GPS antenna to the receiver

OBSERVABLES

PROBLEM 21

What is the minimum number of satellites that must be available for a differential solution of latitude, longitude, height, and time?

(A) 1

(B) 2

(C) 3

(D) 4

PROBLEM 22

What is a pseudo-range measurement in GPS?

(A) a distance measured between a GPS satellite and receiver based on a time shift that depends on the correlation of the codes each generates to reveal the biased time delay between their respective clocks

(B) a distance measured between a GPS satellite and receiver that depends on the observed phase difference between the carrier signals each generates to reveal the number of wavelengths between the two standards

(C) a measurement of a time series of coordinates using GPS satellites, one stationary receiver, and one roving receiver that depends on the corrected range misclosures between the two receivers

(D) a measurement of the accuracy—that is, the standard deviation—of positions derived from GPS satellites in a particular configuration

PROBLEM 23

What is a carrier beat phase measurement in GPS?

(A) a distance between two satellites determined from the comparison of the Doppler-shifted carriers received from both satellites

(B) a measurement of a time series of coordinates using GPS satellites, one stationary receiver, and one roving receiver that depends on the corrected range misclosures between the two receivers

(C) a distance determined by comparing the difference in phase of the carriers generated by two different satellites

(D) a distance between the satellite and receiver determined from the differencing of the reference carrier signal generated in the receiver with the Doppler-shifted carrier signal received from the satellite

PROBLEM 24

Which of the following statements describes the elements of a single difference solution in GPS?

(A) Two receivers track the same two satellites simultaneously over several epochs.

(B) Two receivers track two satellites during the same epoch.

(C) Two receivers track the same satellite during the same epoch.

(D) All of the above are true.

PROBLEM 25

Double differencing is the prevailing method of baseline determination. Which of the following statements is true of double differencing?

(A) A double difference is the difference of two single differences.

(B) A double difference involves one satellite tracked at two stations over several epochs.

(C) Two receivers tracking two satellites at the same time is a double difference observation.

(D) Both A and C are true.

PROBLEM 26

What is the meaning of DOP in GPS?

(A) dilution of precision—a measurement of the geometric strength of the satellite configuration at a particular moment

(B) an abbreviation used to represent the Doppler shift

(C) dynamic operator position—determination of a trajectory with GPS

(D) the standard deviation of the timing errors from the combination of satellite and receiver clocks

GPS SURVEYING

PROBLEM 27

Considering all the components involved, which of the following is not a factor that affects the accuracy of carrier phase static GPS surveying in the relative mode?

(A) receiver clock bias

(B) modeling of atmospheric refraction

(C) accuracy of satellite orbital data

(D) selective availability

PROBLEM 28

Generally, what category of precision can be routinely achieved in the measurement of baselines by GPS static relative positioning techniques?

(A) 1 part in 20,000 parts

(B) 1 cm + 2 ppm

(C) 1 part in 100,000 parts

(D) 1 cm + 10 ppm

PROBLEM 29

The heights determined by uncorrected GPS measurements refer first to which of the following?

(A) the geoid

(B) a reference ellipsoid

(C) NGVD

(D) mean sea level

PROBLEM 30

Which of the following formulas most closely describes the relationship between ellipsoidal height, h, geoidal height, H, and geoidal undulation, N?

(A) $h = H - N$

(B) $H = \dfrac{h}{N}$

(C) $N = h + H$

(D) $h = H + N$

SOLUTION 1

Neither the TRANSIT system nor GPS requires a line-of-sight between stations for successful satellite positioning. However, GPS has many improvements over the TRANSIT system. For example, the signals from TRANSIT satellites are more likely to be affected by local gravity fields and atmospheric drag than those of GPS satellites. The submeter accuracy of the TRANSIT system has been outstripped by GPS, and observations using GPS technology are more efficient than those made with the TRANSIT system.

The answer is (D).

SOLUTION 2

The GPS satellites are named NAVSTAR, an acronym for Navigation System by Timing And Ranging. The NAVSTAR number, such as NAVSTAR 8, indicates the launch sequence of the space vehicle to which it is assigned.

The answer is (C).

SOLUTION 3

The full constellation of GPS satellites is designed to provide four satellites above the horizon everywhere on the earth's surface, 24 hours a day. An approximate 26,000 km orbit provides an orbital period of nearly 12 hours, though each satellite will rise four minutes earlier each day, and the orbit will be nearly circular.

The answer is (D).

SOLUTION 4

The signal from NAVSTAR satellites must pass through a thicker layer of the atmosphere when the satellites are close to the observer's horizon. Variations of the signal, called *tropospheric errors*, are especially unpredictable within the mask angle area, the signals are generally not considered reliable in that region.

The answer is (D).

SOLUTION 5

The two frequencies broadcast by GPS satellites for positioning are 1575.42 MHz, known as L1, and 1227.60 MHz, known as L2. Both are derived from the fundamental frequency 10.23 MHz. L1 is created by the multiplier 154 and L2 by the multiplier 120.

The answer is (C).

SOLUTION 6

The PRN codes are known as the *P code*, the *C/A code*, the *M code*, and the *L2C code*. GPS codes consist of positive and negative ones (+1 and −1). Both carriers, L1 and L2, are modulated by the P, or precision, code and the M, or military, code; the C/A code, or coarse/acquisition code, is found only on L1. L2C is found on L2. The C/A code and L2C are emitted at a frequency of 1.023 MHz; the P code and the M code have frequencies of 10.23 MHz.

The answer is (C).

SOLUTION 7

The C/A code is designed for rapid acquisition by a receiver. The repetition cycle of the C/A code is very short: 1 ms. Many GPS receivers use only the C/A code. The P code is used to assign a PRN number to each GPS satellite; this number differs from the NAVSTAR, or launch sequence number. The PRN number indicates the weekly code sequence assigned to a particular satellite. For example, PRN 12 would consistently be assigned to the 12th week of the 38 week-long P code sequence. This system allows for the transmission of mutually exclusive P code sequences from each space vehicle.

The answer is (A).

SOLUTION 8

The legacy navigation code (NAV), or satellite message, is transmitted at a frequency of 50 Hz, and has information on the system status, clock behavior, GPS time, the ephemerides of the satellites, etc. The full message is 1500 bits long with 25 frames, and is divided into five subframes of 300 bits each. Each subframe consists of 10 words, with 30 bits per word. Every subframe starts with the telemetry word (TLM), which is followed by the handover word (HOW). The HOW provides time information (seconds of the GPS week), allowing the receiver to acquire the week-long P(Y)-code segment.

The answer is (B).

SOLUTION 9

GPS is a passive system, meaning that the satellites do not receive any transmissions from the receivers of GPS users. The only information received by the satellites is transmitted by the official government GPS control ground stations, or upload stations.

The answer is (D).

SOLUTION 10

Every Block IIA satellite has four atomic high-precision oscillators, two cesium and two rubidium, the former being the more accurate and the latter the less expensive. Only one of these four clocks is used at a time; the others provide redundancy. Most GPS receivers employ oven-controlled quartz crystal oscillators as their time standard. The oscillators depend on the piezoelectric effect for their capabilities. They are inexpensive, compact, and consume very little power.

The answer is (B).

SOLUTION 11

The control segment monitors the satellites and produces corrected navigation messages that it uploads on a daily basis. However, it is not the responsibility of the control segment to provide post-computed precise ephemeris data to the civilian-user community. Several organizations, both public and private, provide this service.

The answer is (C).

SOLUTION 12

The Block IIR(M) satellites broadcast the L2C, as well as the military code known as the M-code. L2C is actually composed of the L2CM and L2CL codes, and is more sophisticated than the older coarse/acquisition (C/A) code.

The answer is (C).

SOLUTION 13

The L5 carrier signal provides GPS with a third carrier signal that is higher powered and has a greater bandwidth than previous satellites. L5 also has improved signal design. Unlike L2C, L5 is broadcast in a radio band designated by the International Telecommunication Union (ITU) for the Aeronautical Radio Navigation Services (ARNS). As such, L5 is not prone to interference with ground-based navigation aids and is available for aviation applications.

The answer is (D).

SOLUTION 14

The Master Control Station is in Colorado Springs, Colorado. Monitoring stations are located at all of the other places listed, and there are upload facilities at all but Hawaii.

The answer is (D).

SOLUTION 15

Some GPS antenna types are more susceptible than others to interference from GPS signals that bounce off nearby objects or the ground. The reflected signals, called *multipath*, present a confusing contradiction to the signals received directly from the satellite. The ground plane of a GPS antenna reduces the effect of multipath by blocking most of these reflected signals below the antenna.

The answer is (D).

SOLUTION 16

Channels are dedicated to particular satellites at particular times within a GPS receiver. Through the use of software, hardware, or a combination of both, the separate satellites' signals can be distinguished by their unique codes or their particular Doppler shifts.

The answer is (B).

SOLUTION 17

The time delay of a signal passing through the ionosphere is proportional to the inverse of the square of the frequency of that signal. The higher the frequency, the shorter the time delay. In other words, L1 passes through the ionosphere faster than L2, allowing fairly accurate calculation and near elimination of the ionospheric delay using dual-frequency GPS receivers. This is perhaps the main advantage of a dual-frequency GPS receiver over a single-frequency receiver.

The answer is (A).

SOLUTION 18

Multipath, the reflection of the satellite signal, and errors of the receiver's on-board oscillator will cause errors in GPS measurements attributable to the receiver.

The answer is (D).

SOLUTION 19

A cycle slip is caused by a brief obstruction of the signal from a satellite. The receiver is said to "lose lock," and while the measured phase may be the same when the signal from the satellite is reacquired, a number of cycles will have been lost.

The answer is (B).

GPS

SOLUTION 20

An epoch is an instant of time. The term is most frequently used to describe the time-tagged moments of data reception.

The answer is (C).

SOLUTION 21

The solution of all four variables requires observation of at least four satellites simultaneously.

The answer is (D).

SOLUTION 22

Nearly all civilian receivers that use pseudo-ranges do so with the C/A code. The code is generated in the receiver, derived from its own clock, and is correlated with the same code received from the satellites. The objective of this correlation is to determine the transit time of the signal from the satellites to the receiver. This time multiplied by the speed of light can provide the distance between the satellite and the receiver. The process is complicated by the fact that the clock of the receiver is not synchronous with the satellites' clocks, nor are the satellites' clocks synchronous with one another. These timing errors cause the derived distance to differ somewhat from the actual distance, giving rise to the term *pseudo-range*.

The answer is (A).

SOLUTION 23

The carrier beat phase is the phase remaining when the received satellite carrier signal is differenced (beat) with the reference frequency generated in the receiver. This is the fundamental tool of the codeless GPS measurement since it often involves squaring the channel to create the second harmonic of the carrier. The codes cannot be retained with this technique, and the method presents a cycle ambiguity that must be solved.

The answer is (D).

SOLUTION 24

Single difference has the desirable effect of removing clock errors associated with the satellites. Receiver clock errors remain, however.

The answer is (C).

SOLUTION 25

Double differences are a combination of two single differences and can eliminate the effects of the clock errors in both the satellites and the receivers.

The answer is (D).

SOLUTION 26

Dilution of precision is a term that predates GPS and refers to a measurement of geometric strength. For example, a GPS position determined when all the observed satellites are bunched together in one part of the sky would not have a good geometric dilution of precision (GDOP). A very good GDOP would be available if one satellite were directly overhead and three others were 15° above the horizon and 120° apart in azimuth at the moment of observation.

The answer is (A).

SOLUTION 27

With station spacings under 100 km, selective availability had no adverse effects. Double differencing techniques can virtually eliminate the satellite clock errors that are the foundation of selective availability. Therefore, selective availability did not affect the accuracy of the relative positions determined by a GPS survey. All of the other aspects mentioned do affect accuracy. Selective availability was turned off May 2, 2000.

The answer is (D).

SOLUTION 28

The precision that may be attained with any surveying equipment is dependent on the methods used, and GPS is no exception. However, with proper data processing and well-planned observation strategies, precisions of 1 cm + 2 ppm are routinely available. Even this level of precision is somewhat conservative; GPS is capable of 0.3 cm + 0.01 ppm with precise ephemeris information.

The answer is (B).

SOLUTION 29

The heights determined by GPS measurements are originally ellipsoidal. These heights are then generally converted to orthometric heights, also known as *elevations*, by modeling the differences between the geoid and reference ellipsoid at the newly observed stations, using connections to control points with known orthometric heights.

The answer is (B).

SOLUTION 30

The ellipsoidal height is the sum of the geoidal height and the separation between the ellipsoid and the geoid, also called the *geoidal undulation*. The calculation does presume there is no significant deviation of the vertical. In the coterminus U.S., the geoid is actually below the ellipsoid, giving N a negative value.

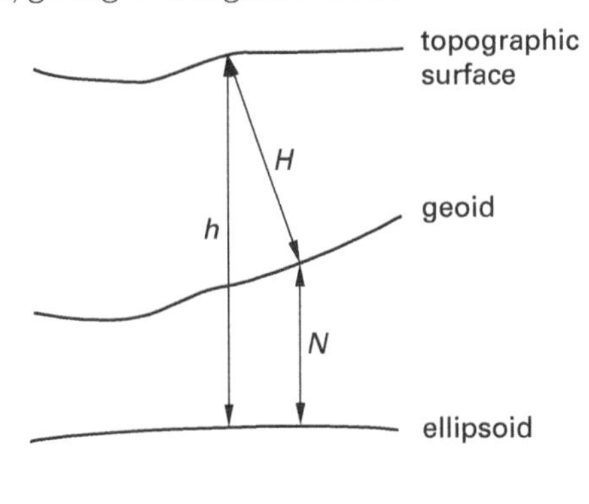

$$h = H + N$$

The answer is (D).

12 Project Management

CLIENT RELATIONS

PROBLEM 1

If a surveyor agrees to locate a particular property line for a client and the client assures the surveyor that an approximate location will be adequate, which of the following statements characterizes the surveyor's best course of action?

(A) The surveyor should be certain the client understands that the monuments he establishes should not be relied upon to locate any improvements.

(B) The surveyor should word the certification of his plat in such a way as to eliminate any liability for reliance on the location he establishes.

(C) The surveyor should be mindful that people will rely on his monuments and either locate the boundary to the best of his ability or not accept the job.

(D) The surveyor should mark the line well inside his client's property to allow for a margin of safety.

PROBLEM 2

Which of the following statements concerning fees is NOT correct?

(A) The liability incurred by a surveyor is proportional to the fee that the client is charged.

(B) In determining the fee to be charged, it is proper to consider whether the work is for a regular client.

(C) In determining the fee to be charged, it is proper to consider the amount of liability involved.

(D) In determining the fee to be charged, it is proper to consider the usual charges for similar work.

PROBLEM 3

Which of the following statements describes a service that a client CANNOT reasonably expect from a surveyor?

(A) A surveyor will avoid provoking disputes with the client's adjoiners.

(B) A surveyor will maintain the confidentiality of communications with the client.

(C) A surveyor will not agree to perform services for which he or she is not qualified.

(D) A surveyor will hold the interests of the client paramount if they come into conflict with the public interest.

PROBLEM 4

If a regular client of one surveyor asks another surveyor to review a property survey that is currently being done by the first surveyor, which of the following questions should the potential reviewer be certain to ask?

(A) What does the client suspect is wrong with the survey?

(B) What was the fee charged by the other surveyor?

(C) How much is the client prepared to pay?

(D) Is the other surveyor aware that the client is requesting such a review?

PROBLEM 5

A regular client has brought to a surveyor's office a plat made by another surveyor. The client wants the surveyor to recertify the plat. The client is anxious to save money and wishes to have the recertification without incurring the cost of any more field work. What should the surveyor do?

(A) Charge a fee for the recertification that is large enough to compensate him for any unforeseen liability.

(B) Call the surveyor who did the work and ask to examine the field notes from the project before recertifying the plat.

(C) Recertify the plat for a fee commensurate with the time and effort required to do so.

(D) Cordially explain to the client the reasons for refusing to do as he asks.

PROBLEM 6

What is a lien?

(A) a contract describing services agreed upon between two surveyors

(B) a court order allowing a surveyor the right of entry onto private property for the performance of a survey without the owner's consent

(C) a defamation of another expressed in writing

(D) an encumbrance placed on property to secure payment

PROBLEM 7

What is the standard traditionally adopted by the courts to determine what constitutes negligence by a land surveyor when providing service to a client?

(A) whether a surveyor exercised the degree of care that a surveyor of ordinary skill and prudence would exercise under similar circumstances

(B) whether a surveyor made errors in judgment that resulted in damage to others

(C) whether a surveyor's work was reviewed and found to fall short of absolute correctness

(D) whether a surveyor exercised the degree of care that the client expected

PROBLEM 8

A client and the client's neighbor wish to execute a property line agreement concerning their common boundary. The client has asked a surveyor to help them execute the agreement. What should the surveyor do?

(A) The surveyor should be sure that there is uncertainty on the part of both parties concerning the exact location of the property line.

(B) The surveyor should reduce the terms of the agreement to writing, describing the line and the monuments by which it may be determined, to avoid any misunderstanding as to the line agreed upon.

(C) The surveyor should first lay out the line described in the deed and show both parties the results of the measurements before they agree on any other line.

(D) Both A and B are appropriate.

PROBLEM 9

Which of the following methods may work to discharge a contract written expressly for surveying services?

(A) mutual agreement of the parties involved

(B) impossibility of performance of the contracted work

(C) the bankruptcy of one of the parties to the contract

(D) all of the above

PROBLEM 10

Which of the following statements describes a disadvantage inherent in a fixed-price contract for surveying services?

(A) Errors or ambiguities in the project's plans that become apparent after the contract has been executed can lead to disagreements.

(B) The contract may seriously impair the incentive to work efficiently and economically since every increase in cost benefits the surveyor.

(C) The actual cost of performing the work may be substantially more than the amount bid.

(D) Both A and C are true.

PROBLEM 11

Which of the following is NOT required for a contract to be valid?

(A) mutual agreement

(B) written form

(C) lawful purpose

(D) mentally competent parties

PROBLEM 12

Creating a written agreement with a client before surveying services are performed offers several advantages to both the surveyor and the client. One type of written agreement is a fixed price contract. A fixed-price contract should NOT include which of the following?

(A) the scope and the cost of the service

(B) other duties as assigned

(C) an estimate of the time of completion

(D) a payment schedule

ETHICS AND LIABILITY

PROBLEM 13

Which of the following activities is NOT prohibited by most codes of ethics for surveyors?

(A) advertising in a self-laudatory manner

(B) practicing in an area of the profession where the surveyor is not proficient

(C) accepting remuneration other than for services rendered

(D) expressing an opinion as to who owns the land when testifying as an expert witness in court

PROBLEM 14

In retracing the work of another surveyor, a surveyor discovers what appears to be a significant error in the position of a particular monumented corner that controls two property lines. The land involved is quite valuable and the client has recently constructed improvements in reliance on the corner. What should be the first action the surveyor takes?

(A) The surveyor should file charges against the other surveyor with the state board of registration.

(B) The surveyor should inform the client that the improvements he has constructed encroach on the adjoining property.

(C) The surveyor should check his own work.

(D) The surveyor should call the other surveyor and share his findings.

PROBLEM 15

Why should a surveyor maintain a comprehensive filing system that includes all the work he performs?

(A) It is mandated by nearly all state laws that regulate land surveying.

(B) It is a professional responsibility.

(C) It is an essential strategy in limiting the surveyor's liability.

(D) It can be an excellent source of income.

PROBLEM 16

A developer is about to begin a large subdivision and has asked a surveyor to bid on the surveying necessary for the first phase. During a subsequent meeting, the developer mentions that he has received a lower estimate from one of the surveyor's competitors. What should the surveyor do?

(A) Offer the developer a lower price than the competitor.

(B) Remind the developer that he will get what he pays for.

(C) Suggest a joint venture between the surveyor and the competitor on this project.

(D) none of the above

PROBLEM 17

A surveyor with special expertise in riparian boundaries is asked to testify as an expert witness at a trial in St. Louis, Missouri. The offer includes the surveyor's travel expenses and a fee. The surveyor has another client in the same vicinity who would like a consultation for a separate project. What should the surveyor do?

(A) Inform both parties of the two objectives and divide the travel expenses between them.

(B) Accept no remuneration for testifying.

(C) Personally pay the travel expenses to St. Louis.

(D) There is no reason that either of the two parties need know anything of the other.

PROBLEM 18

A contract for the surveying services necessary to construct a residential subdivision contains the following language. Which stipulation would be most likely to cause the surveyor to risk unnecessary liability or unethical practice?

(A) All surveying services will be billed per the agreed unit prices.

(B) The surveyor agrees to prepare all necessary drawings and certifications for presentation to the appropriate governmental agencies.

(C) The surveyor agrees to satisfy all state and local regulations regarding monumentation of property corners.

(D) Payment for surveying services will be rendered only after all such work has been performed to the satisfaction of the property owner and developer.

PROBLEM 19

A surveyor is asked to perform a survey of public lands under contract to the Bureau of Land Management (BLM). However, the surveyor is not licensed in the state where the work must be done. What should the surveyor do?

(A) Decline the project.

(B) Find a licensed surveyor in the state who will assume responsible charge for the project.

(C) Accept the project.

(D) Investigate whether the surveyor may acquire a license in the state by comity.

PROBLEM 20

A former client discovers that the lot corner monuments a surveyor established eleven years ago in the client's subdivision were not placed in the position shown on the surveyor's plat. The surveyor has been summoned to appear in court to respond to the client's claim of a faulty survey. The surveyor is aware that a three-year statute of limitations for such action is codified in the state's laws, but not a ten-year statute of repose. The surveyor will most likely be liable. Why?

(A) The court will likely consider the case in the light of the discovery rule.

(B) The client is an attorney.

(C) There is no ten-year statute of repose in the state.

(D) both A and C

PROBLEM 21

An employee reports what is believed to be fraudulent work conducted by a surveyor's firm to the lawful authorities. The surveyor is certain that the report is groundless. Which of the following actions is most likely to be prohibited?

(A) termination of the employee for making the report

(B) revealing the identity of the employee who made the report

(C) immediate investigation into the veracity of the report

(D) none of the above

PROBLEM 22

To minimize the risk associated with a post-construction drawing, which of the following names for such drawings is frequently preferred by a surveyor's insurance carriers?

(A) record

(B) red-line

(C) as-built

(D) measured

PROBLEM 23

Which of the following types of professional services contracts provides the least risk mitigation for a surveyor?

(A) oral agreement

(B) purchase order

(C) letter agreement

(D) written contract

PROBLEM 24

Which language would be LEAST effective in preventing improper changes or reuse by a client or others if included in a transfer agreement of an electronic surveying document, such as a CAD drawing?

(A) "Hard copies retained by the surveyor control over subsequent changes made by others."

(B) "Reuse without the surveyor's verification is at the client's risk."

(C) "The document is copyrighted."

(D) "The client indemnifies the surveyor from any claim or liability for injury or loss resulting from reuse."

ECONOMICS

PROBLEM 25

Joe Foster is paid an hourly wage of $18.50. He has worked 40 hr this week at the regular rate and 5.75 hr of overtime at time and a half. His net pay for the week was $638.69. What percentage of his gross wages was deducted?

(A) 18%

(B) 25%

(C) 29%

(D) 30%

PROBLEM 26

The Remler Surveying Company has eight employees and a monthly payroll of $35,520.00. The firm pays 11% of the gross wages into a pension plan and $35.00 per employee per month for life insurance premiums. What is the weekly cost of these fringe benefits to the nearest dollar?

(A) $974

(B) $1996

(C) $2006

(D) $3046

PROBLEM 27

An investment of $12,000 was made at an interest rate of 7.5% per annum. What is the value of the investment after three years?

(A) $12,930.35

(B) $14,169.20

(C) $14,700.00

(D) $14,907.56

PROBLEM 28

A surveyor landed a lucrative government contract and needed additional equipment and personnel on short notice. The surveyor borrowed $20,000.00. The loan was discharged five years later with a total payment of $30,772.48. What was the interest on the loan if it was compounded annually?

(A) 8.0%

(B) 8.5%

(C) 9.0%

(D) 9.5%

PROBLEM 29

A lot was purchased in January 1988 for $10,000.00. The following end-of-year taxes were assessed: $300.00 in 1988, $350.00 in 1989, $400.00 in 1990, and $500.00 in 1991. The taxes for 1988 were paid, but the taxes for the other three years were not. At the end of 1991 the lot was sold for $20,000.00, and the owner, now the seller, paid the back taxes at an interest rate of 9% compounded annually, as well as a 10% commission to the agent. To the nearest dollar, what portion of the $20,000.00 selling price remained?

(A) $12,495

(B) $14,771

(C) $15,382

(D) $16,648

PROBLEM 30

What is the meaning of the term *depreciation* in financial matters?

(A) the amortizing of debt

(B) the quarterly compounding of interest

(C) the process of increasing the value of equipment

(D) the decrease in value caused by the deterioration of an asset over time

PROBLEM 31

What is the meaning of the term *sinking fund* in financial matters?

(A) It is an interest-bearing account into which periodic deposits are made.

(B) It is synonymous with a depletion allowance.

(C) Establishing a sinking fund is a method of postponing a balloon payment.

(D) It involves uniformly increasing cash flow with a uniform gradient factor.

PROBLEM 32

An EDM purchased for $14,000 is expected to provide service for nine years and has an estimated salvage value of $3200. What is the annual depreciation for the instrument using the straight-line method?

(A) $1200

(B) $1350

(C) $1431

(D) $1500

PROBLEM 33

What is meant by the term *discount rate*?

(A) the rate which determines the value of an investment prior to its maturity

(B) the rate a bank is likely to charge its best customers for loans

(C) the rate of depreciation

(D) the change expected in the interest rate over a particular time period

PROBLEM 34

A surveyor holds a promissory note that will have a value of $3000.00 at maturity two years hence. The surveyor has an immediate need for cash and can obtain the money by assigning the note to another party. This party will pay the surveyor today the sum that is the equivalent of the $3000.00 two years from now. At an interest rate of 10%, most nearly what amount will be paid if the note is discounted today?

(A) $1590

(B) $2480

(C) $2700

(D) $3300

PROBLEM 35

Which of the following is NOT required to have a federal employer identification number (EIN) for its business related dealings with the Internal Revenue Service (IRS)?

(A) a partnership

(B) a new owner of an existing corporation whose previous owner had a valid federal EIN

(C) a sole proprietor without employees

(D) a sole proprietor with employees

PROBLEM 36

In an open, competitive market, the most probable price a real property would realize from a well-informed and prudent buyer that has no special relationship with the seller, is known as

(A) value-in-use

(B) market value

(C) insurable value

(D) investment value

SOLE PROPRIETORSHIPS

PROBLEM 37

Which of the following business structures is NOT a separate legal entity from the owner so that any income or liability incurred by the business is actually incurred by the owner?

(A) corporation

(B) sole proprietorship

(C) partnership

(D) none of the above

OPERATION ANALYSIS AND OPTIMIZATION

PROBLEM 38

Properly implemented operation analysis

(A) models existing processes

(B) provides the information needed to optimize procedures

(C) reveals bottlenecks

(D) all of the above

CRITICAL PATH ANALYSIS

PROBLEM 39

A disadvantage of using critical path analysis (CPA) to monitor the achievement of project goals is that CPA does NOT

(A) show how activities are linked sequentially

(B) clearly establish the relation of tasks-to-time in non-routine projects.

(C) identify the minimum length of time needed to complete a project

(D) apply well to analyzing complex projects

TESTIMONY

PROBLEM 40

A deposition is an oral testimony under oath in response to questions from the opposing attorney and is transcribed by a court reporter. Which of the following should an expert witness NOT do during a deposition?

(A) Accept the judgment of one's attorney as to whether or not to answer a question.

(B) Avoid estimating time or distance.

(C) Present a new document to back up an answer.

(D) Answer the question, "Have you discussed your testimony with anyone previously?"

PROBLEM 41

In the testimony of a land surveyor serving as an expert witness, which of the following is an example of evidence that will most likely NOT be accepted as an exception to the hearsay rule?

(A) proof of boundaries by "reputation in a community"

(B) records of regularly conducted activity made at the time by a person with direct knowledge

(C) recounting a remembered statement made by a property owner, now deceased, about the owner's objection to the staking of a property line

(D) recorded public records

PROBLEM 42

A surveyor giving expert testimony may correctly do which of the following?

(A) Refuse to answer a question because answering will cause civil liability.

(B) Accept a fee for testifying whose payment is contingent on the outcome of the litigation.

(C) Answer the question asked without volunteering information.

(D) Answer a question not fully understood.

LAND DEVELOPMENT PRINCIPLES

PROBLEM 43

A county's comprehensive plan for development is NOT

(A) a formal statement of the community's goals

(B) legally binding like the zoning ordinances

(C) the result of both the public and private sectors' needs and desires

(D) the result of economic, social, and demographic analysis

PROBLEM 44

A typical purpose for a land use map is to

(A) provide absolute information that is used to restrict use to only those purposes set out on the map

(B) include labels on outlined districts that correspond to zoning ordinances in the jurisdiction

(C) provide a guide for future growth and development to achieve the policies set out in the comprehensive plan

(D) represent what the community looked like in the past

PROBLEM 45

A jurisdiction specified a floor area ratio (FAR) of 2.0 on its zoning map and text for a district. On a 50,000 ft^2 parcel, which of the following buildings would be at the allowable limit?

(A) one-story building that occupied the entire parcel

(B) two-story building that occupied one-half of the parcel

(C) four-story building that occupied one-quarter of the parcel

(D) eight-story building that occupied one-quarter of the parcel

PROBLEM 46

Jurisdictions impose subdivision regulations on land developments to

(A) ensure the continuity of systems and new infrastructures will be compatible with the development of the larger community

(B) prohibit residents and businesses from occupying the buildings before all of the land development improvements are completed

(C) allow the release of performance bonds before the project is accepted by the public authorities

(D) delay the approval of submitted subdivision development plans

PROBLEM 47

Which of the following is common among both zoning regulations and restrictive covenants?

(A) They are imposed and enforced by governmental authorities.

(B) They are subject to public hearing and notice requirements when a particular property requests a change.

(C) They limit the use of the property.

(D) They are written in the deed.

PROBLEM 48

The plan of a subdivision is being prepared. The client, who is a developer and not a builder, wishes to have some control over the builders during the construction work and the subsequent marketing and sales efforts. Which of the following is the best course of action for a surveyor to take?

I. Refuse to advise the client on such issues.

II. Share information with the client concerning the general bounds of covenants and restrictions on subdivision plans.

III. Inform the client that only government restrictions can properly be shown on a subdivision plat.

IV. Suggest the client consider covenants and mention that an attorney should review them and write the final draft.

(A) I only

(B) III only

(C) I and III

(D) II and IV

PROBLEM 49

As a project progresses from the preliminary to the final stages, a final subdivision plan takes shape. Which of the following are typical of a final subdivision plan?

I. Streets, drainage, and public utilities have either been completed, or there is a performance bond in place to guarantee their completion before the final subdivision plan's approval.

II. The final subdivision plat will be approved and recorded after the sale of lots has begun.

III. Buildings cannot be constructed before the final subdivision plat has been approved and recorded.

IV. Approved construction documents are not usually part of the final subdivision plan.

V. The final subdivision plat is part of the final subdivision plan.

(A) I, II, and III

(B) I, III, and IV

(C) I, III, and V

(D) II, III, and IV

PROBLEM 50

Sketch, preliminary, and final subdivision plans are submitted to various agencies for their approval. When addressing comments for subsequent submissions, there are guidelines to keep in mind. Which of the following statements is correct?

I. Once the agencies have approved the plans, the liability rests with them instead of the engineer and/or surveyor.

II. All the reviewer's comments are written directly on the plans.

III. After the reviewer's comments are completed, the plans are returned for corrections.

IV. A good rapport with the reviewers is important and valuable.

(A) IV only

(B) III and IV

(C) I, II, and IV

(D) II, III, and IV

PROBLEM 51

A sketch plan and a preliminary plan of a proposed subdivision have some similarities and some differences. Which of these statements is correct?

I. A sketch plan usually does not contain the same level of detailed engineering plans as does a preliminary plan.

II. The sketch plan usually indicates, at a conceptual level, the feasibility and some design characteristics of the development

III. Sometimes the sketch plan and the preliminary plan may be the same document.

IV. The approved preliminary plan should be followed throughout the project's life as closely as possible; altering it may necessitate beginning the whole process anew.

V. A properly approved preliminary plan gives the developer the right to begin construction.

(A) I, III, and V

(B) II, IV, and V

(C) I, II, III, and IV

(D) I, II, III, IV, and V

PROBLEM 52

What is a typical difference between the regulations governing zoning and subdivision?

(A) Zoning provides the comprehensive plan for land use; subdivision regulations govern the creation of lots, blocks, streets, easements, etc.

(B) Zoning is concerned with the details of specific land development projects and related services, such as water, sewer, gas, and electricity; subdivision regulations specify the allowable type and density of development on parcels of land by district.

(C) Zoning governs the process for dividing land into lots; subdivision regulations divide a locality into more general areas, i.e., agriculture, industry, commercial, etc.

(D) Zoning governs the creation of lots, blocks, streets, easements, etc.; subdivision regulations provide the comprehensive plan for land use.

PROBLEM 53

Which of the following statements about a master deed is NOT correct?

(A) It is an instrument by which a homeowner's association or condominium association lists limits on the utilization of the properties covered.

(B) It is an instrument by which a condominium developer converts a single property of multiunit buildings into individually owned units that share ownership in common areas.

(C) It is an instrument that sets forth the percentage interest in the common elements of a condominium attributable to each individual unit.

(D) It is an instrument that conveys real property from one person to another.

PROBLEM 54

Subdivision regulations typically govern which of the following divisions of land?

(A) those created by order of a court of record

(B) those that create an interest in oil, gas, minerals, or water severed from the surface ownership

(C) those that create lots for residential purposes

(D) those created in accordance with land use planning

PROBLEM 55

What is the purpose of an environmental site assessment?

(A) to address the potential effects on the environment of a proposed project

(B) to determine if endangered species are affected by a proposed project

(C) to identify environmental contamination liabilities from past or present activities on a property

(D) to verify soil compaction testing

PROBLEM 56

A planned unit development (PUD) does NOT typically include which of the following?

(A) single-family homes

(B) condominiums

(C) commercial spaces

(D) agricultural land

PROBLEM 57

What services are NOT supplied by the surveyor in a typical land development process?

(A) boundary surveying

(B) topographic surveying

(C) final horizontal alignment of the infrastructure and buildings

(D) determination of land ownership

PROBLEM 58

Which of the following would NOT be a topographic design consideration for a land development project?

(A) swales

(B) inundation areas

(C) steep slopes

(D) traffic flow patterns

PROBLEM 59

Subdivision regulations typically do NOT specifically address which of the following water-related issues?

(A) floodplain boundaries and elevation information

(B) conformance of lot building sites with floodplain zoning regulations

(C) dedication of easements for watercourse or drainage crossings

(D) wetland regulation

PROBLEM 60

Zoning standards are adopted to achieve the comprehensive plan and typically do NOT address which aspects of new land development projects?

(A) land use

(B) location, sizes, and heights of buildings

(C) drawing of subdivision plats

(D) division of a jurisdiction into districts

PROBLEM 61

Why is the survey of site topography an important step in the land development process?

(A) It reveals drainage and floodplain issues and informs the design of site grading.

(B) It locates property boundaries.

(C) It creates lots by protraction.

(D) It eliminates the need for site grading.

PROBLEM 62

Which of the following easements might be required to preserve the historical, architectural, archaeological, or cultural aspects of real property?

(A) conservation easement

(B) easement in gross

(C) easement by estoppel

(D) prescriptive easement

PROBLEM 63

Which of the following questions would NOT typically be a pertinent when considering access to a land development project from an existing public road?

(A) Will there be acceptable sight distances at the intersection?

(B) Can the intersection of the land development entrance and the public road be built within the width and grade requirements?

(C) Will the existing public road be able to handle the increased traffic?

(D) Was the right-of-way of the public road dedicated by deed or not?

SOLUTION 1

The surveyor must be aware that the public will accept the correctness of his or her monuments without question. Location of a boundary should be undertaken with the intention of providing its correct position, or else it should not be attempted at all.

The answer is (C).

SOLUTION 2

Potential liability may be considered in fixing the fee; however, the liability will not be proportional to the fee charged.

The answer is (A).

SOLUTION 3

The surveyor is bound to hold the public interest above the interests of the client if the two come into conflict.

The answer is (D).

SOLUTION 4

The first consideration should be whether the other surveyor has been informed of the request for a review. In order to avoid impropriety, such a review should be undertaken only if the other surveyor has been informed and has given his consent.

The answer is (D).

SOLUTION 5

The recertification of a plat requires that the surveyor accept full responsibility for all the information shown on the document. Clearly, a prudent surveyor is not likely to do such a thing without fully examining the entire project. A client may well be unaware of the burden assumed by a surveyor recertifying a plat. The surveyor should carefully explain his refusal to recertify the plat.

The answer is (D).

SOLUTION 6

Most state laws allow a surveyor to place a mechanic's lien on the property of a client who refuses to pay for services rendered. The lien creates a nonpossessory claim against the property as security until the debt is satisfied.

The answer is (D).

SOLUTION 7

The general standard, reinforced by many court cases, has been whether the surveyor performed the work with the same skill and care that is exercised by others in the profession.

The answer is (A).

SOLUTION 8

A property line agreement in writing is sometimes the best solution for two adjoining property owners who are uncertain as to the location of their common boundary. However, a surveyor who insists on laying out the line described in the deed prior to such an agreement may destroy the force of the agreement itself. The courts have consistently held that the parties may decide the location of their common boundary as long as they both are uncertain of its original location and are both satisfied and clear about the line they settle upon.

The answer is (D).

SOLUTION 9

A contract may be discharged by performing the work to the satisfaction of the parties involved, by exercising the right to terminate the contract (where the contract expressly confers such a right), and by several other methods. All of the methods listed will discharge a contract, including the impossibility of performing the contracted work. For example, a violent storm that washes away a parcel of coastal land would terminate a contract to survey that land.

The answer is (D).

SOLUTION 10

Usually awarded on the basis of competitive bidding, a fixed-price contract has the benefit of a high degree of certainty in cost for the client. However, there are drawbacks for the contractor, including extra work that may arise from uncertainties in plans and specifications.

The answer is (D).

SOLUTION 11

A written form is not a requirement for a contract to be valid. Oral contracts are valid and enforceable; however, they are almost never well-advised. When the terms of an agreement remain unwritten and something goes wrong, contention is likely.

The answer is (B).

SOLUTION 12

An open-ended clause, such as "other duties as assigned," is an invitation for scope creep and is not compatible with a fixed-price contract.

The answer is (B).

SOLUTION 13

A surveyor may express an opinion when testifying as an expert witness; in some situations, the court has allowed such opinions to involve questions of ownership. Most codes of ethics for surveyors do not prohibit such testimony.

The answer is (D).

SOLUTION 14

The first action a surveyor should take when finding what appears to be an error in the work of others is to thoroughly check his own work. If this does not resolve the discrepancy, the next step should be to consult with the other surveyor involved. It is quite possible that the first surveyor has information that the second is not aware of.

The answer is (C).

SOLUTION 15

An all-inclusive filing system is not required by law and will not limit the extent of a surveyor's liability for the work performed, but it is a professional responsibility. For example, a client should be able to return to the surveyor who began his project some time ago and find complete records of the work that was performed.

The answer is (B).

SOLUTION 16

Professions such as surveying do not compete on the basis of price. A client who tries to obtain an advantage by pitting surveyors against each other on the basis of their fees should find them unwilling to play along.

The answer is (D).

SOLUTION 17

The best course is to inform both parties fully and divide the necessary travel expenses between them. It is certainly ethical to accept remuneration for expert testimony, but it is not ethical to accept any pay under what might be construed as false pretenses.

The answer is (A).

SOLUTION 18

A surveyor must adhere to standards set by state regulatory agencies and its code of ethics. If these rules contradict the wishes of the client where contractual language specifies that the client must be satisfied, an irresolvable conflict may develop.

The answer is (D).

SOLUTION 19

A surveyor working in public lands exclusively under the auspices of the BLM is not required to be licensed in the state where the work is done.

The answer is (C).

SOLUTION 20

The discovery rule indicates that the statute of limitations does not begin to run until the client discovers the alleged error. Absent the discovery rule, a surveyor could argue that under the statute of limitations, liability for an error made in a survey would expire after the stipulated period codified in the state—in this case three years. However, many courts consider it unfair for a plaintiff to lose the right to sue because the statute of limitations has run its course before the error is discovered. On the other hand, if the state had enacted a ten-year statute of repose, the actual cause of the action would be extinguished after that period regardless of when the alleged error was discovered.

The answer is (D).

SOLUTION 21

Most whistleblower legislation mandates a "no retaliation" policy. Prior to the Sarbanes-Oxley Act of 2002, whistleblower litigation often arose when there was not a clear policy prohibiting retaliation against employees who complained about suspected law violations. It is important to note that whistleblower statutes do not usually require a whistleblower's complaint be correct to be protected. Whistleblower complaints should be addressed in a consistent, thoughtful, and rapid manner.

The answer is (A).

SOLUTION 22

Surveyors are frequently advised by their insurance carriers to use the term "record drawing" for post-construction drawings. The term "as-built" is commonly used; however, since the data in an as-built often includes data from several different sources, some of which are outside the control of the surveyor, it is usually difficult to verify that everything on the drawing is truly shown as it was built, so insurance carriers advise against it. The term "red-line" for post-construction drawings is attributable to the common practice of making changes to plans in red ink, but is not preferred by insurance carriers. A "measured drawing" is a type of post-construction drawing, typically created from on-site measurements to inform upcoming renovation or to provide historic documentation.

The answer is (A).

SOLUTION 23

Oral agreements are unreliable, as they depend on the memories of the parties to the agreement and typically lack clarity on the scope of services, terms, and conditions. When disputes arise, there is usually no method of amicably resolving them. A written executed contract, including the scope and accuracy of services, along with the payment terms, is the best type of contract for mitigating risk.

The answer is (A).

SOLUTION 24

Clear protection and enforcement are afforded by language stipulating the need for the surveyor's verification for changes or reuse, that the surveyor's hard copy controls over other versions, and that the client holds the surveyor harmless for any claim or liability for injury or loss resulting from reuse. The protection and enforcement afforded by copyright is not as clear. However, there could be debate over the applicability of copyright protection for some types of information and data.

The answer is (C).

SOLUTION 25

The wages may be calculated as follows.

$$\begin{aligned} (\$18.50)(40\text{ hr}) &= \quad \$740.00 \\ (\$27.75)(5.75\text{ hr}) &= \underline{+\$159.56} \\ \text{gross wages} &= \quad \$899.56 \end{aligned}$$

The amount received was $638.69.

Project Management

$$
\begin{aligned}
\text{gross wages} &= \$899.56 \\
\text{net wages} &= \underline{-\$638.69} \\
\text{amount deducted} &= \$260.87
\end{aligned}
$$

$$\frac{\$260.87}{\$899.56} = 0.29$$

The deduction was 29%.

The answer is (C).

SOLUTION 26

The monthly cost of the pension plan is

$$(\$35{,}520.00)(0.11) = \$3907.20$$

The monthly cost of the life insurance premium is

$$(\$35.00)(8) = \$280.00$$

Together they are

$$\$3907.20 + \$280.00 = \$4187.20$$

The weekly cost can be found by dividing the result by 4.3.

$$\frac{\$4187.20}{4.3} = \$973.77 \quad (\$974)$$

The answer is (A).

SOLUTION 27

The formula used to solve this problem is

$$P_n = P_o(1+i)^n$$

P_n is the principal at the end of the interest period, P_o is the principal at the beginning of the interest period, i is the interest, and n is the number of years of the interest period.

$$
\begin{aligned}
P_n &= (\$12{,}000.00)(1+0.075)^3 \\
&= (\$12{,}000.00)(1.075)^3 \\
&= (\$12{,}000.00)(1.2423) \\
&= \$14{,}907.56
\end{aligned}
$$

The answer is (D).

SOLUTION 28

Calculate the interest using the formula

$$P_n = P_o(1+i)^n$$

P_n is the principal at the end of the interest period, P_o is the principal at the beginning of the interest period, i is the interest, and n is the number of years of the interest period.

$$
\begin{aligned}
\$30{,}772.48 &= (\$20{,}000.00)(1+i)^5 \\
\frac{\$30{,}772.48}{\$20{,}000.00} &= (1+i)^5 \\
1.5386 &= (1+i)^5 \\
(1.5386)^{1/5} &= \left((1+i)^5\right)^{1/5} \\
1.09 &= 1+i \\
1.09-1 &= 1+i-1 \\
i &= 0.09
\end{aligned}
$$

The interest was 9.0%.

The answer is (C).

SOLUTION 29

The formula for compound interest is

$$P_n = P_o(1+i)^n$$

P_n is the principal at the end of the interest period, P_o is the principal at the beginning of the interest period, i is the interest, and n is the number of years of the interest period.

In this problem, the taxes for 1989 and 1990 must be paid with interest. The taxes for 1988 were paid at the time, and the taxes for the current year have no interest.

For 1989 taxes (with interest),

$$
\begin{aligned}
P_n &= (\$350.00)(1+0.09)^2 \\
&= (\$350.00)(1.09)^2 \\
&= (\$350.00)(1.188) \\
&= \$415.83
\end{aligned}
$$

For 1990 taxes (with interest),

$$
\begin{aligned}
P_n &= (\$400.00)(1+0.09)^1 \\
&= (\$400.00)(1.09)^1 \\
&= (\$400.00)(1.09) \\
&= \$436.00
\end{aligned}
$$

Project Management

For 1991 taxes (no interest),

$$\$500.00$$

For payment to the agent,

$$\begin{aligned} (\$20{,}000.00)(0.10) &= \quad \$2000.00 \\ \text{selling price} &= \quad \$20{,}000.00 \\ \text{1989 taxes with interest} &= \quad -\$415.83 \\ \text{1990 taxes with interest} &= \quad -\$436.00 \\ \text{1991 taxes} &= \quad -\$500.00 \\ \text{commission to agent} &= \quad \underline{-\$2000.00} \\ & \quad\;\; \$16{,}648.17 \quad (\$16{,}648) \end{aligned}$$

The answer is (D).

SOLUTION 30

The value of all physical assets tends to decrease with use, changes in economic conditions, and technological advances. This deterioration is known as *depreciation.*

The answer is (D).

SOLUTION 31

When a firm anticipates a need for a particular sum of money at a future date for purchase of expensive equipment, it may establish a sinking fund. Regular deposits made to this interest-bearing fund are calculated to yield the particular sum needed on a particular date. A sinking fund represents a special kind of annuity.

The answer is (A).

SOLUTION 32

Under the straight-line, or fixed-percentage, method, depreciation (the wearing value of the asset) is distributed evenly across its life according to the formula

$$D = \frac{C_o - L}{n}$$

D is the annual depreciation, C_o is the original cost, L is the salvage value, and n is the life of the asset in years.

$$\begin{aligned} D &= \frac{\$14{,}000 - \$3200}{9} \\ &= \frac{\$10{,}800}{9} \\ &= \$1200 \end{aligned}$$

The answer is (A).

SOLUTION 33

The discount rate is used to calculate the value of an invested sum of money at some time prior to its maturation. For example, if a promissory note has a mature value of $8000.00 and the holder wishes to know its value one year before it matures, the rate used to calculate that lesser sum of money would be the discount rate.

The answer is (A).

SOLUTION 34

The formula for compound interest is

$$P_n = P_o(1 + i)^n$$

P_n is the current value of the note, P_o is the value of the given sum of money at the date of maturity, i is the interest, and n is the number of years of the interest period. For future delivery, n is negative; the value decreases.

$$\begin{aligned} P_n &= (\$3000.00)(1 + 0.10)^{-2} \\ &= (\$3000.00)(1.10)^{-2} \\ &= (\$3000.00)(0.82645) \\ &= \$2479.34 \quad (\$2480) \end{aligned}$$

The answer is (B).

SOLUTION 35

A sole proprietorship with employees, a partnership, or a corporation must obtain a federal employer identification number (EIN). This number is necessary for business-related forms sent to the Internal Revenue Service (IRS). It identifies the business's tax accounts on all federal and state tax forms. A new owner of an existing business that is required to use an EIN must obtain a new EIN—the former owner's EIN cannot be used. A sole proprietor without employees does not need to obtain an EIN. The proprietor's social security number may be used instead.

The answer is (C).

SOLUTION 36

The market value of real property is the price for which it would trade in an arms-length arrangement.

The answer is (B).

SOLUTION 37

A sole-proprietorship is not a separate legal entity from the owner and any income and/or liability incurred by the business is actually incurred by the owner. Furthermore, all profits are taxed at personal tax rates. In contrast, an incorporated company is a separate legal entity from the owner.

The answer is (B).

SOLUTION 38

Operation analysis, which is modeled after existing processes, is now assisted by software tools so that managers of a firm can analyze performance, find bottlenecks, and optimize processes more efficiently.

The answer is (D).

SOLUTION 39

Critical path analysis (CPA) has been used in large civil and military projects since the 1950s and is especially useful in analyzing complex projects. It facilitates analysis of linked sequential activities, as well as revealing which activities can take place in parallel. CPA helps determine when remedial action is needed and also the minimum length of time needed to complete a project. However, sometimes the tasks-to-time relationship is not as clear in CPA as it might be with other analysis methods. Especially in projects that are not routine, there is often a good deal of uncertainty in the completion times of particular tasks. This uncertainty may limit the usefulness of CPA.

The answer is (B).

SOLUTION 40

During a deposition, it is best to accept the judgment of one's attorney as to whether or not to answer certain questions. Estimation is not a good idea in testimony. It is proper and expected that a witness would discuss testimony with an attorney before a deposition, and if done, should be admitted if asked about. However, it is not a good idea to offer any new documentation in a deposition, especially if it is new to one's attorney.

The answer is (C).

SOLUTION 41

Hearsay is evidence that does not proceed from the direct knowledge of the land surveyor testifying, but from repetition of what that surveyor has heard others say. Hearsay evidence is usually not admitted in court proceedings primarily because it is not presented under oath and before the judge or jury by those who have personal knowledge of the matter. However, there are exceptions to this rule.

Examples of valid exceptions under the Federal Rules of Evidence, Article 1 General Provisions, Rule 803, are proof of boundaries by reputation in a community, records of regularly conducted activity made at the time by a person with direct knowledge, and recorded public records. However, statements recounting a now-deceased property owner's objections to the staking of a property line will most likely not be accepted as an exception to the hearsay rule.

The answer is (C).

SOLUTION 42

A witness may refuse to answer a question when the answer may be self-incriminating, but may not refuse to answer a question when the answer may result in civil liability. It is not acceptable to accept a fee for testifying whose payment is contingent on the outcome of the litigation, and one should not answer a question not fully understood. However, during testimony it is advisable to answer only the questions asked and not to volunteer information.

The answer is (C).

SOLUTION 43

While a county's general, or comprehensive, plan is the foundation of land use decisions; it is usually considered a guide and not a legally binding document.

The answer is (B).

SOLUTION 44

A land use map provides recommendations that are intended to ensure a community will have housing, employment, and recreation in the future. Unlike a zoning map, a land use map does not represent the actual final zoning of the land. Rather, public officials, property owners, and citizens use the land use plan map as a tool to make decisions about future development.

The answer is (C).

SOLUTION 45

The FAR is the floor area ratio that controls the size of buildings. When the FAR is multiplied by the parcel, it produces the maximum allowable amount of floor area a building may occupy. For example, on a 10,000 ft^2 parcel in a district with a maximum FAR of 1.0, the floor area of a building cannot exceed 10,000 ft^2. In that case, one story covering the entire parcel, two stories covering half of the parcel, or four stories covering a quarter of the parcel would meet the limit. However, if the FAR is 2.0, then a building whose footprint covered a quarter of the lot could have eight stories.

The answer is (D).

SOLUTION 46

Jurisdictions impose subdivision regulations on land developments to ensure the continuity of systems and new infrastructures will be compatible with the development of the larger community.

Most jurisdictions allow businesses and residents to occupy buildings before all of the land development improvements are completed. Most often, performance bonds are not released until the project is completed, occupied, and accepted by the public authorities. Most subdivision ordinances have maximum time limits for the consideration of submittals during which period the government must accept or deny it.

The answer is (A).

SOLUTION 47

Both zoning regulations and restrictive covenants do limit the use of the property; however, zoning regulations are imposed and enforced by governmental authority, whereas restrictive covenants are the subject of a private contract. For example, a land developer may impose aesthetic requirements, minimum setbacks, house size restriction, etc., in the language of the deed conveying title to a property. These conditions may bind both the subsequent owner and also every future owner. While a request for a change in zoning of a particular property typically requires public notice and hearings, that is not the case with a restrictive covenant. Zoning regulations are not written in deeds. Restrictive covenants are also known as Covenants, Conditions & Restrictions (CC&Rs).

The answer is (C).

SOLUTION 48

Correctly advising the client is the best course of action. A surveyor should share information with the client concerning the general bounds of covenants and restrictions on subdivision plans. Properly written covenants within the bounds of real estate laws are legally binding, and an attorney should be involved in their final drafting.

The answer is (D).

SOLUTION 49

Streets, drainage, and public utilities have either been completed, or there is a performance bond in place to guarantee their completion before the final subdivision plan is approved. Buildings cannot be constructed before the final subdivision plat has been approved, and the final subdivision plat is part of the final subdivision plan.

Typically, the final subdivision plat must be approved and recorded before the sale of lots can begin. Approved construction documents are usually part of the final subdivision plan.

The answer is (C).

SOLUTION 50

After the reviewer's comments are completed, the plans are returned for corrections. It is always valuable to have a good rapport with the reviewers.

Even though the reviewing agencies usually do finally approve the plans, the liability ultimately rests with the engineer and/or land surveyor. Not all the reviewers' comments appear on the plans, though some do. They are usually accompanied by other narrative comments on other documents.

The answer is (B).

SOLUTION 51

A sketch plan does not contain the same level of detailed engineering plans as does a preliminary plan. It does, at a conceptual level, usually indicate the feasibility and some design characteristics of the development. Sometimes the sketch plan and the preliminary plan may be the same document. The approved preliminary plan should be followed throughout the project's life as closely as possible; altering it may necessitate beginning the whole process anew.

An approved preliminary plan does not give the developer the right to begin construction.

The answer is (C).

SOLUTION 52

The comprehensive plan for land use specifies the allowable type and density of development on parcels of land by district and dividing a locality into more general areas, i.e., agriculture, industry, commercial, etc.; it is the purview of zoning. Subdivision regulations are concerned with the details of specific land development projects; the creation of lots, blocks, streets, easements, etc.; and related services, such as water, sewer, gas, and electricity.

The answer is (A).

SOLUTION 53

A master deed is often used in the establishment of condominiums. It is used to convert a single property into individually owned units that share ownership in common areas and to establish the percentage interest in the common elements attributable to each unit. It also typically lists limits on the utilization of the properties covered. A master deed deals with the usage of property; however, it does not convey real property from one person to another.

The answer is (D).

SOLUTION 54

Subdivision regulations do typically govern the division of land to create lots for residential purposes. However, division of land created by order of a court of record; to create an interest in oil, gas, minerals, or water severed from the surface ownership; or in accordance with land use planning; are not typically covered by subdivision regulations.

The answer is (C).

SOLUTION 55

An environmental site assessment (ESA) is a report that identifies potential or existing environmental liabilities on a property. This information is in the Phase I ESA report and includes an opinion concerning the release of hazardous substances or petroleum products at the property in the past or present. The American Society of Testing and Materials (ASTM) establishes standards for ESAs in commercial real estate transactions. An environmental impact statement (EIS) is a report that addresses the potential effects on the environment of a proposed project. Endangered species are not the subject of ESAs. Soil compaction testing is a construction concern, unrelated to ESAs.

The answer is (C).

SOLUTION 56

Planned unit developments (PUDs) can include a mixture of single-family homes, condos, or town homes, along with retail and other commercial spaces within the development. However, a PUD does not typically include agricultural land.

The answer is (D).

SOLUTION 57

The land development process typically depends on the surveyor for the boundary and topographic surveying, as well as the final horizontal alignment of the infrastructure and buildings. The surveyor does create, locate, and describe land divisions, but determining the ownership of land is not within a surveyor's purview.

The answer is (D).

SOLUTION 58

Governmental entities regulate the locations and grades of utilities, foundations, and drainage, among other things. It follows that the topography of a site influences the design of its development directly. Traffic flow patterns could well be a factor in a land development design but are not a topographic consideration.

A swale is low place or shallow channel, and an inundation area is a region subject to flooding. These, as well as steep slopes, would be topographic design consideration for a land development project.

The answer is (D).

SOLUTION 59

Subdivision regulations typically do require inclusion of floodplain boundaries and elevation information on plats and usually will not approve lots without a building site whose use is consistent with floodplain zoning regulations. They usually do specifically address the dedication of public easements for watercourse crossings. However, while most subdivision regulations forbid building where conditions might create a health or safety hazard, they do not specifically regulate wetlands.

The answer is (D).

SOLUTION 60

The location, sizes, and heights of buildings are subjects typically addressed by zoning regulations, along with land use and the division of a jurisdiction into districts. However, the standards by which subdivision plats are drawn and approved are the purview of subdivision regulations.

The answer is (C).

SOLUTION 61

A proper design of cuts, fills, erosion control, excavation, etc., can only be accomplished if it is based on a complete and accurate representation of the topography of the site that includes knowledge of floodplain issues, drainage patterns, drainage infrastructure, etc.

The answer is (A).

SOLUTION 62

A conservation easement is also known as a preservation easement. It is a legal agreement designed to protect a significant historic, archaeological, or cultural resource through a non-possessory interest in real property. An easement in gross attaches to an individual, not to a particular parcel of land (e.g., an easement that grants an individual a right to hunt on a parcel of land). An easement by estoppel is an easement created by a bar that precludes a person from asserting rights they otherwise would have because of their conduct. A prescriptive easement arises from open, notorious, hostile, adverse, and continuous use of the property of another for the required number of years (e.g., for a pathway or driveway). A prescriptive easement often arises when a fence is built on the wrong side of a boundary line.

The answer is (A).

SOLUTION 63

Traffic volume, sight distance, width, and grade are important considerations when planning access to a land development project from an existing public road. However, the exact nature of the dedication of the right-of-way of the public road is not typically of direct concern.

The answer is (D).

13 Hydrography

FLOW MEASUREMENT

PROBLEM 1

Which of the following discharge rates and periods of time will most nearly accumulate 1 ac-ft?

(A) 1 ft^3/sec for 12 hr

(B) 25 gal/min for 24 hr

(C) 30 miner's in for 6 hr

(D) 50 gal/min for 8 hr

PROBLEM 2

What is the meaning of the word *stage* as it is used in hydrography of streams?

(A) The stage of a stream refers to the amount of water flowing past a given section at a given time.

(B) The stage of a stream refers to the velocity of the water flowing past a given section at a given time.

(C) The stage is the height of the water surface above a given datum.

(D) The stage refers to the location of a gauging station on a stream bank.

PROBLEM 3

Of what use is a stage-discharge curve for a stream?

(A) A stage-discharge curve is an indispensable tool for predicting the occurrence of floods.

(B) The computation of a stage-discharge curve is not necessary for streams with a steady flow.

(C) A stage-discharge curve correlates the height of the water and the flow of a stream.

(D) A stage-discharge curve correlates the velocity of the water and discharge of a stream.

PROBLEM 4

Which of the following is NOT used to measure the level of the water in a stream?

(A) staff gauge

(B) hook gauge

(C) wire weight gauge

(D) vane gauge

PROBLEM 5

Which of the following instruments is equipped with a mercury manometer assembly?

(A) optical current meter

(B) Haskell meter

(C) Parshall flume

(D) bubble gauge

PROBLEM 6

What government agency is responsible for compiling stream flow data in the U.S.?

(A) Bureau of Reclamation

(B) National Geodetic Survey

(C) National Ocean Service

(D) U.S. Geological Survey

PROBLEM 7

The discharge of a stream is the amount of water, in cubic feet per second, flowing past a given section in a given time. The discharge is the product of which of the following factors?

(A) the pulsation of the stream and the depth of the main channel

(B) the cross-sectional area and the mean velocity of the water

(C) the datum of the stage measurements and the average staff gauge readings

(D) the pulsation of the stream and the mean velocity of the water

PROBLEM 8

A gaging site should NOT be chosen along a river

(A) where very low flows may be successfully recorded

(B) where there are high stable banks clear of brush and where there is a uniform bottom

(C) where there is a natural constriction providing relatively constant stage and discharge readings

(D) on a gradually curving section of the river near a confluence

PROBLEM 9

Which sort of gauge is incremented after being placed in position?

(A) staff gauge

(B) float gauge

(C) wire weight gauge

(D) inclined staff gauge

PROBLEM 10

Which of the following current meters is most frequently used in the U.S.?

(A) Ott

(B) Hoff

(C) Price

(D) Haskell

PROBLEM 11

The Price current meter is equipped with a buzzer or sounder that may be set differently depending on the conditions. Which of the following statements best characterizes the available settings?

(A) The electrical contact may be set to sound with every revolution or every other revolution, depending on the velocity of the water.

(B) The meter may be set to emanate a mechanical click with every revolution in slow currents, or an electrical buzz in fast water.

(C) The meter may be set to sound with every revolution in slow currents or with every fifth revolution in fast water.

(D) The faster the current, the more frequently the meter sounds, sounding with every third revolution in slow currents and every revolution in faster water.

PROBLEM 12

The following four problems refer to the illustration shown.

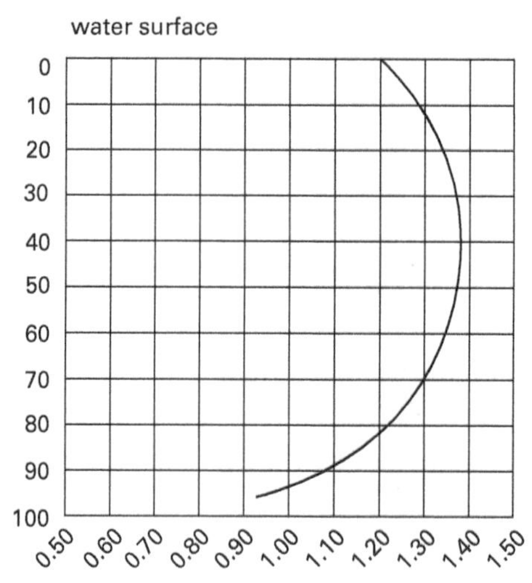

12.1. If the arc in the illustration is considered to be a typical vertical-velocity curve, what units are being expressed by the numbers across the bottom of the chart?

(A) depth of the water in tens of feet

(B) velocity of the water in feet per second

(C) quantity of water in cubic feet

(D) quantity of water in miner's inches

12.2. Which of the following shapes does the vertical-velocity curve most closely approximate?

(A) parabola

(B) cycloid

(C) helicoid

(D) circular curve

12.3. Which of the following elements usually affects the shape of a vertical-velocity curve?

(A) the direction of the wind

(B) whether the water is covered with ice or not

(C) the distance from the banks

(D) all of the above

12.4. Which of the following methods would yield the most accurate indication of the mean velocity expressed by the vertical-velocity curve?

(A) the two-point method

(B) the subsurface method

(C) the six-tenths method

(D) division of the area between the curve and the ordinate by the length of the ordinate

PROBLEM 13

Which of the following statements best describes the vertical integration method of current measurement?

(A) The meter is placed at 20% of the total depth of the stream, and the mean velocity is calculated with a previously established coefficient.

(B) The meter is placed at 60% of the total depth of the stream, and the observed value is taken as the current's mean velocity.

(C) The meter is lowered at a slow and uniform rate to the bottom of the stream and then raised at the same speed. The number of rotations of the meter is recorded and the mean velocity found by using the meter's rating table.

(D) The meter is lowered to 50% of the total depth of the stream in the same position at several different times. The mean velocity is taken as the average of all readings.

PROBLEM 14

Which of the following characteristics of a vertical-velocity curve would be indicative of the flow pattern created in an ice-covered stream?

(A) The shape of the curve would more closely approximate a circular curve than a parabolic curve.

(B) The curvature would be sharper near the surface of the water.

(C) The mean velocity would occur near 75% of the total depth instead of 60%.

(D) Fewer observations would be necessary due to the flatness of the curve.

PROBLEM 15

When calculating the average velocity represented by a vertical-velocity curve using the two-point method, at what depths should readings be taken?

(A) 10% and 90% of the total depth

(B) 20% and 80% of the total depth

(C) 30% and 70% of the total depth

(D) 40% and 60% of the total depth

PROBLEM 16

Which of the following circumstances makes the application of the two-point method inappropriate for finding the mean current velocity of a stream section?

(A) a stream that is less than 2 ft deep

(B) high current velocities

(C) the use of a horizontal-axis current meter

(D) both A and B

PROBLEM 17

A current meter is rated

(A) by the maximum and minimum current velocities it can measure

(B) according to its tolerance to various conditions of silt, current velocities, and methods of suspension

(C) to correlate its number of revolutions per unit of time to specific current velocities

(D) as mechanical, optical, or electrical

PROBLEM 18

Which of the following statements describes a typical purpose of a weir?

(A) A weir is used to stabilize a sounding line.

(B) A weir is a type of station-rating curve.

(C) A weir is a control structure used in canals, shallow streams, and sewers.

(D) A weir is a small portable flume with a throat of 12 in or less.

PROBLEM 19

What sort of weir is used primarily to measure water flow?

(A) rectangular weir

(B) trapezoidal weir

(C) triangular weir

(D) sharp-crested weir

PROBLEM 20

What does the term *head* indicate when referring to a weir?

(A) the width of the crest of the weir

(B) the cross-sectional shape of the wier

(C) the width of the channel approaching a weir

(D) the depth of the flow over a weir

PROBLEM 21

Neglecting the approach velocity of the flow into a 90° sharp-crested V-notch weir, if the head is 6 in, what is the discharge in cubic feet per second?

(A) 0.45 ft^3/sec

(B) 0.83 ft^3/sec

(C) 1.03 ft^3/sec

(D) 1.22 ft^3/sec

PROBLEM 22

What sort of gauge is most often used to measure the head of water flowing over a weir?

(A) a wire weight gauge

(B) a crest-stage gauge

(C) an inclined staff gauge

(D) a hook gauge

HYDROGRAPHY AT SEA

PROBLEM 23

A bathymetric survey is a survey

(A) conducted by sampling material on the seabed

(B) of the depth of water

(C) conducted by sampling seawater

(D) of the salinity of seawater

PROBLEM 24

Mapping the seabed by echo sounding involves the transmission of acoustic waves through seawater. What is the most serious source of error in this procedure?

(A) the width of the acoustic beam

(B) the angle of incidence at the target

(C) spurious echoes

(D) the velocity of the signal

PROBLEM 25

Which of the following affects the velocity of sound through seawater?

(A) salinity

(B) temperature

(C) density

(D) all of the above

PROBLEM 26

A depth determined by echo sounding may be checked with a thermometric depth. Which of the following best describes a thermometric depth?

(A) a probe giving continuous telemetry on salinity, temperature, and sound velocity lowered through the water column

(B) the isothermal determination of the steric anomaly at 1 atmosphere

(C) the depth determined by lowering a thermometer subject to hydrostatic pressure along with a thermometer protected from hydrostatic pressure

(D) the depth calculated by comparing the temperature from a thermistor at the surface with a similar instrument lowered at depth

PROBLEM 27

Which of the following devices involved in echo sounding can be categorized as transducers?

(A) projectors

(B) hydrophones

(C) platen

(D) both A and B

PROBLEM 28

What is the frequency range usually employed in echo sounding?

(A) 1–300 KHz

(B) 100–300 MHz

(C) 1200–1500 MHz

(D) 1500–3000 KHz

PROBLEM 29

Which of the following affects the detection capabilities of echo-sounding equipment?

(A) the nature of the target and the angle of incidence of the wavefront with respect to the target

(B) the pulse length, duration, and the beam width of the transmission

(C) the recording medium

(D) all of the above

PROBLEM 30

As the frequency of an acoustic wave is increased, what are the attendant effects for a given transducer size?

(A) The wavelength is shortened, and the beam width is widened.

(B) The wavelength is lengthened, and the beam width is widened.

(C) The wavelength is lengthened, and the beam width is narrowed.

(D) The wavelength is shortened, and the beam width is narrowed.

PROBLEM 31

What is the relationship between the speed of sound through the atmosphere and the speed of sound through seawater?

(A) The speed of sound through the atmosphere is nearly four-and-one-half times slower that it is in seawater.

(B) The speed of sound through the atmosphere is nearly two times slower that it is in seawater.

(C) The speed of sound through the atmosphere is nearly the same as it is in seawater.

(D) The speed of sound through the atmosphere is nearly three times faster than it is in seawater.

PROBLEM 32

Which of the following statements best describes a bar check of echo-sounding equipment?

(A) The echo sounder is lowered with a small metal plate set at a known distance below the transducer. The sounder is then adjusted to read the calibration distance at each depth.

(B) A metal target is lowered on marked lines below the echo sounder. As the target is lowered to various depths, the echo sounder is adjusted to read the correct distance to the target.

(C) The stylus speed on the recorder is adjusted to correspond with the speed of the pulse from the echo sounder to the seafloor and back.

(D) The motor speed of the recorder is adjusted to correspond with the speed of the vessel in relation to the seafloor.

PROBLEM 33

Obstructions to navigation are sometimes so small as to escape soundings. Which of the following devices is designed to detect such hazards?

(A) backscatter meter

(B) vibro-corer

(C) wire sweep

(D) sediment dredge

PROBLEM 34

Which of the following are examples of seabed samplers?

(A) reversing bottle, the swallow float, and the dredge

(B) grab sampler, the corer, and the dredge

(C) sheave, the sweep, and the corer

(D) pocket scow, the T-bar, and the coaming

PROBLEM 35

Which of the following statements is NOT true of modern recording current meters used in measuring the flow of seawater?

(A) The direction of the flow is determined by an on-board magnetic compass.

(B) The rotor is protected from debris, and the instrument body is watertight.

(C) The recording current meter may be left unattended for long periods.

(D) The recording current meter is designed for use at several depths.

PROBLEM 36

What is a messenger as the term is used in hydrographic instrumentation?

(A) a drogue chute with a sinker used to retard the movement of a float-type current meter

(B) an electronic meter that measures the salinity of seawater by its ability to transmit visible light

(C) a weight that falls down a line and is used to trigger instruments

(D) an electronic interface used to monitor a direct-reading current meter

HYDROGRAPHIC MAPS

PROBLEM 37

When soundings are shown on a hydrographic map, what is the exact position of the measurement?

(A) The position of the sounding is represented by a + sign.

(B) The position of the sounding is represented by the decimal point.

(C) Every sounding on the map is connected with contour lines.

(D) The position of the sounding is represented by the letter *s*.

PROBLEM 38

Which of the following dimensions is equal to 1 fathom?

(A) 1 yd

(B) 1 rod

(C) 4 ft

(D) 6 ft

PROBLEM 39

What should distinguish the low-water line on a hydrographic map?

(A) It is the heaviest line on the map.

(B) It is indicated by a dashed line.

(C) It is the second heaviest line on the map.

(D) It is indicated by a dash-dot line.

PROBLEM 40

Where the datum of a hydrographic map is mean low water, how may soundings that are below the datum be easily distinguished from those that are above?

(A) Soundings above the datum are usually circled, while those below the datum are not.

(B) Soundings that are below the datum are usually circled, while those above the datum are not.

(C) Soundings above the datum are usually expressed to the nearest foot, while those below the datum are expressed to the nearest tenth of a foot.

(D) Soundings below the datum are usually in black, while those above the datum are in some other color.

PROBLEM 41

The standard hydrographic symbols used by the USC&GS (now the National Geodetic Survey) in the preparation of nautical charts include which of the following?

(A) Land areas are in yellow, while marsh or beach areas are in green.

(B) Contour lines on land are shown in red, while underwater contours are in black.

(C) Depths of water are shown in fathoms, while elevations on land are shown in feet.

(D) The contour interval is 10 ft on the land but 5 ft under the water.

PROBLEM 42

The most convenient method of plotting soundings on a map is dependent on the procedures used to locate them in the field. Which of the following plotting techniques is best suited to soundings located by resection—that is, measuring two angles between three known points from the survey vessel?

(A) polar protraction

(B) use of two tangent protractors

(C) use of a three-armed protractor

(D) tracing-cloth method

SOLUTION 1

An acre-foot is the quantity of water required to cover an acre to a depth of 1 ft, or 43,560 ft^3. A flow of 1 ft^3/sec over 12 hr, or 43,200 sec, comes closest to accumulating 1 ac-ft.

The answer is (A).

SOLUTION 2

The stage of the water in a stream is its height above a given datum.

The answer is (C).

SOLUTION 3

The primary use of a stage-discharge curve for a particular stream at a particular section is to estimate the flow of the water by measuring the stage, or height, of the water surface.

The answer is (C).

SOLUTION 4

Each of the gauges listed is used to measure the stage of water in streams, except for the vane gauge.

The answer is (D).

SOLUTION 5

A manometer is an instrument used to measure the pressure of the fluid in a tube, and is an integral part of a bubble gauge. Nitrogen gas is discharged through a tube at a fixed elevation in a stream below its minimum stage, and the pressure at the orifice is measured by the manometer. The pressure in the tube varies directly with the height of the water.

The answer is (D).

SOLUTION 6

The Water Resources Division of the U.S. Geological Survey publishes an annual series of water supply papers concerning the surface water in the U.S. Data on stream flow may be obtained from the district offices of the U.S. Geological Survey.

The answer is (D).

SOLUTION 7

The discharge of a stream is the product of the cross-scctional area at right angles to the flow and the mean velocity of the water.

The answer is (B).

SOLUTION 8

A gaging site near a confluence or on a curve in the river should not be chosen because it would invite variable readings in the stage measurements. The location of a gauging site should be on a straight section of the river far from any confluence.

The answer is (D).

SOLUTION 9

An inclined staff gauge is usually graduated after being set in a permanent foundation since it follows the slope of the bank.

The answer is (D).

SOLUTION 10

The Price current meter, developed by the U.S. Geological Survey, is usually the meter of choice in the U.S. It has the advantages of being useful over a wide range of velocities, being repairable in the field, and having bearings that stand up well to silt.

The answer is (C).

SOLUTION 11

The pentacount setting on the Price current meter causes the instrument to sound with every fifth revolution and is used in water with a current exceeding 6 ft/sec.

The answer is (C).

SOLUTION 12

12.1. The typical arrangement of a vertical-velocity curve is to express the depth of each reading as a percentage of the distance from the water surface to the bottom of the ordinate. The velocity of the water is expressed in feet per second along the abscissa.

The answer is (B).

12.2. The vertical-velocity curve most closely resembles a parabola, and for most calculations is assumed to be one.

The answer is (A).

12.3. Each of the conditions listed has some influence over the shape of the vertical-velocity curve for a particular section. Another element that could be included would be the depth of the channel.

The answer is (D).

12.4. The assumption that the vertical-velocity curve is a parabola allows for several methods of approximation of the mean velocity expressed by a particular curve. The most accurate method is the most time-consuming and can be summarized as finding the axis of the parabola. Dividing the area between the curve and the ordinate by the length of the ordinate yields the length of a particular axis, which will usually occur between 10% and 30% of the total depth.

The answer is (D).

SOLUTION 13

A horizontal-axis current meter is preferred in the integration method of flow measurement; a vertical-axis current meter should not be used. The integration method is based on the assumption that all horizontal and vertical current forces have equal influence on the current meter.

The answer is (C).

SOLUTION 14

Due to the friction created by the underside of the ice, the velocity of the current is significantly decreased near the water surface, giving the vertical-velocity curve a sharper curvature near the top.

The answer is (B).

SOLUTION 15

The average of readings taken at 20% and at 80% of the depth from the water surface will closely approximate the average velocity represented by a vertical-velocity curve. This method is widely used and yields good results in most cases. It also avoids the time-consuming process of finding the average area included within a particular vertical-velocity curve.

The answer is (B).

SOLUTION 16

The use of a horizontal-axis current meter, such as a Hoff meter, does not prevent the application of the two-point method. However, very high current velocities, depths of less than 2 ft, and high levels of debris in the water can seriously degrade the results.

The answer is (D).

SOLUTION 17

A current meter is rated by towing it through still water at various speeds while counting its revolutions. This is necessary because each meter has slightly different bearing frictions and other physical attributes that affect the rate at which it revolves.

The answer is (C).

SOLUTION 18

A weir is built much like a dam; however, water is allowed to flow over a weir, usually through a notch. A weir is often used to measure flow.

The answer is (C).

SOLUTION 19

A broad-crested weir may function as a water-measuring structure, but that is usually not its primary purpose. However, the exclusive function of a sharp-crested weir is water measurement.

The answer is (D).

SOLUTION 20

The head is the height of the water that flows over the crest of the weir. The head is usually measured from the top of the crest to the surface of the water pooled upstream of the weir. In calculating the discharge, the measurement is made at a distance that is three times the length of the head upstream from the crest of the weir and is then corrected for the velocity of the approaching current where the channel is contracted.

The answer is (D).

SOLUTION 21

The general formula for the calculation of the discharge over a 90° triangular weir, neglecting the velocity, which is generally quite low with such control, is

$$Q = 2.49H^{2.48}$$

Q is the discharge in cubic feet per second, and H is the height of the head (6 in) converted to feet.

$$\begin{aligned} Q &= (2.49)(0.5)^{2.48} \\ &= (2.49)(0.1792) \\ &= 0.45\ \text{ft}^3/\text{sec} \end{aligned}$$

The answer is (A).

SOLUTION 22

A hook gauge is most often used to measure the head of a flow over a weir. The gauge is mounted upstream, at a distance from the crest that is three times the head, with the point of the hook coincident with the water surface.

The answer is (D).

SOLUTION 23

A bathymetric survey involves the measurement of the depth of oceans and other large bodies of water. The term may also refer to the mapping of the topography of the seabed based on soundings.

The answer is (B).

SOLUTION 24

The velocity of the acoustic signal through seawater is the most important source of error in echo sounding.

The answer is (D).

SOLUTION 25

The density of the water has a large impact on the speed of the acoustic signal. The temperature, the salinity, and the density vary with the depth, but like the atmospheric corrections applied to the signal from an EDM, these elements are difficult to quantify.

The answer is (D).

SOLUTION 26

The measurement of the temperature-pressure gradient with a pair of thermometers, one protected from and one exposed to hydrostatic pressure, can determine depth within 5 m.

The answer is (C).

SOLUTION 27

An instrument that transforms one kind of energy to another is a transducer. The projectors in echo-sounding equipment are driven by electrical power, but produce an acoustic signal. The hydrophones receive that signal after it has been reflected and transform it back into an electrical impulse.

The answer is (D).

SOLUTION 28

The frequency range most often used by echo-sounding equipment is 1–300 KHz.

The answer is (A).

SOLUTION 29

Resolution is usually limited to half of the pulse length, unless the beam incidence is not normal to the target. The recording medium may not have the capability of displaying the echoes received if resolution is necessary in real time.

The answer is (D).

SOLUTION 30

For a given transducer size, as the frequency increases, the wavelength is shortened and the beam width is narrowed.

The answer is (D).

SOLUTION 31

The speed of sound through the atmosphere at sea level is approximately 1087 ft/sec. The speed of sound through seawater is approximately 4900 ft/sec, or nearly four-and-one-half times faster.

The answer is (A).

SOLUTION 32

A bar check involves lowering a metal bar or disk to various depths while the echo sounder is adjusted to correctly record the known depths to the target.

The answer is (B).

SOLUTION 33

The wire sweep, or drag, is used to locate pinnacle rocks, wrecks, and other hazards in channels that have been selected as safe routes at particular depths. The drag consists of wires or rods held up by buoys, which are then towed by two boats at the required depth. When an obstruction is encountered, the buoys describe two lines that converge at the hazard.

The answer is (C).

SOLUTION 34

The corer is used to obtain a sample from several feet into the seabed, the dredge is designed to bring up loose material lying on top of the seabed, and the grab sampler penetrates the seabed.

The answer is (B).

SOLUTION 35

Most current meters are designed for use at specific depths. Several different types of meters may be arrayed simultaneously to determine flow at various depths.

The answer is (D).

SOLUTION 36

A messenger is a small weight with a slot that allows it to be attached to a line holding a hydrographic instrument at depth. At the appropriate moment, the weight is allowed to fall down the line to trigger the instrument.

The answer is (C).

SOLUTION 37

The position of the decimal point indicates the point at which the sounding was made.

The answer is (B).

SOLUTION 38

A fathom is equal to 6 ft.

The answer is (D).

SOLUTION 39

The heaviest line on a hydrographic map is the high-water line. The second heaviest line is the low-water line.

The answer is (C).

SOLUTION 40

The usual practice is to show soundings below mean low water in black, while soundings above the datum are lettered in some other color.

The answer is (D).

SOLUTION 41

The usual convention is to show marshes and beaches in green and the upland areas in yellow.

The answer is (A).

SOLUTION 42

A three-armed protractor can be used to set off the angles measured at the point of the sounding and then moved across the plan until the three arms coincide with the three control stations. However, this technique cannot provide a unique solution when the three control points are on the circumference of a circle that also includes the position of the sounding.

The answer is (C).

Geographic Information System (GIS)

DATA DEFINITIONS

PROBLEM 1

The term *topology*, as used in a geographic information system (GIS), is

(A) the complete coverage of a plane or a volume by its division into contiguous spatial elements

(B) a means of organizing data into columns and rows

(C) the relationships, such as adjacency and connectivity, between points, lines, and polygons

(D) a commonly used raster file format

PROBLEM 2

In GIS, which of the following is usually NOT considered spatial data?

(A) points

(B) lines

(C) polygons

(D) time

PROBLEM 3

In GIS, a TIN is

I. an irregular tessellation

II. a type of digital map illustrating specific subjects or topics

III. a three-dimensional surface formed from a series of irregularly shaped triangles

IV. a model that may be used as an alternative to the regular grid of a raster that comprises a digital elevation model (DEM)

V. a triangulated irregular network

VI. a format for spatial data developed by the U.S. Bureau of the Census

(A) II and IV

(B) I, II, IV, and VI

(C) I, III, IV, and V

(D) III, IV, V, and VI

PROBLEM 4

Which of the following is NOT used in a GIS vector data model?

(A) cells

(B) points

(C) lines

(D) polygons

PROBLEM 5

Which of the following statements about slope and aspect in GIS is NOT correct?

(A) Slope is expressed as a rate of change.

(B) Aspect is indicated by a direction.

(C) Usually both slope and aspect are attributes displayed over areas instead of at points.

(D) Usually both slope and aspect are attributes displayed at points instead of in areas.

PROBLEM 6

Which of the following is NOT an advantage of vector data compared with raster data?

(A) compact data structure

(B) topology can be completely described with network linkages

(C) retrieval of graphics and attributes is clear and concise

(D) easy spatial analysis within polygons

PROBLEM 7

Which of the following is NOT an advantage of raster data compared with vector data?

(A) simple data structure

(B) convenient layer overlays

(C) easy spatial analysis

(D) informational integrity and clearly defined network linkages

METADATA CONCEPTS

PROBLEM 8

What is metadata?

(A) data used to determine the relative abundance of energy ranges

(B) a dynamic link library file

(C) data about data

(D) a large amount of data

PROBLEM 9

Which of the following organizations defined the standard for digital geospatial metadata?

(A) FGDC

(B) PEIMS

(C) EDSC

(D) NCICB

RELATIONAL DATABASE SYSTEMS

PROBLEM 10

A tuple, or relation, that is part of a relational data structure corresponds to which of the following?

(A) row

(B) column

(C) degree

(D) foreign key

PROBLEM 11

Which of the following is NOT a GIS database structure?

(A) hierarchical

(B) network

(C) relational

(D) choropleth

PROBLEM 12

A relational database management system (RDBMS) is based on several straightforward principles. Which of the following is NOT one of those principles?

(A) The data are in tables that have rows and columns.

(B) All rows in a particular table have identical columns.

(C) Each column has a data type, int, float, decimal, etc.

(D) Classes are used to associate rows from one table to another.

ATTRIBUTE VALUES

PROBLEM 13

Which of the following is NOT a category of attributes used in GIS?

(A) nominal

(B) ordinal

(C) interval

(D) cellular

PROBLEM 14

Which of the following is another way to characterize *attribute data* as the term is used in GIS?

(A) spatial data

(B) key

(C) feature class data

(D) aspatial data

OVERLAY

PROBLEM 15

What is the objective of a point-in-polygon overlay?

(A) find which points fall inside the polygon

(B) interpolate between points

(C) a least-squares-adjustment of the control point data

(D) the reclassification of the polygon feature

PROBLEM 16

In a GIS, what type of overlay would be most appropriate to determine the length of a portion of state highway that crosses a particular city's limits?

(A) line-in-polygon

(B) point-in-polygon

(C) polygon-on-polygon

(D) point-on-line

BUFFERING

PROBLEM 17

In GIS, buffering

I. is the creation of a zone of a specified width around a point, line, or polygon

II. can be done at a constant or variable distance

III. can be used to support proximity analysis

IV. is a data compression technique

V. is a piecewise form of kriging

(A) I, II, and III

(B) I, II, IV, and V

(C) III, IV, and V

(D) IV and V

PROBLEM 18

When buffering is done on a point, what is the characteristic shape of the resulting area?

(A) linear

(B) circular

(C) polygonal

(D) rectangular

CODING STANDARDS

PROBLEM 19

Which of the following statements about SQL is correct?

(A) It is a language that provides an object-oriented mapping solution.

(B) Its statements are always defined by tuple variables.

(C) It is a language that provides an interface to relational database systems.

(D) It is a deductive system based on a Datalog query evaluator.

PROBLEM 20

Which of the following represents the Open Geospatial Consortium's (OGC's) XML-based data standard?

(A) SDTS

(B) HTML

(C) DNF

(D) GML

SPATIAL DATA ACCURACY STANDARDS

PROBLEM 21

The Federal Geographic Data Committee (FGDC) has developed standards for reporting methodology; geodetic control networks; spatial data accuracy; and architecture, engineering, construction, and facilities management, among others. Part 3 of these standards is known as the National Standard for Spatial Data Accuracy (NSSDA). According to the NSSDA, what is the minimum number of check points required to estimate points on a map and in digital geospatial data with a valid positional accuracy?

(A) 10

(B) 20

(C) 30

(D) 40

PROBLEM 22

According to the National Standard for Spatial Data Accuracy (NSSDA), the root-mean-square error (RMSE), as it is used to estimate positional accuracy, means the

(A) positive square root of the variance of the sampling distribution of a statistic

(B) square root of the average of the set of squared differences between dataset coordinate values and coordinate values from an independent source of higher accuracy for identical points

(C) standard deviation of a random variable divided by the mean

(D) difference between an average of a set of values and an estimate thereof, derived from a random sample

PROBLEM 23

According to National Standard for Spatial Data Accuracy (NSSDA) specifications, the test points used to determine the accuracy of a data set should have which of the following characteristics?

(A) The accuracy of the test points and data set points should be equal.

(B) The positions of the test points and data set points should be determined by the same means.

(C) The test points should be well defined, easy to locate, and easy to measure.

(D) The test points and data set points should be dissimilar.

PROBLEM 24

National Standard for Spatial Data Accuracy (NSSDA) specifications stipulate that positional accuracy values for ground distances should be reported in which unit of measure?

(A) meters

(B) feet

(C) the same unit of measure (meters or feet) used in the independent test points

(D) the same unit of measure (meters or feet) used in the data set being tested

PROBLEM 25

National Standard for Spatial Data Accuracy (NSSDA) specifications state that the number of significant digits for positional accuracy values should be

(A) equal to the number of significant digits for the data set point coordinates

(B) one more than the number of significant digits for the data set point coordinates

(C) equal to the number of significant digits for the independently determined test points

(D) equal to four digits

SOLUTION 1

In a general sense, topology is the study of properties of geometric figures that do not change when such figures are distorted. However, in a geographic information system (GIS), the term *topology* is used to describe the continuity of space and spatial properties like adjacency, connectivity, and containment. For example, lines can be said to connect to points, just as polygons are connected lines. Polygons can be adjacent to each other and/or connected to each other to form a network. These relationships are discussed in GIS as topological properties.

The answer is (C).

SOLUTION 2

Time is usually not considered a spatial component.

The answer is (D).

SOLUTION 3

In GIS, a TIN is a triangulated irregular network. It is a representation of a continuous three dimensional surface built from triangular facets that connect heights, and non-overlapping triangles are connected along their edges. A TIN is also an alternative to the regular grid of a raster that comprises a digital elevation model (DEM). It has the advantage of allowing the representation of complex areas with densely spaced information, while sparse information can be used to represent simpler areas.

The answer is (C).

SOLUTION 4

A GIS vector data model uses three basic geometrical entities: points, lines, and polygons. A raster model uses cells.

The answer is (A).

SOLUTION 5

In GIS, slope and aspect are attributes displayed over areas, not points. Slope may be expressed as a rate of rise or fall per a horizontal distance, and as a rate of change, which may be a percentage, tangent of an inclination angle, decimal, or ratio. Aspect is the direction of a face of a slope, usually expressed as an azimuth from north or in a particular quadrant.

The answer is (D).

SOLUTION 6

When compared with the raster model, a disadvantage of using the vector model is that spatial analysis within polygons is difficult. Additional disadvantages include its complex data structures; map overlay can cause difficulties and other modeling can be difficult because each unit has a different topological form; and display and plotting can be burdensome, especially cross-hatching.

The answer is (D).

SOLUTION 7

Some of the advantages of the raster structure are its simple data structure and the capacity to overlay and combine mapped data with remotely sensed data easily. Spatial analysis is also easy with raster data, as is modeling because each spatial unit has the same size and shape.

The disadvantages of the raster model include the large volume of the data and its storage. The use of large cells to reduce data volumes means that feature structures can be lost. In short, information is sometimes lost in the raster model and network linkages are sometimes difficult. Transformation from one map projection to another is also difficult.

The answer is (D).

SOLUTION 8

Metadata is data about data. It is structured information that is used to describe some characteristics of data and is sometimes used to speed up data querying and management by saving users from complex search operations.

The answer is (C).

SOLUTION 9

The Federal Geographic Data Committee (FGDC) defined the standard for digital geospatial metadata in *Content Standard for Digital Geospatial Metadata* (CSDGM). By Executive Order 12906, signed on April 11, 1994, all federal agencies "...shall document all new geospatial data it collects or produces...using the standard under development by the FGDC." FGDC finalized a revised version of its standard in 1998.

The answer is (A).

SOLUTION 10

A relation in a relational data structure is made up of tuples. Each tuple is comprised of a row in that table.

The answer is (A).

SOLUTION 11

A choropleth is not a GIS database structure; it is a type of map in which each area is shaded according to the data type. Hierarchical, network, and relational are all database structures. In a hierarchical database, the foundation of the structure is the parent, and those a step below are children of this foundation. The parent-child relationship is repeated down the hierarchy, and each parent may have many children, but each child can have only one parent. In a network database structure, a child can have more than one parent. Network databases have less redundancy than do hierarchical databases. The structure used most often in GIS is the relational database. In a relational database, the data is stored in tables and the tables are related to one another by keys.

The answer is (D).

SOLUTION 12

In a relational database management system (RDBMS), data are represented in tabular form, all rows in a particular table have identical columns, and each column has a data type. The relational approach supports queries that involve several tables by the use of links across tables. Primary keys, not classes, are used to associate rows from one table to another.

The answer is (D).

SOLUTION 13

Cellular is not a category of attributes used in GIS. GIS usually has four categories of attributes: nominal, ordinal, interval, and ratio. The nominal category includes attributes that differentiate one item from another. Nominal attributes do not usually include the order or size of item. The ordinal category includes attributes that rank entities by order into hierarchies. However, the ordinal category is not normally concerned with the magnitude of the difference between features. Interval attributes are also concerned with the order of items and the magnitude of the differences between them, though the intervals in this category are not necessarily proportional as they are with ratio attributes. Ratio attributes are those concerned with a well-defined starting point and the logical distribution of the magnitude between features.

The answer is (D).

SOLUTION 14

An attribute is a characteristic of a spatial feature that is usually described by numbers or letters. Attributes are often stored in tabular format and are linked by keys to the feature in a relational database. A key in GIS is a unique identifier assigned to each record in a table that helps link it to others. Attributes are aspatial, or nonspatial, data. Feature classes are collections of the same type of entities that share the same spatial and attribute data.

The answer is (D).

SOLUTION 15

The objective of a point-in-polygon overlay is to find which points fall inside a polygon.

The answer is (A).

SOLUTION 16

A line-in-polygon overlay would be most appropriate to determine the length of a portion of state highway that crosses a particular city's limits. This method is used to find the common space between a line feature class and a polygon feature class. The line limited to the area of polygon allows the computation of a "contained in" relationship.

The answer is (A).

SOLUTION 17

Buffering is the specification of a constant or variable width around a point, line, or polygon and can be used to support proximity analysis.

The answer is (A).

SOLUTION 18

The characteristic shape of a buffer around a point is circular.

The answer is (B).

SOLUTION 19

The structured query language (SQL), pronounced "sequel," is a language that provides an interface to relational database systems. It is a special-purpose, nonprocedural query language used for, among other things, requesting information from a database. It was developed by IBM in 1974–75 and was originally called SEQUEL for "structured English query language." Oracle Corporation introduced SQL as a commercial database system in 1979 and it has now become a de facto standard. Some superset of ANSI SQL is built into nearly every relational database management system (RDBMS). While it is an ISO and ANSI standard, many database products support SQL with proprietary extensions to the standard language. SQL queries take the form of a command language to select, insert, update, locate data, etc. SQL is compatible with HTML and XML.

The answer is (C).

SOLUTION 20

The Open Geospatial Consortium (OGC) developed the Geography Markup Language (GML). It is part of the OGC web services architecture that facilitates automated searches for spatial data and spatial services on the web. GML separates content from presentation so data presentation can be under program control. With GML, some problems with incompatible data models can be solved. The Ordnance Survey of Great Britain and the U.S. Census Bureau's Topologically Integrated Geographic Encoding and Reference system (TIGER) use GML.

The answer is (D).

SOLUTION 21

The National Standard for Spatial Data Accuracy's (NSSDA's) specifications require that horizontal and vertical accuracy be tested. This test is done by comparing planimetric coordinates and elevations of well-defined points in the dataset with coordinates of the same points from an independent source of higher accuracy. The minimum number of check points is 20. They should be distributed to reflect the geographic area of interest and the distribution of error in the dataset.

The answer is (B).

SOLUTION 22

The National Standard for Spatial Data Accuracy (NSSDA) defines root-mean-square error (RMSE), which is used to estimate positional accuracy, as the square root of the average of the set of squared differences between dataset coordinate values and coordinate values from an independent source of higher accuracy for identical points.

The answer is (B).

SOLUTION 23

The accuracy of a data set is tested by comparing the coordinates of test points to the coordinates of corresponding points in the data set. The test points should be independent of and more accurate than the data set being evaluated. The test points should also be well defined, easy to locate, and easy to measure.

The answer is (C).

SOLUTION 24

NSSDA specifications state that positional accuracy values for ground distances should be reported in the same unit of measure as that in the data set being tested.

The answer is (D).

SOLUTION 25

According to NSSDA specifications, the number of significant digits for positional accuracy values should equal the number of significant digits for the data set point coordinates.

The answer is (A).

15 Written Communication

SENTENCE STRUCTURE

PROBLEM 1

What must be added to the phrase "The instrument with tracking capability" to most simply make it a sentence?

(A) a verb

(B) a subject

(C) an adjective

(D) a gerund

PROBLEM 2

Determine which of the following sentences is correct.

Which of the crew is leaving the job site?

Which of the crew are leaving the job site?

(A) The first sentence is correct; the second sentence is incorrect.

(B) The second sentence is correct; the first sentence is incorrect.

(C) Both sentences are correct.

(D) Neither sentence is correct.

PROBLEM 3

Which of the following sentences is the most grammatically correct?

(A) The job could be profitable.

(B) More favorable conditions than over there.

(C) I don't want to do that project, it's too far away.

(D) If that work could be done efficiently this late in the year and it was not so far from here I might consider doing it but as it is, I can't.

PROBLEM 4

What must be added to the phrase "tracked satellites for two hours" to most simply make it a sentence?

(A) a preposition

(B) a verb

(C) a subject

(D) a conjunction

PROBLEM 5

There are four types of sentences, and each is comprised of clauses. Which of the following correctly pairs the type of clause with its appropriate sentence type?

(A) simple: one independent clause

compound: one independent clause and at least one dependent clause

complex: two independent clauses

compound-complex: at least two dependent clauses and at least one dependent clause

(B) simple: one independent clause

compound: two independent clauses

complex: one independent clause and at least one dependent clause

compound-complex: at least two independent clauses and at least one dependent clause

(C) simple: one independent clause

compound: two independent clauses

complex: at least two dependent clauses and at least one dependent clause

compound-complex: one independent clause and at least one dependent clause

(D) simple: one independent clause

compound: at least two dependent clauses and at least one dependent clause

complex: one independent clause and at least one dependent clause

compound-complex: two independent clauses

PROBLEM 6

The sentence "Although we usually use laser scanning, close-range photogrammetry is specified" is composed of which of the following?

(A) one independent clause

(B) one independent and dependent clause

(C) two independent clauses

(D) two dependent clauses

PROBLEM 7

An independent clause

(A) contains a subject and a verb

(B) cannot stand alone as a sentence

(C) is most often introduced with a subordinating conjunction

(D) cannot be a complete sentence without the addition of a dependent clause

PROBLEM 8

A dependent clause

(A) cannot stand alone as a sentence

(B) does not contain a subject and a verb

(C) cannot be introduced with a relative pronoun

(D) cannot be a complete sentence without the addition of an independent clause

PROBLEM 9

Which of the following sentences correctly uses "sitting" as a gerund?

Kevin did not like the man sitting beside him on the train.

Kevin did not like the man's sitting beside him on the train.

(A) the first sentence

(B) the second sentence

(C) neither sentence

(D) both sentences

GRAMMAR

PROBLEM 10

Which of the following sentences is written correctly?

(A) If I except a offer, it will affect my job.

(B) I will only accept an offer that will not effect my other work.

(C) I never except an offer that has any affect on my free time.

(D) Accepting such an offer will affect my health.

PROBLEM 11

Which of the following sentences is written correctly?

(A) Among you and me, the figure he sites are explicit exaggerations.

(B) The idea we sighted between us was implicit.

(C) Between you and me, the sources cited were explicit and answered all objections.

(D) Among the two of us we site rules implicitly.

PROBLEM 12

Which of the following sentences is written correctly?

(A) I lay the pencil on the desk.

(B) I lied one paper on the other.

(C) If you're tired, lay on the couch.

(D) Don't lie the candle on the radiator.

PROBLEM 13

Which of the following is NOT written correctly?

(A) The city amended its subdivision regulations. He emended the text of the legal description.

(B) Their assumption has not been tested and may prove to be false. His presumption is clearly based on the location of the fence and the corner monuments.

(C) He just said that to illicit a response. The contract was just the latest maneuver in his elicit scheme.

(D) I want to assure you that your survey will be completed on time. We want to ensure that you are satisfied with our performance and your company will have no reason for concern when you insure the title.

PROBLEM 14

Which of the following is written correctly?

(A) I imply its further from home and yet takes less time.

(B) I infer its good from your furthering our interests with fewer cost.

(C) I infer from the demonstration that less effort is required now to move its bulk farther.

(D) It's implied that less numbers of passengers want to travel further.

PREPARING PROPOSALS

PROBLEM 15

A surveyor is writing a proposal and has an approach that may be better than the method described in the request for proposal (RFP) documents. How should the surveyor write the proposal?

(A) Do not include the alternative approach.

(B) Present only the alternative approach.

(C) Present the alternative approach, but only after describing how to carry out the work using the method described in the RFP.

(D) Present the alternative approach and explain why the method described in the RFP documents is inferior.

PREPARING REPORTS

PROBLEM 16

Which of the following best describes the purpose of an abstract in a technical report?

(A) It is a summary that avoids technical concepts and focuses on managerial concerns.

(B) It lists each heading and its beginning page number.

(C) It is an informative or descriptive summary for the technical readers.

(D) It contains the supporting information at the end of the report.

PUNCTUATION

PROBLEM 17

Which of the following lists is most correct in punctuation and format?

(A) The surveying departments will be evaluated in six categories

Personnel	Equipment
Record Keeping	Profitability
Experience	Technological Sophistication

(B) To save files–

1. input your username and password and log into the system

2. activate the download application

3. find and choose the correct directory in the structure

(C) The data logger default settings

Mask Angle 15°.

SNR 6.

PDOP 4.

(D) Write three descriptions:

1. to illustrate the proper use of metes and bounds

2. to distinguish between the use of radians and degrees in curves

3. to illustrate the difference between exceptions and reservations

PROBLEM 18

Which of the following has the correct punctuation?

(A) The invoice was paid by Lusk & Co..

(B) Herbert et al.., in *Photogrammetry*, explains the process.

(C) What did the judge mean when he said, "The claim of estoppel is not supported by the evidence"?

(D) "What is your opinion?," asked Mark.

PROBLEM 19

Which of the following uses the appropriate dash?

(A) The description was on pages 14–16 in the title commitment documents. (An en dash was used.)

(B) The convention was held January 10—January 14 at the University of Colorado. (An em dash was used.)

(C) The party chief was frustrated–his crew had not arrived–as he set out the stakes himself. (En dashes were used.)

(D) It was the mechanic—not the service manager—that noticed the brake line was leaking. (Em dashes were used with a space before and after each of them.)

BIBLIOGRAPHIC REFERENCING

PROBLEM 20

Using the *Chicago Manual of Style's* bibliographic format, how should a book be cited if it has three authors?

(A) in the order they appear on the title page

(B) alphabetically

(C) the first author followed by et al.

(D) any author followed by etc.

PROBLEM 21

How should references be organized in a bibliography?

(A) in the order they appear in the book

(B) alphabetically by the author's surname

(C) sequentially by year of publication

(D) alphabetically by title

PROBLEM 22

Which of the following should one NOT do when citing quotations from other sources in text?

(A) Correct the spelling, capitalization, and internal punctuation of the original text.

(B) Use ellipses points to show any omission from the original text.

(C) Alter grammatical forms, the tense of a verb, and the person of a pronoun to conform the original quotation to a text's prose, so as alterations are set off by brackets.

(D) Long quotations should be placed in a separate paragraph and indented from the left and right margins of the main text.

SOLUTION 1

Adding a verb to the phrase "the instrument with tracking capability" would most simply make it a sentence. For example, replacing "with" with "has" would turn the phrase into the sentence "The instrument has tracking capability."

The answer is (A).

SOLUTION 2

Relative pronouns such as "which" can either be singular or plural depending on the antecedent—the words to which they refer. For example, the first sentence is correct if the antecedent (in this case the crew) refers to one member. However, the second sentence is correct if the antecedent refers to more than one member of the crew. Because the number of crew members leaving the job site is unknown, both of the sentences are correct.

The answer is (C).

SOLUTION 3

The sentence "The job could be profitable," is simple, but the most grammatically correct. The sentence "More favorable conditions than over there" is a fragment. The sentence "I don't want to do that project, it's too far away" contains a comma splice. The sentence, "If that work could be done efficiently this late in the year and it was not so far from here I might consider doing it but as it is, I can't" is a run-on.

The answer is (A).

SOLUTION 4

Adding a subject to the phrase "tracked satellites for two hours" would most simply make it a sentence. For example, the addition of "The receiver" would turn the phrase into the sentence "The receiver tracked satellites for two hours."

The answer is (C).

SOLUTION 5

A simple sentence contains one independent clause (e.g., "The surveyor found the error"). A compound sentence contains two independent clauses (e.g., "The surveyor found the error, but it was systematic"). A complex sentence contains one independent clause and at least one dependent clause (e.g., "Although the surveyor tried to prevent the error, it occurred"). A compound-complex sentence contains at least two independent clauses and at least one dependent clause ("Although the surveyor tried to prevent the error, it occurred, and he had to repeat the entire project").

The answer is (B).

SOLUTION 6

The sentence "Although we usually use laser scanning, close-range photogrammetry is specified" is a complex sentence. It has one dependent clause that is prefaced by a subordinating conjunction and is joined to an independent clause.

The answer is (B).

SOLUTION 7

An independent clause contains both a subject and a verb and it can stand alone as a sentence.

The answer is (A).

SOLUTION 8

A dependent clause contains a subject and a verb, but it cannot stand alone as a sentence. Dependent clauses are usually introduced by a subordinating conjunction or a relative pronoun. They can be combined with either independent clauses or other dependent clauses to make a sentence.

The answer is (A).

SOLUTION 9

A gerund is a verbal noun that ends in "-ing." The sentence "Kevin did not like the man's sitting beside him on the train" correctly uses "sitting" as a gerund. The sentence implies Kevin dislikes that the man is sitting next to him, not that he dislikes the man himself. The first sentence "Kevin did not like the man sitting next to him" implies that Kevin dislikes the man regardless of where the man is sitting.

The answer is (B).

SOLUTION 10

The sentence "Accepting such an offer will affect my health" is written correctly. The proper article to use before a word that begins with a consonant sound is an "a." For a word that begins with a vowel sound, use the article "an." "Accept" means to agree to take something. "Except" means to exclude. An "affect" is something that influences or produces a change. An "effect" is something that is produced by a cause; it is a consequence.

The answer is (D).

SOLUTION 11

The sentence "Between you and me, the sources cited were explicit and answered all objects" is written correctly. It would not be correct to write "between you and I," because "between" is a preposition. Pronouns that come after prepositions are in the accusative case, not the nominative case.

"Between" should be used when speaking about two subjects, and "among" should be used when referring to more than two subjects. "Citing" means to refer to. A "site" is a location, and a "sight" is something that is seen. "Explicit" means clearly expressed, but when something is implied it is "implicit."

The answer is (C).

SOLUTION 12

According to the *Chicago Manual of Style*, lay is a transitive verb (it always takes a direct object), while lie is an intransitive verb (it never takes a direct object).

In the answer choices given, only the sentence "I lay the pencil on the desk" is written correctly. The word lay is used in the present tense when an object is set down. (One could also use the past tense form, laid.) While the word "lie" can be used as an intransitive verb to refer to reclining (e.g., lie down and sleep—notice there is no direct object), the sentence "Don't lie the candle on the radiator" is incorrect because it has a direct object (the candle).

The answer is (A).

SOLUTION 13

The verb *elicit* means to draw out information and ought to be used in the sentence "He just said that to *elicit* a response." *Illicit* is an adjective and ought to be used in the sentence, "The contract was just the latest maneuver in his *illicit* scheme."

Regarding the use of *amend* and *emend*, *amend* is the more general term of the two. *Emend* is usually more specific and implies correction of text in preparation for publication of some kind.

An *assumption* is a hypothesis based on supposition, not evidence. A *presumption* is founded on probable evidence and may support a decision.

Ensure is a term meaning to make certain or guarantee something will, or won't occur. *Insure* is used in the context of underwriting financial risk. An *assurance* is an earnest declaration.

The answer is (C).

SOLUTION 14

The sentence "I infer from the demonstration that less effort is required now to move its bulk farther" is correct.

Farther implies a greater distance, but *further* refers to a greater extent or degree. *Fewer* is the proper word to indicate the physical numbers of items. *Less* refers to matters of degree or value. An *implication* is a suggestion, but an inference is a conclusion. The possessive form of *it* is *its*. *It's* is a contraction of *it is*.

The answer is (C).

SOLUTION 15

If a surveyor has an alternative solution to the project described in the request for proposal (RFP) and would like to propose it, the surveyor should first describe how to perform the work using the method described in the RFP. A surveyor may offer a solution as an alternative, but not as a substitute for the original method described in the RFP documents.

The answer is (C).

SOLUTION 16

In a technical report, an abstract is an informative or descriptive summary for the readers. An executive summary avoids technical concepts and focuses on managerial concerns. A table of contents lists each of the report's headings and the page number they begin on. An appendix contains the supporting information at the end of the report.

The answer is (C).

SOLUTION 17

According to the *Chicago Manual of Style*, a vertical list is best introduced with a complete sentence followed by a colon. Periods should be used at the end of complete sentences. When the list is numbered, the number is followed by a period and the items should begin with a capital letter.

The answer is (D).

SOLUTION 18

The *Chicago Manual of Style* states that when an expression that normally requires a period ends a sentence (e.g., Lusk & Co.), no additional period is needed. Furthermore, neither a question mark nor an exclamation mark should be accompanied by a comma or a period.

The answer is (C).

SOLUTION 19

An en dash has the width of a typesetter's "N" and the em dash the width of the typesetter's "M." A hyphen is shorter than both the en and em dash. The en dash means "through" and is correctly used in choice (A) to indicate the description was on pages 14 through 16. Em dashes are usually used in pairs to break up a sentence, or to set off a phrase or clause. Spaces should not be used before or after either an en or an em dash. Choice (D) would be correct if there were no spaces before and after the em dashes.

The answer is (A).

SOLUTION 20

According to the *Chicago Manual of Style*, when a book has two or more authors, the names should be listed in the order they appear on the book's title page. Only the first author's name is inverted (last name first), and a comma must appear both before and after the first author's name. For books by four to ten authors, all names are usually given in a bibliography.

The answer is (A).

SOLUTION 21

A bibliography should be organized alphabetically by an author's surname.

The answer is (B).

SOLUTION 22

When citing quotations, it is generally not acceptable to alter the spelling, capitalization, and internal punctuation of the original text; it should be reproduced exactly.

Ellipses points should be used to show any omission from the original text. It is acceptable to alter grammatical forms, the tense of a verb, and the person of a pronoun to conform the original quotation to a text's prose when these changes are set off by brackets. While the meaning of a long quotation is defined differently for different publishing styles, it is correct to place these quotations in a separate, indented paragraph.

The answer is (A).

16 Computer Operations

ORDER OF ARITHMETIC OPERATIONS

PROBLEM 1

Considering the rules of precedence used by computers, what solution will be generated for the calculation shown?

$$9 + (2 \times 6) \times 5^4 \div 10$$

(A) 650
(B) 750.9
(C) 759
(D) 1312.5

PROBLEM 2

A computer's compiler uses the rules of precedence to specify the order in which operators should be evaluated. Given that the symbol * indicates multiplication, what solution will the computer generate for the calculation shown?

$$6 + 8 * 3$$

(A) 30
(B) 42
(C) 70
(D) 72

PROBLEM 3

Using the rules of precedence, what solution will a computer generate for the calculation shown?

$$10 + 7 \times (9 - 6)$$

(A) 31
(B) 51
(C) 67
(D) 72

BITS AND BYTES

PROBLEM 4

A megabyte contains how many bytes?

(A) 2^8
(B) 2^{10}
(C) 2^{20}
(D) 2^{30}

PROBLEM 5

What is a terabyte?

(A) 2^{20} bytes
(B) 2^{30} bytes
(C) 2^{40} bytes
(D) 2^{50} bytes

GRAPHICAL USER INTERFACES

PROBLEM 6

Which of the following best describes a computer's user interface that utilizes a mouse to click on buttons, drop-down menus, icons, and check boxes that activate text-based hyperlinks?

(A) GUI
(B) CLI
(C) HTML
(D) HTTP

PROGRAMMING CONCEPTS

PROBLEM 7

Surveying computer programs often require the manipulation of data types or variables that include decimals. Which of the following data types CANNOT include decimals?

(A) float

(B) double

(C) char

(D) real

PROBLEM 8

What are the permissible values for a Boolean variable?

(A) true and false

(B) −9,223,372,036,854,775,808 to −9,223,372,036,854,775,807

(C) 0 to 255

(D) 0 to 65,535

PROBLEM 9

Regarding computers, a sampling rate is

(A) the frequency, usually in hertz, that an analog signal is checked during the creation of its digital representation

(B) the speed at which the smoothing of a digital representation's roughness caused by aliasing takes place

(C) the generation of a false frequency in parallel with the creation of a correct frequency in the creation of an analog representation of a digital signal

(D) all of the above

OPERATING SYSTEMS

PROBLEM 10

A computer's operating system does which of the following?

(A) manages all other programs in a computer

(B) defines data types and their interactions with one another

(C) creates a set of computer files where documents can be created or edited

(D) finds and removes viruses

PROBLEM 11

Which of the following is NOT the name of an operating system?

(A) Windows

(B) XML

(C) UNIX

(D) Linux

COMPUTER ARCHITECTURE

PROBLEM 12

Which type of computer connection will communicate data the fastest?

(A) USB 3

(B) Thunderbolt

(C) FireWire

(D) USB 2

PROBLEM 13

A computer's typical central processing unit (CPU) contains which of the following?

I. arithmetic and logic unit

II. internal cache

III. control unit

IV. decode unit

V. prefetch unit

VI. bus

VII. registers

VIII. instruction set

IX. main memory, RAM

X. secondary storage

(A) II, IV, VI, and IX

(B) VI, VII, VIII, and X

(C) I, II, IV, V, VI, VII, IX, and X

(D) I, II, III, IV, V, VI, VII, and VIII

DATA FLOW

PROBLEM 14

A stack collection is created in the following order: 3, 9, 6, 2, 8, 1, 7, 5, 4. What is the order of the collection's removal?

(A) 1, 2, 3, 4, 5, 6, 7, 8, 9

(B) 3, 9, 6, 2, 8, 1, 7, 5, 4

(C) 4, 5, 7, 1, 8, 2, 6, 9, 3

(D) 9, 8, 7, 6, 5, 4, 3, 2, 1

PROBLEM 15

Which of the following statements about lossy data compression methods is NOT correct?

(A) decompressed data may not match original data exactly

(B) most codecs are lossless, not lossy

(C) they are used in internet applications

(D) repeatedly compressing and decompressing data will degrade the data's quality

PROBLEM 16

Which of the following are used as measures, or indicators, of computer performance?

I. MIPS

II. IODE

III. clock speed

IV. KPI

V. FLOPS

VI. OBD

VII. MOPS

VIII. PAIS

(A) I, II, III, and IV

(B) I, III, V, and VII

(C) II, IV, VI, and VIII

(D) all of the above

LOOPING

PROBLEM 17

Doing the same thing over and over with a piece of computer code can be called iteration or looping. There are several kinds of loops. Which of the following loops will always run at least once?

(A) WHILE

(B) DO WHILE

(C) both WHILE and DO WHILE

(D) none of the above

PROBLEM 18

Why must a computer be capable of floating point math operations?

(A) Computer operations require fractions expressed as decimals or other non-integers.

(B) Graphic-intensive programs and computer-aided design (CAD) need floating point math.

(C) Most arithmetic and logic units (ALU) only perform integer mathematics.

(D) all of the above

INTERNET

PROBLEM 19

Which of the following is a network that uses common protocols to connect many geographically dispersed telecommunications networks, as opposed to the applications that run on that network?

(A) world wide web

(B) internet

(C) FTP

(D) Telnet

PROBLEM 20

Which of the following communicate interactively by hypertext transfer protocol (HTTP)?

(A) file servers

(B) database servers

(C) web servers

(D) transaction servers

SOLUTION 1

The rules of precedence, or order of operations, can be recalled using the mnemonic, "please excuse my dear Aunt Sally." This mnemonic stands for parentheses, exponents, multiplication, division, addition, and subtraction.

To find the solution for the calculation $9 + (2 \times 6) \times 5^4 \div 10$, first multiply the 2 and 6 together to get 12. The calculation becomes $9 + 12 \times 5^4 \div 10$.

Then, evaluate the exponent 5^4. The calculation becomes $9 + 12 \times 625 \div 10$. Next, multiply 12 and 625, so that the calculation becomes $9 + 7500 \div 10$. Divide 7500 by 10 to get 750. Finally, add 9.

Therefore, $9 + (2 \times 6) \times 5^4 \div 10$ is equal to 759.

The answer is (C).

SOLUTION 2

According to the rules of precedence, multiplication takes precedence over addition. Multiply the 8 and 3 together, and then add 6.

Therefore, $6 + 8 \times 3$ is equal to 30.

The answer is (A).

SOLUTION 3

The rules of precedence require operations within parentheses to be simplified first, so subtract six from nine. The calculation becomes $10 + 7 \times 3$. Then, because multiplication precedes addition and subtraction, multiply seven by three to get 21. Finally, add 10.

Therefore, $10 + 7 \times (9 - 6)$ is equal to 31.

The answer is (A).

SOLUTION 4

Abbreviated MB, a megabyte contains 1,048,576 bytes, which can be expressed as 1024^2, or 2^{20}.

The answer is (C).

SOLUTION 5

A terabyte is equal to 1024 gigabytes, one trillion bytes, or 2^{40} bytes.

The answer is (C).

SOLUTION 6

GUI stands for graphical user interface, which is an interface in which a user interacts with a computer by manipulating graphical elements such as buttons, drop-down menus, icons, and checkboxes. CLI is an abbreviation for command line interface, an older method of controlling a computer whereby the user types in commands in response to a prompt. HTML is not an interface. It stands for hypertext markup language and is used to create documents on the world wide web. HTTP stands for hypertext transfer protocol, which is a set of rules for exchanging files between a web browser and a web server.

The answer is (A).

SOLUTION 7

All the data types listed can include decimals except char. Char includes single characters such as individual letters, digits, punctuation marks, and other symbols. It cannot contain decimals.

The answer is (C).

SOLUTION 8

The only permissible values for a Boolean variable are true and false. This data type is named for George Boole (1815–1864), an English mathematician.

The answer is (A).

SOLUTION 9

When an analog signal is converted to a stream of numbers (that is, when it is digitized) every number is a sample. The frequency, or the number of samples per second in hertz, is often known as the sampling rate.

The answer is (A).

SOLUTION 10

Abbreviated OS, an operating system manages a computer's various programs after being loaded on to the computer by the boot program. The OS also receives input from the mouse and keyboard, sends information to the computer screen, and stores directories and files directly on the computer.

The answer is (A).

SOLUTION 11

Windows, UNIX, and Linux are all operating systems. XML, on the other hand, is an abbreviation for extensible markup language. XML is a subset of SGML, or standard generalized markup language (ISO 8879). This language describes a class of data objects called XML documents that are stored on computers. It also partially describes programs that process these objects.

The answer is (B).

SOLUTION 12

A USB 2 connection has a maximum transfer rate of 480 mbps. USB 3 has a speed of 5 gbps. FireWire can transfer data at two speeds—400 mbps and 800 mbps. Thunderbolt's data transfer speed is 5.4 gbps.

The answer is (B).

SOLUTION 13

The central processing unit (CPU) is a computer's brain and core. The CPU is composed of an integrated circuit with logic gates that carry out the Boolean operations and other instructions the CPU receives from software. Typical CPUs contain the following components.

- *Arithmetic logic unit* (ALU): The ALU controls a computer's arithmetic, adding, subtracting, multiplying, and dividing functions. It carries out the Boolean logic operations, sums values and increments counters, and sends the intermediate results to the CPU's registers. The ALU is also known as the math coprocessor.
- *Internal cache*: Of the two types of cache memory, L1 and L2, L1 is the internal cache. It is a small amount of static random-access memory, also known as SRAM. SRAM is most used in cache memory and is extremely fast, which is important as a CPU is the fastest of a computer's components. The internal cache is one of the components that help prevent the processor from having to wait for instructions.
- *Control unit*: The control unit has four jobs—fetch, decode, execute, and store. Machine language from the instruction cache is translated into binary code in the decode unit. This translation is necessary for the ALU to understand the instructions. Based on the instructions it receives from the decode unit, the control unit tells the ALU and the registers what to do and where to put their results.
- *Prefetch unit*: The prefetch unit holds upcoming instructions in its small amount of memory and compares the instructions to the cache and the main memory. It is another component that minimizes the time the CPU waits for instructions.
- *Bus*: The bus moves data from place to place in the chip's logic structure.
- *Registers*: These are small bits of memory in the CPU that hold temporary data and are used to increase processing speed. Using the registers, the CPU does not have to expend time accessing normal memory.
- *Instruction set*: Permanent machine code is stored on the CPU in the instruction set. There are two types of instruction sets, CISC (complex instruction set computer), and RISC (reduced instruction set computer). RISC is the faster of the two sets.

Though the main memory is closely associated with the CPU, it is separate from it. Random-access memory, or RAM, is where the CPU sends instructions, and where these instructions wait until the CPU needs to retrieve them. However, RAM is not actually a part of the CPU. Secondary storage is used to store data that is not in active use. It is not directly accessible to the CPU. Secondary storage devices, such as DVDs and flash memory, have larger storage capacities than the previously mentioned memory devices, but they operate more slowly.

The answer is (D).

SOLUTION 14

A stack is a last-in, first-out collection, often abbreviated LIFO. It is much like an actual physical stack of objects—the last item is placed on top of those that came before it and becomes the first item to be removed.

The answer is (C).

SOLUTION 15

A *codec*, which is short for coder-decoder, is a device or program used to encode and decode data streams or analog and digital signals. Most codecs use lossy compression, not lossless compression.

In a *lossy data compression method*, decompressed data may not match the original data exactly when restored. Many internet applications, such as streaming of MP3 files, use lossy data compression to achieve greater reduction in data size; however, repeatedly compressing and decompressing data degrades the data's quality. (This drawback is also known as generation loss.) Unlike lossy compression, a *lossless data compression method* reduces data size less but restores compressed data back to its original form exactly.

The answer is (B).

SOLUTION 16

MIPS, clock speed, FLOPS, and MOPS are used to measure computer performance. MIPS, or millions of instructions per second, is a measure of a computer's processing speed. In some cases, MIPS is a better indicator of computer performance than the more regularly used clock speed standard. (However, critics of MIPS often change the acronym to stand for meaningless indication of processor speed.) Clock speed, or frequency, is a measure of the number of cycles the CPU completes during each tick of the computer's clock. It is usually discussed in terms of megahertz or gigahertz.

MOPS, which stands for millions of operations per second, is another standard of computer performance. FLOPS, also known as floating point operations per second, can be useful in computer applications where floating point operations are a critical component of regularly used applications.

However, it is important to note that using a benchmark such as speed to compare computer performance is not well suited to a one-size-fits-all solution. Computer performance is dependent on a number of factors, including the particular functions a computer is expected to perform.

The answer is (B).

SOLUTION 17

A WHILE loop is based on a condition, usually a Boolean expression, and a block of code. Its function may be summarized as, "While this condition is true, run." A WHILE loop will test the condition before it begins. If the condition is true, it will run and stop only when the condition is false. However, if the condition is initially false it will not run at all. On the other hand, a DO WHILE loop tests the condition only after it has completed its first run. Therefore, a DO WHILE loop will always run at least once.

The answer is (B).

SOLUTION 18

A computer must be capable of floating point math operations because computer operations require fractions expressed as decimals or other non-integers; most arithmetic and logic units (ALUs) only perform integer mathematics; and graphic-intensive programs need floating math. Computer-aided design (CAD), graphic programs, and 3D work are just a few of the computer applications that require more than the integer-only capability of the ALU in the computer's CPU. Therefore, the floating point unit (FPU) is necessary to fill the gap and provide the fraction capability.

The answer is (D).

SOLUTION 19

The only network listed in the options provided is the internet. The internet is composed of geographically dispersed networks that are connected by communication devices and common protocols. Applications that run on the internet include the world wide web (www), which supports a hypertext transfer protocol (HTTP). The file transfer protocol (FTP) and Telnet are protocols.

The answer is (B).

SOLUTION 20

Web servers communicate with hypertext transfer protocol (HTTP) over the internet. File servers respond to client requests for records and return these records to the client across the network. A database server processes client queries and after the requested data is found, passes the results back to the client. A client invokes a remote procedure that executes a transaction at a transaction server and the results are returned back to the client over the network.

The answer is (C).

17 Basic Sciences

LIGHT AND WAVE PROPAGATION

PROBLEM 1

Which of the following lists the colors of the visible spectrum in a sequence from the shorter to the longer wavelength?

(A) violet, blue, green, yellow, orange, red

(B) violet, blue, yellow, green, red, orange

(C) red, orange, yellow, green, blue, violet

(D) orange, red, green, yellow, blue, violet

PROBLEM 2

Electronic distance measurements (EDMs) rely on an accurate estimation of the speed of light through the atmosphere. The ratio between the speed of light in a vacuum and the speed of light in the atmosphere is known as the refractive index of air. Derivation of the correct refractive index of air depends on knowing the wavelength under consideration, the temperature, the humidity, and which of the following?

(A) absorption

(B) atmospheric pressure

(C) wind speed

(D) emissivity

PROBLEM 3

Which of the following lists the electromagnetic spectrum's regions in a sequence from the shorter to the longer wavelength?

(A) radio, microwave, infrared, visible, ultraviolet, x-rays, gamma rays

(B) gamma rays, x-rays, ultraviolet, visible, infrared, microwave, radio

(C) gamma rays, x-rays, visible, ultraviolet, infrared, microwave, radio

(D) radio, microwave, visible, infrared, ultraviolet, x-rays, gamma rays

PROBLEM 4

Which list of symbols correctly describes the following equation as applied to waves of the electromagnetic spectrum?

$$\lambda = \frac{c}{f}$$

(A) λ = wavelength, f = speed of light in a vacuum, c = frequency

(B) λ = speed of light in a vacuum, f = wavelength, c = frequency

(C) λ = wavelength, f = frequency, c = speed of light in a vacuum

(D) λ = speed of light in a vacuum, f = frequency, c = wavelength

PROBLEM 5

One hertz is a full wavelength that takes one second to cycle through how many degrees of phase?

(A) 45°

(B) 90°

(C) 180°

(D) 360°

REFRACTION

PROBLEM 6

Which of the following is NOT a result of refraction?

(A) objects on the horizon appearing higher than they actually are

(B) the colors of a rainbow

(C) magnification by lenses

(D) the apparent increase in the frequency of the signal from a GPS satellite as it rises

PROBLEM 7

A ray of light enters glass from air at 45° as measured from the normal; that is, the imaginary line perpendicular to the boundary between the two materials at the point of entry. Given that the index of refraction for air is 1.0003 and the index of refraction for the glass is 1.5198, the ray's change in direction in the glass will be

(A) toward the normal 17.264°

(B) away from the normal 31.475°

(C) toward the normal 0.465°

(D) away from the normal 23.873°

BASIC ELECTRICITY

PROBLEM 8

A lighted lamp with a resistance of 7 Ω is powered by a battery through a connecting circuit carrying 2 A. What is the voltage provided by the battery?

(A) 6 V

(B) 9 V

(C) 12 V

(D) 14 V

PROBLEM 9

Six 2 V, 60 A-hr batteries are connected in series; that is, the positive terminal of the first battery is connected to the negative terminal of the second battery, the positive terminal of the second is connected to the negative of the third, and so on. Which of the following represents the voltage and amperage of the resulting combination?

(A) 2 V, 60 A-hr

(B) 2 V, 360 A-hr

(C) 12 V, 60 A-hr

(D) 12 V, 360 A-hr

DENDROLOGY

PROBLEM 10

Which of the following trees is a closed-seed, flowering plant?

(A) slash pine

(B) fir

(C) oak

(D) juniper

PROBLEM 11

A surveyor in a central Rocky Mountain state is searching for the corner monument of Sections 25, 26, 35, and 36 in a Public Land Survey System 125 years after the original notes were recorded. The notes mentioned four bearing trees that were marked less than one chain from the monument when it was set, and were described as silver spruce, with a diameter breast height (dbh) of 5 in. When the surveyor arrives at the position where he expects to find the monument, there is a stand of pyramidal shaped trees with diameters of approximately 15 in. The evergreen needles on the trees are about 1 in long and sharp. The trees' bark is gray to red-brown and scaly, and their cones are 2 in to 4 in long. What type of trees are these?

(A) red spruce

(B) white spruce

(C) blue spruce

(D) Norway spruce

BIOLOGY

PROBLEM 12

What is the average pulse and respiration rate of a healthy adult person at rest?

(A) 12 beats and 6 breaths per minute

(B) 32 beats and 10 breaths per minute

(C) 72 beats and 17 breaths per minute

(D) 112 beats and 29 breaths per minute

GRAVITY

PROBLEM 13

Which of the following is NOT affected by gravity?

(A) an orthometric elevation

(B) an ellipsoidal height

(C) a value measured by a gravimeter

(D) heights of the sea measured at a tide station

PROBLEM 14

Which of the following is the foundation of the theory of continental drift?

(A) plate tectonics

(B) vendian extinction

(C) tidal flow

(D) glaciation

PROBLEM 15

Water will always flow along which of the following?

(A) from a higher ellipsoidal height to a lower ellipsoidal height

(B) from a higher dynamic height to a lower dynamic height

(C) water will not always flow along either A or B

(D) water will always flow along A and B

PROBLEM 16

A *gal* is a unit of measurement concerning

(A) speed

(B) radiance

(C) acceleration

(D) pressure

TEMPERATURE AND HEAT

PROBLEM 17

Which of the following statements about temperature is NOT correct?

(A) Temperature is the measurement of the average of the kinetic energy of molecules.

(B) The pressure of a gas is directly proportional to its temperature.

(C) Temperature and heat are the same.

(D) There is no maximum temperature, but there is a minimum temperature, known as absolute zero, at which all molecular motion stops.

FORCES AND MECHANICS

PROBLEM 18

On the Apollo 15 mission in 1971, astronaut Dave Scott brought a falcon feather from the Air Force Academy's mascot with him on the mission to the moon. There he conducted an experiment originally suggested by Joe Allen. He dropped the feather and a hammer at the same time. What happened?

(A) The feather hit the surface first.

(B) The hammer hit the surface first.

(C) The hammer and the feather hit the surface at the same time.

(D) none of the above

PROBLEM 19

Which of the following statements about an object with a mass of 1 kg is NOT true?

(A) Its mass is the numerical measure of its inertia.

(B) If the object has a mass of 1 kg on the earth, it would also have a mass of 1 kg on the moon.

(C) Its mass is how much it resists changes in its motion.

(D) Its mass is a measure of the gravitational attraction force on the object.

PROBLEM 20

Which of the following statements is true for the pressure on an object under water, presuming that the water is equally dense?

(A) The pressure under 35 ft of water in a 4 ft wide well is less than the pressure under 35 ft of water in a 50 mi wide lake.

(B) The pressure under 35 ft of water in a 4 ft wide well is more than the pressure under 35 ft of water in a 50 mi wide lake.

(C) The pressure under 35 ft of water in a 4 ft wide well is the same as the pressure under 35 ft of water in a 50 mi wide lake.

(D) none of the above

PROBLEM 21

When a pulley system is used to lift an engine out of a truck, which of the following happens?

(A) The work output from the pulley system is more than the work input.

(B) The pulley system increases both the magnitude and the distance of the input force.

(C) The pulley system reduces the magnitude of the input force.

(D) The work output from the pulley system is almost exactly the same as the work input.

PROBLEM 22

Which of the following correctly describes a difference between force and energy?

(A) force is vector quantity; energy is a scalar quantity

(B) energy is a push or pull on a body; force is the ability to do work

(C) unbalanced energy produces acceleration; force changes only if velocity changes

(D) force flows from higher temperature to lower temperature; energy is a push or pull on a body

PROBLEM 23

What is the meaning of the term *specific gravity*?

(A) the dimensionless ratio of the weight of a material to that of the same volume of water

(B) the density of a material in units of mass per volume

(C) the ratio of the volume of water that a given body of rock will hold against the pull of gravity to the volume of the body itself

(D) the ratio of the total surface of a substance to its volume-surface area per unit mass

SOLUTION 1

Even though there are no distinct boundaries between colors in the visible spectrum, the order is clear. The wavelengths from shortest to longest are as follows: violet, 380–450 nm; blue, 450–495 nm; green, 495–570 nm; yellow, 570–590 nm; orange, 590–620 nm; and red, 620–750 nm.

The answer is (A).

SOLUTION 2

The speed of light in a vacuum is approximately 299,792,458 m/s. However, in the earth's atmosphere at sea level, the speed of light is approximately 299,702,532 m/s—a difference of about 300 ppm. This speed is affected by the wavelength of the light, and the temperature, humidity, and pressure of the atmosphere. An electronic distance measurement (EDM) is usually standardized at a specific refractive index, usually about 12°C and 1013.25 mbar. It is said that a change of approximately 1°C, or 3.5 mbar, from the standard will result in approximately 1 ppm error.

The answer is (B).

SOLUTION 3

The electromagnetic spectrum is a continuum from gamma rays with very short wavelengths of approximately 0.1 Å and less, to radio waves with long wavelengths of approximately 10 m and more.

The answer is (B).

SOLUTION 4

The speed of light in a vacuum (c), divided by the frequency (f) of an electromagnetic wave, equals the wavelength (λ).

The answer is (C).

SOLUTION 5

One hertz is a full wavelength that takes one second to cycle through 360° of phase.

The answer is (D).

SOLUTION 6

Refraction is responsible for objects on the horizon appearing higher than they actually are, the colors of a rainbow, and magnification by lenses. It is *not* responsible for the apparent increase in the frequency of a GPS signal as the GPS satellite moves closer to the observer. This phenomenon is due to the Doppler effect.

The answer is (D).

SOLUTION 7

To determine the change in the ray's direction in the glass, use Snell's law; where N_i is the refraction index of the first medium, N_r is the refraction index of the second medium, θ_i is the incident angle with respect to the normal, and θ_r is the refracted angle with respect to the normal.

$$\begin{aligned}
N_i \sin\theta &= N_r \sin\theta_r \\
1.0003 \sin 45^\circ &= 1.5198 \sin\theta_r \\
\theta_r &= \sin^{-1}\frac{(1.0003)(0.7071)}{1.5198} \\
&= \sin^{-1} 0.4654 \\
&= 27.736^\circ \\
\theta_i - \theta_r &= 45.000^\circ - 27.736^\circ \\
&= 17.264^\circ
\end{aligned}$$

According to Fermat's principle, a ray of light's path between two points is that which is traversed in the least time. Therefore, when light travels from a high index of refraction to a low index of refraction, its velocity increases and it bends away from the normal. When traveling from a low index of refraction to a high index of refraction, light's velocity decreases and bends toward the normal.

The answer is (A).

SOLUTION 8

To determine the voltage provided by the battery, use Ohm's law; where I is the current expressed in amperes (A), E is the voltage expressed in volts (V), and R is the resistance expressed in ohms (Ω).

$$\begin{aligned}
E &= IR \\
&= (2\text{ A})(7\ \Omega) \\
&= 14\text{ V}
\end{aligned}$$

The answer is (D).

SOLUTION 9

When batteries are connected in series, the voltage of the combination is equal to the sum of the voltages of the individual batteries and the amperage is unchanged. When batteries are connected in parallel (that is, the batteries are connected + to + and – to –) the voltage of the combination is unchanged and the amperage of the combination is equal to the sum of the amperages of the individual batteries. A typical car battery is six 2 V battery cells in series.

The answer is (C).

SOLUTION 10

An oak tree is an example of a closed seed, flowering plant (also known as an angiosperm). Angiosperm trees have flattened or broad leaves and their seeds are enclosed in a fruit covering.

The answer is (C).

SOLUTION 11

The trees are not red spruces. Red spruces are located primarily in the Maritime Provinces of Canada west to Maine, southern Quebec, and southeastern Ontario; and south into central New York, eastern Pennsylvania, northern New Jersey, and Massachusetts. The needles of the red spruce generally do not reach 1 in and their cones are smaller than 2 in. The trees are also not white spruces or Norway spruces. White spruces have a range from Newfoundland and Labrador west across Canada along the northern limit of trees, to the Hudson Bay, the Northwest Territories, and the Yukon. The needles of the white spruce are not sharp and their cones do not reach 4 in. Norway spruces are native to the European Alps, the Balkan Mountains, and the Carpathians Mountains extending north to Scandinavia. The cones of the Norway spruce are 4 in to 6 in long.

Therefore, the trees are blue spruce, also known as the silver spruce—a species discovered in 1861. Blue spruce trees are long lived and grow slowly. For example, in the 125 years since the notes were taken, the trees would gain approximately 10 in in diameter. Their needles are about 1 in long and sharp, their bark is gray to red-brown and scaly, and their cones are 2 in to 4 in long.

The answer is (C).

SOLUTION 12

A pulse of 60 to 100 beats per minute is normal, though it can fluctuate. A typical respiration rate for an adult person at rest ranges from 15 to 20 breaths per minute. Over 25 breaths per minute, or fewer than 12 breaths per minute, is not normal.

The answer is (C).

SOLUTION 13

An ellipsoidal height is not influenced by gravity. On the other hand, an orthometric elevation is measured in relation to the geoid, the equipotential surface defined by gravity. Tidal variation is also directly affected by gravity, as are the measurements of a gravimeter.

The answer is (B).

SOLUTION 14

The plate tectonic model posits that the earth's surface is composed of rigid plates and that these plates are moving.

The answer is (A).

SOLUTION 15

Water will not necessarily flow from a higher ellipsoidal height to a lower ellipsoidal height because these heights are not based on gravity. However, water will definitely flow from a higher dynamic height to a lower dynamic height—this is one reason that the International Great Lakes Datum 1985 (IGLD85) is expressed as dynamic heights.

The answer is (B).

SOLUTION 16

The acceleration of gravity (the rate of change in the velocity of a falling body) can be quantified in *gals*. The namesake for this unit of measurement is Galileo. If an object falling at 1 cm per second at the end of the first second of its fall has accelerated to a rate of 2 cm per second at the end of the next second, then it has accelerated 1 gal. In other words, 1 gal is an acceleration of 1 cm per second squared.

The answer is (C).

SOLUTION 17

Temperature is the measurement of the average of the kinetic energy of molecules. Therefore, the pressure of a gas is indeed directly proportional to the temperature (Boyle's law). Heat on the other hand, is thermal energy transferred from a hot to a cold region. Heat and temperature are not the same. If they were the same, adding heat to a boiling pot of water would raise its temperature; but it doesn't.

The answer is (C).

SOLUTION 18

On Earth, the feather falls more slowly than the hammer because of air resistance. On the Moon, there is insignificant atmosphere and virtually no air resistance. The feather fell at the same rate of acceleration as the hammer, and they hit the surface at the same time. Galileo was proven correct.

The answer is (C).

SOLUTION 19

Mass is not the same as weight. An object's weight is a measure of the force of gravity on it. Mass, on the other hand, is inertia, or resistance to changes in motion. A mass of 1 kg on Earth remains a mass of 1 kg on the Moon; however, the weight of the object would only be about 1/6 of its weight on Earth.

The answer is (D).

SOLUTION 20

The water's depth and density are the contributors to the pressure on an object under a column of water; therefore, the pressure under 35 ft of water in a 4 ft wide well is the same as the pressure under 35 ft of water in a 50 mi wide lake.

The answer is (C).

SOLUTION 21

Ignoring friction and other small inefficiencies, the work output is approximately the same as the work input. The work, or energy, output from a simple machine cannot be more than the work, or energy, input. Though the pulley system increases the magnitude of the input force (i.e., there was a mechanical advantage), no machine can increase the magnitude and the distance of the force input.

The answer is (D).

SOLUTION 22

Energy has magnitude but not direction. It is a scalar quantity. On the other hand, force is a vector. Force has magnitude and direction. As specified by the first law of thermodynamics, energy and mass are not created nor destroyed, but force is not conserved. (For example, a simple machine like a lever can increase input force but it cannot increase input energy.)

The answer is (A).

SOLUTION 23

Specific gravity is the dimensionless ratio of the weight of a material to that of the same volume of water. Most common minerals have specific gravities between 2 and 7. Gold has a higher specific gravity (19.3) than does magnetite (5.2). For example, 1 cm^3 of magnetite will be 5.2 times heavier than 1 cm^3 of water.

The answer is (A).